Groundwater Ecology

AQUATIC BIOLOGY Series

Series Editor

James H. Thorp
Department of Biology
University of Louisville
Louisville, Kentucky

Plate 1   *Upper:* The natural bridge of Daxiaocaokou, 100 m high (Zhigin wunty, Guizhou province, southern China). Photograph courtesy of P. Audra. *Lower:* The underground river of Longdong Cave in Wufeng wunty (Hubei province, central China). Photograph courtesy of J. Bottazi.

Groundwater
Ecology

Edited by

Janine Gibert
Laboratoire d'Hydrobiologie et Ecologie Souterraines
U.A./C.N.R.S. 1451
Ecologie des Eaux Douces et des Grands Fleuves
Université Claude Bernard-Lyon 1
Villeurbanne
France

Dan L. Danielopol
Limnological Institute
Austrian Academy of Sciences
Mondsee
Austria

Jack A. Stanford
Flathead Lake Biological Station
The University of Montana
Polson, Montana

ACADEMIC PRESS
San Diego   New York   Boston   London   Sydney   Tokyo   Toronto

Cover illustration: The interstitial habitat and some of the subterranean-dwelling organisms. See Chapter 8, Figure 1.

This book is printed on acid-free paper. ∞

Academic Press, Inc.
A Division of Harcourt Brace & Company
525 B Street, Suite 1900, San Diego, California 92101-4495

United Kingdom Edition published by
Academic Press Limited
24-28 Oval Road, London NW1 7DX

Library of Congress Cataloging-in-Publication Data

Groundwater ecology / edited by Janine Gibert, Dan L. Danielopol, Jack Stanford.
     p.   cm. -- (Aquatic ecology series)
    Includes bibliographical references and index.
    ISBN 0-12-282110-6 (case)
    1. Groundwater ecology. 2. Groundwater. 3. Aquifers.
  I. Gibert, Janine. II. Danielopol, Dan l.   III. Stanford, Jack Arthur, date. IV. Series.
  QH541.5.G76G76   1994
  574.5' 2632--dc20   94-17095
                                   CIP

PRINTED IN THE UNITED STATES OF AMERICA
94 95 96 97 98 99 MM 9 8 7 6 5 4 3 2 1

In honor of Professor René Ginet

This book brings together numerous studies of groundwater ecology by researchers from all over the world. For many years Professor Ginet has contributed to and encouraged this kind of research at the University Claude Bernard (Lyon 1) in France where he created the Groundwater Hydrobiology and Ecology Laboratory in 1960. He was the guiding spirit for many of the researchers who have written chapters in this book.

An international specialist in hypogean Amphipod Crustaceans, he is the author of (among numerous volumes and papers) a remarkable monograph on *Niphargus* and of a book (in collaboration) called *Introduction to Underground Biology and Ecology* which has been responsible for awakening many a vocation.

This book gives us the opportunity to thank him for all he has done for groundwater research, and we should like to present it as a mark of our gratitude and affection.

Janine Gibert

Contents

SECTION *ONE*

HYDRODYNAMICS AND GEOMORPHOLOGY OF GROUNDWATER ENVIRONMENTS

2 Karst Hydrogeology

A. Mangin

3 Porous Media and Aquifer Systems

L. Zilliox

4 Water Geochemistry: Water Quality and Dynamics

M. Bakalowicz

SECTION *TWO*

*BIOLOGICAL ORGANIZATIONS AND
CONSTRAINTS IN GROUNDWATER*

7   *Microbial Ecology of Groundwaters*

A. M. Gounot

8 Adaptation of Crustacea to Interstitial Habitats: A Practical Agenda for Ecological Studies

D. L. Danielopol, M. Creuzé des Châtelliers, F. Moeszlacher, P. Pospisil, and R. Popa

9 Biotic Fluxes and Gene Flow

T. C. Kane, D. C. Culver, and J. Mathieu

10 Species Interactions

D. C. Culver

20   *Ecological Basis for Management of Groundwater in the United States: Statutes, Regulations, and a Strategic Plan*

C. A. Job and J. J. Simons

EPILOGUE

Conclusions and Perspective   543

Jack A. Stanford and Janine Gibert

Contributors

Numbers in parentheses indicate the pages on which the authors' contributions begin.

M. *Bakalowicz* (97), Laboratoire Souterrain du CNRS, Moulis, 09200 Saint-Girons, France

J.-P. *Bravard* (157), Université Lyon III, Département de Géographie, 69239 Lyon, France

M. *Creuzé des Châtelliers* (157, 217, 313), Université Lyon I, U.R.A. CNRS 1451, Ecologie des Eaux Douces et des Grands Fleuves, Laboratoire d'Hydrobiologie et Ecologie Souterraines, 69 622 Villeurbanne Cedex, France

D. C. *Culver* (245, 271, 451), Department of Biology, American University, Washington, DC 20016

D. L. *Danielopol* (217), Limnological Institute, Austrian Academy of Sciences, Gaisberg, 116, 5310, Mondsee, Austria

M.-J. Dole-Olivier (7, 313), Université Lyon I, U.R.A. CNRS 1451, Ecologie des Eaux Douces et des Grands Fleuves, Laboratoire d'Hydrobiologie et Ecologie Souterraines, 69 622 Villeurbanne Cedex, France

B. K. Ellis (367), Flathead Lake Biological Station, The University of Montana, Polson, Montana 59860

D. W. Fong (451), Department of Biology, American University, Washington, DC 20016

J. Gibert (7, 425, 543), Université Claude Bernard-Lyon 1, U.A./C.N.R.S. 1451, Ecologie des Eaux Douces et des Grands Fleuves, Laboratoire d'Hydrobiologie et Ecologie Souterraines, 69 622 Villeurbanne Cedex, France

A. M. Gounot (189), Université Lyon I, Ecologie Microbienne, U.R.A. CNRS 1450, 69 622 Villeurbanne Cedex, France

C. A. Job (523), U. S. Environmental Protection Agency, Groundwater Protection Division, Washington, DC 20460

W. K. Jones (451), Karst Waters Institute, Charleston, West Virginia 25414

T. C. Kane (245, 451), Department of Biological Sciences, University of Cincinnati, Cincinnati, Ohio 45221

R. Laurent (425), Université Lyon I, U.R.A. CNRS 1451, Ecologie des Eaux Douces et des Grands Fleuves, Laboratoire d'Hydrobiologie et Ecologie Souterraines, 69 622 Villeurbanne Cedex, France

R. Maire (129), Laboratoire Environnement, Centre d'Etudes de Géographie Tropicale, CNRS, Domaine Universitaire de Bordeaux; and Laboratoire CIBAMAR, Géologie-Recherche, Université Bordeaux I, 33 405 Talence Cedex, France

F. Malard (425), Université Lyon I, U.R.A. CNRS 1451, Ecologie des Eaux Douces et des Grands Fleuves, Laboratoire d'Hydrobiologie et Ecologie Souterraines, 69 622 Villeurbanne Cedex, France

A. Mangin (43), Centre National de la Recherche Scientifique, Laboratoire Souterrain, Moulis, 09200 Saint Girons, France

J. Margat (505), BRGM, Avenue de Concyr, 45060 Orléans Cedex 2, France

P. Marmonier (313), Université de Savoie, Département d'Ecologie Fondamentale et Appliquée, 73011 Chambéry, France

D. Martin (313), Université Lyon I, U.R.A. CNRS 1451, Ecologie des Eaux Douces et des Grands Fleuves, Laboratoire d'Hydrobiologie et Ecologie Souterraines, 69 622 Villeurbanne Cedex, France

J. Mathieu (245), Université Lyon I, U.R.A. CNRS 1451, Ecologie des Eaux Douces et des Grands Fleuves, Laboratoire d'Hydrobiologie et Ecologie Souterraines, 69 622 Villeurbanne Cedex, France

F. Moeszlacher (217), Limnological Institute, Austrian Academy of Sciences, Gaisberg, 116, 5310, Mondsee, Austria

J. Notenboom (477), Laboratory of Ecotoxicology, National Institute of Public Health and Environmental Protection, 3720 BA Bilthoven, The Netherlands

S. Plénet (477), Université Lyon I, U.R.A. CNRS 1451, Ecologie des Eaux Douces et des Grands Fleuves, Laboratoire d'Hydrobiologie et Ecologie Souterraines, 69 622 Villeurbanne Cedex, France

D. Poinsart (157), Université Lyon III, Département de Géographie, 69239 Lyon, France

S. Pomel (129), Laboratoire Environnement, Centre d'Etudes de Géographie Tropicale, CNRS, Domaine Universitaire de Bordeaux; and Laboratoire CIBAMAR, Géologie-Recherche, Université Bordeaux I; 33 405 Talence Cedex, France

R. Popa (217), Institute of Speology, "E. G. Racovitza", Rumanian Academy, 11 RO-78109, Bucuresti, Rumania

P. Pospisil (217, 347), Institute of Zoology, University of Vienna, 1090 Vienna, Austria

J.-L. Reygrobellet (425), Université Lyon I, U.R.A. CNRS 1451, Ecologie des Eaux Douces et des Grands Fleuves, Laboratoire d'Hydrobiologie et Ecologie Souterraines, 69 622 Villeurbanne Cedex, France

J. J. Simons (523), U. S. Environmental Protection Agency, Groundwater Protection Division, Washington, DC 20460

J. A. Stanford (7, 367, 543), Flathead Lake Biological Station, The University of Montana, Polson, Montana 59860

D. L. Strayer (287), Institute of Ecosystem Studies, Millbrook, New York 12545

M.-J. Turquin (477), Université Lyon I, U.R.A. CNRS 1451, Ecologie des Eaux Douces et des Grands Fleuves, Laboratoire d'Hydrobiologie et Ecologie Souterraines, 69 622 Villeurbanne Cedex, France

Ph. Vervier (425), Centre d'Ecologie des Ressources Renouvelables, CNRS, 31055 Toulouse Cedex, France

N. J. Voelz (391), Department of Biological Sciences, St. Cloud State University, St. Cloud, Minnesota 56301

J. V. Ward (7, 367, 391), Department of Biology, Colorado State University, Fort Collins, Colorado 80523

L. Zilliox (69), Institut Franco-Allemand de Recherche sur l'Environnement, 67037 Strasbourg Cedex, France

Introduction

When we consider the Earth's hydrosphere, we are always impressed by the extent of oceanic waters, which dominate the surface of the Earth. However, 97% of all freshwater is subsurface; lakes and rivers represent less than 2% (L'Volich, 1974). The continents' groundwater constitutes a subsurface hydrosphere (Pinneker, 1983). Complex linkages between surface water and groundwater are determined by geologic and climatic conditions existing within drainage basins, and organisms traverse this geohydraulic continuum in association with fluxes of energy and matter. This is the essence of groundwater ecology.

For a very long time ecological studies of groundwater, compared with other research (e.g., limnology and marine ecology), remained undeveloped. This was partly because of the technical difficulties associated with sampling the underground aquatic environment. But a complete revolution in groundwater ecological research has occurred. Investigations moved from descriptive natural science to a modern ecological approach; descriptive study of the "strange organisms" in caves is now complemented by population, community, and process-response perspectives in many different groundwater ecosystems. Research is focused on the dynamics and the functioning of these ecosystems at various scales, from the study of individuals and their microhabitats to the complex interactions occurring within large aquifers or parts of subsurface drainage basins.

Groundwater ecology is now a relevant domain of research, not only because the groundwaters represent the most extensive array of ecosystems on our planet (except perhaps in the oceans), but also because much of the world's human population depends on subsurface sources of uncontaminated water. Indeed, about 75% of the inhabitants of the European Community are supplied with groundwater (RIVM/RIZA, 1991), and in the United States 50% (more than 90% in rural areas) of the potable resource is groundwater (General Accounting Office, 1991).

I. OBJECTIVES OF THE BOOK

Although marine, lake, and stream research is widely published, groundwater ecology has not been synthesized to convey progress in the field. The present book is intended to fill this gap. The subject matter and authors were chosen to describe the multidisciplinary nature of modern groundwater studies. Environmental change is pervasive and increasing as a consequence of human activities. Ecologists must be informed and must react creatively to better understand and manage groundwater resources. It is hoped that this volume will play a seminal role in the development of groundwater ecology as an essential scientific discipline.

II. AUDIENCE

Everyone has a stake in aquatic ecosystems, and this book aims at a very wide audience. We hope that it will be useful not only to the student directly involved in groundwater ecological research, but also to those dealing with general topics in ecology. Both should be interested in the unique and newly discovered aspects of biological and physical pattern and process in groundwater, as well as how those features pertain to or elaborate general ecological principles. Ecologists who study landscape boundaries and the contact zones between surface water and groundwater environments (i.e., ecotones), or more precisely river–aquifer interactions (i.e., the hyporheic zone), will also find interesting information in this book. Geohydrologists and other scientists classically interested in aspects of groundwater quantity and quality may find relevence in coupling their work with ecological concerns, as described herein (Nachtnebel and Kovar, 1991). When planning new groundwater protection strategies, water managers may find the case studies summarized herein to be useful. Finally, the book may be used as a text for courses in groundwater ecology, which are now being contemplated or taught at the university level.

III. CONTENT AND STRUCTURE OF THE BOOK

Although in the past most books dealing with subterranean ecology relied heavily on cave organisms (Vandel, 1964; Mohr and Poulson, 1966; Ginet and Decou, 1977; Culver, 1982), this volume presents groundwater ecology in its full complexity; i.e., it considers both the ecology of the karst groundwaters (Biospeology sensu lato) and that of the porous media of unconsolidated rocks (Phreatobiology sensu lato), as well as the interaction zones between surface water and groundwater environments [Naiman and Décamps, 1990 (ecotone concept); Gibert *et al.*, 1990 (surface water/ groundwater ecotone)].

In some ways the book reflects the historical development of groundwater ecology (Fig. 1). Original research was mainly initiated by zoologists, and this tradition has survived. That is why this book contains more on the investigation of groundwater *Metazoa* than on microbiology and geochemistry recently reviewed by Chapelle (1993) and for the data of the deep subsurface microbiology see Frederickson and Hicks (1987) and the special issue of *Geomicrobiology* (1989). Much of the ecology of the subsurface environment is studied by researchers interested in basic research in evolutionary ecology, community ecology, ecosystem functioning, or a combination thereof. But the need for applied research, such as ecotoxicology and environmental chemistry, is now a priority for groundwater studies in many countries.

The book is composed of 20 chapters grouped in four sections. As a general introduction, the first chapter presents the characteristics of groundwater ecosystems and outlines some prospects of future groundwater ecological research. The physical and chemical characteristics of aquifers in various areas of the world are synthesized in the Chapters 2–6. Hydrogeology and geomorphology provide the template for ecological organization and community structure discussed in later chapters in the book.

The second section deals mainly with the relationships between subsurface organisms and their environment. The papers are topical because we wanted to summarize developments in groundwater ecology and review subjects that were seldom dealt with in the groundwater ecological literature, such as the problem of ecological constraint (i.e., limits to biological distributions, biotic fluxes and species interactions in groundwater, and adaptations of groundwater organisms).

The third section includes six chapters that involve case histories of

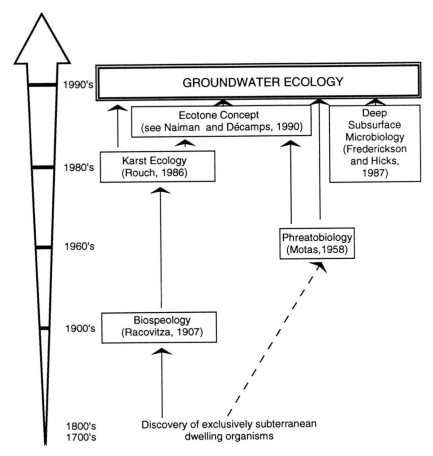

FIGURE 1   Diagrammatic representation of the merging of research perspectives in ground-water ecology [for more details, see the reviews by Danielopol (1982), Rouch (1986), Gibert (1992), and Camacho (1992)].

particular ecosystems, including large alluvial aquifers, karstic systems, and chains of aquifers along a large river. Several well-known research groups in groundwater ecology were asked to summarize their research. The papers emphasize the dynamics of animal communities in combination with hydrological and geomorphological processes, especially with respect to the contemporary view that groundwaters are open, interactive, and hierarchical networks with vital biological and physical connections to other environments.

The last section of the book discusses aspects of applied ecological research. One of the chapters deals with ecotoxicology at the organism level as well as at the ecosystem level. The main interest of this chapter is to show the ecological consequences of groundwater pollution and to discuss how our knowledge can be applied to groundwater management. The final two chapters discuss groundwater management with respect to protection and enhancement of the function of aquifers. Pure drinking water is a matter of national security; we cannot do without sustainable supplies. In this context, the book shows the urgency for managers and water agencies to incorporate the concepts of ecosystem science into new statutes for conservation and sustainable exploitation of aquifers and attendant biota. All in all, we believe that this book supports the idea that the problems of groundwater ecology are sufficiently progressive to be further supported by material and human capital.

ACKNOWLEDGMENTS

It is our pleasure to acknowledge the great cooperation of the contributors and reviewers who examined each chapter. The help of Academic Press in publishing this book is also acknowledged.

REFERENCES

Camacho, A. I., ed. (1992). "The Natural History of Biospeleology," Monogr. Mus. Nac. Cienc. Nat., 7, C.S.I.C., Madrid.

Chapelle, F. H. (1993). "Ground-Water Microbiology and Geochemistry." Wiley, New York.

Culver, D. C. (1982). "Cave Life: Evolution and Ecology." Harvard Univ. Press, Cambridge, MA.

Danielopol, D. L. (1982). Phreatobioloy reconsidered. *Pol. Arch. Hydrobiol.* **29**(2), 375–386.

Frederickson, J. K., and Hicks, R. J. (1987). Probing reveals many microbes beneath earth's surface. *A.S.M. News,* **53,** 78–79.

General Accounting Office. (1991). "Water Pollution: More emphasis needed on prevention in EPA's efforts to protect groundwater." USGAO. GAO/RCED-92-47. 57 p.

Geomicrobiology. (1989). Special Issue on Deep Subsurface Microbiology, 7, 1/2, 131 p.

Gibert, J. (1992). Groundwater ecology from the perspective of environmental sustainability. *In* "Ground Water Ecology" (J. A. Stanford and J. J. Simons, ed.) pp. 3–33. *Am. Water Resourc. Assoc.,* Bethesda, MD.

Gibert, J., Dole-Olivier, M. J., Marmonier, P., and Vervier, P. (1990). Surface water-groundwater ecotones. *In* "The Ecology and Management for Aquatic-Terrestrial Ecotones" (R. J. Naiman and H. Décamps, eds.), Man & Biosphere Ser. pp. 199–225. Parthenon Publ., London.

Ginet, R., and Decou, V. (1977). "Initiation à la biologie et à l'écologie souterraines." Delarge, Paris.

L'Vovich, M. I. (1974). "World Water Resources and their Future" (Engl. Transl. A.G.U.). Mysl' P. H., Moscow.

Mohr, C. E., and Poulson, T. L. (1966). "The Life of the Cave." McGraw-Hill, New York.

Motas, C. (1958). Freatobiologia, o noura ramura a limnologiei (Phreatobiology, a new field of limnology). *Natura (Bucharest)* **10,** 95–105.

Nachtnebel, H. P., and Kovar, K. (1991). Hydrological basis of ecological sound management of soil and groundwater. *IAHS Publ.* eds. **202.**

Naiman, R. J., and Décamps, H., eds. (1990). "The Ecology and Management of Aquatic-Terrestrial Ecotones," Man & Biosphere Ser., Vol. 4. Parthenon Publ., London.

Pinneker, E. V. (1983). "General Hydrogeology." Cambridge Univ. Press, Cambridge, UK.

Racovitza, E. G. (1907). Essai sur les problèmes biospéologiques. Biospeologica I. *Arch. Zool. Exp. Gén.* **4,** 371–488.

RIVM/RIZA (1991). "Sustainable Use of Groundwater: Problems and Threats in the European Communities." RIVM/RIZA, The Netherlands.

Rouch, R. (1986). Sur l'écologie des eaux souterraines dans le karst. *Stygologia* **2,** 352–398.

Vandel, A. (1964). "Biospéologie: La biologie des animaux cavernicoles." Gauthier-Villars, Paris.

1

Basic Attributes of Groundwater Ecosystems and Prospects for Research

J. Gibert,* J. A. Stanford,† M.-J. Dole-Olivier,*
and J. V. Ward‡

*Université Claude Bernard-Lyon I
U.A./C.N.R.S. 1451
Ecologie des Eaux Douces et des Grands Fleuves
Laboratoire d'Hydrobiologie et Ecologie Souterraines
69622 Villeurbanne Cedex, France

†Flathead Lake Biological Station
The University of Montana
Polson, Montana 59860

‡Department of Biology
Colorado State University
Fort Collins, Colorado 80523

I. INTRODUCTION

Understanding of the structure and function of the world's ecosystems changed considerably during recent years as additional dimensions were progressively taken into account. Modern theories on pattern and process hierarchization (O'Neill *et al.*, 1986; Kolosa, 1989; Kotliar and Wiens, 1990), patch dynamics (Pickett and White, 1985; Townsend, 1989), and ecotones and boundary dynamics (Naiman and Décamps, 1990; Hansen and di Castri, 1992) have demonstrated the nonlinear nature of biotic and abiotic processes and stressed the need for multiscale (space and time) studies. The view of the factors controlling the structure and operation of aquatic ecosystems has expanded and gradually evolved from a unifactor to a multifactor orientation (Minshall, 1988). Moreover, the scale on which a process is examined may influence perception of controlling factors (Auger *et al.*, 1991). This leads to a variety of problems associated with sampling, data acquisition, experimental design, extrapolation of results, and, especially, synthesis of ecosystem structure and function. However, all ecosystems are fundamentally defined by the way the organic matter and biogenic elements are organized in time and space with respect to environmental variability (e.g., Andrewartha and Birch, 1954).

Our objective in this chapter is to describe the attributes of groundwater ecosystems with respect to scalar phenomena and environmental change. We discuss the classical views that initiated groundwater ecology as a science and argue for a more process-based approach. We then elucidate problems and challenges in groundwater ecosystem analysis. Finally, we discuss the relevance to groundwater ecology of some prevalent theoretical concepts in ecology.

II. THE GROUNDWATER REALM: THE CLASSICAL VIEW

Ecological descriptions of the underground environment were initially developed with reference to surface systems. Table I emphasizes the most obvious differences between the two environments and gives a general description of the phenomena that characterizes them. Such a representation does not, however, demonstrate the great heterogeneity and the real diversity occurring in groundwater systems (e.g., components, fluxes and organismal responses) (Rouch, 1986). More complete descriptions of groundwater systems are available in former works, such as Delamare-Deboutteville (1960), Vandel (1964), Mohr and Poulson (1966), Ginet and Decou (1977), and Culver (1982).

TABLE I Biological and Physical Organization of the Groundwater Environment in Contrast to Surface Waters

	Underground system	Surface system
Environment	Constant darkness	Light (alternance day/night)
	Habitats: restricted variety (no vegetation) and small size	Habitats: high diversity
	Environmental fluctuations	
	Physical inertia, predictability	Frequent and large variations, low predictability
Organisms	Morphological, physiological, and behavioral specializations to underground environment	No adaptation to underground environment (except ecological one for ubiquitous species)
	Special classification of organisms defined by their relation to the underground habitat	
Communities Populations	A selection dominant	r or K selection dominant
	Mobility toward a nondominant strategy in relation to habitat predictability (or unpredicability) and favorableness (or unfavorableness)	
Biocenosis	Richness, diversity, and density: low and variable	Richness, diversity, and density: generally high and variable
	Biotic constraints: low	Biotic constraints: generally higher
	Cenotic strategy i	Cenotic strategy i or s
	Population concentration in interstitial habitat near contact points with surface one	Different spatial distribution
Functional characteristics	Heterotrophy and allotrophy	Autotrophy
	With or without rigorous resource filtration by the habitat	Optional resource filtration
	Low trophic resource polyphagous diet (detritus feeders dominant)	Very important trophic resources
	Short and simple food webs	High diversity of diets and possibility of very complex food webs
	System with low productivity	System with higher productivity

A. Physical Environment

The primary characteristic of the physical environment is permanent darkness; the disappearance of the nycthemeron is without any doubt the fundamental property, and almost all the biotrophic attributes are influenced by the constancy of total darkness. The only exception occurs in the immediate vicinity of portals in "open" karstic systems, where light may have an attenuated influence.

Another fundamental characteristic is general reduction in the diversity of habitats in both alluvial and karstic environments and reduction of available space within interstitial environments. In the case of karsts, reduced habitat diversity is due essentially to the absence of vegetation. However, physical heterogeneity still exists due to diversity of void shape and size (see Chapter 2). The interstitial environment is even more unfavorable. Vegetation is absent, and physical variety is reduced to a minimum because there is only one habitat type—the pore space—which is considerably reduced in size relative to surface environments. However, pore space has considerable heterogeneity (i.e., the arrangement of grains, the void size, and the physical and chemical characteristics of aquifers) and determines habitability of the interstitial environment by metazoan animals.

Physical inertia increases with depth in interstitial environments; hydrologic and chemical variation, which is often dramatic in surface waters, usually is not very evident in interstitial substrata many meters below the surface (Castany, 1985). Karst is much more heterogeneous, varying from "inertial" systems, which store large water reserves with long retention time (years or longer), to "noninertial" karstic systems, which store very little water and whose flux is rapid (days or shorter) (Mangin, 1983, 1986). With respect to physical stability of the karst environment, it is necessary to recognize the degree to which the pattern is predictably repeated in time as water passes through the karst aquifer system. In many instances in which the hydrology is well documented, discharge patterns may not vary over long time periods owing to the moderating effect of aquifer storage. Rouch (1980, 1986) and Turquin (1981) refer to the predictability of underground systems. Predictability of inertial karstic systems is due to constancy, whereas the predictability of noninertial karstic systems results largely from contingency. Hence, the notion of stability is a relative one, but stability is usually much greater in groundwaters than surface waters. Ramifications for organisms are profound.

B. Groundwater Fauna

1. Biological Characteristics

Hypogean organisms display biological, morphological, physiological, or behavioral characteristics that appear to be linked to the physical limitations of the environment ("troglomorphic features"; Christiansen, 1962, 1992). Three morphological characteristics of "archetypal" groundwater animals that may be linked to the absence of light are a general lack of pigmentation, an ocular regression (microphtalmy or anophtalmy), and hypertrophy of sensory organs. Appendages tend to be long and numerous, and highly developed chemical and mechanical receptors are usually present.

In addition many authors have noted a decrease or increase in size compared with that of epigean relatives and a general convergence of body shape for different taxa. Organisms as different as the Annelida, Planaria, or Crustacea tend to become longer and thinner (vermiform). This characteristic is mainly true for interstitial animals, because of the reduced size of the available living space. Subterranean fauna also are generally distinguished by relatively slow metabolic rates. This is expressed by a longer period needed for ontogenetic processes, a lengthening of each stage of the life cycle, an increase in longevity, and less frequent reproduction. Primary ethological characteristics are reduced motor and reproductive output and unique behaviors such as stereotropism, thigmotropism, thigmotactism, or various other ethophysiological strategies (Ginet and Decou, 1977; Holsinger, 1988).

2. Ecological Characteristics

Strategic ecological pathways followed by the organisms to adapt to underground life are numerous and quite varied. Moreover, some epigean organisms move in and out of groundwaters and some groundwater forms are occasionally found at the surface. Yet, others are found only in groundwater. This led some authors to draw up special nomenclatures, which are based on morphological, behavioral, and ecological adaptations of animals to underground life and on their presence or absence in caves [troglofauna; see review in Camacho (1992)] or in various types of groundwater habitats (stygofauna; Ginet and Decou, 1977; Botosaneanu, 1986). In addition, stream ecologists introduced for porous alluvia the notions of occasional and permanent hyporheos (Williams, 1984) and the terms of stygobionts and amphibionts (Stanford and Ward, 1993). These terms are more complementary than exclusive. We propose the *sensu lato* (stygo) as the most ecologically descriptive of groundwater fauna, and we applied this classification to porous alluvia (see Fig. 1). However, the Schiner–Racovitza classification scheme (troglofauna) is currently used by many authors (see, for example, Chapter 9).

Stygoxenes are organisms that have no affinities with the groundwater systems, but occur accidentally in caves or alluvial sediments. Nevertheless, stygoxenes can influence processes in groundwater ecosystems, for example, functioning as either predators or prey. Some vertebrates are mentioned as stygoxenes in karst systems (*e.g.*, fishes and salamanders; Ginet and Decou, 1977). Some planktonic groups (Copepoda: Calanoida) or species (Cladocera) and a variety of benthic crustacean and insect species may passively infiltrate alluvial sediments.

Stygophiles have greater affinities with the groundwater environment than do stygoxenes, because they appear to actively exploit resources in the groundwater system, actively seek protection from unfavorable situations in the surface environment resulting from biotic or stochastic processes, or both. In porous aquifers, stygophiles are subdivided into three categories:

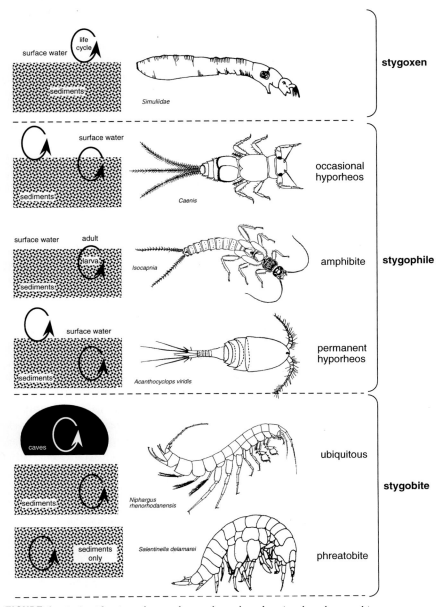

FIGURE 1 A classification of groundwater fauna based on its phenology and its presence or absence in various groundwater environments.

(1) the occasional hyporheos, (2) the amphibites, and (3) the permanent hyporheos. The occasional hyporheos consists of larvae, especially of aquatic insects, of most of the benthos. Eggs may hatch, and early instar larvae

reside in this zone during their early development, whereas later life stages predominate in the benthos (Williams, 1984; Pugsley and Hynes, 1985, 1986). In this case individuals of the same species could either spend all their lives in the surface environment or spend a part of their lives in the surface environment (in addition to the obligate aerial stage) and a part in groundwater (early stages for protection from predators or later stages for more favorable conditions). Amphibite species are a particular kind of stygophile because their life cycle necessitates use of both surface water and groundwater systems. A taxonomically variable group of *Plecoptera* species, including members of the *Isocapnia, Paraperla, and Kathroperla,* lives in the total darkness of the deep hyporheic zone of the Flathead River (Montana) for one or more years before returning to the river to emerge as terrestrial adults (Stanford and Gaufin, 1974; Stanford and Ward, 1988, 1993). In this case, the interstitial stage is predominant, epigean life being restricted to emergence and aerial adult stage (Berthelemy, 1968). The permanent hyporheos consists of many species (both immature and adult) of nematodes, oligochaetes, mites, copepods, ostracods, cladocerans, and tardigrades that may be present during all life stages either in groundwater or in benthic habitats. In this case, an aerial epigean stage is not necessary to complete the life cycle.

Stygobite organisms are specialized subterranean forms, obligatory hypogean as described above. Some are widely distributed in all types of groundwater systems (both karst and alluvia), and sometimes they are found very close to the surface as in forest canals under dead leaves, where they occur in great numbers (Ginet and David, 1963). A good example of such an ubiquitous stygobite is provided in the Rhône River basin by the amphipod *Niphargus rhenorhodanensis*. In contrast, phreatobite species are stygobites that are restricted to the deep groundwater substrata of alluvial aquifer (phreatic water).

3. Biotic Assemblages

In response to habitat properties, the population usually adopts a demographic strategy corresponding to the adversity–selection model or A-type strategy of Greenslade (1983). This strategy is particularly descriptive of many subterranean organisms (Rouch, 1986), which are characterized by a long life, late maturity, and low rates of development, fecundity, and growth. Likewise, the A-type strategy usually occurs in physically predictable, but ecologically unfavorable, environments, as described above for hypogean systems.

Reproductive strategy often responds to an extreme situation that may be modulated by changing environmental conditions; for example, whenever more favorable conditions eventuate an increase in available food resources, some populations begin to reflect K-type strategies (e.g., high diversity). If the environment becomes less predictable, such as in systems with highly stochastic hydrology, the populations may take on r-type strategies (e.g., fecundity closer to that of many surface organisms, higher growth rates, variable population density, and so on). This variability expresses the partial overlapping of the three strategies, r, K, and A, taken into account by Greenslade's definition (1983) according to gradients of predictability and favorableness.

Underground biocenoses are less speciose and populations are less dense in the majority of epigean environments; therefore the different biotic pressures, such as predation and interspecific competition, are correspondingly less severe. Principles of community organization were developed by Blandin *et al.* (1976), who recognized two types of cenotic strategies. The i-type strategy is based on the ability of individuals to react when the environmental conditions changed, whereas the s-type strategy is based on species richness and functional redundancy (Table II). According to Rouch (1980), the cenotic strategy of hypogean animals is probably of the i type. The author bases his statement on the reduced diversity and richness of hypogean biocenoses rather than on the genetic properties of natural populations.

Another important characteristic of subterranean biocenoses is their nonuniform spatial distribution . For instance, higher biodiversity and concentrations of organisms occur in or near the transition zones between surface water and groundwater rather than deeper in the groundwater system.

TABLE II   Comparison of Attributes Describing Systems in Which Environmental Variation and Biodiversity Are low with Those Describing Systems in Which Environmental Variation and Biodiversity Are Higher (from Blandin et al., *1976)*

Biocenotic strategy	
s-Strategy species	i-Strategy individual
High ⟨ Species richness / Biodiversity / Quantity of energy	Low ⟨ Species richness / Biodiversity / Quantity of energy
Constant energy fluxes	Discrete energy fluxes: simple and short ways for the circulation of matter and energy
Potential functional redundancy of species	Dominance of one species
Modification of the environment:	
Substitution of extirpated species by others sustaining the same function in the ecosystem	Ecosystem survival depends on the survival of all the species and on the capacity of populations to produce genetically new individuals

4. Energy within the Groundwater Environment

Lacking light energy, groundwater food webs are almost entirely hetero-trophic. Bioproduction is largely dependent on transport of resources from the surface (allotrophy). Materials entrained from the surface include bio-mass and detritus. Quantity depends on the possibilities that exist on the surface, on the dynamics of the water, and also on the filtration potential of the underground environment. Most unpolluted groundwaters are charac-terized by scarce trophic resources. Food webs are generally simple, charac-terized by few trophic links, which limits energy loss from one component to the next and ensures a high energy efficiency. Groundwater microbes, including bacteria, fungi, and protozoans, are the primary consumers and may themselves be consumed by micro- and macroinvertebrates. But inverte-brate diets are not specialized; most are polyphagous [Ginet and Decou, 1977; see examples in Culver (1985)]. Some studies of cave fauna also examined trophic relationships (Culver *et al.*, 1991; see also Chapter 10). The adaptive response to food-poor waters by aquatic cave animals is displayed by their resistance to starvation and their feeding efficiency (Cul-ver, 1985).

III. GROUNDWATER ECOSYSTEMS

A. The Conceptual Foundation

From the above discussion it is apparent that the classical view of groundwater ecology emphasized low biodiversity and bioproductivity. Or-ganisms are highly adapted to life in dark, physically stable environments, in which energy resources are generally limited. However, owing to the dependency on advective fluxes and elucidation of the existence of complex biological and physical interactions, groundwater environments are now being synthesized from an ecosystem perspective.

It is in karsts that the roots of the concept of the groundwater ecosystem are found. Karst was defined as a discontinuous and anisotropic system with topographical and hydrogeological features corresponding to a physical entity (Castany and Margat, 1977). Hence, the former simple notion of the cave environment changed to recognize the karst aquifer with two fundamen-tal superimposed zones, a vadose zone and a saturated zone. Caves and their habitats were thus considered as parts, or subunits, of a broader and coherent physical unit. Mangin (1974, 1975) defined a karstic system as a functional unit, involving organized flow pathways forming a drainage unit. This definition emphasizes physical structure (void organization) and func-tion (hydrological aspects, organization of water flows, and importance of storage). Rouch (1977) demonstrated that physical subunits of karst corresponded to biological components, hence defining for the first time the concept of karstic ecosystem. This concept, developed for the Baget system (France) (Mangin, 1986; Bakalowicz, 1986; Rouch, 1980, 1982), was confirmed on the Dorvan karst system, whose structure and dynamics were studied for more than 10 years (Gibert *et al.*, 1978; Gibert, 1986, 1989). In the same way, Danielopol (1982) presented porous and riverine aquifers as conceptual ecosystems, open systems with input—output zones and finite configurations, which can be visualized from a process—response perspective. As in the karst system, different subunits of porous aquifers were based on hydrological and geomorphologic (*i.e.*, physical) features, which yield interactive biological components (Creuzé des Châtelliers, 1991a). We recognize that groundwater systems should not be viewed as semidesert areas, colonized by rare relict forms, but, rather, as dynamic ecosystems responding to a plethora of ecological processes similar to occur within the surface realm (Rouch, 1977).

B. Hierarchical Approaches and Keystone Processes in Groundwater Ecosystem Dynamics

Complex processes determine the structure and function of groundwater ecosystems. As in epigean stream ecosystems (Minshall, 1988), these processes are expressed at different spatial and temporal scales: the microhabitat or microscale (10^{-1}–10^0 years; 10^{-2}–10^0 m^3), the habitat (aquifer sector) or mesoscale (10^0–10^2 years; 10^0–10^2 m^3), the aquifer or macroscale (10^2–10^4 years; 10^2–10^5 m^3), and the regional/continental drainage or megascale ($>10^4$ years; $>10^5$ m^3). These domains are a nested series of spatial and temporal configurations, each integrating all the patterns and processes ongoing at lower levels within the hierarchy and each linked by the next larger scale (Allen and Starr, 1982; O'Neill *et al.*, 1986; Delcourt and Delcourt, 1988). Thus, the main biological and physical factors responsible for the structure and operation of a groundwater ecosystem must be viewed at different levels of resolution (Fig. 2). Moreover, if we can identify the main processes within the dimensions of these domains, it is obvious that processes can cross the arbitrary boundaries and be expressed on more than one scale, underscoring the complexity of processes and responses in an ecosystem context.

The megascale domain ranges from regional, through continental, to global and concerns the majority of geologic time during which plate tectonics have changed the configuration of continent basins. Thus, geology (lithography, stratigraphy, structure, and formation of the geological deposits) and climate, including glaciation during the Quaternary period, marine regressions, and marine transgressions, have exerted a determining control on the nature and distribution of aquifers. Orogeny, for example, the uplifting of the Alps, produced compartments within aquifers and a multiplicity of geomorphologic evolutions (Maire, 1990). Basin filling by nonindurated

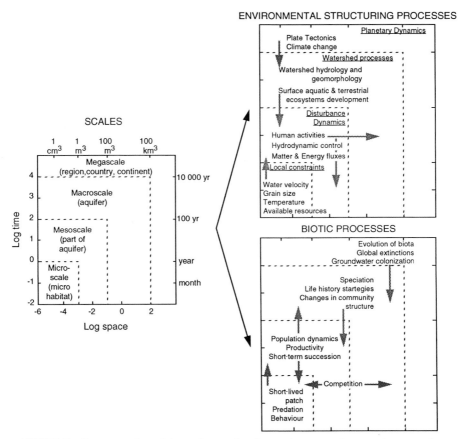

FIGURE 2 Representation of the various scales that are important in groundwater ecology and the biological and physical processes that influence ecosystem processes at those various scales.

deposits, such as alluvium, has occurred in nearly all regions of the planet and resulted in storage of important sources of groundwater (Freeze and Cherry, 1979). In limestone areas, karstification (Gèze, 1965; Bakalowicz 1979; see also Chapters 2 and 5) produces voids in the limestone matrix and controls water infiltration and storage processes.

All these processes have influenced biotic colonization of the underground realm. Extant animals living in groundwaters have evolved from surface water ancestors that lived either in the sea or in freshwater bodies. Whatever the origin and mode of colonization (*i.e.*, active dispersion of generalist animals or passive invasion), groundwater biota gradually became geologically isolated from ancestral marine and littoral populations. Hence, zoogeography of groundwater biota involves the entire biosphere, and the species distributions of stygobites is a shifting mosaic of communities through the geological ages (Rouch, 1986).

At the macroscale, the hydrologic and geomorphologic processes of the catchments determine aquifer properties, the distribution of voids, and the water circulation characteristics. The surface environment, represented by lands (with various vegetation types), rivers, and lakes, also exerts a primary control over the underground circulation of matter and energy. Aquifers receive, from the surface, climatic pulses and organic matter that are essential elements for their life and evolution. Exchanges take place, and different degrees of connectivity can change potential capacities and characteristics of aquifers (Gibert *et al.*, 1990). These processes create mosaics of various aquifers at the scale of regions or countries. Flux of water and suspended and dissolved materials directly or indirectly determine the habitability of the geohydraulic milieus and associated changes in community structure. Large spaces in karstic channels enable a large size range of organisms to occur; therefore, larger species are able to range widely in the deep saturated zones (Rouch *et al.*, 1968; Gibert, 1986). In contrast, very tiny spaces in sandy aquifers create an impassable barrier for many organisms. Hence, the aquifer pore space influences not only biological and physical fluxes but also speciation and life history strategies (Ginet and Decou, 1977; Juberthie, 1984; Williams, 1991).

The mesoscale domain encompasses processes expressed as hydrodynamic controls, matter and energy fluxes, and the human impact. Aquifers are influenced to a greater or lesser extent by different types of natural disturbances, floods, drought, and inundation, and to a lesser extent by sedimentation and erosion. Most disturbances are not extreme and have no long-term effect. Hence, the system remains in a stable quasiequilibrium condition (Schumm and Lichty, 1956) that may persist over a time scale of 10^2 years. At this scale the impact of humans can be very important by overexploitation of groundwater for drinking water supply or for irrigation or industrial needs. Over time, the vast subsurface reservoir of freshwater, which was relatively unblemished by human activities a few decades ago, is gradually becoming degraded. Degradation of groundwater quality often requires long periods of time before the true extent of the degradation is readily detectable (Freeze and Cherry, 1979). The consequences are eutrophication, clogging of interstices, altered productivity, and loss of ecosystem resiliency. On this scale of 10^2 years and from a biological point of view, short-term succession occurs in response to changes in environmental gradients and predominant disturbance regimes. Floods influence the life in groundwater systems by (1) enrichment of energetic resources and modification of the organic matter stored, (2) dispersing nutrients and allochthonous fauna into the aquifer, (3) restructuring groundwater habitats and populations, and (4) extirpating autochthonous species (Barr, 1968; Gibert, 1986, 1989; Marmonier and Dole, 1986). Hence, an ecosystem in the mesoscale domain is quite dynamic, but may be viewed in the context of a quasiequilibrium state (naturally resilient and long-term stable); unless it is damaged by human pollution or the rare occurrence of some other very catastrophic disturbance, in which case the system proceeds to a new quasiequilibrium state possibly characterized by very different biological and physical conditions.

The microscale domain encompasses events occurring over the year and on the spatial scale of the pore, fissure, or channel. The processes take place generally during the annual hydrological cycle. The local constraints on biota include water velocity, grain size, temperature, and available energy resources. These processes generate short-lived patches and promote colonization by epigean animals.

What are the significant scale considerations in groundwater ecology? For many years ecological studies focused mainly on the microscale (Vandel, 1964; Poulson and White, 1969; Delamare-Deboutteville, 1971). In the past 20 years the view has expanded to encompass larger spatial scales and long-term studies (mesoscale and macroscale) (Rouch, 1986; Danielopol, 1992). This evolution is due to (1) better understanding of aquifer hydrodynamics, (2) use of new techniques and new sampling methods (Mathieu *et al.*, 1991), and (3) improvements in data processing in the field and laboratory. Many ecological patterns and processes of direct relevance to the relatively new discipline of landscape ecology are resolvable between the microscale and the mesoscale domains. Good examples are given in this volume.

It is not enough to merely recognize the existence of processes at different scales; we also have to consider interactive processes that transfer across scales. For example, autochthonous microbial metabolism in pristine groundwaters appears to be low, limited by the paucity of dissolved and particulate organic matter derived from external (catchment) sources, and influenced by water flux rates. Especially in interstitial systems, redox gradients that locally influence oxidation pathways are interactive with porosity and entrainment of allochthonous organic matter. Moreover, food web interactions at the microhabitat scale, such as the microbial loop involving consumption of bacteria by protozoa, are directly influenced by microbial metabolism. Hence, elementary food web formation is influenced by processes that interactively are derived from catchment to pore size scales, and they probably are manifested in time rather randomly in relation to climatic variations. This aspect of functional hierarchy is one of the most important unresolved problems in contemporary groundwater ecology. A few new mathematical tools, such as chaos theory, fractals, and complexity theory (Naiman *et al.*, 1988), provide new insights but they have only recently been used in hydrogeological models (Mangin, 1986; Curl, 1986; Laverty, 1987; Lenormand, 1985) and have not been examined in an ecological context in groundwaters.

C. Heterogeneity of Groundwater Systems, Boundaries, and Ecotones

Heterogeneity is produced by patchiness and boundaries that define patches, and both are scale dependent. Within groundwater systems, at megascale and macroscale, the "grain" size is large and the "extent" is small (Kotliar and Wiens, 1990); detection of some boundaries is difficult because the discontinuities are "averaged out" by inexplicit measures or inappropriate units. At mesoscale and microscale the aggregation rate is high and heightens patch boundaries. For example, patterns of allopatry and sympatry in populations of cave crustaceans may be determined by the intensity of competition for resources (Culver, 1982).

However, at all scales of investigation, the transition zones between surface and underground ecosystems is recognized as a critically important boundary or ecotone (Gibert *et al.*, 1990; Vervier *et al.*, 1992). Within this ecotone, ecological contrast, amplitude of change of biological and physical variables, and stochasticity of time-series pattern are the highest relative to the geohydraulic continuum. Rates and directions of matter and energy flux change dramatically in the ecotone. As further elaboration of the scale transfer problem noted above, the surface water–groundwater ecotone may control the metabolism of the underground systems because it may function as a temporary or permanent sink of inorganic and organic matter derived from the catchment (Pinay *et al.*, 1990). Hence, ecotones may also function as natural filtering and buffering systems, thereby improving water quality. This concept has been used to design artificial groundwater recharge schemes (Vanek, 1994).

In the surface water–groundwater transition zone community dynamics are a function of the contiguity between the environmental conditions and the ecological status of organisms (*i.e.*, life-history traits). Community dynamics reflect the collective responses of the set of conditions present in the space–time window of interest. The heterogeneity of the contact zones can be represented considering different types of exchanges, from clogging zones to permanent and continuous relations in space and time through different types of aquifers. That is why it is often difficult to precisely define the ecotone.

For example, the definition of the hyporheic zone of alluvial rivers has been variously interpreted in biological (penetration of amphibites) or physical (penetration of river water) terms (Bretschko, 1992; Pennak and Ward, 1986; Stanford and Gaufin, 1974; Triska *et al.*, 1989, 1993; Stanford and Ward, 1988, 1993), and, although the problem is somewhat semantic, misinterpretations may occur. Conceptually, the seminal definition[1] (Orghi-

---

[1] "Cet habitat est constitué par les cavités interstitielles imbibées d'eau des graviers qui tapissent le lit des rivières. Les eaux interstitielles proviennent soit uniquement des eaux de la rivière, par le phénomène nommé "underflow", soit des eaux phréatiques des rives, qui s'écoulent vers la rivière . . . le biotope hyporhéique constitue une zone intermédiaire entre les eaux épigées et les eaux phréatiques, les valeurs de ces facteurs étant également intermédiaires . . ." (Orghidan, 1955).

dan, 1955) restricted the hyporheic zone to the interstitial spaces constituted by the sediments of the stream bed (involving lateral limits of this zone), which is considered a transition zone between surface water and groundwater (involving vertical limits). This construct takes explicitly into account the spatial and temporal variability of the transition zone, because interstitial water can be derived from the surface stream only (*i.e.*, underflow or downwelling), from the phreatic zone only (*i.e.*, drainage area or upwelling), or from a mixing of surface water and groundwater (with intermediate values for physicochemical parameters). Many biotic and abiotic tracers can be used in attempt to delineate the hyporheic zone, but they provide many different responses (Gibert *et al.*, 1990). Hence, considerable spatial and temporal variations may exist, and it is impossible to find a unanimous definition for the exact limits of this zone. Consequently, European investigators rarely use the term hyporheic zone, and others, especially in the United States, usually use it but too often do not parameterize. Consequently, we recommend that, in each case, the parameter used for delimitation of the hyporheic zone be clearly stated. In any case, these problems of terminology must not hold up progress toward a better understanding of the processes within the hyporheic zone. We conclude that the solution is to think in terms of a dynamic ecotone or transition zone (Vervier *et al.*, 1992; Stanford and Ward, 1993).

IV. RELEVANCE OF SOME PREVALENT CONCEPTS IN ECOLOGY TO GROUNDWATER ECOLOGY

It is important for the advance of the discipline that groundwater ecology should rest on a solid theoretical foundation and that intellectual exchange with other fields be encouraged (Danielopol, 1980; Culver, 1982). Most major concepts developed from general and aquatic ecology (Cummins *et al.*, 1984; Minshall *et al.*, 1985; Stanford and Covich, 1988; Ward and Stanford, 1991; Carpenter *et al.*, 1992; Naiman, 1992) are also applicable to groundwater systems, although few have been intensively investigated in a hypogean context. If we consider only the groundwater milieu, some topics are reasonably well studied at the organismal/population level (colonization, speciation, ecological adaptations, and limiting factors), at the community level (food webs, species diversity, and zonation), and at the ecosystem level (energy flow, island biogeography, and role of disturbance).

In the discussion that follows, the relationships of several of these concepts to groundwaters are briefly considered under the following topics: (1) biodiversity, (2) stability and disturbance, (3) trophic dynamics, and (4) spatio–temporal pathways. Because we agree with Hynes (1983) that groundwaters must be incorporated into stream ecosystem theory, key interactions between groundwaters and lotic ecosystems are included in the ensuing material.

A. Biodiversity

Modern perspectives of biodiversity recognize the importance of diversity at genetic, species, ecosystem, and landscape levels of organization (Noss, 1990) as well as the importance of processes that influence biota at all levels of organization (*i.e.,* the functional diversity, Marmonier *et al.,* 1993). Groundwaters constitute a diverse array of habitats that contain a reservoir of biodiversity, which has been largely or totally ignored in calculations of global biodiversity. It has been estimated that the actual number of extant epigean species in the tropics is 10 to 20 times greater than the number currently known to science (Cairns, 1988). There is no denying that tropical epigean organisms are poorly known, but it is likely that on a global scale the groundwater biota is virtually unknown. It appears that groundwater biotopes contain highly diverse faunas with high levels of local and regional endemism (Marmonier *et al.,* 1993). In addition stygophiles may also be found in suitable epigean biotopes. As noted above, many truly epigean species colonize, or even depend upon, near-surface hypogean waters, especially when immature, which dramatically increases groundwater biodiversity.

What factors might lead to faunal diversification in groundwaters? On an evolutionary scale, great age and the relative constancy and predictability of habitat conditions would seem conducive to speciation. Although groundwater systems are interconnected and as noted above many stygobites are ubiquitously distributed and genetically vagile, habitat fragmentation may be sufficient to promote reproductive isolation (Barr and Holsinger, 1985). Ancestral species of ancient freshwater lineages were fragmented by plate tectonics. For example, crangonyctid amphipods and asellid isopods probably originated in Laurasia prior to the separation of North America and Eurasia (Barr and Holsinger, 1985), and bathynellaceans may have evolved in freshwaters prior to the breakup of Pangaea (Schminke, 1981). Groundwater faunas are comparatively protected from major climatic changes and other catastrophes that have led to massive extinctions of the epigean biota. For example, hypogean animals may persist in groundwater refugia beneath glaciers (Botosaneanu and Holsinger, 1991). In addition, climate changes may facilitate the evolutionary invasion of groundwaters by epigean forms. Thermophilic relicts colonized groundwaters during glaciation because of the winter-warm conditions, whereas psychrophilic relicts colonized groundwaters as the climate became warmer and drier in postglacial periods. Marine regression or tectonic uplift, resulting in the gradual freshening of coastal interstitial and crevicular habitats, has facilitated the evolutionary invasion of inland groundwaters by marine animals (Stock, 1980; Holsinger, 1988; Notenboom, 1991).

Notenboom (1991) considers marine regressions "as vicariant events in the evolutionary history of thallasoid stygobiont lineages" because populations may be stranded in a new environment as sea level falls and groundwaters freshen. In groups with low vagility, small founder populations are isolated in brackish/freshwater coastal stygohabitats, which leads to peripatric speciation (Mayr, 1982).

On an ecological time scale, spatial and temporal heterogeneity tends to promote biodiversity. Lateral migrations of rivers across their floodplains produce a mosaic of habitat patches, disturbance levels, and successional sequences; these phenomena may be responsible for the high biodiversity of the (epigean) Amazon biota (Salo *et al.*, 1986). Floodplain dynamics also create a diversity of habitats in alluvial aquifer systems (Boulton *et al.*, 1992; Danielopol and Marmonier, 1992; Stanford and Ward, 1993; Ward *et al.*, 1994), thereby potentially enhancing the diversity of hypogean forms. Owing to physical heterogeneity in karst systems, the diversity of habitats and species is often high, not only at the scale of the flow pathway (drain), but also at the aquifer scale (Rouch *et al.*, 1968; Culver, 1982; Gibert, 1986).

B. Environmental Stability and Disturbance

We began this chapter with a discussion of early views of groundwater ecology and noted several times that both karst and interstitial systems are relatively benign (nonvarying) compared with surface environments. But it is now recognized that all groundwater ecosystems are characterized by some amount of physical variation. As a consequence, distribution and abundance of groundwater biota may be examined as for surface biota, from the perspective of patch dynamics (Pickett and White, 1985) or gradient analysis (Hall *et al.*, 1992). In addition, from an ecosystem perspective, aquifer systems are progressively more variable or "open" the closer they are to the surface, the greater the pore space, or both.

The role of disturbance is a central theme of ecology in general and stream ecology in particular (Resh *et al.*, 1988). Disturbance undoubtedly serves as a major structuring force in groundwater environments, where cause and effect are manifested over a range of scales (Fig. 2). The effect of disturbance, such as spates, is modified as it propagates into aquifers in relation to the filtering effect of the transition zone and the structure of the aquifer. Indeed, Mangin (Chapter 2) eloquently described propagation of flood waves through a complex karst aquifer and showed that physical heterogeneity is a primary property of karst that cannot be described by simple linear relations. Long-term data sets have been used to analyze flood disturbance and recovery in karstic systems and their biological responses: variations of drift density and rate, changes of community structure, and species demography (Rouch, 1977, 1980; Turquin, 1981; Gibert, 1986; see also Gibert *et al.*, this volume). They are influenced by year-to-year variations in flow patterns, long-term trends of the dry and wet weather cycle, and lags in the response of short- and long-lived organisms. Indeed, month after month, flood after flood, biological responses can change in some parts of the karst system. However, if we consider the karst system as a whole, then we assume that most communities are likely to be resilient and have reached equilibrium. Some parts of the system probably have a high "physical stability" comparable to that encountered in porous aquifers, which authorized the development of permanent and diversified communities. In ecotonal conditions, floods may temporarily disrupt normal vertical distribution patterns of biota, although this may partly reflect spatial shifts in ecotonal conditions (Marmonier and Dole, 1986; Gibert *et al.*, 1990; Palmer, 1990; Marmonier and Creuzé des Châtelliers, 1991; Vervier and Gibert, 1991).

The intermediate disturbance hypothesis, initially developed to explain diversity patterns in tropical forests and coral reefs (Connell, 1978) and subsequently applied to the stream biota (Ward and Stanford, 1983), may also be applicable to the biodiversity of groundwater biota. The essence of this hypothesis is that biodiversity will be low under conditions of extreme disturbance in which only fugitive species can survive; it will also be low in the absence of disturbance because superior competitors exclude other species and will attain the highest diversity at intermediate levels of disturbance at which fugitive species and superior competitors coexist. Dole (1983) compared the fauna of three reaches corresponding to different physical stability levels in the alluvial plain of the Rhône river. The highest number of taxa occurred at the site with intermediate physical stability, thereby providing preliminary support for the intermediate disturbance hypothesis. In addition to changes in biodiversity, the composition of assemblages was modified in relation to physical stability. Sites that were unstable contained a fauna dominated by epigean forms well adapted to low physical stability, whereas sites far from the river, were characterized by a diverse fauna of stygobites intolerant of disturbance (Marmonier *et al.*, 1993). Other examples are discussed by Dole-Olivier *et al.* (Chapter 12).

C. Trophic Dynamics

The basic food resource in most groundwater ecosystems is organic matter from external sources. Hypogean organisms have at their disposal potential food of complex composition. Organic matter may be transported underground by streams that enter caves through sinkholes and other openings, especially during floods, or by percolating groundwater that flows into karst through crevices and fissures or into alluvial systems through soil and underflow (Ginet and Decou, 1977; Grimm and Fisher, 1984; Holsinger, 1988; Fiebig and Lock, 1991). For example, percolation waters of the upper part of the Mammoth Cave (Kentucky) contain about 7–8 mg/l^{-1} of total organic carbon (TOC) during summer, 2–3 mg/l^{-1} in winter, and 10–15 mg/l^{-1} in autumn at the first flood (Barr, 1981). Delay (1978) demonstrated that infiltration waters of two pyrenean caves (France) contains amino acids from 2 to 66 $\mu g/l^{-1}$ and free glucids of about 2.5 mg.l^{-1}. Stanford and Ward (1988) observed that dissolved organic carbon concentrations were consistently low in the pristine, unpolluted groundwaters of the Flathead River alluvial aquifers. Moreover, at the same site Ellis *et al.* (1994) found that entrainment of riverine particulate organic matter (POM) was rapidly attenuated with distance from the river channel, whereas dissolved nitrogen and phosphorus increased (see Chapter 14). These observations suggested that a rapid turnover of available particulate carbon forms was occurring and that microbial production could be limited by paucity of dissolved organic matter (DOM). Unfortunately, little is known about microbial respiration and rates of microheterotrophic production in aquifers, except that they are low, perhaps paradoxically low with respect to the abundance of metazoans that are often present. It is apparent, however, that organic pollution can greatly stimulate microbial activity and, in severe cases, eventuate in oxygen stagnation and steep changes in redox gradients that can greatly affect the distribution and abundance of hypogean metazoans (Chapters 14 and 18). Indeed groundwater contains generally dissolved nutritive substances in low concentrations.

In the absence of phototrophs, fungi and bacteria form the cellular components of the biofilm in groundwater systems (Dickson and Kirch, 1976; see also Chapter 7). Except for the photosynthetic algae, which are absent (or present in very low abundances), it is likely that the biofilm attached to substrate particles in groundwater functions in a manner similar to that of the epilithon of rivers. Indeed, Ellis *et al.* (1994) showed that most of the bacteria and fungi growing in the interstitial alluvium of the Flathead River was epilithic (Barlocher and Murdoch, 1989).

River epilithon consists of a polysaccharide matrix, secreted by the microorganisms, in which the cellular components are embedded. The matrix entrains dissolved, colloidal, and particulate organic matter from the water, and organic compounds are held by adsorption and diffusional resistance. High-molecular-weight compounds are degraded into compounds available to the bacteria and fungi of the biofilm by enzymes and exoenzymes held within the matrix. The stimulatory effects of experimental nutrient additions on the heterotrophic components of river biofilm are, however, mediated through algae (Peterson *et al.*, 1985), a component that is absent from the biofilm in groundwater environments. Nutrient dynamics and fluxes within hypogean biofilms, between biofilms and the surrounding groundwater, and from epigean to hypogean microbial communities are poorly understood and provide a fruitful avenue of research.

The biofilm, by incorporating DOM and thereby making it available to higher trophic levels, is an important part of the "microbial loop" first described for pelagic ocean waters (Pomeroy, 1974). In subterranean systems, biofilms and associated primary consumers, such as protozoa, presumably form similar microbial loops, but evidence is limited (Ellis *et al.*, 1994). Detailed data on the functional roles of microbial loops in carbon and nutrient dynamics have been available for some time for biofilms in porous media used to treat sewage (Taylor and Jaffe, 1990; Wilderer and Characklis, 1989), which, given the strong vertical light gradient, may serve as an appropriate analogue to direct such investigations in groundwater environments.

In any case, the biofilm and microbial loop provide food for grazing groundwater invertebrates (Barlocher and Murdoch, 1989), which form complex communities (Fig. 3) involving many species from many taxa. As noted above, most groundwater animals are polyphagous because food resources usually are scarce. Studies have examined distributions of the invertebrate fauna in alluvial systems from which inferences about trophic

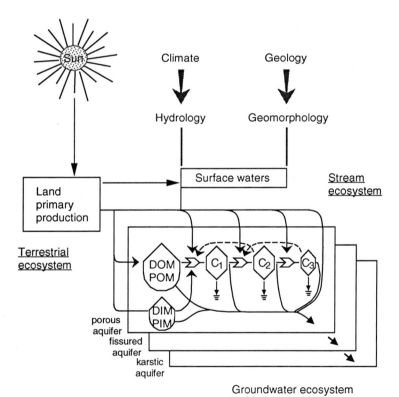

FIGURE 3   Conceptualization of energetics in the three major groundwater ecosystems that can exist in the geohydraulic continuum, showing the flux of dissolved and pariculate inorganic (DIM and PIM) and organic (DOM and POM) matter, from the terrestrial environments through the food web contained in the aquifer. Modified from Benke *et al.* (1988). The less complex food webs that generally occur in groundwaters in comparison with stream ecosystems are emphasized.

structure can be made (Pennak and Ward, 1986; Boulton *et al.*, 1992; Bretschko, 1992; Danielopol, 1989). Trophic relationships within groundwater systems were mainly studied in karsts. A synthesis is provided by Culver (Chapter 10).

The trophic cascade hypothesis, initially applied to the limnetic zone of lakes (Carpenter *et al.*, 1985; Carpenter and Kitchell, 1988, 1993), proposed that high consumers indirectly influence phytoplankton composition and productivity through interactions that cascade down the food web. This concept of a top-down control of lake productivity and community structure recognizes that food web interactions as well as bottom-up control by nutrients can influence trophic dynamics. Evidence supporting the concept of cascading trophic interactions has been demonstrated for stream environments (Power and Matthews, 1983; Power, 1990). We do not know to what extent and under what circumstances high consumers influence lower trophic levels and biofilm dynamics in groundwaters.

However, it is apparent that speciose food webs exist in groundwaters, and it is highly likely that they interact with geochemical processes associated with redox gradients and hydrodynamics to influence material transformations and fluxes between surface water and groundwaters. Use of stable isotopes to elaborate trophic relationships is promising in surface waters and should be used in groundwater studies (Hershey *et al.*, 1993; Peterson *et al.*, 1985). Release of dissolved forms of nitrogen and phosphorus in the groundwater can create bursts of autotrophic bioproduction in which groundwaters upwell to the surface (Hill, 1990; Valett *et al.*, 1992), thereby providing a vital stimulus to inland surface waters and wetlands (Stanford and Ward, 1993). Clearly, additional work on food web dynamics and influences on material flux and transformations in groundwater and surface water ecotones is badly needed.

D. Resource Limitation and Stress Tolerance

Resources to support the biological components of groundwater ecosystems appear to be limiting in bioproduction, and biota are organized around the physical structure of the aquifer and the movement of water and allochthonous materials through it. Owing to the relative stability of many groundwater systems, native biota are highly evolved and therefore sensitive to environmental change (Chapter 18). We assume that food webs and microbially mediated chemical processes are likewise variable resilient. Groundwater ecosystems are not all exposed in the same way to anthropogenic perturbations (Gibert *et al.*, 1991). Deep groundwater covered by almost impermeable rock is naturally well protected, but once it is damaged it is subject to very long recovery times. Usually, shallow groundwater such as river underflow, or large karst drains directly influenced by epigean dynamics, was considered to have a low resistance to environmental change caused by human perturbations. However, karst as well as porous systems present different zones with different types of "physical, chemical, and biological stability." Thus resistant or protected zones, such as floodplain margins or annex drains, can attenuate the effect of pollutants. Hence, groundwater ecosystems appear now more resistant to anthropogenic perturbations than previously recognized (Chapter 16).

When the perturbation takes place in a groundwater system (stress reaction), in some cases we can follow restoration processes. For example, Ginet and Turquin (1984) observed the restoration of a groundwater ecosystem in only a few decades after cultivation and intensive visits to cave were suspended. These observations are echoed in many of the chapters included in the present volume and underscore the need to better understand the process–response relationships in groundwater ecosystems.

Many authors, such as Odum *et al.* (1979) and Kolosa and Pickett (1992), have based evaluation of stress on the model of system organisation, its integrity, and its persistence. This supports a holistic, multidisciplinary approach, rather than a reductionist approach. Clearly, we do not have a detailed knowledge in many groundwater ecosystems.

E. Spatial Pathways in Stream Ecology

The groundwater system may be envisioned as domains in a nested series of spatial and temporal configurations (Fig. 2). On a global scale and in an evolutionary time frame, all aquatic habitats (surface waters; marine water and freshwater; psammolittoral; and porous, fissured, and karstic aquifers) are interconnected (Delamare-Deboutteville, 1960; Hutchinson, 1967; Pennak, 1968; Husmann, 1971; Ward and Palmer, 1994). In this section we provide a broad spatial perspective to demonstrate the importance of integrating groundwaters into stream ecosystem theory (Hynes, 1983). We generally focus on alluvial river–aquifer systems (Gibert *et al.,* 1990; Vervier and Gibert, 1991).

A conceptual model of the four-dimensional nature of lotic ecosystems (Amoros *et al.,* 1986; Ward, 1989a) includes interactions between rivers and adjacent groundwaters along the vertical dimension. As initially formulated, however, the other spatial dimensions (longitudinal and lateral) in the model dealt only with epigean pathways. In the ensuing discussion, each of the three spatial dimensions is briefly examined from a largely hypogean context.

1. Longitudinal Patterns

On a large scale, stream ecology has a long and continuing tradition of analyzing changes along the longitudinal dimension, including the altitudinal gradient, as embodied in zonation schemes (Illies and Botosaneanu, 1963) and the continuum model (Vannote *et al.,* 1980). On the catchment scale, the hyporheic corridor concept (HCC) is a physical theoretical construct

(Stanford and Ward, 1993) that incorporates groundwater dynamics along the longitudinal stream profile. A basic premise of the hyporheic corridor concept is that reaches of aggraded alluvium (floodplains) alternate with bedrock-constrained canyon segments. River water downwells at the upstream end of aggraded segments, moves through the unconfined floodplain aquifer, and returns to the channel as upwelling at the downstream end (Fig. 4). In canyon segments interstitial flow is normally limited to gravel bars and thin bed sediments. However, this new concept must be discussed in other hydrographic contexts. Indeed, on the Rhône river, longitudinal patterns were presented in a rather different way by Creuzé des Châtelliers (1991b). He considers the hyporheic layer as "a sequence of subsystems forming a biological discontinuum (Perry and Schaeffer, 1987) determined by different geomorphological and hydrogeological mechanisms."

Another biological study (Chapter 15) has examined the groundwater fauna over an extensive elevation gradient (2000 altitudinal and 475 longitudinal km). In stark contrast to the macrobenthic stream fauna (Ward, 1986), the longitudinal distribution of hypogean faunal assemblages in the South Platte River, Colorado, did not correspond to the longitudinal gradient. This reflects, in part, the disjunct distributions of some stygobiontic forms (*e.g.*, the archiannelid *Troglochaetus* and the syncarid *Bathynella*) along the river's course and suggests that the groundwater fauna is to some extent decoupled from the altitudinal gradient. It appears that site-specific geomor-

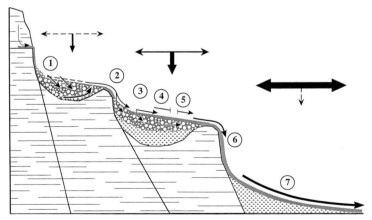

FIGURE 4 A cross-sectional representation of the hyporheic corridor. Surface and interstitial flows are represented by arrows, and numbers refer to various segments in the river continuum: (1) headwaters, (2) headwater transition, (3) montane floodplain, (4) montane transition, (5) piedmont valley floodplain, (6) piedmont transition, and (7) coastal plain. Vectors above the floodplains represent the relative volume of annual overland (horizontal arrows) versus interstitial (vertical arrows) flow for these floodplains [from Stanford and Ward (1993)].

phic and hydrologic features override the influence of direct correlates of elevation (*e.g.*, temperature) in structuring hypogean faunal assemblages. On a smaller scale, Creuzé des Châtelliers and Reygrobellet (1990) observed that interstitial faunal communities of the Rhône River clearly reflect the longitudinal effect of geomorphic and hydrologic interactions. Longitudinal faunal patterns have also been recorded at gravel bars. In a 1200-m-long gravel bar of the Rhône River the distribution of interstitial animals reflected the relative contributions of surface water versus groundwater (see Chapter 12).

2. Lateral Patterns

Lateral interactions between river channels and the surrounding landscape form an important part of stream ecosystem theory (Fisher and Likens, 1973; Hynes, 1975; Welcomme, 1979; Cummins *et al.*, 1984; Amoros and Roux, 1988; Junk *et al.*, 1989; Ward, 1989b; Gregory *et al.*, 1991). Stream ecologists should be aware that the lateral dimension also has a hypogean component (Freeze and Cherry, 1979; Stanford and Ward, 1992) that may be massive in some rivers.

An example of channel–aquifer interactions on the floodplain scale involves the Kalispell Valley, Montana (Stanford and Ward, 1988; see also Chapter 14). The valley contains an expansive unconfined alluvial aquifer that is hydraulically interactive with the Flathead River. As noted above, influent groundwater provides an important nutrient subsidy to the oligotrophic river, which contains a diverse plecopteran fauna, eight species of which are amphibites, specialized for a hypogean existence. Analysis of interstitial crustacean distribution patterns across the floodplain indicated the heterogeneous nature of the alluvial aquifer (Ward *et al.*, 1994). It appears that the patchy lateral distribution of crustaceans relates to geomorphic and hydrologic heterogeneity. The unconfined aquifer consists of a subterranean lattice work of alluvial-filled paleochannels of high hydraulic conductivity (Stanford and Ward, 1993). Porosity and exchange properties exhibit strong spatial gradients, but do not exhibit a discernible pattern on the floodplain scale. Examples on several aquifers such as the Rhône River aquifer can be provided (Chapter 12).

The lateral distribution of interstitial animals has also been examined on a scale of meters. Faunal densities progressively declined along a transect from hyporheic (underflow), through shore (2 m from channel), to phreatic (20 m from channel) habitats in a Rocky Mountain river (Pennak and Ward, 1986). Although there was considerable faunal overlap, each habitat contained a distinctive faunal assemblage. Seyed-Reihani (1980) observed similar distributions on the Rhône River. In the Sava River (Yugoslavia), Mestrov *et al.* (1983) studied faunal distribution along transects varying from several meters to several hundred meters from the channel, and Williams (1989), for a Canadian stream, distinguished three faunal assemblages (bank, margin, and stream bed) along a 5-m transect.

3. Vertical Patterns

The vertical distribution of interstitial assemblages results directly from the structure and dynamics of the surface water–groundwater ecotone, which varies in space and time. This was illustrated for the Rhône River (Fig. 5), where five different assemblages were found, ranging from ecotonal specialists (*Niphargopsis casparyi, Fabaeformiscandona wegelini,* and *Pseudocandona albicans,* which mainly colonized the interaction zone) to benthic or phreatic specialists (*e.g., Herpetocypris reptans, Heptagenia* sp., *Microcharon reginae,* and *Bathynella* sp., which were situated above or below the interaction zone). Both epigean and hypogean forms possessing greater ecological tolerances colonized the interaction zone and layers situated above and below, respectively (*e.g., Cypria ophthalmica, Leuctra fusca, Niphargus kochianus,* and *Pseudocandona zschokkei*).

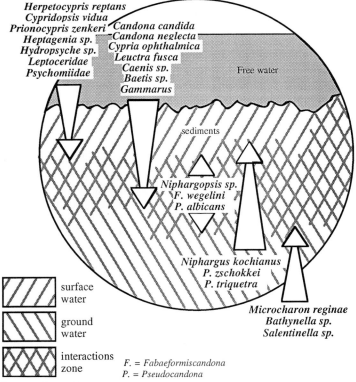

FIGURE 5 Specificity of groundwater organisms in the interaction zone between surface and hypogean components of an alluvial aquifer of the Rhône River, France. The interaction zone fluctuates as a consequence of the changing flow pattern and associated piezometric gradient in the stream channel.

Considerable research has been conducted mainly by stream ecologists investigating vertical distribution patterns of macroinvertebrates in superficial bed sediments (Coleman and Hynes, 1970; Radford and Hartland-Rowe, 1971; Bishop, 1973; Williams and Hynes, 1974; Poole and Stewart, 1976; Morris and Brooker, 1979; Bretschko, 1981; Pennak and Ward, 1986). Temperature range, current, and light rapidly decline with depth below the stream bed. The extent to which epigean animals penetrate bed sediments depends in part on substrate porosity and oxygen levels. In a Canadian river, interstitial waters 30 cm below the bed surface were only 5% saturated with oxygen (Williams and Hynes, 1974), whereas there was no depletion of oxygen down to 60 cm in an Austrian mountain brook (Bretschko, 1981). In most situations stream macroinvertebrates penetrate at least 30 cm into coarse substratum. For example, Coleman and Hynes (1970) found comparable numbers of animals in each of four layers (extent, 7.5 cm) in the top 30 cm of a stream bed. Only black flies (Simuliidae) were confined to the top layer. Even fish occurred within the substrate interstices; seven darters (*Etheostoma flabellae*) were collected from the bottom layer.

Phreatobiological studies have examined faunal depth distribution patterns deeper than 1 m below the stream bed or in alluvium located laterally from river channels (Husmann, 1975; Danielopol, 1976, 1991; Dole, 1985). In a Rocky Mountain river, densities of interstitial animals increased markedly with depth (15, 30, and 50 cm) in the stream bed; density maxima occurred at 2 m from the water's edge, at a depth of 30 cm; and densities were similar at the three depths at a site 20 m from the river (Pennak and Ward, 1986).

Dole (1985) found marked differences in the vertical structure (-0.5 m to -3.0 m) of interstitial animals in three hydrographic units (active channel, side arm, and old meander) that differed in physical stability and the relative influence of surface water–groundwater. There was a tendency for stygobiontic forms to occur at greater depths beneath the active channel. In the alluvial sediments of a backwater of the Danube, Danielopol (1991) found true hypogean forms inhabiting superficial layers in which environmental conditions fluctuated widely.

The four-dimensional perspective of undisturbed riverine–aquifer systems provides an appropriate framework for understanding and evaluating anthropogenic perturbations. According to this perspective, human-induced disturbances are viewed as severed interactive pathways.

V. CONCLUSION

The ecosystem concept in groundwater is now far from universally accepted. For far too long research has been focused on the organisms according to an analytical approach. The concept of the ecosystem has greatly changed the conception of the groundwater realm and the study of organisms. By considering adequate spatial and temporal scales and using habitat dynamics as a template for the study of biological and physical interactions, theoretical ecological models may provide new insights into the processes that contribute to ecosystem organization.

Groundwater ecology taken in an ecosystem context is a young and fertile discipline. The conceptual and methodological bases for research into groundwater are subject to spurts of growth. More is still to be done in developing a dynamic model and in elaborating the component forces. We hope this synthesis and others in this book will provide a foundation for linking new research, emphasizing groundwaters and dynamic ecosystems, with management to protect and conserve groundwater resources for future generations.

REFERENCES

Allen, T. F. H., and Starr, B. T. (1982). "Hierachy: Perspectives for Ecological Complexity." Univ. of Chicago Press, Chicago.

Amoros, C., and Roux, A.-L. (1988). Interaction between water bodies within the floodplains of large rivers: Function and development of connectivity. *In* "Connectivity in Landscape Ecology" (K. F. Schreiber, ed.), Münstersche Geogr. Arb. 29, pp. 125–130.

Amoros, C., Roux, A.-L., Reygrobellet, J. L., Bravard, J. P., and Pautou, G. (1986). A method for applied ecological of fluvial hydrosystems. *Regulated Rivers* 1, 17–36.

Andrewartha, H. G., and Birch, L. C. (1954). The distribution and abundance of animals. Thesis, Univ. of Chicago Press, Chicago.

Auger, P., Baudry, J., and Fournier, F. (1991). "Hiérarchies et échelles en écologie." Naturalia Publications. Paris.

Bakalowicz, M. (1979). Contribution de la géochimie des eaux à la connaissance de l'aquifère karstique. Thèse, Univ. P. et M. Curie, Paris-6.

Bakalowicz, M. (1986). De l'hydrogéologie en karstologie. *Journ. Karst Euskadi*, pp. 105–128.

Barlocher, F., and Murdoch, J. H. (1989). Hyporheic biofilm—a potential food source for interstitial animals. *Hydrobiologia* 184, 61–67.

Barr, T. C. (1968). Cave ecologie and the evolution of troglobites. *Evol. Biol.* 2, 35–102.

Barr, T. C. (1981). Biological tour of Mammoth cave, Kentucky. *Proc. Int. Speleol., 8th, Bowling Green*, pp. 1–7.

Barr, T. C., and Holsinger, J. R. (1985). Speciation in cave faunas. *Annu. Rev. Ecol. Syst.* 16, 313–337.

Benke, A. C., Hall, C. A. S., Hawkins, C. P., Lowe-McConnell, R. H., Stanford, J. A., Suberkropp, K., and Ward, J. V. (1988). Bioenergetic considerations in the analyses of stream, ecosystems. *J. North Am. Benthol. Soc.* 7, 480–502.

Berthelemy, C. (1968). Contribution à la connaisance des Leuctridae (Plécoptères). *Ann. Limnol.* 4, 175–198.

Bishop, J. E. (1973). Observations on the vertical distribution of the benthos in a Malaysian stream. *Freshwater Biol.* 3, 147–156.

Blandin, P., Barbault, R., and Lecordier, C. (1976). Réflexions sur la notion d'écosystème: Le concept de stratégie cénotique. *Bull. Ecol.* 7(4), 391–410.

Botosaneanu, L., ed. (1986). "Stygofauna mundi." E. J. Brill, Dr. W. Backhuys, Leiden.

Botosaneanu, L., and Holsinger, J. R. (1991). Some aspects concerning colonization of the subterranean realm—especially of subterranean waters: A response to Rouch and Danielopol, 1987. *Stygologia* **6**, 11–39.

Boulton, A. J., Valett, H. M., and Fisher, S. G. (1992). Spatial distribution and taxonomic composition of the hyporheos of several Sonoran Desert streams. *Arch. Hydrobiol.* **125**(1), 37.

Bretschko, G. (1981). Vertical distribution of zoobenthos in an Alpine brook of the Ritrodat-Lunz study area. *Verh. Int. Ver. Limnol.* **21**, 873–876.

Bretschko, G. (1992). Differentiation between epigeic and hypogeic fauna in gravel streams. *Regulated Rivers* **7**, 17–22.

Cairns, J., Jr. (1988). Can the global loss of species be stopped? *Speculations Sci. Technol.* **11**, 189–196.

Camacho, A. I., ed. (1992). "The Natural History of Biospeology," Monogr. Mus. Nac. Cienc. Nat., 7, C.S.I.C., Madrid.

Carpenter, S. R., and Kitchell, J. F. (1988). Consumer control of lake productivity: Large-scale experimental manipulations reveal complex interactions among lake organisms. *BioScience* **38**(11), 764–769.

Carpenter, S. R., and Kitchell, J. F., eds. (1993). "The Trophic Cascade in Lakes." Cambridge Univ. Press, Cambridge, UK.

Carpenter, S. R., Kitchell, J. F., and Hodgson, J. R. (1985). Cascading trophic interactions and lake productivity. *BioScience* **35**, 634–639.

Carpenter, S. R., Fisher, S. G., Grimm, N. B., and Kitchell, J. F. (1992). Global change and freshwater ecosystems. *Annu. Rev. Ecol. Syst.* **23**, 119–139.

Castany, G. (1985). Liaisons hydrauliques entre les aquifères et les cours d'eau. *Stygologia* **1**, 1–25.

Castany, G., and Margat, J. (1977). "Dictionnaire français d'hydrogéologie." Bureau de Recherches Géologiques et Miniéres, Orléans.

Christiansen, K. (1962). Proposition pour la classification des animaux cavernicoles. *Spelunca* **2**, 76–78.

Christiansen, K. (1992). Biological processes in space and time: Cave life in the light of modern evolutionary theory. *In* "The Natural History of Biospeology" (A. I. Camacho, ed.), pp. 456–478. Monogr. Mus. Nac. Cienc. Nat., C.S.I.C., Madrid.

Coleman, M. J., and Hynes, H. B. N. (1970). The vertical distribution of the invertebrate fauna in the bed of a stream. *Limnol. Oceanogr.* **15**, 31–40.

Connell, J. H. (1978). Diversity in tropical rain forests and coral reefs. *Science* **199**, 1302–1310.

Creuzé des Châtelliers, M. (1991a). Dynamique de répartition des biocénoses interstitielles du Rhône en relation avec des caractéristiques géomorphologiques (secteurs de Brégnier-Cordon, Miribel-Jonage et Donzère-Mondragon). Thèse, Univ. Lyon.

Creuzé des Châtelliers, M. (1991b). Geomorphological processes and discontinuities in the macrodistribution of the interstitial fauna. A working hypothesis. *Verh. Int. Ver. Limnol.* **24**, 1607–1612.

Creuzé des Châtelliers, M., and Reygrobellet, J.-L. (1990). Interactions between geomorphological processes, benthic and hyporheic communities: First results on a bypassed canal of the Frech upper Rhône River. *Regulated Rivers* **5**, 139–158.

Culver, D. C. (1982). "Cave Life: Evolution and Ecology." Harvard Univ. Press, Cambridge, MA.

Culver, D. C. (1985). Trophic relationships in aquatic cave environments. *Stygologia* **1**(1), 43–53.

Culver, D. C., Fong, D. W., and Jernigan, R. W. (1991). Species interactions in cave stream communities—Experimental results and microdistribution effects. *Am. Midl. Nat.* **126**(2), 364–379.

Cummins, K. W., Minshall, G. W., Sedell, J. R., Cushing, C. E., and Petersen, R. C. (1984). Stream ecosystem theory. *Verh. Int. Ver. Limnol.* **22**, 1818–1827.

Curl, R. L. (1986). Fractal dimensions and geometries of caves. *Math. Geol.* **18**(8), 765–783.

Danielopol, D. L. (1976). The distribution of the fauna in the interstitial habitats of riverine sediments of the Dnube and Piesting (Austria). *Int. J. Speleol.* **8**, 23–51.

Danielopol, D. L. (1980). The role of the limnologist in ground water studies. *Int. Rev. Gesamten Hydrobiol.* **65**, 777–791.

Danielopol, D. L. (1982). Phreatobiology reconsidered. *Pol. Arch. Hydrobiol.* **29**, 375–386.

Danielopol, D. L. (1989). Groundwater fauna associated with riverine aquifers. *J. North Am. Benthol. Soc.* **8**, 18–35.

Danielopol, D. L. (1991). Spatial distribution and dispersal of interstitial Crustacea in alluvial sediments of a backwater of the Danube at Vienna. *Stygologia* **6**, 97–110.

Danielopol, D. L. (1992). La phréatobiologie de Constantin Motas dans la perstective des développements actuels de l'écologie aquatique souterraine. *Trav. Inst. Spéol. "Emile Racovitza"* **31**, 11–20.

Danielopol, D. L., and Marmonier, P. (1992). Aspects of research on groundwater along the Rhône, Rhine and Danube. *Regulated Rivers* **7**, 5–16.

Delamare-Deboutteville, C. (1960). "Biologie des eaux souterraines littorales et continentales." Hermann, Paris.

Delamare-Deboutteville, C. (1971). "La vie dans les grottes," P.U.F. Colloq. Que sais-je? Presses Univ. de Paris.

Delay, B. (1978). Milieu souterrain et écophysiologie de la reproduction et du développement des Coléoptères Bathysciinae hypogés. *Mém. Biospéol.* **5**, 1–349.

Delcourt, H. R., and Delcourt, P. A. (1988). Quaternary landscape ecology: Relevant scales in space and time. *Landscape Ecol.* **2**(1), 23–44.

Dickson, G. W., and Kirch, P. W. (1976). Distribution of heterotrophic microorganisms in relation to detritivors in Virginia caves (with supplemental bibliography on cave micology and microbiology). In "The Distributional History of the Biota of the Southern Appalachians. Part IV. Algae and Fungi" (B. C. Parker and M. K. Roane, eds.), pp. 205–226. Univ. Press of Virginia, Charlottesville.

Dole, M.-J. (1983). Le domaine aquatique souterrain de la plaine alluviale du Rhône à l'Est de Lyon: Ecologie des niveaux supérieurs de la nappe. Thèse, Univ. Lyon.

Dole, M.-J. (1985). Le domaine aquatique souterrain de la plaine alluviale du Rhône à l'Est de Lyon. 2. Structure verticale des peuplements des niveaux supérieurs de la nappe. *Stygologia* **1**, 270–291.

Ellis, B. K., Stanford, J. A., and Ward, J. V. (1994). Ground water microbiol ecology of the alluvial aquifers of a large gravel-bed river. *J. North Am. Benthol. Soc.* (in press).

Fiebig, D. M., and Lock, M. A. (1991). Immobilization of dissolved organic matter from groundwater discharging through the stream bed. *Freshwater Biol.* **26**, 45–55.

Fisher, S. G., and Likens, G. E. (1973). Energy flow in Bear Brook, New Hampshire: An integrative approach to Stream ecosystem metabolism. *Ecol. Monogr.* **43**, 421–439.

Freeze, A. R., and Cherry, J. A. (1979). "Groundwater." Prentice-Hall, Englewood Cliffs, Nd.

Gèze, B. (1965). "La spéléologie scientifique." Seuil, Paris.

Gibert, J. (1986). Ecologie d'un système karstique jurassien. Hydrogéologie, dérive animale, transit de matières, dynamique de la population de Niphargus (Crustacé-Amphipode). *Mém. Biospéol.* **13**, 1–379.

Gibert, J. (1989). Functional sub-units of an exsurgence karstic system, and exchanges with the surface environment. Reflections on the characterization of natural aquatic groundwater ecosystems. *Verh. Int. Ver. Limnol.* **23**(2), 1090–1096.

Gibert, J., Laurent, R., Bourne, J. D., and Ginet, R. (1978). L'écosystème karstique du Massif de Dorvan (Torcieu, Ain, France). Présentation de l'environnement physique et le peuplement animal souterrain. *Actes 6 Congr. Suisse Spéléol.*, 16–18 September 1978. *Porrentruy*, pp. 37–68.

Gibert, J., Olivier, M.-J., Marmonier, P., and Vervier, P. (1990). Surface water-groundwater ecotones. In "Ecology and Management of Aquatic-Terrestrial Ecotones" (R. J. Naiman and H. Décamps, eds.), Man & Biosphere Publ., pp. 199–225. Parthenon Publ., London.

Gibert, J., Marmonier, P., Turquin, M.-J., and Martin, D. (1991). Anthropogenic disturbance of surface landscape: Consequences on groundwater ecosystems. *In* "Terrestrial and Aquatic Ecosystems: Perturbation and Recovery" (O. Ravera, ed.), pp. 310–319. Ellis Horwood, London.

Ginet, R., and David, J. (1963). Présence de *Niphargus* (Amphipode Gammaridae) dans certaines eaux épigées des forêts de la Dombes (département de l'Ain, France). *Vie Milieu* **14**(2), 299–310.

Ginet, R., and Decou, V. (1977). "Initiation à la biologie et à l'écologie souterraines." Delarge, Paris.

Ginet, R., and Turquin, M.-J. (1984). "Evolution réciproque des biocénoses et des activités humaines dans les réserves naturelles: Réserve Naturelle de la Grotte de Hautecourt." Conf. Perm. Réserves Nat., Ministère de l'Environnement, Paris.

Greenslade, P. J. M. (1983). Adversity selection and the habitat templet. *Am. Nat.* **122**(3), 352–365.

Gregory, S. V., Swanson, F. J., McKee, W. A., and Cummins, K. W. (1991). An ecosystem perspective of riparian zones. *BioScience* **41**, 540–551.

Grimm, N. B., and Fisher, S. G. (1984). Exchange between surface and interstitial water: Implications for stream metabolism and nutrient cycling. *Hydrobiologia* **111**, 219–228.

Hall, C. A. S., Stanford, J. A., and Hauer, F. R. (1992). The distribution and abundance of organisms as a consequence of energy balances along multiple environmental gradients. *Oikos* **65**, 377–390.

Hansen, A. J., and di Castri, F. (1992). Landscape boundaries. Consequences for biotic diversity and ecological flows. *Ecol. Stud.* **9**, 4–452.

Hershey, A. E., Pastor, J., Peterson, B. J., and Kling, G. W. (1993). Stable isotopes resolve the drift paradox for *baetis* mayflies in an arctic river. *Ecology* **74**(8), 2315–2325.

Hill, A. R. (1990). Ground water flow paths in relation to nitrogen chemistry in the near-stream zone. *Hydrobiologia* **206**, 39–52.

Holsinger, J. R. (1988). Troglobites: The evolution of cave-dwelling organisms. *Am. Sci.* **76**, 146–153.

Husmann, S. (1971). Ecological studies on freshwater meiobenthon in layers of sand and gravel. *Smithson. Contrib. Zool.* **76**, 161–169.

Husmann, S. (1975). Versuch zur Erfassung der verticalen Verteilung von Organismen und chemischen Substanzen im Grundwasser von Talauen und Terrassen; Methoden und erste Befunde. *Int. J. Speleol.* **6**, 271–302.

Hutchinson, G. E. (1967). "A Treatise on Limnology," Vol. 2. Wiley, New York.

Hynes, H. B. N. (1975). The stream and its valley. *Verh.—Int. Ver. Theor. Angew. Limnol.* **19**, 1–15.

Hynes, H. B. N. (1983). Groundwater and stream ecology. *Hydrobiologia* **100**, 93–99.

Illies, J., and Botosaneanu, L. (1963). Problèmes et méthodes de la classification et de la zonation écologique des eaux courantes, considérées surtout du point de vue faunistique. *Mitt.—Int. Ver. Theor. Angew. Limnol.* **12**, 1–57.

Juberthie, C. (1984). La colonisation du milieu souterrain; théories et modèles, relations avec la spéciation et l'évolution souterraine. *Mém. Biospéol.* **11**, 69–102.

Junk, W. J., Bayley, P. B., and Sparks, R. E. (1989). The flood pulse concept in river-floodplain systems. *Can. J. Fish. Aquat. Sci., Spec. Publ.* **106**, 110–127.

Kolosa, J. (1989). Ecological systems in hierarchical perspective: Breals in community structure and other consequences. *Ecology* **70**(1), 36–47.

Kolosa, J., and Pickett, S. T. A. (1992). Ecosystem stress and health: An expansion of the conceptual basis. *J. Aquat. Ecosyst. Health* **1**, 7–13.

Kotliar, N. B., and Wiens, J. A. (1990). Multiple scales of patchiness and patch structure: A hierarchical framework for the study of heterogeneity. *Oikos* **59**, 253–260.

Laverty, M. (1987). Fractals in karst. *Earth Surf. Processes Landforms* **12**, 475–480.

Lenormand, R. (1985). Invasion percolation in an Etched network: Measurement of a fractal dimension. *Phys. Rev. Lett.* **54**, 2226–2229.

Maire, R. (1990). La haute montagne calcaire. *Karstol. Mém.* **3**, 1–731.

Mangin, A. (1974). Contribution à l'étude hydrodynamique des aquifères karstiques (première et deuxième parties). *Ann. Spéléol.* **29**(3), 283–332; (4), 495–601.

Mangin, A. (1975). Contribution à l'étude hydrodynamique des aquifères karstiques. (troisième partie). *Ann. Spéléol.* **30**(1), 21–124.

Mangin, A. (1983). L'approche systémique du karst, conséquences conceptuelles et méthodologiques. *Réun. Monogr. Karst-Larra, 1982*, pp. 142–157.

Mangin, A. (1986). Réflexion sur l'approche et la modèlisation des aquifères karstiques. *Journ. Karst Euskadi, Donostia-San Sébastian, Spain* **2**, 11–31.

Marmonier, P., and Creuzé des Châtelliers, M. (1991). Effects of spates on interstitial assemblages of the Upper Rhône River. Importance of spatial heterogeneity. *Hydrobiologia* **210**, 243–251.

Marmonier, P., and Dole, M.-J. (1986). Les Amphipodes des sédiments d'un bras court-circuité du Rhône: logique de répartition et réaction aux crues. *Rev. Fr. Sci. Eau* **5**, 461–486.

Marmonier, P., Vervier, P., Gibert, J., and Dole-Olivier, M.-J. (1993). Biodiversity in ground waters: A research field in progress. *Trends Ecol. Evol.* **8**(11), 392–395.

Mathieu, J., Marmonier, P., Laurent, R., and Martin, D. (1991). Récolte du matériel biologique aquatique souterrain et stratégie d'échantillonnage. *Hydrogéologie* **3**, 217–223.

Mayr, E. (1982). Processes of speciation in animals. *In* "Mechanisms of Speciation" (C. Barigozzi, ed.), pp. 1–19. Liss, New York.

Mestrov, M., Stilinovic, B., Habdija, I., Lattinger, R., Maloseja, Z., Kerovec, M., and Cicin-Sain, L. (1983). The ecological characteristics of interstitial groundwaters in relation to the water of the River Sava. *Acta Biol. Knj.* **9**, 1, 5–33.

Minshall, G. W. (1988). Stream ecosystem theory: A global perspective. *J. North Am. Benthol. Soc.* **7**(4), 263–288.

Minshall, G. W., Cummins, K. W., Petersen, R. C., Cushing, C. E., Bruns, D. A., Sedell, J. R., and Vannote, R. L. (1985). Developments in stream ecosystems theory. *Can. J. Fish. Aquat. Sci.* **42**, 1045–1055.

Mohr, C. E., and Poulson, T. L. (1966). "The Life of the Cave." McGraw-Hill, New York.

Morris, D. L., and Brooker, M. P. (1979). The vertical distribution of macro-invertebrates in the substratum of the upper reaches of the River Wye, Wales. *Freshwater Biol.* **9**, 573–583.

Naiman, R. J. (1992). "Watershed Management: Balancing Sustainability and Environmental Change." Springer-Verlag, New York.

Naiman, R. J., and Décamps, H., eds. (1990). "The Ecology and Management of Aquatic-Terrestrial Ecotones," Man & Biosphere Ser., Vol. 4. Parthenon Publ., London.

Naiman, R. J., Décamps, H., Pastor, J., and Johnston, C. A. (1988). The potential importance of boundaries to fluvial ecosystems. *J. North Am. Benthol. Soc.* **7**(4), 289–306.

Noss, R. F. (1990). Indicators for monitoring biodiversity: A hierarchical approach. *Conserv. Biol.* **4**, 355–364.

Notenboom, J. (1991). Marine regressions and the evolution of groundwater dwelling amphipods (Crustacea). *J. Biogeogr.* **18**, 437–454.

Odum, E. P., Finn, J. T., and Franz, E. H. (1979). Perturbation theory and the subsidy-stress gradient. *BioScience* **29**(6), 249–352.

O'Neill, R. V., De Angelis, D. L., Waide, J. B., and Allen, T. F. H. (1986). "A Hierarchical Concept of Ecosystems." Princeton Univ. Press, Princeton, NJ.

Orghidan, T. (1955). Un nou domeniu de viata acvatica subterana: "Biotopul hiporeic." *Bul. Stiint. Sect. Biol. Stiinte Agric. Sect. Geol. Geogr.* **7**(3), 657–676.

Palmer, M. A. (1990). Temporal and spatial dynamics of meiofauna within the hyporheic zone of Goose Creek, Virginia. *J. North Am. Benthol. Soc.* **9**(1), 17–25.

Pennak, R. W. (1968). Historical origins and ramifications of interstitial investigations. *Trans. Am. Microsc. Soc.* **87**, 214–218.

Pennak, R. W., and Ward, J. V. (1986). Interstitial faunal communities of the hyporheic and adjacent groundwater biotopes of a Colorado mountain stream. *Arch. Hydrobiol., Suppl.* **74,** 356–396.

Perry, J. A., and Schaeffer, D. J. (1987). The longitudinal distribution of riverine benthos: A river dis-continuum? *Hydrobiologia* **148,** 257–268.

Peterson, B. J., Hobbie, J. E., Hershey, A. E., Lock, M. A., Ford, T. E., Vestal, J. R., McKinley, V. L., Hullar, M. A. J., Miller, M. C., Ventullo, R. M., and Volk, G. S. (1985). Transformation of a tundra river from heterotrophy to autotrophy by addition of phosphorus. *Science* **229,** 1383–1386.

Pickett, S. T. A., and P. S. White, eds. (1985). "The Ecology of Natural Disturbance and Patch Dynamics." Academic Press, Orlando, FL.

Pinay, G., Décamps, H., Chauvet, E., and Fustec, E. (1990). Functions of ecotones in fluvial systems. *In* "The Ecology and Management of Aquatic-Terrestrial Ecotones" (R. J. Naiman and H. Décamps, eds.), Man & Biosphere Ser., Vol. 4, pp. 141–169. Parthenon Publ., London.

Pomeroy, L. R. (1974). The ocean's food web, a changing paradigm. *BioScience* 24(9), 499–504.

Poole, W. C., and Stewart, K. W. (1976). The vertical distribution of macrobenthos within the substratum of the Brazos River, Texas. *Hydrobiologia* **50,** 151–160.

Poulson, T. L., and White, W. B. (1969). The cave environment. Limestone caves provide unique natural laboratories for studying biological and geological processes. *Science* **165,** 971–981.

Power, M. C. (1990). Effects of fish in river food webs. *Science* **250,** 811–814.

Power, M. E., and Matthews, W. J. (1983). Algae-grazing minnows (*Campostoma anomalum*), piscivorous bass (*Micropterus spp.*), and the distribution of attached algae in a small prairie-margin stream. *Oecologia* **60,** 328–332.

Pugsley, C. W., and Hynes, H. B. N. (1985). Summer diapause and nymphal development in *Allocapnia pygmaea* (Burmeister), (Plecoptera: Capniidae), in the Speed River, Southern Ontario. *Aquat. Insects* 7(1), 53–63.

Pugsley, C. W., and Hynes, H. B. N. (1986). Three-dimensional distribution of winter stonefly nymphs, *Allocapnia pygmaea,* within the substrate of a southern Ontario river. *Can. J. Fish. Aquat. Sci.* 43(9), 1812–1817.

Radford, D. S., and Hartland-Rowe, R. (1971). Subsurface and surface sampling of benthic invertebrates in two streams. *Limnol. Oceanogr.* **16,** 114–119.

Resh, V. H., Brown, A. V., Covich, A. P., Gurtz, M. E., Li, H. W., Minshall, G. W., Reice, S. R., Sheldon, A. L., Wallace, J. B., and Wissmar, R. C. (1988). The role of disturbance in stream ecology. *J. North Am. Benthol. Soc.* 7, 433–455.

Rouch, R. (1977). Considérations sur l'écosystème karstique. *C.R. Hebd. Seances Acad. Sci., Ser.* D 284,1101–1103.

Rouch, R. (1980). Le système karstique du Baget. X. La communauté des Harpacticides. Richesse spécifique, diversité et structures d'abondance de la nomocénose hypogée. *Ann. Limnol.* 16(1), 1–20.

Rouch, R. (1982). Le système karstique du baget. Comparaison de la dérive des Harpacticides à l'entrée et à la sortie de l'aquifère. *Ann. Limnol.* 18(2), 133–150.

Rouch, R. (1986). Sur l'écologie des eaux souterraines dans le karst. *Stygologia* 2(4), 352–398.

Rouch, R., Juberthie-Jupeau, L., and Juberthie, C. (1968). Recherches sur les eaux souterraines. 3. Essai d'étude du peuplement de la zone noyée d'un karst. *Ann. Spéléol.* 23(4), 717–733.

Salo, J., Kalliola, R., Häkkinen, I., Niemelä, Y., Puhakka, M., and Coley, P. D. (1986). River dynamics and the diversity of Amazon lowland forest. *Nature (London)* **322,** 254–258.

Schminke, H. K. (1981). Perspectives in the study of the zoogeography of interstitial crustacea: Bathynllacea (Syncarida) and Parastenocarididae (Copepoda). *Int. J. Spéléol.* **11,** 83–89.

Schumm, S. A., and Lichty, R. W. (1956). Time, space and causality in geomorphology. *Am. J. Sci.* **263,** 110–119.

Seyed-Reihani, A. (1980). Etude écologique du milieu interstitiel lié au fleuve Rhône en amont de Lyon. Thése, Univ. Lyon.

Stanford, J. A., and Covich, A. P. (1988). Community structure and function in temperate and tropical streams. *J. North Am. Benthol. Soc.* 7(4), 261–529.

Stanford, J. A., and Gaufin, A. R. (1974). Hyporheic communities of two Montana rivers. *Science* 185, 700–702.

Stanford, J. A., and Ward, J. V. (1988). The hyporheic habitat of river ecosystems. *Nature* (*London*) 335, 64–66.

Stanford, J. A., and Ward, J. V. (1992). Management of aquatic resources in large catchments: recognizing interactions between ecosystem connectivity and environmental disturbance. *In* "Watershed Management" (R. J. Naiman, ed.), pp. 91–124. Springer-Verlag, New York.

Stanford, J. A., and Ward, J. V. (1993). An ecosystem perspective of alluvial rivers: Connectivity and the hyporheic corridor. *J. North Am. Benthol. Soc.* 12(1), 48–60.

Stock, J. H. (1980). Regression model evolution as exemplified by the genus *Pseudoniphargus* (Amphipoda). *Bijdr. Dierkol.* 50, 105–144.

Taylor, S. W., and Jaffe, P. R. (1990). Biofilm growth and the related changes in the physical properties of a porous medium. 1. Experimental investigation. *Water Rev. Res.* 26(9), 2153–2159.

Townsend, C. R. (1989). The patch dynamics concept of tream community ecology. *J. North Am. Benthol. Soc.* 8(1), 36–50.

Triska, F. J., Kennedy, V. C., Avanzino, R. J., Zellweger, G. W., and Bencala, K. E. (1989). Retention and transport of nutrients in a third-order stream in northwestern California: Hyporheic processes. *Ecology* 70(6), 1893–1905.

Triska, F. J., Duff, J. H., and Avanzino, R. J. (1993). Patterns of hydrological exchange and nutrient transformation in the hyporheic zone of a gravel-bottom stream: Examining terrestrial-aquatic linkages. *Freshwater Biol.* 29, 259–274.

Turquin, M.-J. (1981). Profil démographique et environnement chez une population de *Niphargus virei* (Amphipode troglobie). *Bull. Soc. Zool. Fr.* 106, 457–466.

Valett, H. M., Fisher, S. G., Grimm, N. B., Stanley, E. H., and Boulton, A. J. (1992). Hyporheic–surface water exchange: Implications for the structure and function of desert stream ecosystems. *In* "Ground Water Ecology" (J. A. Stanford and J. J. Simons, eds.), pp. 395–405. Am. Water Resour. Assoc., Bethesda, MD.

Vandel, A. (1964). "Biospéologie: La biologie des animaux cavernicoles." Gauthier-Villars, Paris.

Vanek, V. (1994). Heterogeneity of groundwater–surface water ecotones. *Conf. Int. Ecotones Eaux Souterraines/Eaux Surf., Lyon, France* (J. Gibert, J. Mathieu and F. Fournier, eds.) Cambridge University Press, London. (submitted for publication).

Vannote, R. L., Minshall, G. W., Cummins, K. W., Sedell, J. R., and Cushing, C. E. (1980). The river continuum concept. *Can. J. Fish. Aquat. Sci.* 37, 130–137.

Vervier, P., and Gibert, J. (1991). Dynamics of surface water/groundwater ecotones in a karstic aquifer. *Freshwater Biol.* 26(2), 241–250.

Vervier, P., Gibert, J., Marmonier, P., and Dole-Olivier, M.-J. (1992). A perspective on the permeability of the surface freshwater-groundwater ecotone. *J. North Am. Benthol. Soc.* 11, 93–102.

Ward, J. V. (1986). Altitudinal zonation in a Rocky Mountain stream. *Arch. Hydrobiol., Suppl.* 74, 133–199.

Ward, J. V. (1989a). The four-dimensional nature of lotic ecosystems. *J. North Am. Benthol. Soc.* 8, 2–8.

Ward, J. V. (1989b). Riverine-wetland interactions. *DOE Symp. Ser.* 61, 385–400.

Ward, J. V., and Palmer, M. A. (1994). Distribution patterns of interstitial freshwater meiofauna over a range of spatial scales, with emphasis on alluvial river-aquifer systems. *Hydrobiologia* (in press).

Ward, J. V., and Stanford, J. A. (1983). The intermediate disturbance hypothesis: An explanation for biotic diversity patterns in lotic ecosystems. *In* "Dynamics of Lotic Ecosystems" (T. D. Fontaine and S. M. Bartell, eds.), pp. 347–356. Ann Arbor Sci. Publ., Ann Arbor, MI.

Ward, J. V., and Stanford, J. A. (1991). Research directions in stream ecology. *In* "Advances in Ecology" (J. Menon, ed.), pp. 121–132. Counc. Sci. Res., Trivandrum, India.

Ward, J. V., Stanford, J. A., and Voelz, N. J. (1994). Spatial distribution patterns of Crustacea in the floodplain aquifer of an alluvial river. *Hydrobiologia* (in press).

Welcomme, R. L. (1979). "Fisheries Ecology of Floodplain Rivers." Longman, London.

Wilderer, P. A., and Characklis, W. G. (1989). Structure and function of biofilms. (W. G. Characklis and P. A. Wilderer eds.). *In* "Structure and Function of Biofilms," pp. 5–17. Wiley, New York.

Williams, D. D. (1984). The hyporheic zone as a habitat for aquatic insects and associated arthropods. *In* "The Ecology of Aquatic Insects" (V. H. Resh and D. M. Rosenberg, eds.), pp. 430–455. Praeger, New York.

Williams, D. D. (1989). Towards a biological and chemical definition of the hyporheic zone in two Canadian rivers. *Freshwater Biol.* **22**, 189–208.

Williams, D. D. (1991). Life history traits of aquatic arthropods in springs. *Mem. Entomol. Soc. Can.* **155**, 63–87.

Williams, D. D., and Hynes, H. B. N. (1974). The occurrence of substratum of a stream. *Freshwater Biol.* **4**, 233–256.

HYDRODYNAMICS AND GEOMORPHOLOGY OF GROUNDWATER ENVIRONMENTS

Karst Hydrogeology

A. Mangin

Centre National de la Recherche Scientifique
Laboratoire Souterrain
Moulis
09200 Saint Girons, France

I. INTRODUCTION

The specificity of the karst milieu comes from the dissolution of the rocks under the action of water, following a complex physicochemical process. This phenomenon is called karstification (Davis, 1930; Cvijic, 1960; Gèze, 1965; Thrailkill, 1968; Herak and Stringfield, 1972; Bögli, 1980; Milanovic, 1981; Jennings, 1985; Dreybrodt, 1988; White, 1988; Ford and Williams, 1989). Whatever the origin of preexisting interstices (pores for chalk and certain dolomites or fissures for limestone), the dissolution will produce voids that are occasionally very developed, because, in certain cases, they are penetrable by humans (speleological network). All of these voids are in communication with one another (Mangin, 1974b, 1975, 1983, 1984a) and will induce an organized structure that drains water. Thus we describe karst, all milieus for which, following the phenomena of dissolution, appear to be a morphology constituted essentially by voids displaying a certain organization. In the same way, the word karstification will be used only if dissolution leads to organization. So, all limestone massifs are not properly called karst unless a widespread dissolution producing an organization has occurred. This definition is different from and more restricted than that given by Gèze (1973): "a region made up of carbonated rock, compact and soluble, in which appear forms with both superficial and subterranean characteristics." In this chapter, first the karst characteristics will be examined and then a presentation of the karstic aquifer with different approaches will be discussed.

II. HYDRODYNAMIC CONTROL OF KARSTIFICATION

Water mediates karstification in terms of quality (i.e., acidity permitting the dissolution of rock) and quantity. From a quality point of view, duality exists between the kinetics of chemical reactions tied to water and the speed at which water flows. Indeed, it has been proved (Roques, 1964; Bakalowicz, 1979) that carbonate reactions involving the carbon dioxide content of water and leading to the dissolution of rock are not instantaneous, but last several hours and, in some cases, even several days. Thus, if the flow is too fast, there is not enough time for dissolution to be completed, and if, on the contrary, it is too slow, the water becomes saturated and the dissolution reaction stops. Therefore, an optimal flow velocity exists for creating karst.

From a quantity point of view, the more abundant the water, the greater the dissolution is. But, inversely, the larger the voids, the more the water will be drained. Qualitative and quantitative conditions create an interrelation

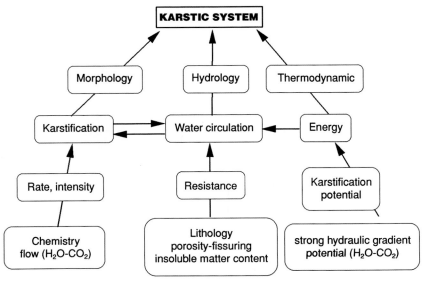

FIGURE 1   Functional scheme of genesis and evolution of karstic systems.

between the structure of voids and that of flow. The result is a complex structure that is hierarchically organized from upstream to downstream (Fig. 1).

Consequently, it is necessary to use a thermodynamic approach to understand how these two processes operate. Chemical and hydraulic mechanisms are considered fluxes that occur according to available energy. This energy constitutes the "karstification potential" with respect to water quantity CO_2 contents and strength of hydraulic gradient. On the basis of available energy, interactions of chemical and hydrologic fluxes define the karst characteristics. Variations of the parameters involved in chemical reactions (nature of the element, acid, CO_2, sulfuric acid, temperature, etc.) do not have any influence on the structure of the flow and, thus, on the type of karst. Indeed, if chemical mechanisms are responsible for the creation of voids, it is the flow that determines the hierarchical structure of karst. Therefore, classifications that do not refer to the structure of flow, notably climatic classifications frequently used by geographers, seem unfounded.

III. THE KARSTIC AQUIFER

The karst definition as discussed above is strongly linked to knowledge of flow. The hydrogeologic study allows the definition not only of the flow characteristics, but also of the genesis and evolution of the karstic milieu. The major problem in karst hydrogeology lies in the heterogeneity of this milieu. Thus, it is necessary to define this heterogeneity and how it belies classical approaches. A new approach that leads to an original karst representation is proposed.

A. Heterogeneity of Karst: Scale Problems

Original porosity linked to the interstice matrix is usually very weak, with the exception of rocks such as chalk or certain dolomites (Murray, 1960; Powers, 1961). Porosity is often increased by fissuration (Tripet, 1972; Borelli and Pavlin, 1967; Paloc, 1964). Nevertheless it remains weak, not more than 2%, and the hydraulic conductivity is also low, 10^{-7} m/sec (Kiraly, 1975; Simeoni, 1976). Subsequently the karstification leads to the appearance of wide voids and gradually increases these two parameters (*e.g.*, porosity can reach 15% in some areas of karst). But karstification induces a very strong heterogeneity determined by an organized structure from upstream to downstream, often leading to a progressive drainage of water toward a single outlet (Mangin, 1974b, 1975, 1984a).

The heterogeneity of hierarchical milieus possesses properties of its own (Cushman, 1990), and this causes scale problems. Indeed, whereas porous and fissured media can be viewed as homogeneous at the macroscopic level (Matheron, 1967), hierarchical milieus cannot be seen in this way (Fig. 2).

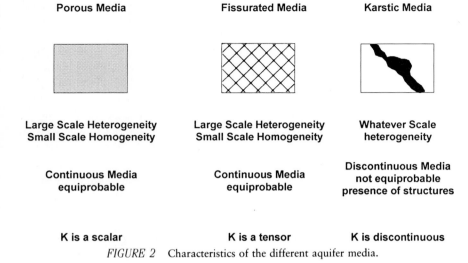

Porous Media	Fissurated Media	Karstic Media
Large Scale Heterogeneity Small Scale Homogeneity	Large Scale Heterogeneity Small Scale Homogeneity	Whatever Scale heterogeneity
Continuous Media equiprobable	Continuous Media equiprobable	Discontinuous Media not equiprobable presence of structures
K is a scalar	K is a tensor	K is discontinuous

FIGURE 2   Characteristics of the different aquifer media.

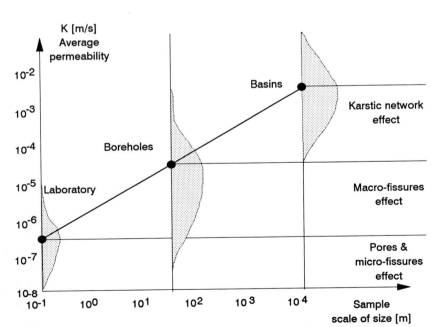

FIGURE 3   Schematic representation of the scale effect on the permeability in karst [after Kiraly (1975)].

Nor can the notion of "representative elementary volume" (Castany, 1982) be used. Theoretically, Darcy's law is invalid because this law, which linearizes the nonlinear Navier–Stokes law, imposes a statistical homogeneity (Marle, 1967). Because the distribution of karstic voids is hierarchical, it does not agree with a random distribution. So variables that characterize this repartition are regionalized variables (Matheron, 1965). Consequently, the parameters defining the karstic aquifer (porosity, hydraulic conductivity, etc.) are themselves structured with respect to the level of scale.

For example, Kiraly (1975) showed that following the change of scale (pores and microfissures, macrofissures, and the karstic network), hydraulic conductivity is a continuous function and is expressed by a straight line in logarithmic coordinates (Fig. 3). Therefore, the properties that are related to a change in scale are invariant and their dimensions are self-similar: the karstic milieu possesses the characteristics of fractal geometry (Mangin, 1986) due to its properties (Mandelbrot, 1975; Feder, 1988; Vicsek, 1989; Gouyet, 1992). This statement is important because it leads to a rejection of the classical approach of hydrogeology based on the validity of Darcy's law and on the linearity of the head flux and other relations. In other words, it is based on an analytical approach conceived from the resolution of differential equations. It is therefore a holistic approach that we are forced to adopt.

B. The Systemic Approach to Karst

1. The Karstic System

The only scale capable of expressing the properties of karstic aquifers is that that corresponds to the hierarchical structure in its entirety. But the discontinuity generated by the structure of voids causes the properties of the milieu to vary considerably from one point to another and makes the notion of an equivalent milieu meaningless for the aquifer in its entirety.

A similar problem can be found in surface hydrology, for which there exists a structure of drainage (streams and rivers) in which reservoirs are incorporated (alluvial ground water and various aquifers); together they constitute a catchment area. In the same way, karst is treated as a catchment area, for which the name the karstic system has been suggested (Mangin, 1974a), and is defined as "the area at the level from which the flow of karstic type becomes organized to constitute a unit of drainage " (Mangin, 1974b, 1975). Hence, the term karstic system is used in the sense defined by systemic analysis (Walliser, 1977), that is to say, a black box that is the center of a dynamic process (flow), determined by input (rain, from a quantitative as well as a qualitative point of view) and output (outlet flow also treated from both a quantitative and a qualitative point of view). The advantage of such a definition is that it promotes a systemic approach as a method for the study of karst (operational research, convolution, etc.).

This implies that the identification of karst and its boundaries rigorously corresponds to a physical definition of the systems.

Whenever the karstic massif is well defined and whenever the flow is supplied only by rainwater that has fallen on the massif itself, the notion of karstic aquifer and that of karstic system are identical. But it is not always the case. Indeed, some karstic massifs drain surface runoff from impermeable terrain, which then becomes part of the system because the organization of underground drainage depends on the organization of surface flow. In this case the notion of a system is no longer equivalent to that of an aquifer. That is the reason why unary (single-phased) karsts (with only the karstic aquifer) and binary karsts (with the participation of surface flow) should be distinguished. Such names refer to thermodynamic concepts, because each part of the system (karst and surface flow) can be interpreted as a different phase; it is particularly important when the evolution of karsts is considered. Therefore, the definition of a karstic system is not obvious and the identification of its characteristics require empirical study.

2. Functional Approach

Analytical approaches used in the study of the flow in aquifers are inappropriate for karst because of the properties already mentioned. Often, the nonlinearity of flow (inadequacy in Darcy's law) is emphasized, but this is not the only limiting factor. For example, nonlinear terms (using Saint Venant's equations) can apply. In the same way, the heterogeneity of the milieu may be taken into account by using the method of finite elements and imposing a geometry that conveys heterogeneity. The problem is essentially the differences in hydraulic conductivity between the different parts of the aquifer in relation to the duration of flow pulses. Indeed, within a distance of a few meters, hydraulic conductivity can change from 10 to 10^{-6} m/sec during a period of rapidly rising water. This creates discontinuities in the integration of differential equations, and Dirichlet's conditions are no longer met. This analytical or structural approach (Mangin, 1985) must be abandoned in favor of a functional or systemic approach, which is based on the study of input–output relations.

With the karstic system entity previously defined and used as a reference, the analysis of input and output functions, and thus of time series, determines the dynamic process from which the properties of functional karst should be identified and characterized. In this approach, the heterogeneity of the aquifer (drains, reservoirs, etc.) is translated by a time-modulated flow at the output. If a drain permits a rapid flow, strong floods will occur; on the contrary, the presence of reserves will modulate the floods. Hence, an accurate, functional description of karst requires an understanding of flow impulse timing in relation to the responses of the karstic system. Karst is then perfectly defined from a function: the unit hydrograph that corresponds to the impulse response of the system.

C. Hydrologic Model of Karst

The proposed model to account for the karstic milieu relies on its hydrologic structure. Consequently, the same distinctions that are found in the aquifer can also be found in the karstic milieu. There are three parts that can be differentiated from the surface to the output (Fig. 4).

1. Nonkarstic Catchment Basin

The karstic aquifer may be connected to part of a nonkarst catchment area. If this nonkarstic area is totally drained by the karstic aquifer, it must be integrated with the system. Such organization may modulate (delay) the input to karst and may mediate change in water quality. If this nonkarstic area is not totally drained by the karstic area, it is not integrated with the system. The inflow by this surface runoff is then considered an input.

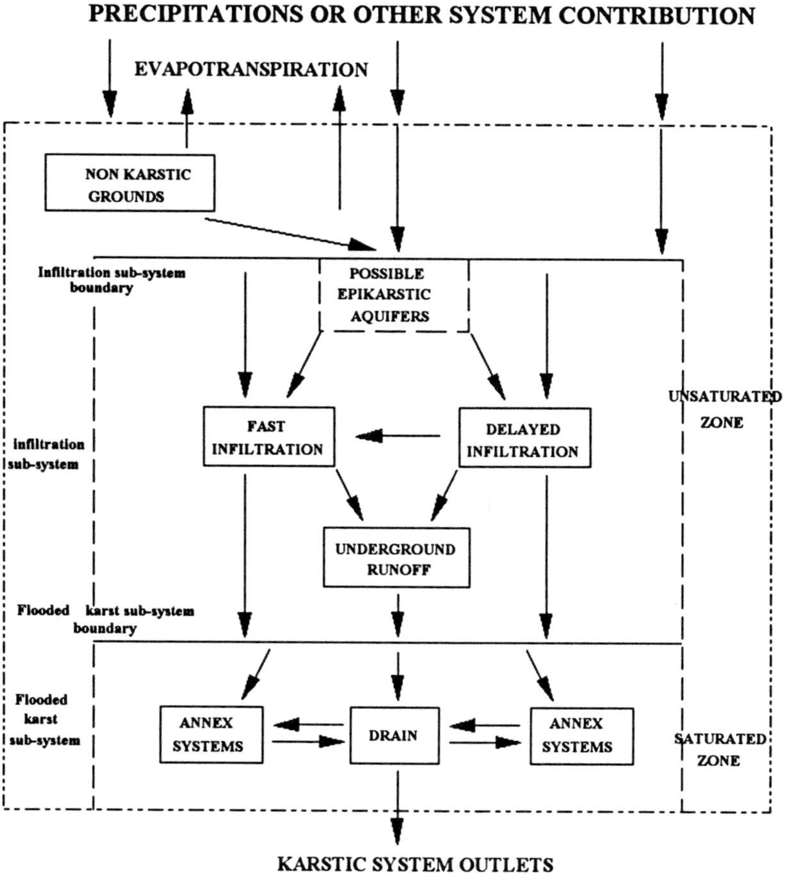

FIGURE 4   Schematization of a karstic system.

2. The Infiltration Zone

The infiltration zone is made up of several parts. Near the surface there exists, sometimes in a discontinuous manner, a weathered zone several meters deep with high porosity and permeability (porosity from 5 to 10%). This epikarstic zone (Mangin, 1974a)—called the subterraneous zone by Williams (1983)—may contain large water reserves (*i.e.*, the epikarstic aquifer). The epikarstic aquifer mediates karstification at depths in the massif or substratum. Otherwise, the low (original) permeability of the terrain would lead to surface runoff with no possibility of infiltration.

Infiltration is carried out following two modes. The first one is runoff through large but not numerous fissures; this is fast infiltration. The second one is diphasic (flow of mixed water and air) through minute fissures; this is delayed infiltration. The whole infiltration zone, with the exception of the epikarstic zone, possesses generally a low porosity (ca. 2%), with a lognormal distribution of voids with depth.

3. The Flooded Zone

The flooded zone (flooded karst) is the saturated zone of the aquifer. It is in this part that the organization of voids and flow is the strongest with drains and annex systems to drainage (ASD). Drains determine the transmissive function of the aquifer; ASD situated downstream of the aquifer and on each side of the drains provide the storage function. It must be noted that "underground river" is not synonymous with "drain," it is just a part of the latter. The porosity of the flooded zone can reach 15%. Depending on the genesis of the aquifers, drains may organize very near the piezometric surface (jurassian type or epiphreatic type) or at a great depth (vauclusian type or deep-phreatic type), according to European terminology. Figure 5 provides a representation of a karstic system.

IV. PROPOSED METHODS FOR STUDYING KARST

A. Hydrodynamic Methods

The characterization of aquifers is generally based on their internal parameters and the conditions at their boundaries. The internal parameters deal with either flux (*e.g.*, permeability, hydraulic conductivity, and transmissivity) or water storage (*e.g.*, porosity and storage coefficient). Conditions at the boundaries modulate flow (*e.g.*, constant flux, variable flux, and zero flux). The heterogeneous nature of the karstic milieu prevents application of such parameters. Alternatively, as a first approach, the aquifer can be treated as a black box and the flux of water through the system can be quantified by input (catchment precipitation) and output (discharge) estima-

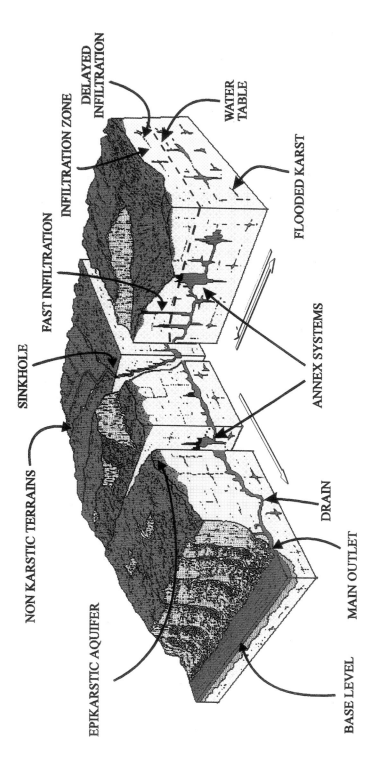

NON KARSTIC TERRAINS

SINKHOLE

FAST INFILTRATION

INFILTRATION ZONE

DELAYED INFILTRATION

WATER TABLE

FLOODED KARST

ANNEX SYSTEMS

EPIKARSTIC AQUIFER

DRAIN

MAIN OUTLET

BASE LEVEL

FIGURE 5   Representation of a karstic system.

tions. The system is characterized by its impulse response (unit hydrograph). Unfortunately, the nonlinearity of karstic systems does not permit flow data to be easily processed (*e.g.*, deconvolution). However, the impulse response may be estimated by recession curve analysis, sorted discharges, and correlation and spectral analysis, procedures that are described below. If data about flow at the outlet are not available, it is nevertheless possible to use piezometric data as indicators of flux (Walliser, 1977). With these methods, it is possible not only to understand the global functioning of the system, but also to identify the elements contributing to this functioning. Thereby they can be used to penetrate into the black box and to study these elements.

1. Analysis of Recession Curves

From physical models, we try to characterize the way the flood wave moves and the way the reserves empty, using numeric parameters. This analysis has been employed for a very long time (Sherman, 1941) using different models with various interpretations (Mangin, 1974b, 1975). Theoretically, this method implies that (1) the rainfall is unitary (which is not the case) and (2) during recession infiltration is separated from depletion (which is impossible). So it is necessary to use some approximations: (1) instead of working with a unit hydrograph, the real hydrograph is used, and (2) instead of using the entire hydrograph, only the recession part of it is studied, and an empirical formulation is applied to distinguish the falling part and the depletion part in the recession. Whereas this method calls upon rigorous physical modeling, it remains semiquantitative.

To study depletion, Maillet's model is used (exponential emptying of a reservoir). For the falling part, an empirical model is adjusted to the form of the decreasing discharges, the chosen formula being a homographic function (Mangin, 1974b, 1975). This function does not strictly account for infiltration, but is a suitable approximation of the way infiltration proceeds. From later considerations, discharge Q will be expressed as follows (Fig. 6),

$$Q = Q_R + q,$$

which will be written

$$Q_{(t)} = Q_{R0}e^{-\alpha t} + q_0 \frac{1 - \eta t}{1 + \varepsilon t},$$

where α, η, and ε, with the dimension T^{-1}, are the parameters of the model.

Due to approximations, the starting point of the curve (t_0) is arbitrarily taken at the highest point of the flood (then $Q_R = Q_{R0}$ and $q = q_0$). Q_R is the depletion discharge. Based on the physical model used, Q_R is >0 only when infiltration has reached zero ($q = 0$ at time t_i in Fig. 6). However, this discharge is extrapolated until time t_0, which is an approximation, and this extrapolated discharge, between t_0 and t_i, joined with the depletion

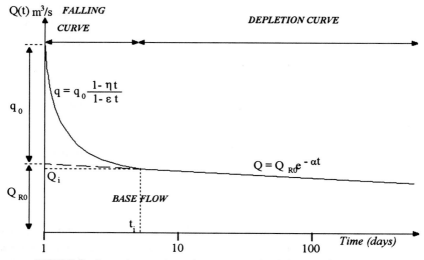

FIGURE 6   Recession curves analysis (see text for definition of parameters).

discharge (*sensu stricto*) is called base flow. With the α coefficient (depletion coefficient) and the totality of the base flow, the proposed model computes the depletion volume according to the formula

$$V = \int_0^\infty Q_{R0}e^{-\alpha t}dt = C\frac{Q_{R0}}{\alpha},$$

where C is a necessary constant (because of the units used) equal to 86,400 when discharge is expressed in m^3/sec and time, in days. This volume, V, has been called dynamic volume. It does not match exactly the volume of the flooded zone (reserves). For this, it would be necessary for the curve to decrease continuously at the same rate until infinity, as is supposed in the formula and which is rarely the case. In fact, only the water volume situated above the level of the outlet can be naturally depleted. The flow also takes into account the volume of moving water situated under the level of the outlet. Inevitably this water is included in the integration of the depletion curve. Hence, the depletion curve cannot continue to decrease at the same rate until infinity. Whenever V is less than 100,000 m^3, the flooded karst is considered negligible; for values included in the 10,000,000- to 100,000,000-m^3 range, the flooded karst is important.

An estimation of the falling curve (q) is obtained by subtracting base flow from decreasing stream flow. It reaches a zero value (time t_i) when infiltration is equal to zero, as predicted by the model. q does not perfectly match with infiltration for two reasons: (1) because, between t_0 and t_i, the depletion of the flooded karst has been extrapolated, and (2) because q

corresponds to the infiltration after the transit through the flooded karst. The η coefficient gives exact information about the infiltration duration, because it is the inverse of t_i. The ε coefficient accounts for the curve concavity and, therefore, for the infiltration heterogeneity.

2. Sorted Discharges

This is another method that makes it possible to establish the characteristics of the input–output relation, hence, to get information about the unit hydrograph. It has been demonstrated (Max, 1980) that, because the discharge series can be assimilated to a random variable, a probability density function describes the input–output relation (*i.e.*, unit hydrograph properties).

To obtain a sorted discharge curve, a certain number of classes of discharge must be defined, and, for each one, the frequency of occurrence must be established. The result is a histogram of sorted discharges on which a function is adjusted. In order to ease the graphical adjustment of the model, the obtained function is linearized by taking the probability Y ordinate and working with cumulated values. The function $F_{(x)}$ has been experimentally described (Mangin, 1974b, 1975),

$$F_{(x)} = \frac{2}{\sqrt{2\pi}} \int_0^x e^{\frac{-u^2}{2}} du,$$

with $x = a\,(Q - Q_0)$ or $x = a\,(\log Q - \log Q_0)$, an entity greater than 0 in both cases. In the former it means that the system is poorly karstified; in the latter, it means that the system is well karstified. When the aquifer is heterogeneous (plurimodal or truncated unit hydrograph), the resulting graph will be a sequence of straight lines. So the method clearly shows the discharge thresholds at which the heterogeneity appears and the modifications in the flow characteristics that result. For example, overflows or extra inputs can be detected and quantified, as well as the discharge values at which they occur. This method differs noticeably from those typically used in hydrology.

3. Correlation and Spectral Analysis

Compared with the two previous methods, correlation and spectral analysis is much more general. It deals with the totality of output time series in relation with the totality of input time series. Consequently it offers much more information about the impulse response of the system (unit hydrograph) and also about the structure of the time series and multiple input–output links.

This method is similar to time series analysis (Box and Jenkins, 1976; Jenkins and Watts, 1968; Yevjevich, 1972) or geostatistic analysis (Matheron, 1970) because of the tools it uses. It differs by excluding all inferential aspects (ARMA process and kriging), which lead to certain restrictions (for example, stationary state of series). The flow, rain, or piezometric series are analyzed from a descriptive point of view in order to establish their structure: trend, periodic components, and random components. The identification of these structures and their isolation after decomposition are used to explain the processes that are responsible for them. Thus, no hypothesis on the series to be analyzed is imposed and no pretreatment is necessary. However, the series must be long enough to give prominence to the structures they express and, above all, must not be incomplete. The underscoring of structures expressed in the time series is carried out with the use of three tools.

a. The Correlogram   The correlogram shows how events are linked to each other for larger and larger time intervals. Its slow decrease conveys the existence of a trend. Values close to zero show the independence of events and, thus, their random nature. As for periodic phenomena, they are marked by amplified periodicities. The correlogram is obtained using the following formula, proposed by Jenkins and Watts (1968):

$$r_k = \frac{C_k}{C_0} \quad \text{with} \quad C_k = n^{-1} \sum_{1}^{n-k} (x_i - \bar{x})(x_{1+k} - \bar{x}).$$

The time series, being defined by its x values, has an average of $\bar{x}$; r is the value of the correlogram with k as the lag, k varying from 0 to m, which is called the truncation point. "The memory effect," which is the time during which the correlogram reaches the value 0.1–0.2, is one important parameter deduced from the correlogram. Therefore, it characterizes the decreasing speed of autocorrelation. If this decrease is rapid, the response is quasirandom, but, if it is slow, the response is very well structured. The appearance of a structure in the time series reveals the inertial nature of the system, which is often related to large reserves. Thus, a correlogram that decreases rapidly characterizes a well-karstified (well drained) system.

b. The Variogram   The variogram expresses the evolution of the total variance when the events are independent of one another. We can demonstrate (Mangin, 1984b) that the variogram, from a descriptive point of view, is perfectly symmetrical to the correlogram; it starts at a zero value (whereas the correlogram starts at 1) and reaches a maximum value of variance when the correlogram has a zero value. Then, its use is the same as that of the correlogram. It is obtained using the following formula:

$$\gamma_k = \frac{1}{2} \sum_{1}^{n-k} (x_i - x_{i+k})^2.$$

Due to its properties, the variogram is interpreted in the same way as the correlogram.

c. The Spectral Density Function   The spectral density function corresponds with the change from a time mode (time series space) to a frequency mode by changing variables (Fourier's transformation). This transformation is interesting because very often it permits a better understanding of the time series components, because they are then well separated. It has been demonstrated (Ventsel, 1973) that the transformation of the time series in the frequency mode corresponds with the decomposition of the variance of the time series components expressed in frequency. It has also been demonstrated (Max, 1980; Réfrégier, 1993) that this decomposition of the variance is obtained by Fourier's transformation of the correlogram (Wiener–Kinschine's theorem). The formula used is that proposed by Jenkins and Watts (1968),

$$S_F = 2\left[1 + 2\sum_1^m D_k r_k \cos 2\pi Fk\right],$$

where k is the step and $F = j/2m$.

D_k is a window that is necessary to ensure that the S_F estimated values are not biased. After experimentation with several hydrologic problems (Mangin, 1984b), it appears that the best windows are those of Tukey,

$$D_k = \frac{\left(1 + \cos\dfrac{\pi k}{m}\right)}{2},$$

and of Parzen,

$$D_k = 1 - 6\frac{k^2}{m^2}\left(1 - \frac{k}{m}\right) \quad \text{for} \quad 0 \le m \le \frac{n}{2}$$

$$D_k = 2\left(1 - \frac{k}{m}\right)^3 \quad \text{for} \quad \frac{n}{2} \le k \le n.$$

Tukey's window filters much less than that of Parzen. It is therefore preferable, because less information is lost. Its drawback is that sometimes it contains peaks that are artifacts. If there is any doubt about the interpretation of a peak after the use of Tukey's window, Parzen's window can be used to clear up the uncertainty.

When high values of density are situated near zero frequencies, the existence of a trend is indicated (note that some trends may be due to a periodic phenomenon with a period longer than the time series studied). Each peak conveys the presence of periodic phenomena, the interpretation of which leads to the characterization of the system's behavior. The intensity of the spectrum can be negligible in high frequencies, indicating a lack of usable information in this range. The frequency from which this condition appears is called the cutoff frequency. The lower its value, the more inertial the system is and so the less karstified it is.

To characterize a system, physicists use the band width notion (spectrum area divided by its maximum value), which characterizes the frequency interval (in the frequency mode) in which the input–output relation intervenes. We shall use the inverse of this band width, called regulation time, which corresponds with the duration of the impulse function. Due to this method, the area is equal to 2, and, thus, the regulation time is obtained by dividing the maximum value of the spectrum by 2.

In correlation and spectral analysis, the choice of lag (k) and truncation (m) is fundamental because it determines the observation window (time interval in which the analysis is carried out). Indeed, all information less than $2k$ long (Shannon's sampling theorem; Max, 1980; Réfrégier, 1993) cannot be noticed. However, if a structure exists in the time series during times inferior to $2k$, the correlogram and the variogram will indicate it (nugget effect, conveyed by a thrust fault in the graph, near low abscissa), as well as the spectral density function ("aliasing," conveyed by an uplift of the spectrum near the 0.5 frequency). As for m, it must always be inferior to $n/2$ (n being the length of the time series). In order to have data that can be easily processed, it is better that m be inferior to $n/3$. Finally, the analysis can be carried out in two different ways.

— A simple analysis: in this case, the time series is supposed to be the response of the system to a random function at the input (white noise). Because of this simplification, the analysis leads to the identification and description of the components of the time series (trend, periodicity, and randomness).

— A cross-analysis: in this case, the time series is considered a response of the system to a time series at the input (causal relation).

The cross-correlogram is obtained following a formula similar to that of the "simple correlogram":

$$r_{+k} = r_{x,y(k)} = \frac{C_{x,y(k)}}{S_x S_y} \quad \text{with} \quad C_{x,y(k)} = n^{-1} \sum_{1}^{n-k} (x_i - \bar{x})(y_{i+k} - \bar{y})$$

$$r_{-k} = r_{x,y(k)} = \frac{C_{x,y(k)}}{S_x S_y} \quad \text{with} \quad C_{y,x(k)} = n^{-1} \sum_{n1}^{n-k} (y_1 - \bar{y})(x_{i+k} - \bar{x})$$

$$\text{and with} \quad S_x^2 = n^{-1} \sum_{1}^{n} (x_i - \bar{x})^2 \quad \text{and}$$

$$S_y^2 = n^{-1} \sum_{1}^{n} (y_i - \bar{y})^2.$$

The cross-correlogram establishes the input–output relation. If the input function is considered to be random (and that is the case of a rainfall within a period of several days to several weeks), the cross-correlogram corresponds with the impulse response of the system. This leads, by definition, to the characterization of karstic systems. When the input cannot be considered a quasirandom function, the cross-correlogram can still provide information on the response of the system: causal or noncausal relation between input and output, kind of relation (directly or inversely proportional), importance of the relation, etc.

Contrary to simple analysis, the cross-correlogram is not symmetrical. Due to this fact, the cross-correlogram provides a complex function based on a real function (the cospectrum) and an imaginary function (the quadrature spectrum), defined, respectively, in the following way:

$$K_{x,y(F)} = 2\left[r_{x,y(0)} + \sum_{1}^{m} (r_{x,y(k)} + r_{y,x(k)})D_k \cos 2\pi Fk \right]$$

$$q_{x,y(F)} = 2\left[r_{y,x(0)} + \sum_{1}^{m} (r_{x,y(k)} - r_{y,x(k)})D_k \sin 2\pi Fk \right].$$

In these formulas D_k is the window of simple analysis.

In polar coordinates, the cospectrum and quadrature spectrum provide an amplitude function and a phase function, which are easier to use and are expressed, respectively, using the following formulas:

$$S_{x,y(F)} = \left[K^2_{x,y(F)} + q^2_{x,y(F)} \right]^{1/2}$$

$$\theta_{x,y(F)} = \arctan\left[\frac{q_{x,y(F)}}{K_{x,y(F)}} \right].$$

The amplitude function gives information about the existence or lack thereof of a relationship between the input and the output, with regard to frequencies. This function is important because all of the other functions (described later) are interpreted in relation to it. This function must not have a zero value, because the properties of the transfer between the input and the output can be examined only when a transfer exists.

Frequency after frequency, the phase function defines the lag between the input and the output. In the time mode, this lag is expressed by the relation:

$$\tau = \frac{\theta}{2\pi F}.$$

Through its intermediary, the response time of a system can be evaluated.

In the frequency mode, two other functions can be calculated: the coherence and gain functions.

The coherence function is written as

$$C_{x,y(F)} = \frac{S_{x,y(F)}}{(S_{x(F)}S_{y(F)})^{1/2}},$$

for which $S_{x,y(F)}$ corresponds with the amplitude function and $S_{x(F)}$ and $S_{y(F)}$ correspond, respectively, to the simple spectra of input (X) and output (Y).

The coherence function conveys the linearity of a system. A system is linear when the increase or the decrease of the output function is proportional to the variation of the input. Nonlinearity often means not only that the output function is related to the studied input, but also that other phenomena must also be taken into account. Linearity is an important characteristic of systems.

The gain function is written as

$$g_{x,y(F)} = \frac{S_{x,y(F)}}{S_{x(F)}},$$

with the same notations as above. The gain function expresses, depending on frequency, either the amplification or the attenuation of the output signal compared with the input signal. The attenuation in high frequencies is often offset by an amplification in low frequencies. In karst, for example, this phenomenon corresponds with the storage process during periods of floods and depletion during periods of low water level. Thus the gain function allows us to evaluate the dynamics of the reserves.

Once the different components of the time series are identified (trend, periodic function, and random function), it is necessary to isolate them in order to identify the mechanisms that have generated them. To do so, a selective filter needs to be used. In most cases, recursive linear filters are sufficient. Two types of filters are used.

— The first-order differentiation filter, which leads to the elimination of all trends. Periodic and random functions are kept as residuals. This filter replaces the time series with its variations,

$$\hat{x}_{(t)} = (x_t - x_{t-1}),$$

$\hat{x}$ being the filtered times series.

—The equally weighted moving average (EWMA) filter, which eliminates all periodic functions with a period equal to a multiple of the amplitude of the filter. The random component is divided by the amplitude of the filter, and all other components remain intact. The terms of the filtered time series are obtained in the following way,

$$\hat{x}_{(t)} = \sum_{-m}^{+m} a_k x_{t+k} \quad \text{with} \quad a = \frac{1}{2m},$$

where formula $2m$ is called the amplitude of the filter.

The value $\hat{x}$ is situated at $t + 1/m$, that is to say, at one point of the nonfiltered time series if the amplitude is an even number or between two points if the amplitude is an odd number. So we notice that the filtered time series is shorter than the original time series with $2m$ data less. This filtering is used, above all, to study the trend. Indeed, to study the residual, precautions must be taken because of the possible presence of periodicities introduced by the filter itself, known as Slutsky—Yule's effect (Barbut and Fourgeaud, 1971).

B. Chemical, Isotopic, Thermal, and Biological Methods

These are indirect approaches that should be coupled with hydrodynamic analysis. Indeed, this section emphasizes that analysis of the karstic system responses is fundamentally multidisciplinary.

1. The Chemical Responses

The chemical responses, founded on the works of Roques (1964) and completed by those of Miserez (1973) and Bakalowicz (1970, 1975a,b, 1977, 1979; see also Chapter 4), are powerful tools. The study of the variations of chemical composition of water with time, during a hydrogeologic cycle or during floods, provides data about the hydrologic functioning of the system: identification of the degree of organization of the aquifer, different modes of infiltration, mechanisms of karstification, etc. (Puyoo, 1976; Fleyfel, 1979; Kuhfuss, 1981; Muller *et al.*, 1982; Lepiller, 1980; Fabre, 1983; Viéville, 1983; Botton, 1984; Muet, 1985; Martin, 1991; Marchet, 1991; Bouchaala, 1991).

2. The Isotopic Responses

Although precautions for interpretation are needed (Margrita *et al.*, 1983) due to the heterogeneity of the reservoir, which does not allow a good mixing of isotopes within the karstic reservoir, and to the existence of a response during the transient regime, the isotopic responses (tritium, deuterium, oxygen 18, and carbon 13 and 14) give important complementary information on rate of water renewal, heterogeneity of the aquifer, and so on (Bakalowicz *et al.*, 1974; Eberentz, 1975; Fleyfel, 1979; Fleyfel and Bakalowicz, 1980; Blavoux *et al.*, 1979).

3. The Thermal Responses

The continuous study of the water temperature at the outlet (Andrieux, 1972, 1976, 1978) provides information about the way the aquifer functions: distinction between energy transfer and mass transfer, size of the reserves, etc. (Pasquier, 1975; Marjolet and Salado, 1976; Lacas, 1976; Tissot and Tresse, 1978; Lescaut, 1980; Botton, 1984).

4. The Biological Responses

The study of animal drift during floods (Rouch, 1968, 1970, 1977, 1980, 1982, 1986) gives information about the functioning of flooded karst between reserves and drains, the existence of large water reserves in the infiltration zone, and the role of surface runoff for binary karst (Lescher Moutoué, 1973; Bertrand, 1974; Rouch, 1982; Turquin, 1981; Moeschler *et al.*, 1982; Gibert *et al.*, 1978; Gibert, 1986).

V. CLASSIFICATION AND EVOLUTION OF KARSTIC SYSTEMS

A. Proposition for a Classification

Karst is a hierarchical structure of voids and flows. Classification of karstic systems should be founded on the degree of organizational structure, which is linked to karstification, and, therefore, every case can be observed between poorly drained media (poorly karstified systems) and, at the other extreme, very well drained media (well-karstified systems). Classification of karst based on hydrogeologic criteria encompasses this milieu, which is not possible by direct observation. Using spectral and correlation analysis, different types of karstic systems may be defined (Fig. 7).

These types (Fig. 7) were given the names of four simple and well-studied European karst systems. The first three are from the Pyrenees and the fourth is from southern Spain (Mangin, 1982). This proposed classification differentiates between two extremes. On the one hand, well drained aquifers are noninertial systems (the "Aliou" type). Their memory effect is poor (the

TYPES	MEMORY EFFECT (r = 0.1-0.2)	SPECTRAL RANGE (truncation frequency)	REGULATION TIME	UNIT HYDROGRAPH
ALIOU (Well Drained)	POOR (5 d)	VERY WIDE (0.30)	10-15 d	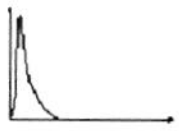
BAGET	SMALL (10-15 d)	WIDE (0.20)	20-30 d	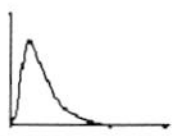
FONTESTORBES	LARGE (50-60 d)	NARROW (0.10)	50 d	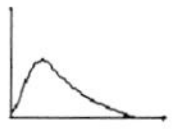
TORCAL (Poorly Drained)	EXTENSIVE (70 d)	VERY NARROW (0.05)	70 d	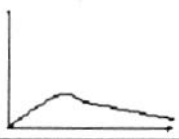

FIGURE 7   Classification of karstic systems from the results of correlation and spectral analysis done on four karst systems that have been studied in detail and are now taken as references (see text for definition of the memory effect, spectral range, regulation time, and unit hydrograph).

value 0.1–0.2 for the correlogram is reached rapidly—5 days). The spectral range is very broad (a good filtering of the observed variations is obtained only for the frequency of 0.3). The regulation time is 10 to 15 days. The shape of the impulse response is very sharp and not spread out. This type of system is characteristic of well-karstified systems having a functional karstic structure. On the other hand, poorly drained systems are inertial (the "Torcal" type). Their memory effect is extensive (the value 0.1–0.2 for the correlogram is reached slowly—70 days). The spectral range is very narrow (after a value frequency of 0.05, the intensity of the spectrum is zero). The regulation time is 70 days. The shape of the impulse response is rounded and well spread out. This is the case for fissured, poorly karstified systems. We shall see later that it is also the case for systems that were karstified, but are no longer functional. Between these two extremes, two intermediary types are defined: "Baget" and "Fontestorbes," which provide a scale for classification.

Correlation and spectral analysis (the multimodal cross-correlogram, anomalies on the simple correlogram and variogram, and the nested effect) may also provide evidence for the complexity of systems: unary or binary systems, simple or composite ones, etc. The sorted discharge method defines complex systems, with the advantage of being able to quantify this complexity: value of thresholds and concerned volumes. Recession curve analysis may complement previous methods. It accounts for the way the different zones of the system are configured. It also makes it possible to quantify the extent of some of these zones. When correlation and spectral anlysis is not applicable (chronicles with gaps), this method can give some information about the organizational degree of the system. A classification was set up (Mangin, 1974b, 1975) from the two variables, k and i, where k is the ratio between the maximum dynamic volume obtained during the longest observation period and the annual average transit volume obtained during the same period and where i corresponds to the value of the function $y = (1 - \eta t)/(1 + \varepsilon t)$, for t = two days.

Five principal types of system have been distinguished (Fig. 8), from the study of 60 different systems:

(I) $k < 0.1$, $i < 0.25$, well-karstified downstream system, and well-developed speleological network
(II) $0.1 < k < 0.5$, $i < 0.25$, and a well-karstified upstream system leading downstream to a large flooded karst (case of systems that have evolved)
(III) $k < 0.5$, $0.25 < i < 0.5$, and a system more karstified upstream than downstream with a delay in its input (case of binary karst)
(IV) $k < 0.5$, $i > 0.5$, and a zone of complex systems (composites)
(V) $k > 0.5$ and nonkarstic systems

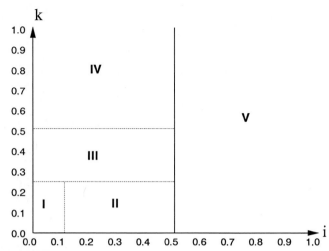

FIGURE 8   Classification of karstic systems from the results of recession curves analysis (see text for definition of i and k and of different classes—I to V).

This classification, less complete than the previous one, gives some similar information but also some more specific data (e. g., dynamic volume of flooded karst).

B. Evolution of Karstic Systems

The genesis and evolution of karst have been inferred from the structure of karsts of various origins and ages (Davis, 1930; Cvijic, 1960; Herak and Stringfield, 1972; Ford and Williams, 1989). Such comparative analysis ignores the dynamic functioning of karst, on which the approach given herein is founded (*i.e.,* structure is inferred from function). Theoretically, this assume that there is a good relation between the organization of voids and flows. However, the dynamics of any system evolve over time. For example, if an organization of drainage is established at the local base level (i.e., the level of the spring changes), a new organization will progressively develop. Void structures will be abandoned and will provide evidence of previous hydrologic conditions. These are the dry caves so often seen in the karstic massif. If the hydrodynamic functioning of the karstic system is set in a thermodynamic context, it will be possible to follow its evolution. The concepts used are those of the thermodynamics of irreversible processes, sometimes far from equilibrium. With this approach, the karstic system corresponds to an open system because it transfers mass and energy beyond its limits: the karst is said to be functional. Both mass and energy are provided by water. When water flows in a low hydraulic gradient and does not have any power of dissolution (absence of CO_2), only mass transfer is taken into account. On the contrary, when the hydraulic gradient is high and dissolution is possible, water intervenes as a flux of energy, which, necessarily, leads to mass transfer. From a thermodynamic point of view, a system is in equilibrium when its state remains constant, that is to say, when its structure does not change. For an open system (thus functional), this equilibrium is not exactly a thermodynamic equilibrium, but is rather a stationary state, because it still transfers energy to its surrounding environment. The degree of structure (or entropy) of such a system depends, at the same time, on its position related to its stationary state and on the amount of energy exchanged. Any variation in environmental conditions (modification of the base level and climatic change) will affect the state of the system and its entropy.

If the flux of energy changes, there are two eventualities. (1) With a decreasing flux, the voids become oversized, and, when the flux stops, the karst becomes nonfunctional (closed system). That is the case of karst found in Florida and in certain areas of Cuba. When the system is binary (see Section III. B. 1), voids are sealed with sediments (*i.e.,* sealed karst). (2) With an increasing flux, erosion and dissolution resume, resulting in an enlargement of voids or an unsealing of previously sealed pipes. When all flux of mass and energy ceases, the karstic voids are abandoned (isolated system) and they then give rise to a paleokarstic morphology (the term "paleokarstic" is sometimes used to describe very old and sealed karst, but the term "fossil karst" seems more appropriate in this case).

Because karstic structure is obviously linked to the flux of energy and of mass, which varies continuously due to the external conditions of the system, most karsts are constantly changing. This is true for all climatic variation and all change in the base level. The speed of karstic system formation may be rapid, on the order of 10,000 to 20,000 years (Bakalowicz, 1975b; Mangin, 1983). Because of this, karst is extremely sensitive to the slightest change in external conditions. The result is that most karsts, if not all, are polyphased (in this last word, "phase" has only temporal significance, with no reference to thermodynamics).

A thermodynamic context permits an unambiguous view of karst evolution (*i.e.,* dynamic response to change in base level, producing the concept of functional or nonfunctional karst and paleokarst) and allows a more synthetic characterization of karst (*i.e.,* the notion of dynamic structure and configuration as inferred in Figs. 7 and 8). But most importantly, an energetics perspective is essential to derivation of a more predictive understanding of karst, including responses to tectonic forces, changes in climate pattern, and human-mediated changes in water chemistry.

ACKNOWLEDGMENT

I thank R. Rouch, D. D'Hulst, and J. Stanford for valuable comments and helpful reviews of the manuscript

REFERENCES

Andrieux, C. (1972). Le système karstique du Baget. I. Sur la thermique des eaux au niveau des exutoires (Note préliminaire). *Ann. Spéléol.* **27**(3), 525–541.

Andrieux, C. (1976). Le système karstique du Baget. 2. Géothermie des eaux à l'exutoire principal selon les cycles hydrologiques 1974 et 1975. *Actes Colloq. Hydrol. 2nd, Pays Calcaires,* Besançon, pp. 1–26.

Andrieux, C. (1978). Les enseignements apportés par la thermique dans les karsts. *In* "Réunion AGSO," pp. 49–63. Sepmast, Toulouse.

Bakalowicz, M. (1970). Géochimie des eaux d'aquifères karstiques. I. Relation entre minéralisation et conductivité. *Ann. Spéléol.* **29**(2), 167–173.

Bakalowicz, M. (1975a). Variations de la conductivité d'une eau en fonction de la température et précision des mesures. *Ann. Spéléol.* **30**(1), 3–6.

Bakalowicz, M. (1975b). Géochimie des eaux karstiques et karstification. *Ann. Spéléol.* **30**(4), 581–589.

Bakalowicz, M. (1977). Etude du degré d'organisation des écoulements souterrains dans les carbonates par une méthode hydrogéochimique nouvelle. *C.R. Hebd. Seances Acad. Sci.* **284**, 2463–2466.

Bakalowicz, M. (1979). Contribution de la géochimie des eaux à la connaissance de l'aquifère karstique et de la karstification. Thèse, Univ. Paris-6.

Bakalowicz, M., Blavoux, B., and Mangin, A. (1974). Apports du traçage isotopique naturel à la connaissance du fonctionnement d'un système karstique. Teneur en oxygène 18 de trois systèmes des Pyrénées (France). *J. Hydrol.* **23**, 141–158.

Barbut, M., and Fourgeaud, C. (1971). "Eléments d'analyse mathématique des chroniques." Hachette Université, Paris.

Bertrand, J. Y. (1974). Recherches sur l'écologie de *Faucheria faucheri* (Crustacés, Cirolanides). Thèse, Univ. Paris-6.

Blavoux, B., Burger, A., Chauve, P., and Mudry, J. (1979). Utilisation des isotopes du milieu à la prospection hydrogéologique de la chaîne karstique du Jura. *Rev. Géol. Dyn. Géogr. Phys.* **21**, 295–306.

Bögli, A. (1980). "Karst Hydrology and Physical Speleology." Springer-Verlag, Berlin.

Borelli, M., and Pavlin, B. (1967). Approach to the problem of underground water leakage from the storage in karst regions. *A. I. H. Publ.* **73**, 120–138.

Botton, R. (1984). Etude de certaines modalités du fonctionnement de l'aquifère karstique (zone d'infiltration et zone saturée) sur un champ de forages nord-Montpellierain. Thèse, Univ. Montpellier.

Bouchaala, A. E. (1991). Hydrogéologie d'aquifères karstiques profonds et relations avec le thermalisme. Exemple de la partie occidentale du massif de Mouthoumet (Aude, France). Thèse, Univ. Franche-Comté, Besançon.

Box, G. E. P., and Jenkins, G. M. (1976). "Time Series Analysis, Forecasting and Control." Holden-Day, San Francisco.

Castany, G. (1982). "Principes et méthodes de l'hydrogéologie." Dunod Université, Paris.

Cushman, J. H. (1990). "Dynamics of Fluids in Hierarchical Porous Media." Academic Press, London.

Cvijic, J. (1960). La géographie des terrains calcaires. *Monogr.—Acad. Serbe Sci. Arts* **341**, 1–212.

Davis, W. M. (1930). Origin of limestone caverns. *Bull. Geol. Soc. Am.* **41,** 475–628.

Dreybrodt, W. (1988). "Processes in Karst Systems: Physics, Chemistry and Geology." Springer-Verlag, Berlin.

Eberentz, P. (1975). Apport des méthodes isotopiques à la connaissance de l'aquifère karstique. Thèse, Univ. Paris-6.

Fabre, J. P. (1983). Etude hydrogéologique de la partie sud-ouest du Causse de Martel (Quercy). Thèse, Univ. Toulouse.

Feder, J. (1988). "Fractals." Plenum, New York and London.

Fleyfel, M. (1979). Etude hydrologique, géochimique et isotopique des modalités de la minéralisation et du transfert du carbone dans la zone d'infiltration d'un aquifère karstique (le Baget, Pyrénées ariègeoises). Thèse, Univ. Paris-6.

Fleyfel, M., and Bakalowicz, M. (1980). Etude géochimique et isotopique du carbone minéral dans un aquifère karstique. *Colloq. Carbonates,* Bordeaux, pp. 231–245.

Ford, D. C., and Williams, P. W. (1989). "Karst Geomorphology and Hydrology." Unwin Hyman, London.

Gèze, B. (1965). "La spéléologie scientifique." Seuil, Paris.

Gèze, B. (1973). Lexique des termes français de spéléologie physique et de karstologie. *Ann. Spéléol.* **28,** 1–20.

Gibert, J. (1986). Ecologie d'un système karstique jurassien: Hydrogéologie, dérive animale, transits de matières, dynamique de la population de *Niphargus* (Crustacé, Amphipode). *Mém. Biospéol.* **13**(40), 1–380.

Gibert, J., Laurent, R., Bourne, J. D., and Ginet, R. (1978). L'écosystème karstique du "Massif de Dorvan" (Torcieu, Ain, France). *Actes Congr. Suisse Spéléol., 6th,* Porrentruy, pp. 37–53.

Gouyet, J. F. (1992). "Physique et structures fractales." Masson, Paris.

Herak, H., and Stringfield, V. T. (1972). "Karst. Important Karst Regions of the Northern Hemisphere." Elsevier, Amsterdam.

Jenkins, G. M., and Watts, D. G. (1968). "Spectral Analysis and its Applications." Holden-Day, San Francisco.

Jennings, J. N. (1985). "Karst Geomorphology." Basil Blackwell, Oxford.

Kiraly, L. (1975). Rapport sur l'état actuel des connaissances dans le domaine des caractères physiques des roches karstiques. *Int. Union Geol. Sci., Hydrogeol. Karstic Terrains B* **3,** 53–67.

Kuhfuss, A. (1981). Géologie et hydrogéologie des Corbières Méridionales, région de Bugarach-Rouffiac des Corbières. Thèse, Univ. Toulouse.

Lacas, J. L. (1976). Introduction à la méthodologie d'étude et d'utilisation des champs hydro-thermiques des aquifères karstiques d'après l'exemple du site de l'exsurgence de la source du Lez (Hérault-France). Thèse, Univ. Montpellier.

Lepiller, M. (1980). Contribution de l'hydrochimie à la connaissance du comportement hydrogéologique des massifs calcaires. Etude de quelques systèmes karstiques du massif du Semnoz et de la région d'Annecy (Savoie, Haute Savoie, France). Thèse, Univ. Grenoble.

Lescaut, M. T. (1980). "Etude des profils thermiques en forages dans les aquifères lorrains," Mém, D.E.A. Univ. Lille I.

Lescher-Moutoué, F. (1973). Sur la biologie et l'écologie des copépodes cyclopides hypogés (Crustacés). *Ann. Spéléol.* **28**(3), 429–502; (4), 581–674.

Mandelbrot, B. (1975). "Les objets fractals, forme, hasard et dimension." Flammarion, Paris.

Mangin, A. (1974a). Notion de systèmes karstiques. *Spelunca Mém.* **8,** 65–68.

Mangin, A. (1974b). Contribution à l'étude hydrodynamique des aquifères karstiques. *Ann. Spéléol.* **29**(3), 283–332; (4), 495–601.

Mangin, A. (1975). *Ann. Spéléol.* **30**(1), 21–124.

Mangin, A. (1982). Mise en évidence de l'originalité et de la diversité des aquifères karstiques. *Colloq. Hydrol., 3rd, Pays Calcaires,* Neuchâtel, pp. 159–172.

Mangin, A. (1983). L'approche systémique du karst, conséquences conceptuelles et méthodologiques. *Karst-Larra Reun. Monogr. Karst-Larra, 1982*, pp. 141–157.

Mangin, A. (1984a). Ecoulement en milieu karstique. *Ann. Mines* 5/6, 135–142.

Mangin, A. (1984b). Pour une meilleure connaissance des systèmes hydrologiques à partir des analyses corrélatoire et spectrale. *J. Hydrol.* 67, 25–43.

Mangin, A. (1985). Progrès récents dans l'étude hydrogéologique des karts. *Stygologia* 1, 239–257.

Mangin, A. (1986). Réflexion sur l'approche et la modèlisation des aquifères karstiques, *Jorn. Karst Euskadi, Donostia-San Sebastian, Spain* 2, 11–31.

Marchet, P. (1991). Approche de la structure et de l'evolution des systèmes aquifères karstiques par l'analyse de leur fonctionnement: Application au NW du Causse de Martel (Quercy-France). Thèse, Univ. Toulouse.

Margrita, R., Guizerix, J., Corompt, P., Gaillard, B., Calmels, P., Mangin, A., and Bakalowicz, M. (1983). Réflexions sur la théorie des traceurs. Applications en hydrogéologie isotopique. *Colloq. Int. Hydro Isot. Mise Valeur Ressources Eau*, Vienne, IAEA-SM-270-84, pp. 1–27.

Marjolet, G., and Salado, J. (1976). Contribution à l'étude de l'aquifère karstique de la source du Lez (Hérault). I. Aspect hydrochimique. II. Aspect hydrodynamique. Thèse, Univ. Montpellier.

Marle, C. (1967). Ecoulements monophasiques en milieu poreux. *Rev. Inst. Fr. Pét.* 22, 1471–1509.

Martin, P. (1991). Hydromorphologie des géosystèmes karstiques des versants nord et ouest de la Sainte Baume (B. du Rh. France). Etude hydrologique, hydrochimique et de vulnérabilité à la pollution. Thèse, Univ. Aix-Marseille III.

Matheron, G. (1965). "Les variables régionalisées et leur estimation." Masson, Paris.

Matheron, G. (1967). "Eléments pour une théorie des milieux poreux." Masson, Paris.

Matheron, G. (1970). La théorie des variables régionalisées et ses applications. *Cah. Cent. Morphol. Math. Fontainebleau*, pp. 1–212.

Max, J. (1980). "Méthodes et techniques de traitement du signal et applications aux mesures physiques," 2nd ed. Masson, Paris.

Milanovic, P. T. (1981). "Karst Hydrogeology." Water Resour. Publ., Littleton.

Miserez, J. J. (1973). Géochimie des eaux du karst jurassien. Thèse, Univ. Neuchâtel.

Moeschler, P., Muller, I., and Schotterer, U. (1982). Les organismes vivants, indicateurs naturels dans l'hydrodynamique du karst, confrontés aux données isotopiques, chimiques et bactériologiques, lors d'une crue de la source de l'Areuse (Jura Neuchâtelois, Suisse). *Beitr. Geol. Schweiz, Hydrol.* 281, 213–224.

Muet, P. (1985). Structure, fonctionnement et évolution de deux systèmes aquifères karstiques du nord du Causse de Martel (Corrèze). Thèse, Univ. Orléans.

Muller, I., Schotterer, U., and Siegenthaler, U. (1982). Etude des caractéristiques structurales et hydrodynamiques des aquifères karstiques par leurs réponses naturelles et provoquées. *Eclogae Géol. Helv.* 75, 65–75.

Murray, R. C. (1960). Origin of porosity in carbonate rocks. *J. Sediment. Petrol.* 30, 59–84.

Paloc, H. (1964). Caractéristiques hydrogéologiques des dolomies de la région languedocienne. *Mém. C.E.R.H., Fac. Sci.*, Montpellier, pp. 123–127.

Pasquier, C. (1975). Contribution à l'étude des bassins karstiques de la région de Champlive (Doubs). Thèse, Univ. Besançon.

Powers, R. W. (1961). Arabian upper jurassic carbonate reservoir rocks. *In* "Classification of Carbonate Rocks: A Symposium" (W. E. Ham, ed.), Vol. 1, pp. 122–192.

Puyoo, S. (1976). Etude hydrogéologique du massif d'Arbas (H. G.). Thèse, Univ. Paris-6.

Réfrégier, P. (1993). "Théorie du signal. Signal, information, fluctuation." Masson, Paris.

Roques, H. (1964). Contribution à l'étude statique et cinétique des systèmes gaz carbonique-eau-carbonate. *Ann. Spéléol.* 19(2), 255–484.

Karst Hydrogeology   67

R. (1968). Contribution à la connaissance des Harpacticides hypogés (Crustacés-Copépodes). *Ann. Spéléol.* **23**(1), 5–167.

Rouch, R. (1970). Le système karstique du Baget. 1. Le phénomène d'"hémorragie" au niveau de l'exutoire principal. *Ann. Spéléol.* **25**(3), 665–709.

Rouch, R. (1977). Considérations sur l'écosystème karstique. *C.R. Hebd. Seances Acad. Sci.* **284**, 1101–1103.

Rouch, R. (1980). Les Harpacticides, indicateurs de l'aquifére karstique. *Mém. Hors-Sér. Soc. Géol. Fr.* **11**, 109–116.

Rouch, R. (1982). Les structures de peuplement des Harpacticides dans l'écosystème karstique. *Int. J. Crustacean Res.* **7**, 360–368.

Rouch, R. (1986). Sur l'écologie des eaux souterraines dans le karst. *Stygologia* **2**(4), 352–398.

Sherman, L. K. (1941). The unit hydrograph and its application. *Bull. Assoc. State Eng. Soc.* **17**, 4–22.

Simeoni, G. P. (1976). Etude de la perméabilité des formations calcaires du Jura neuchâtelois. *Bull. Cent. Hydrogéol. (Neuchâtel)* **1**, 9–18.

Thrailkill, J. V. (1968). Chemical and hydrologic factors in the excavation of limestone caves. *Geol. Soc. Am. Bull.* **79**, 19–46.

Tissot, G., and Tresse, P. (1978). Etude des systèmes karstiques du Lison et du Verneau, région de Nans-sous-Sainte Anne (Doubs). Thèse, Univ. Besançon.

Tripet, J. P. (1972). Etude hydrogéologique du bassin de la source de l'Areuse. Thèse, Neuchâtel.

Turquin, M. J. (1981). Profil démographique et environnement chez une population de *Niphargus virei*. *Bull. Soc. Géol. Fr.* **106**(4), 457–466.

Ventsel, M. (1973). "Théorie des probabilités." Mir, Moscou.

Vicsek, T. (1989). "Fractal Growth Phenomena." World Scientific, Singapore.

Viéville, J. (1983). Etude hydrogéologique du massif de l'Etang de Lers (Pyrénées ariégeoises. Thèse, Univ. Montpellier.

Walliser, B. (1977). "Systèmes et modèles. Introduction critique à l'analyse des systèmes." Seuil, Paris.

White, W. B. (1988). "Geomorphology and Hydrology of Carbonate Terrains." Oxford Univ. Press, Oxford.

Williams, P. W. (1983). The role of the subcutaneous zone i karst hydrology. *J. Hydrol.* **61**, 45–67.

Yevjevich, V. (1972). "Stochastic Processes in Hydrology." Water Resour. Publ., Fort Collins, CO.

3

Porous Media and Aquifer Systems

L. Zilliox

Institut Franco–Allemand de Recherche sur l'Environnement
Campus du CNRS—Université Louis Pasteur
BP 20
67037 Strasbourg Cedex, France

I. THE POROUS MEDIUM AS A COMPLEX FIELD OF FLUID FLOW

A. Definition and Observation Scales

A porous medium is a type of reservoir in which the study of the equilibrium and the movement of fluids is particularly complicated when compared with the knowledge of fluid mechanics in pipes or channels. For example, it is difficult to access the internal "geometry" of a porous medium. In the laboratory samples, it may be examined by penetrating optic fibers into the medium and transmitting images directly on to the analyzer or by observation under a microscope of thin lamina of the material obtained after injection of resins into the pores. The field analysis of soundings and loggings may yield knowledge of the complex structure of the interconnected pores of the medium. But irrespective of the scale of the observation, it has not been possible to give a detailed quantitative description of a "porous reservoir." A natural porous reservoir is represented by the alluvia—the mixture of pebbles, gravel, sand, and clays—which any river deposits in its channel's floodplain. These alluvial masses accumulated at different periods, in formations of variable thickness in valleys enclosed to varying extents, constitute aquifers with interstitial porosity. The observation at different scales of such a granular, unconsolidated massif accentuates several "heterogeneity levels" of the porous medium.

In Fig. 1, level 1 defines heterogeneity on the "granulometric" scale, whereas levels 2 and 3 show successively the heterogeneity of the "sedimentary deposit" and the heterogeneity of the "geological structure." Under certain circumstances in which it is important to take into consideration "elementary lengths of distance covered" across a "granular bed" (of the

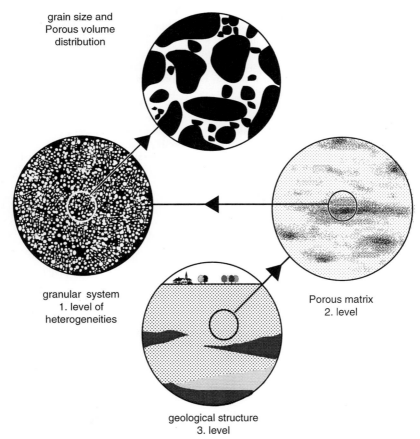

grain size and
Porous volume
distribution

granular system
1. level of
heterogeneities

Porous matrix
2. level

geological structure
3. level

FIGURE 1   Observation scales of heterogeneities in a sand and gravel aquifer [after Spitz (1985)].

"reactor" type in chemical engineering, for example), several "pore sub-scales" can be identified (Fig. 2).

In order to study the flow of water through a complex system composed of the interconnected interstices of a porous medium of the alluvial aquifer type, the experimenter needs to adopt a perspective on a very much larger scale than that of the pores in order to consider the permeable reservoir in its entirety. Consequently, Fig. 3 gives a schematic representation of an aquifer, from the soil surface to the impermeable bedrock in a vertical section according to the general direction of the underground water flow. The part of the aquifer saturated with water is called groundwater. In the case of Fig. 3 the groundwater is said to be free because its surface is in contact with the atmospheric air through the unsaturated area.

B. The Hydrodynamic Global Approach

The medium can be considered as a continuous medium when the objective is to describe the phenomena taking place on the scale of the reservoir during the study of the movement of a free groundwater defined

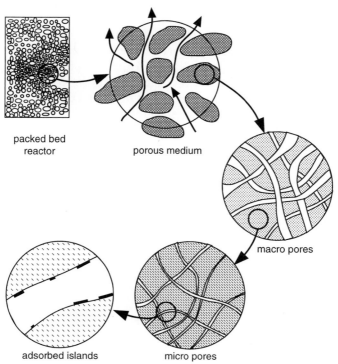

FIGURE 2 Multiple length scales in a packed-bed reactor (from porous disordered medium to micropores and sorption sites) [after Quintard and Whitaker (1993)].

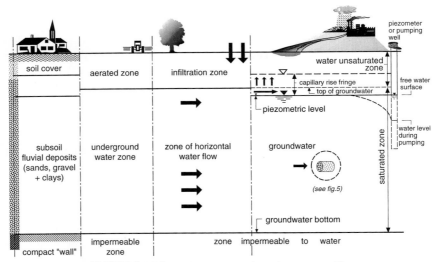

FIGURE 3   Schematic representation of porous aquifer.

by the field of the hydraulic charge. Then it may be possible to apply the theory of potential movement using all of the methods, mathematical, numerical, analogical, or a combination thereof.

The porous structure of the reservoir does not need to be considered when the interest is limited to the productive capacities of the alluvial groundwater as "quantity of water." Neither the nonuniform composition of the water nor the permeability heterogeneities of the medium are taken into account in the analysis globally based on the network formed of current lines and equipotentials.

In the case of movements with potential it is then possible to apply the "superposition principle": if the movement toward a well [Fig. 4 (top)] is known in the horizontal plane (x, y) and in a straight line [Fig. 4 (middle)], the characteristics of the flow of a well placed in the rectilinear current [Fig.4 (bottom)] are determined by superposition. By calling the uniform upstream velocity of the rectilinear current V and the discharge of the well Q, the equations of the current lines, ψ = resultant movement constant, are obtained by addition of the current functions, $\psi_c = -Vy + $ constant (rectilinear current) and $\psi_p = (Q/2\pi)\theta + $ constant (well), i.e., $\psi_c + \psi_p = \psi = -Vy + (Q/2\pi)\theta + $ constant $[\theta = \arctan (y/x)]$.

The representation of Fig. 4. (bottom) is used in the definition of the protection perimeter of a well supplying drinking water, which takes into account in particular the line of separation ψ_0 corresponding to the limit of influence of the well. This is essential when examining the transport of a pollutant toward a catchment field for groundwater. In most cases, the behavior of the fluids contained in the reservoir—they may be waters con-

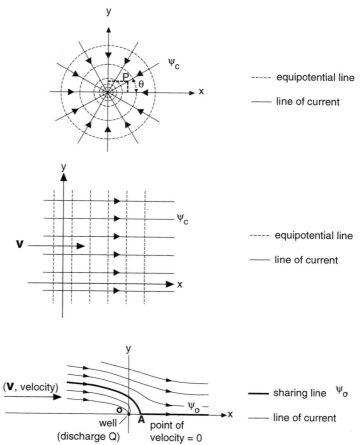

FIGURE 4   Visualization of field of flow in a continuous medium (movements in the horizontal plane in the three cases). (top) flow toward the well; (middle) rectilinear flow; and (bottom) well placed in a rectilinear flow (symmetry/Ox).

taminated physically (thermal effects), chemically (dissolved substances), or biologically (microorganisms) or other liquids immiscible with water or even a gas—results from mechanisms brought into play at the level of the pores, and it is necessary to acquire a good understanding of the phenomena that will be manifested on the scale of the entire reservoir.

Between the scale of the reservoir and that of the pore space, an intermediate level of the "representative elementary volume"(REV) of the porous medium should be considered. On this scale, the objective is to establish the so-called "macroscopic" laws by taking mean values (in the sense of probabilities) of the laws established in fluid mechanics in a continuous medium and applying them to an elementary flow duct (i.e., through the pore space). In this way one changes from a minute geometric characterization to a block scale sufficiently large to include a considerable number of empty and filled spaces such that a mean effect emerges; it will be possible to treat it as in a continuous medium and to define locally "macroscopic dimensions." This concept, which establishes an equivalence between the actual dispersed medium and an idealized continuous medium, makes it difficult to grasp the spatial variability of the natural media. In order to treat the problem of the scale change in porous medium, Matheron (1967) used a probalistic approach relating DARCY's empirical law (filtrometer experiment) to the equations of Navier–Stokes established in fluid mechanics in continuous media. DARCY's law is the equation for the movement of a fluid (water, for example) at the REV scale, in which the peculiarities at the granulometric level (scale of the pores) are no longer taken into consideration. In the case of a continuous medium, this law simply expresses the relationship between the "macroscopic" idealized velocity (or unit discharge) q and the hydraulic gradient (or piezometric slope) J, by $q = KJ$, with K = DARCY coefficient (or permeability coefficient or also hydraulic conductivity).

The REV scale is usually that that applies to the sample of porous medium in the laboratory. But any part of the natural aquifer exhibits heterogeneities in which dimensions range from that of the empty spaces (also existing in the REV) to that of the distance, for example, between a supply point and a catchment point or between injection wells and a direction-finding piezometer during an *in situ* tracer test. In these cases, in view of the extent of the field of observation, it will be necessary to define sizes characteristic of this scale greater than a heterogeneity level of order 2 or even 3. They will no longer be local values of a REV but will integrate the "regional variations" and will need to be determined with the aid of statistical methods.

C. Characteristic Parameters of a Porous Reservoir

When an alluvial aquifer is used to supply drinking water, it is necessary to treat two concerns in parallel. The first relates to the quantitative study of the water resource, in which movement is treated without regard to composition. The second relates to the quality of the water. In the first case, the capacity of the porous reservoir to contain the water and its capacity to transmit a water flux are evaluated. In the second case, the capacity of the reservoir to select elements (physical, chemical, or biological) transported by the water and its capacity to disperse these elements (dissolved, in suspension, or both) will be estimated.

Accordingly, particular attention needs to be paid to three parameters of the reservoir: the porosity, the permeability, and the dispersivity (Fig. 5). Their importance is in relation to the content (stock of water), the flux (circulation of the water), and the dispersion (transport by water), respectively.

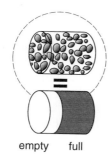

POROSITY, necessary
"to determine the available **volume**
for the water between the grains"
(defines the **content** of the
reservoir)

empty   full

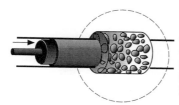

PERMEABILITY, necessary
"to determine the pressure to be
exerted to overcome the resistance
offered by the solid matrix to the
movement of the water"
(defines the **flux** through the
reservoir)

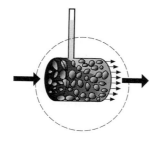

DISPERSIVITY, necessary
"to determine the capacity of the
matrix to generate mixing of waters
of different compositions"
(defines the **dispersion** in the
reservoir)

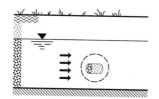

THE AQUIFER : a "filter"
(porous medium) ?

THE UNDERGROUND WATER :
a transport "vector"

FIGURE 5   Parameters characterizing a porous medium ("reservoir").

1. Porosity

During tests on a sample of a porous medium practically saturated with water, the volume of gravitational water collected compared with the total volume of the sample indicates the value of its effective porosity (corresponding to the drainable water content). Whereas this effective porosity is obtained as a ratio of two volumes, the kinematic porosity is obtained as the ratio of two velocities, the DARCY velocity and the mean macroscopic velocity. Is this kinematic porosity (corresponding to the mobile water content) a function of the hydraulic load gradient? Very few measurements are known to answer this question.

Is it not risky to want to attempt to determine this kinematic porosity help by *in situ* tracing at the level of a portion of reservoir, knowing that the heterogeneities of the medium will cause the average macroscopic velocity to be overestimated because of predominate transit through the most permeable zones? A tracing test would be more relevant for estimating kinematic porosity, by comparison with the value for the total porosity of the medium.

2. Permeability

Recall from DARCY's law that the permeability coefficient K, or hydraulic conductivity with dimensions LT^{-1}, depends not only on the sedimentological characteristics of the porous reservoir but also on the physical properties of the circulating fluid. Thus, permeability k, a magnitude specific for porous matrix, with dimensions L^2 and also called intrinsic permeability, is linked to K by the relationship $K = k \cdot \rho g / \mu$, where g is the acceleration due to gravity and ρ and μ are the density and the dynamic viscosity of the circulating fluid, respectively. In the alluvial substrata values of the horizontal permeability are always much higher than those of the vertical permeability. In fact, in such an aquifer volume the permeability is usually considered as a tensorial property.

Knowledge of the local values of the permeability at different depths of a reservoir enables the heterogeneity to be estimated. It is, however, not certain that it is always possible to take advantage of this knowledge to forecast the overall behavior of the reservoir. In some cases the concept of transmissivity of the groundwater can be used. Transmissivity corresponds, in a vertical plane, to the product of the mean horizontal hydraulic conductivity and the thickness of the groundwater (saturated zone).

Using the composition of the local permeabilities to determine a regional permeability (i.e., average distribution of the heterogeneities of an aquifer) works well only in simple cases of a stratified reservoir with a succession of layers parallel or perpendicular to the flow of water. In most cases stochastic approaches must be used along with deterministic models more common in hydrology. This methodology is being developed in hydrogeology in order to take into account the random spatial variability of characteristic parameters. The same analysis of the heterogeneity of a reservoir as that performed on the basis of local and regional permeabilities is found in the characterization of the dispersive properties of the reservoir.

3. Dispersivity

In an alluvial aquifer the transport of a tracer (i.e., a substance miscible with water without modifying the physical properties of the water and noninteractive with the alluvia) from an emission well to a locating well allows estimation of the dispersion coefficient from the restitution curve, giving the distribution of the tracer concentrations with time in the measuring well (the mapping methods used in the field are found in the abundant literature on this subject). Taken according to the direction of the mean flow of the underground water, this coefficient, denoted D_L and called the longitudinal dispersion coefficient (dimensions $L^2 T^{-1}$) is usually expressed by the relationship $D_L = \alpha_L \cdot u^m$, where u is the mean macroscopic velocity of flow (ratio of the unit flow rate to the kinematic porosity of the substratum), m is a constant taken to be equal to 1 under the conditions of water movement in an alluvial aquifer, and α_L is an intrinsic longitudinal dispersion coefficient called longitudinal dispersivity. D_T and α_T, similar parameters corresponding to the transversal dispersive effects (in a plane perpendicular to the mean direction of flow in the reservoir), are introduced in the same manner.

Just like the permeability parameter, the dispersivity is representative of the scale of measurement. It corresponds, at the local level, to the variability of the velocities in a porous space and, at the regional level, to the heterogeneities of the field of macroscopic velocities representative of the sedimentological nature of the reservoir. It is the parameter whose determination in the field is the most difficult compared with the measurement of porosities and permeabilities.

D. Heterogeneous Flow Conditions

Overall knowledge of the contours of the aquifer and the determination of parameters at some sites do not remove uncertainties concerning the variability of the magnitudes in space. In this context it is also necessary to enlarge the concept of heterogeneity beyond the effects due to the sedimentological structure of the reservoir (generator of preferential flow channels and of areas of delayed diffusion, for example); it is also necessary to take into consideration the effect of areas with differential saturations (capillary effects in polyphasic flow and water trapped in aerated aquifer, in particular) as well as the incidence of density contrasts (at the origin of the development of interference instabilities in the "miscible displacements" of soft and salt waters, for example).

Conditions of "heterogeneous movement" are shown schematically in Fig. 6 according to various observation levels of water flow and the transport of elements dissolved in the water. With respect to the concepts of homogeneity and isotropy, a porous aquifer is homogeneous if it exhibits at every point, in a given direction, the same resistance to the movement of water; if this resistance is the same irrespective of the direction, the aquifer is isotropic.

(1) dispersive effects of flow observed at the pore-volume scale
(first level of heterogeneity)

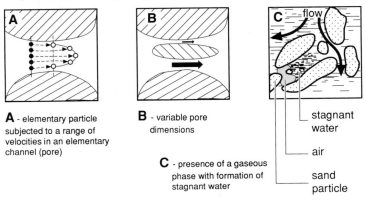

A - elementary particle
subjected to a range of
velocities in an elementary
channel (pore)

B - variable pore
dimensions

C - presence of a gaseous
phase with formation of
stagnant water

stagnant
water

air

sand
particle

(2) mixing effects of multipermeability structures (second level)

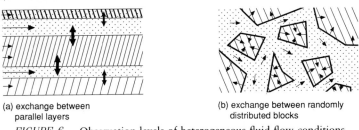

(a) exchange between
parallel layers

(b) exchange between randomly
distributed blocks

FIGURE 6   Observation levels of heterogeneous fluid flow conditions.

II. MULTIPHASE FLOW IN POROUS MEDIA

Whether it is a question of water movement in the unsaturated area of a porous aquifer (biphasic, water–air) or the movement of a petroleum product accidentally infiltrated into the soil cover of an aquifer (triphasic, oil–water–air), one must, in both cases, confront the problem of polyphasic flow in a porous substratum. Whether interest is focused on understanding the role of surface water/groundwater ecotones or on the study of a problem of pollution of the alluvial groundwater by a fluid immiscible with water, a certain number of elements are necessary in order to analyze the behavior at the interfaces of the gases, liquids, and solids simultaneously present in the porous aquifer (Zilliox, 1980).

A. Interfacial Phenomena in the Fluid–Solid System

The study of interfacial phenomena provides a useful insight into the relative movement of several fluid phases (immiscible gases and liquids) in a porous medium. Figure 7 presents a schematic view of the contact between two immiscible fluids adjacent to a solid face. The fluid–fluid interface forms an angle,θ, with the solid face called the " contact angle"; this angle is directly proportional to the "wettability" of the solid by one of the fluids in the presence of the other.

Usually, wettability can be defined as the adherence energy between the solid and the liquid due to molecular attraction; this energy is often measured by the angle θ , also called the wetting angle (Larde *et al.*, 1965). The fluid on the side on which $\theta < \pi/2$ (Fig. 7) is called the "wetting phase"; the other fluid is called the nonwetting or repelled phase. If the interfacial tension between the two fluids is designated σ and the respective interfacial tensions between fluid 1 and fluid 2 and the solid are designated σ_1 and σ_2, respectively, then at equilibrium $\sigma \cos \theta = \sigma_2 - \sigma_1$ (Young's equation). When the fluid–fluid interface moves across the face of the solid, the tensions and wettability (which determine the curvature of the fluid–fluid interface) change with the direction of movement; this pheomenon is called wetting hysteresis. The curvature of the interface (Fig.7) indicates the existence of a pressure difference between fluid 1 and fluid 2; this pressure difference is called capillary pressure and is expressed symbolically by $p_c = p_2 - p_1$, p_1 being the pressure in the wetting phase and p_2 being the pressure in the nonwetting phase (the pressure excess is found on the same side as the center of curvature).

The connection between capillary pressure and wettability is demonstrated by the following illustration. Imagine a capillary tube completely filled with wetting fluid (1), placed in contact with a reservoir containing a nonwetting fluid (2) (the pressures being strictly identical on either side of the fluid–fluid interface). In order to cause fluid 2 to penetrate into the tube , it is necessary to exert at the mouth of this tube a pressure that will be a function of the radius (r) of the tube, the contact angle (θ), and the interfacial tension (σ). The minimal pressure (Δp) that must be exerted in

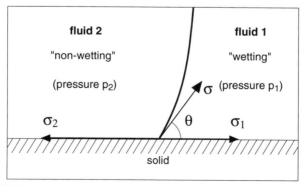

FIGURE 7   Interface between two immiscible fluids in contact with a solid.

order to cause fluid 2 to enter the tube and displace fluid 1 from the tube is none other than the capillary pressure (corresponding to the dimensions of this tube). This relation is expressed by

$$\Delta p = p_c = \frac{2\sigma \cos \theta}{r}.$$

The measurement of the displacement pressure actually constitutes a measurement of the wettability of the solid face. The higher the pressure needed for one liquid to displace another, the more nonwetting that fluid is for the solid. This concept forms the basis of the principle of capillary pressure measurement. If the fluid filling the tube is the less wetting of the two with respect to the solid, the other fluid will enter the tube spontaneously when there is no initial pressure difference, even if the force of gravity is acting in the opposite direction. A well-known consequence of this phenomenon is the capillary rise of water in a porous soil, forming the vadose zone.

B. Distribution of Several Immiscible Fluids

In an aquifer with interstitial porosity in which several nonmiscible fluids occupy the interstitial volume, the capillary forces play an important role. Two parameters prevail in contiguous fluids.

1. Saturation by fluid i is defined as being the fraction of the porous volume occupied by this fluid; it is represented by

$$S_i = \frac{\text{volume of fluid } i}{\text{total porous volume}},$$

and, for all the fluids present, the equation is $\Sigma_i S_i = 1$

2. The mean pressure differences in the two fluids in contact within the porous medium is called capillary pressure (p_c).

In the porous medium, the variability of the pore sizes in three dimensions leads to complex fluid distribution patterns. Hence, between two fluids at equilibrium a clear-cut interface will no longer exist (i.e., with a well-defined curvature). Instead, there will be a transition zone in which the capillary pressure will be a function primarily of the saturation of each of the liquids. A hysteresis phenomenon associated with wettability and the variable form of the interstices exists, and, hence, no single relationship between capillary pressure and saturation can be described.

In the case of two fluids, 1 (wetting) and 2 (nonwetting), the value of S_2, for example, is not sufficient to determine p_c (S_2). In fact, for a given value of S_2, the two fluids can be distributed in the porous medium in several different ways, and it is necessary to know the direction of change in the saturation by fluid 2, which has led to S_2. Drainage and imbition depend on whether the saturation by a nonwetting fluid is increased or decreased, respectively. The displacement of the wetting fluid at complete saturation with this fluid is termed "primary drainage," and the displacement of the nonwetting fluid at complete saturation by this fluid will be called "primary imbibition."

Figure 8 shows typical drainage and imbibition curves for fluids 1 and 2. The extreme saturation zones, in which one of the fluids is no longer capable of being displaced by the other fluid (limit values S_{01} and S_{02}), can be made out. In the intermediate zone, the distributions of the two fluids are simultaneously in a funicular state, and each equilibrium state between the two fluids corresponds to a minimal energy state. Therefore, the wetting fluid will occupy the smallest pores. When the medium contains more than two fluids, the phenomena are qualitatively analogous, but more complicated. We will return to this later when considering flow in three phases: water, air, and oil.

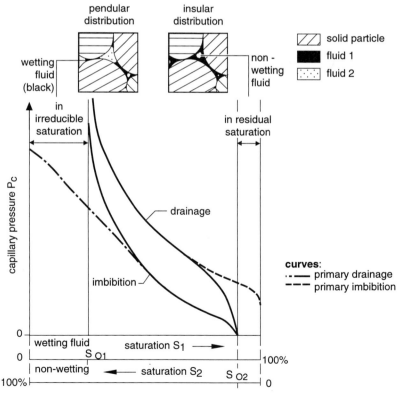

FIGURE 8 Capillary pressure curves in terms of saturation by two immiscible fluids. 1, wetting; 2, nonwetting.

A special case is the study of the water–air boundary in the unsaturated zone of an aquifer. In this zone the effect of capillarity is defined by the difference between the local interstitial water pressure and the standard atmospheric pressure. This negative pressure (depression expressed in terms of water height) is usually called suction and denoted by H. The distribution of water in the unsaturated soil is defined by the volumetric water content (c), which is the volume of water in an apparent unit volume of soil. If n denotes the porosity, then $c = n \cdot S$, where S is water saturation.

The "suction" parameter can be measured directly by means a tensiometer (Fig. 9a). Volumes of water are preferably measured by radioisotopic tracers, because tracers (like gravimetric measures) are not destructive to the medium. The tensiometer consists of a hydraulic circuit connecting a porous cell in contact with the soil to a pressure transducer. The equalization of the pressures on either side of the porous cell is achieved by transfer of water through the cell, which is impermeable to air. Figure 9b illustrates schematically the principle of such a device operating in contact with an unsaturated porous medium. The water content of soil is determined either by gammametry (the method favored in the laboratory) or by the use of a neutron probe (implanted *in situ*). Gammametry is based on the adsorption of a beam of gamma rays emitted by a radioisotope. The water content is determined by comparing the count rate for photons passing through dry soil with the count rate of photons passing through the same thickness of moist soil. Vachaud *et al.* (1970) employed this method using americium 241. The principle of the neutron humidimeter is based on the deceleration of fast neutrons by the hydrogen atoms present in the soil water. The measurement is performed by introducing into the medium a source of fast neutrons and a slow neutron detector, combined in the same probe. The slow neutron count is proportional to the water content. This procedure works well in the field (Mutin and Soeiro, 1969).

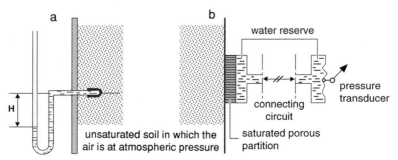

FIGURE 9 Measurement of suction: working principle of the tensiometer. (a) Suction measurement (Bear, 1972); (b) tensiometer principle (Thony, 1970).

C. Simultaneous Flow of Different Fluid Phases

As we have seen above, Darcy' law expresses the relationship between the specific discharge (q) and the hydraulic gradient. It may be written in the generalized form

$$q = -\frac{k}{\mu}\,\text{gradient }\phi,$$

where μ is the (absolute) dynamic viscosity of the fluid, k is the intrinsic permeability (the tensor for an anisotropic medium), and $\phi = p + \rho g z$ is the total potential (sum of the pressure potential and position potential), with p = pressure, ρ = fluid density, g = the acceleration due to gravity, and z = elevation.

In the case in which several fluid phases are present in the porous medium, the following relation, by analogy with Darcy's law, holds for each phase i,

$$q_i = -\frac{k_i}{\mu_i}\,\text{gradient }\phi_i$$

where k_i, "effective permeability" for phase i in the presence of other fluids, is a function of the saturation level of the medium with respect to this phase. Experiment shows that the values of k_i are always lower than that for k because the presence of other fluids hinders the flow of fluid i. The "relative permeabilities" are then defined as $k_{ri} = k_i/k$ (values included between 0 and 1), which depend on how the phases are distributed in the porous volume.

In the case of fluid couple 1 (wetting) and 2 (nonwetting) saturating the medium, the following relationships hold:

$$q_1 = -\frac{k_{r1}\cdot k}{\mu_1}\,\text{gradient }\phi_1$$

$$q_2 = -\frac{k_{r2}\cdot k}{\mu_2}\,\text{gradient }\phi_2$$

with
$$\phi_1 = p_1 + \rho_1 g z$$
$$\phi_2 = p_2 + \rho_2 g z$$
and
$$p_2 - p_1 = p_c(S_1)$$

$$S_1 + S_2 = 1.$$

Figure 10 presents typical relative permeability curves as a function of saturation.

Although in practice it is generally true that relative permeabilities are essentially functions of saturation, it should be pointed out that more detailed study will show that relative permeability curves, like capillary pressure curves, are not determined in a unique manner. Moreover, hysteresis will occur also as a function of the direction of change in the saturation. It is important to note that the relative permeabilities cancel out for saturation values above zero. So long as the saturation by the wetting fluid remains at the stage of pendicular distribution (S_1 below the irreducible saturation: Fig. 8) this phase is immobile; the same is true for the nonwetting fluid in insular distribution ($S_2 < S_{02}$).

A special case exists for water flow in unsaturated soil, in which air is supposed to remain immobile. In order to study water transfer independent readings of the changes in the specific water content (c) and the capillary pressure or suction (H) are taken (see Section II.B). During a drainage

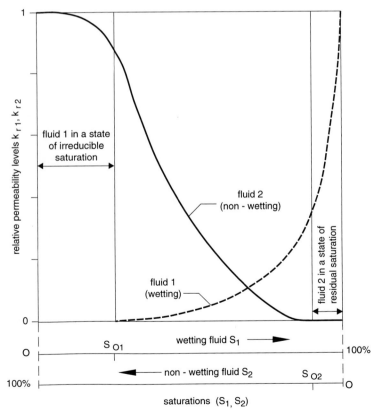

FIGURE 10   Relative permeability curves for two fluids. 1, wetting; 2, nonwetting.

experiment in a vertical column (with Oz axis downward) (Thony, 1970), knowing the instantaneous local measurements of c, it is possible to obtain values for the specific discharge (q) by integration of the equation of continuity

$$\frac{\partial q}{\partial z} = \frac{\partial c}{\partial t}.$$

From measurements of H, the potential gradients ($H - z$) are determined, and, by calculation using Darcy's law, it is possible to obtain the hydraulic conductivity, $K\ (c) = -q/\text{gradient}\ (H - z)$. The movement of three immiscible fluids in a porous medium can be studied formally as biphasic flow; three relative permeabilities that are also dependent on the saturation conditions will be defined (Fig. 11) (after Leverett and Lewis, 1940 see in von Engelhardt, 1965). Every point on the triangular diagram corresponds to a distribution of saturations of the three fluids. The isoplets for air, water, and oil are reported in Fig. 11a; in Fig. 11b are shown the zones of possible three-phase, two-phase, or one-phase flow as a function of the saturations.

These results are obviously limited to a given case of pore geometry, interfacial tensions, and wettabilities. In Fig. 11a the particular pattern of isoperms relating to the oil, which expresses both the wettability of this

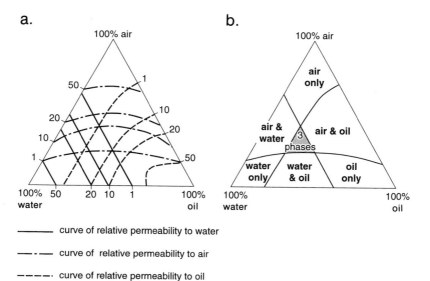

————— curve of relative permeability to water

—·— curve of relative permeability to air

————· curve of relative permeability to oil

FIGURE 11   Degrees of relative permeability (a) and zones of mobility (b) for air, water, and oil in a porous medium, as a function of saturation by each of these three phases. (a) Relative permeability, expressed as a percentage, to each of the three phases, for different levels of saturation by those phases. (b) Areas of possible flow of one, two, or three phases for different levels of saturation.

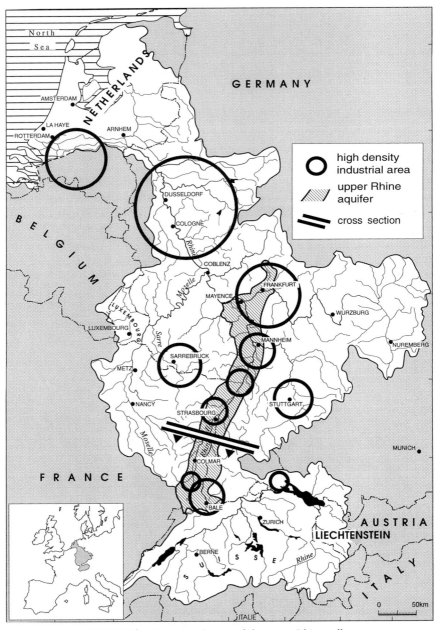

FIGURE 12   The groundwater of the upper Rhine valley.

phase relative to air and its nonwettability relative to water, should be noted (Dracos, 1966). The permeability to water, on the other hand, is uniquely a function of the water saturation (parallel isoperms).

III. TRANSPORT PHENOMENA IN POROUS AQUIFER SYSTEMS

A. The Porous Aquifer: An Interactive Compartment in a River Valley Hydrosystem

A continental hydrosystem, including the physical medium and its biological content, including humans (and their activities), may be illustrated by the cartographic representation in Fig. 12, which shows the porous aquifer of the Rhine valley in the rift basin between the Black Forest and the Vosges. The exploitation and management of the water in such a hydrosystem occur on the scale of the alluvial plain (Ackerer and Zilliox, 1990) and the entire river valley, taking into account the atmospheric water, surface waters, water in the soil, and underground water all at the same time. In order to quantify the transport of chemical substances toward (and into) the underground waters, it is necessary not only to know the capacity of the medium (soil cover, root systems, sedimentary deposits, and alluvia, with their specific biocenoses) to transmit a water flux, but also to know how to take into account the capacity of this porous medium to select the elements transported by the water, to disperse them or to adsorb them, or even to transform them (chemical and biological processes).

The elements below show how the free alluvial groundwater is situated in the overall interactive system. Water exchanges occur between the rivers at the surface and the underground water, which are variable in quantity depending on the particular river segment, season, hydraulicity of the rivers, and relative position of the surface of the groundwater to that of the bottom of the bed of the watercourse. The alluvia in contact with the groundwater are not inert to the passage of pollutants. Thus, the sediments of the water courses and the fine elements (clays and loams) of the aquifer react with toxic contaminants such as traces of heavy metals or pesticides (organic or inorganic micropollutants). Laboratory experiments on porous media have shown that under certain conditions (for example, a change in the water composition), a micropollutant such as mercury can be remobilized, after being bound, and transported by the water with the risk that it will reach a drinking water catchment downstream. Hence it appears that the alluvia may store micropollutants (which is fortunate for the protection of the groundwater) up to a certain threshold (most often unknown); but other elements that modify the composition of the water can be released from these permeable alluvia and pollute the groundwater.

What distinguishes the problems of pollution of the underground waters from those of the surface waters is, first the time scale. In the case of a river for which the velocity at which the water moves is on the order of km/hr, the time of renewal of the water volume contained in its bed is, on average, about a few days and maximally several weeks. In the porous aquifer in which the underground water moves at velocities on the order of km/year, the indicator of the renewal time for the volume of water contained in the reservoir is expressed in decades, even centuries. As a result of its rapid dispersion the pollution of a water course (the effects of which are most often visible) is capable of being quickly detected. Conversely, the pollution of underground water (circulating very slowly and practically invisibly in the underground compartment) may not be observed for several years or more.

Thus, reaction times are of the same order of magnitude in the case of depollution. After the cause of the pollution has been brought under control and decontamination procedures have been implemented, the original quality of the surface water may be regained after a relatively short time. But the pollution in the aquifer reservoir may require decades for restoration. In extreme cases, the situation may even be irreversible. This notion of irreversibility is in itself complex, because it depends on the persistence of various classes of pollutants. Persistence of pollution is a concept associated with the nature and degree of intensity of the interrelations, of variable duration, between hydrological compartments and ecosystems associated with the river basin.

Detailed knowledge of the dynamics of transfer of pollutants toward the groundwaters is usually fragmentary. Such transfer is a function of the specificity of the medium traversed, the type of pollutant, and conditions of its transport by water.   As regards the characterization of the site of transfer toward the groundwater, it is necessary to distinguish two types of

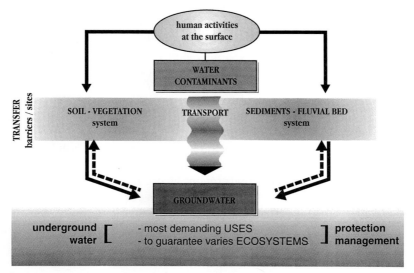

FIGURE 13   Schematic representation of multidisciplinary research program of PIREN/water—Alsace third phase (1989—1993).

screens or biogeochemical barriers: (1) soils, dry (at the surface) or flooded, with or without plant cover, exhibiting very varied water conditions and (2) sediments and silted deposits in the water courses, ponds, quarries, etc., in a state of maximal water saturation. These porous media are most often described in the context of a problem or set of problems of interest to a single scientific discipline and a given scale. The present state of knowledge as to their *in situ* structure and their role as "active" media (internal functioning of heterogeneous ecotones) in the transfer of the pollutants is insufficient. Interdisciplinary research is required (Fig. 13).

A classification of the pollutants by family can be made according to physical (total or partial miscibility with water, volatility, etc.), chemical (organic or mineral nature, solubility, etc.), and biological (level of degradation, degree of toxicity, etc.) characteristics. Although these types of data are known for this or that substance under precise conditions of analyses and measurement, complete knowledge about them under the conditions of circulation in natural porous media is largely unavailable.

Transport by the water comprises hydrodynamics and reaction processes. The contaminating substances (in particulate form or in solution) exhibit differentiated "affinities" toward the solid matrix or fluid phases (water or air) composing the natural medium of transfer. Static experiments in closed reactors usually enable equilibria to be attained with regard to the distribution of an element between fluid and solid phases. Under dynamic conditions, in which the fluxes of pollutants are variable, the exchange occurs differently. Hence, it is useful to develop experimental procedures in order to obtain a better understanding of the exchange mechanisms and functions of the nature and specific behavior of the pollutants.

Various reaction processes (chemical and biological) are influenced by conditions in the ecotone (vadose zone) and are affected by very many "environmental" factors (pH, temperature, speciation, and presence of other substances—colloids, organic matter, etc.). The exchanges may or may not be reversible; the kinetics also are different and linked to the conditions of transport (rates of movement). Finally, the relative position of the surface of the groundwater to the soil surface is crucial because the infiltration rate increases as the porous medium becomes the more saturated. Quantification of transfer requires an understanding of a hierarchy of physical mechanisms and biological and chemical processes, which govern it. Extensive experimental research must be done in order to provide predictive models that can be used by managers of water quality. Identification of the groundwater/surface water ecotones and discontinuities is essential because reaction rates change dramaticaly in short distances.

B. Some Mechanisms

Analysis of pollutant flux is complex. The dynamic nature of pollutants occurs in various spatial and time scales with respect to accumulation of toxic pollutants in all the compartments of an alluvial hydrosystem. Understanding of a certain number of elementary mechanisms and their interactions may permit models to be developed with a view to active prevention. Unfortunatly only curative technologies are applied in many cases after pollutant spills.

Most of the contaminations in alluvial systems are miscible with water. Even in the case of pollution by petroleum products, it is the soluble hydrocarbon fraction that constitutes the essential danger because it is dispersed by the movement of the water. Movement of the pollutant also will be subjected to transport retardants that are the stationary and reactive solid, liquid, or gaseous phases of the porous substrata. The use of geochemical, hydrogeological, and fluid mechanical knowledge is thus necessary.

During its displacement, the pollution will also be under the influence of pollutant biotransformers. The analysis of such perturbations is the concern of ecology and microbiology (or even toxicology as far as the effects on living matter are concerned). Pollutants therefore spiral downstream in response to interactive biophysical processes that differentially hold and then release the pollutant. The relative propensity to retain the pollutant is termed the spiraling length. The approach to as complete a vision of the mechanisms as possible must necessarily be interdisciplinary. In this type of approach, modelization will be useful in two ways: (i) it will help the researcher to view the phenomena as transdisciplinary agents and to construct a finalized representation compatible with the reality (even if it lacks properties judged irrelevant as a function of the choice of objective, levels of observation, availability of data, intensity of calculations, etc.) and (ii) it will help the decision maker by placing at his or her disposition a practical tool designed, depending on the case, to reveal the changing mechanisms of sustained pollution, to take measures to control pollution, or to consider town and country planning and the management of natural resources (including primarily water and soils) by minimizing the risks of pollution.

An example of model mechanisms in the transfer of a contaminant of water under standard conditions of isothermal transport, practically saturated porous medium, and first-order reaction processes is provided. In unidirectional flow, according to Ox (x is the coordinate in the direction of water movement), the expression of the conservation of a solute mass whose flux is affected by mechanisms and processes, numbered (1) to (4), may be expressed as

$$\frac{\partial C_{water}}{\partial t} = \underbrace{-\frac{q}{n}\frac{\partial C_{water}}{\partial x}}_{(1)} + \underbrace{D_a\frac{\partial^2 C_{water}}{\partial x^2}}_{(2)} - \underbrace{\frac{\rho}{n}\frac{\partial C_{soil}}{\partial t}}_{(3)} - \underbrace{\lambda\left(C_{water} + \frac{\rho}{n}C_{soil}\right)}_{(4)}.$$

The "rate of change" of the concentration of the pollutant in the water (C_{water}) is a function of the (hydrodynamic) mechanisms and processes (chemical processes at the interfaces and biotransformation processes) expressed by the terms (1) to (4), with the following definitions.

(1) The advection (mean displacement, also called convective movement), which depends on the unit discharge (q) (DARCY's notional velocity); n is the kinematic porosity of the medium.

(2) The apparent dispersion ("mechanical mixing" and molecular diffusion), which depends on the apparent dispersion coefficient (D_a), which is a function of the rate of movement of the interstitial water. If this "mean actual" rate is zero, only the diffusion phenomenon of the pollutant comes into play.

(3) The interaction with the solid phase, which takes into account the adsorption and desorption phenomena at the surface of the alluvial matter (porous medium), which is in contact with the moving water. C_{soil} represents the quantity of pollutant retained on the porous medium (solid matrix); ρ is the apparent density of the porous medium.

(4) The degradation processes, which take into account the decrease in the concentrations in the liquid phase and on the solid matrix as a result of biological activity. This term depends on a diminution constant λ.

The change in the concentration profiles in the water, starting from an initial signal of concentration (C_0) (i.e., at the entry to the system for a given duration) and under the influence of mechanisms and reactions that can act simultaneously on the transport of a solute in depth, is shown schematically in Fig. 14 (Zilliox and Schenck, 1989). An example of representations of the impact of hydrodynamic mechanisms on the transport in a porous aquifer system may be described in two cases.

1. Visualization of the Effects of a Permeability Heterogeneity in the Vertical Plane

One objective of the Hydrodynamics of Porous Media research team of the Institute of Fluid Mechanics in Strasbourg is to understand the elementary mechanisms of dispersion and transfer that characterize the contamination of an aquifer of the alluvial type with interstitial porosity. The experimental research approach consisted of placing emphasis on significant parameters in order to estimate by means of a simple but representative laboratory model the effect produced on the mechanism under study: thus, concerning the fundamental aspect of hydrodynamic dispersion, the team has studied the influence of density contrasts, exchanges between the solute and the porous medium, and heterogeneous structures of the porous matrix on the change in the mixing zones.

The experiments performed on a vertical two-dimensional physical model have enabled the mechanisms of dispersion to be studied, for example, in the case of a secondary fluid source, miscible with water, placed in the horizontal flow of water. Figure 15 shows how in the case of stable

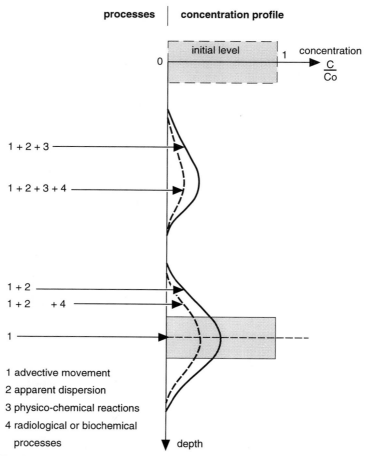

processes | **concentration profile**

initial level   1   concentration
0                           $\frac{C}{Co}$

1 + 2 + 3

1 + 2 + 3 + 4

1 + 2

1 + 2   + 4

1

1 advective movement
2 apparent dispersion
3 physico-chemical reactions
4 radiological or biochemical
  processes   depth

FIGURE 14   Change of the concentrations in water from the entry signal (initial level) in response to the combined action of various mechanisms.

movement—from left to right—the mixing zone is perturbed by the presence of a heterogeneity formed by a vertical wedge of a porous medium, the permeability of which is higher than that of the surrounding medium; it is possible in particular to arrive in this wedge to unstable states (Muntzer and Zilliox, 1978).

Downstream from this perturbation and as a function of the extent of the instabilities created, an extension in depth of the mixing zone, which may vary with time, is observed; it should be noticed that the mean concentration in the mixing zone diminishes under these conditions by dilution–dispersion after passing through the more permeable wedge.

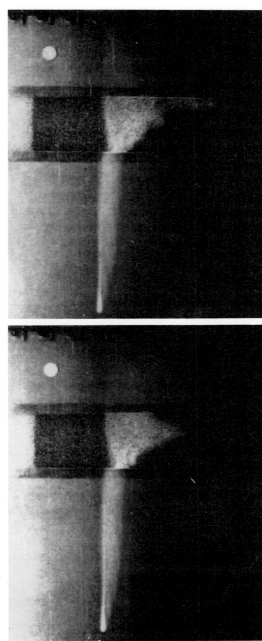

FIGURE 15 Influence of a vertical break in permeability on the change of an initially stable mixing zone, clearly visible in the two photographs (movement in saturated porous medium, from left to right). Left, mixing zone at t. Right, mixing zone at $t_2 > t_1$.

2. Schematic Representation of the Change in Space of a Mixing Zone: Principle of Delimitation of a Contaminated Area Downstream from a Pollution Source

Starting from a point source (point O), where a pollutant discharge occurs under stationary-state conditions, and considering the change of concentrations in three-dimensional space, one determines the surfaces of equal concentration using an equation of partial differentials. Supposing an isotropy of the permeabilities, dispersivities of the medium, and a discharge of constant source in a rectilinear current, one can devise the scheme shown in Fig. 16 (Zilliox *et al.*, 1982).

If u represents the mean real velocity in the direction of flow according to Ox, C, the concentration, D, the dispersion coefficient in the (y,z)plane, and m, the source discharge (in mass of contaminant per unit time), then the spatial variation of the concentration is given by the expression

$$C\,(x,y,z) \;=\; \frac{m}{4\pi Dx}\exp\left(-u\frac{y^2 + z^2}{4Dx}\right),$$

which is the solution to the equation

$$u\frac{\partial C}{\partial x} \;=\; D\left(\frac{\partial^2 C}{\partial y^2} + \frac{\partial^2 C}{\partial z^2}\right).$$

This simplified model corresponds to stationary-state hydromechanical conditions. If a "norm" defines a limit concentration (C_L) for the polluting

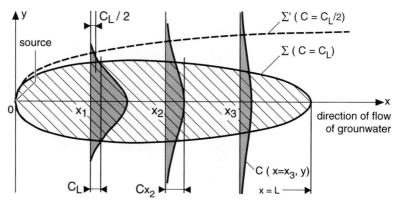

Delimitation of a contaminated area : $\Sigma\,(C = C_L)$

FIGURE 16   Delimitation of a contaminated area: $\Sigma\,(C = c^{te})$ (schematic construction in the median plane (x,y)).

substance concerned (corresponding, for example, to a critical value that must not be exceeded from the ecotoxicological point of view), then the envelope Σ ($C = C_L$) delimiting the actual zone of contamination of the aquifer is determined according to the representation shown in Fig. 16.

On the Ox axis, with $D = \alpha_T u$ (see Section I.C.3.), where α_T is the dispersivity of the medium in the (y,z) plane, the concentration is expressed by the relationship

$$C(x,0,0) = \frac{m}{4\pi\alpha_T u} \cdot \frac{1}{x}.$$

Provided the values of m (mass discharge) and α_T (dispersivity) are known, the maximal possible extent of pollution (Fig. 16) is given by the expression

$$x = L = \frac{m}{4\pi\alpha_T u} \cdot \frac{1}{C_L}.$$

The contaminated area, shown in three dimensions, is generated by rotation round the Ox axis under the conditions of isotropy shown above. The heterogeneous structure of the actual aquifer and the natural potentialities for biochemical transformation (the microbiological activity will be the more intense the slower the rate of movement of the water, for example) will contribute to deform and reduce, respectively, the contaminated area in most cases. (The evaluation based only on the hydrodynamic approach offers maximal security when there is risk of contamination of a water catchment point in the alluvial aquifer.)

The foregoing considerations raise the question of the hierarchy of mechanisms and processes of transfer. Is the necessary knowledge available to allow analysis of the transfer of a pollutant toward the groundwater—under defined conditions of the medium traversed and transport—to yield a representation describing the mechanism(s) or process(es) involved? In other words, can a predictive model be devised and supported by the data according to criteria relating to the incidence and importance of the mechanisms and reactions?

For example, it is known that chloride (a natural, conservative tracer) is transported in water by convective–dispersive processes. But the transport of nitrates (nonconservative) in a reducing medium rapidly results in a blockage (retention). On the other hand, in a saturated aquifer, nitrates may play the role of tracer. Elements such as mercury, may be bound to sediments, and subsequently remobilized as a result of the action of chlorides. Predicting concentration thresholds for the various ions and elements with respect to passage through a porous medium is problematic. Extensive experimental research needs to be carried out to provide a better grasp of this hierarchy. Only then will predictive models be very useful to decision makers.

REFERENCES

Ackerer, P., and Zilliox, L. (1990). Groundwater case study: Upper Rhine Aquifer. *In* "Global Freshwater Quality: A First Assessment" (M. Meybeck, D. Chapman, and R. Helmer, eds.), pp. 253–260. Programme Global Environment Monitoring Systems (GEMS), United Nations Environment Programme (UNEP), and WHO.

Bear, J. (1972). "Dynamics of Fluids in Porous Media." Am. Elsevier, New York.

Dracos, T. (1966). Physikalische Grundlagen und Modellversuche über das Verhalten und die Bewegung von nichtmischbaren Flüssigkeiten in homogenen Böden. *Mitt. Versuchsanstalt Wasserbau Erdbau,* No. 72. E.T.H. Zürich, 48 p.

Larde, Briant, J., Labrid, J., and Marle, C. (1965). Quelques aspects des phénomènes interfaciaux dans le déplacement de l'huile par l'eau en milieu poreux. *Rev. Inst. Fr. Pét.* **20**(2), pp. 253–279. Ed. Technip, Paris.

Matheron, G. (1967). "Eléments pour une théorie des milieux poreux." Masson, Paris.

Muntzer, P., and Zilliox, L. (1978). Rôle des structures hétérogènes sur la propagation de polluants dans la nappe alluviale. *Proc. Symp. Int. I.H.E.S. Ed. CERGH-USTL* Montpellier, pp. 473–483.

Mutin, and Soeiro, F. A. (1969). Infiltration en milieu naturel argileux, interprétation physique des phénomènes. *Houille Blanche* **8–69.**

Quintard, M., and Whitaker, S. (1993). Transport in ordered and disordered porous media. *Chem. Eng. Sci.* **48**(14), 2537–2564.

Spitz, K.-H. (1985). Dispersion in porösen Medien: Einfluss von Inhomogenitäten und Dichteunterschieden. Mitteilungen des Instituts für Wasserbau, Universität Stuttgart. *In* Schadstoffe im Grundwaser/DEG (H. Kobus, ed.), Vol. 1, No. 60. 1992 VCH Verlag, Weinheim, FRG.

Thony, J. L. (1970). Etude expérimentale des phénomènes d'hystérésis dans les écoulements en milieu poreux non saturés. Thèse, Univ. Sci. Méd., Grenoble, France.

Vachaud, G., Cisler, J., Thony, J. L., and De Baker, L. (1970). Utilisation de l'émission gamma de l'Américium-241 pour la mesure de la teneur en eau d'échantillons de sols non saturés. *C.R. Isot. Hydrol. A.I.E.A.,* Vienne, pp. 643–661.

von Engelhardt, W. (1965). Physikalische Grundlagen des Verhaltens von Mineralöl und Mineralölprodukten im Boden. *In* "Gutachten," pp. 7–48. Verhalten von Erdölprodukten im Boden, Bad Godesberg, FRG.

Zilliox, L. (1980). Multiphase flow in porous media. *In* "Aquifer Contamination and Protection: UNESCO Studies and Reports in Hydrology" (R. E. Jackson, ed.), No. 30, pp. 163–175. UNESCO, Paris.

Zilliox, L. (1989). Les relations eaux de surface-eaux souterraines dans le bassin de l'Ill domaniale. Impact des inondations et d'aménagements, transferts de polluants à la nappe. *Rapp. Synth., Programme* PIREN-Eau/Alsace, 124 p., Strasbourg.

Zilliox, L., and Schenck, C. (1989). Quelques réflexions sur le transfert de polluants dans les eaux souterraines. *In* "Qualité et conservation des sols: Devenir des polluants dans les sols," *Rech.-Environ. Collection* Vol. 34, pp. 281–288, SRETIE, Ministère de l'Environnement, Paris.

Zilliox, L., Muntzer, P., and Kresser, W. (1982). Beitrag zur Frage der Grundwasserverschmutzung: Verdrängungsvorgänge in gesättigten porösen Medien. *Oesterr. Ing.-Arch.-Z.* **1,** 127. pp. 1–8.

4

Water Geochemistry: Water Quality and Dynamics

M. Bakalowicz

Laboratoire Souterrain du CNRS
Moulis
09200 Saint-Girons, France

I. INTRODUCTION

Continental waters are the main agents that weather rocks and transform landscapes. Water chemistry is the result of interactions between infiltrating water (rain or surface water) and rocks (Schoeller, 1962; Drever, 1982). As a matter of fact, water plays two different complementary parts (Bakalowicz, 1979): (1) as a chemical reagent, dissolving minerals and organic matter, and (2) as a transport agent of energy and mass. These two parts taken together describe water geochemistry.

Therefore, the geochemical characteristics include chemical information, depending on laws of aquatic chemistry (stoechiometry, thermodynamics, etc.) and hydrogeological information, given by the nature (geochemical and physical characteristics) and by the function (hydrodynamical characteristics) of particular hydrological systems.

In groundwater geochemistry two different approaches can be considered: a descriptive approach, in opposition to the modeling approach, which is considered a predictive one (Hakenkamp *et al.*, 1993). Descriptive concepts consider empiricisms of system structure and function, assumptions proceeding from empirical analyses and general synthesis models (de Marsily, 1981; Castany, 1982; Hubert, 1984). Modeling attempts explicitly simulate groundwater geochemical distribution and/or evolution derived from empirical data (Chapelle, 1993). Artificial tracers are commonly used to describe water movement and aquifer zones for local studies [see discussion by M. Palmer (1993) for the hyporheic zone and by Margrita *et al.* (1984) for aquifers].

This chapter mainly refers to the descriptive approach. The first part describes the factors responsible for chemical concentrations of groundwaters and transformation. The second part gives examples of groundwater geochemistry in practice.

II. WATER CHEMICAL CONTENT: FACTORS INFLUENCING CHEMICAL SPECIES IN GROUNDWATER

A. Factors Responsible for Water Chemistry

All minerals and, consequently, all rocks are soluble to some extent in water, but some of them are more soluble than others. The solubility of a mineral (i.e., the quantity of dissolved minerals in water) depends on its nature, on physical variables (such as temperature and pressure), and on water chemical content. However, mineral solution never is an instantaneous process. That is why time for water–rock contact partly determines the dissolved solids content of water. The longer the water–rock contact, the more mineralized the water. Some factors contribute to increase the solution velocity, such as the surface of water–rock contact and the flow regime.

Therefore, the chemical content of waters results from three factors: (1) the type of rock (*e.g.*, mineralogy and grain size) in contact with the flowing water; (2) the climate and, in general, environmental conditions, determining flow rate, temperature, pressure, etc.; and (3) the flow conditions, determining especially the presence of a gaseous phase and the time of contact between water and rock (*i.e.*, residence time of water in the system).

The first factor imparts the main geochemical characteristics of water. The examples from Table I show that the water chemical content, for exam-

TABLE I   Chemical Compositions of Groundwaters from Various Geological Environments

	1	2	3	4	5	6	7	8	9	10
pH	5.9	5.7	7.0	7.60	7.30	7.85	7.20	6.7	7.65	7.32
C	—	24	172	323	667	2220	3570	15	—	—
TDS	83	45	125	—	—	—	—	—	233	3025
Ca	7.0	1.6	38.2	64.2	111	610	840	6.1	44.0	265
Mg	2.0	0.7	1.5	7.3	7.9	56.0	190	0.7	23.7	98.5
Na	18.0	2.35	1.00	1.35	23.0	45.0	9100	0.9	7.9	644
K	1.5	1.95	0.30	1.25	1.6	30.0	660	0.4	0.3	5.6
HCO_3	24.0	14.7	110	186	350	182	187	19.0	256	412
SO_4	4.0	3.0	8.0	34.0	30.0	1600	3200	2.3	2.5	303
Cl	17.0	2.0	1.85	2.60	39.0	53.0	14450	1.8	7.1	1290
SiO_2	13.0	16.8	6.0	3.5	4.5	2.7	1.9	26.0	—	—
F	—	0.25	0.04	0.03	0.04	0.04	0.03	0.40	—	—
PO_4	—	0.50	—	—	—	—	—	—	—	—

Note. Contents are given in mg/l, except for electric conductivity (C), which is given in μS/cm. 1, granite of Limousin (Massif Central, France); 2, Holocene basalt (Besse-en-Chandesse, Massif Central, France); 3, albo-cenomanian schists and argillaceous sandstones (central Pyrenees, France); 4, limestone (experimental karst system of Le Baget, central Pyrenees, France); 5, limestone (karst aquifer of Lez spring, Montpellier, France); 6, Triassic gypsum (Corbières, France); 7, Triassic halite and gypsum (Corbières, France); 8, albian quartz sands (Basin of Paris, France); 9, Mesozoic dolostone (northern Sicily, Italy); 10, karst aquifer contaminated by a natural sea water intrusion (northern Sicily, Italy).

ple, the specific conductance or the total dissolved solids (TDS), is strongly related to the aquifer rock geochemistry. Salt rocks, halite, and gypsum (Nos. 6, 7, and 10) (Table I) have high TDS, corresponding to high Ca, Mg, and SO_4 or Na, K, and Cl concentrations. On the contrary, waters flowing through granites or basalts (Nos. 1 and 2) or through quartz sands or sandstones (Nos. 3 and 8) present low TDS, with low ion concentrations. Dissolved silica also is higher in these waters than in others. Medium TDS values correspond to waters from carbonate aquifers (Nos. 4, 5, and 9), in which HCO_3 and Ca are dominant in limestones and HCO_3, Ca, and Mg, in dolostones. A peculiar example from northern Sicily (No. 10) shows the effect of seawater intrusion into a karstic aquifer; the water, which contains only 4% seawater, is undrinkable but may be used carefully for watering orange trees.

The second and third factors are particularly responsible for the solution rate of rock, depending on flow rate, and for ion content, depending on residence time and chemical content. Waters with a very long residence time, such as mineral and thermal groundwaters, show higher ion concentrations than surface waters or shallow groundwaters, when in contact with the same rocks. A clear relationship exists between the total dissolved content and the residence time. For example, Fontes *et al.* (1989) showed that salinity increases when residence time increases in low fracture flows in crystalline rocks.

In groundwater systems in which residence time is short, such as some alluvial and karst aquifers, changes of flow conditions are nearly always responsible for strong changes in chemical content. Indeed, alluvial and karst aquifers are known for showing seasonal variations in their total dissolved solids. On the contrary, waters of deep or thermal aquifers do not generally show any seasonal variation in their chemical content, except when they mix with shallow waters.

A part of the dissolved solids may originate outside of the aquifer. Part may be supplied by rainwaters and air dust (*e.g.*, mainly from marine salt and continental dust). In industrial areas and near megapoles, air pollution enriches rainwaters in various salts (*e.g.*, mainly NO_3 and SO_4, which are responsible for the anthropic acidity of rain) and in other chemicals such as heavy metals (Pb) and organic compounds (hydrocarbons). Human land use activities may introduce additional pollutants into water systems. For example, domestic wastewaters may concentrate sodium chloride, potassium chloride, and borate (from washing powder) and also ammonia, nitrate, and nitrite. The latter three are biologically labile and, depending on the oxygen content of groundwater, may react in accordance with redox gradients. Transformations, precipitation, and other processes may eventuate.

B. The Main Chemical Systems Involved in Water Chemistry

Water chemistry in its simplest form involves two-phase and three-phase systems (i.e., water–rock and gas–water–rock systems). Three-phase systems are the more complicated, involving more reactions and equilibria, but they may record conditions for free surface flow or infiltrating water in the vadose zone and phreatic flow for lack of a gas phase, because some of their chemical variables "keep in memory" the relations between solution and gas and between solution and rock. The CO_2–H_2O–carbonate system will be detailed because it is the best known and the most common example.

1. Two-Phase Systems

Two-phase systems concern salts, like chlorides, sulfates, and nitrates. They generally are related to evaporites and to pollutants, many times mediated by natural microbial processes. In polluted waters, Cl^- and NO_3^- are often of domestic origin; SO_4^{2-} and NO_3^- are from fertilizers. Cl^- content can reach >100 mg/l at the outlet of wastewater treatment plants; NO_3^- is then very low, but NO_2^- and NH_4^+ are abundant (some in mg/l).

a. Sulfate as a Natural Tracer   Evaporite origin may be determined by analyzing strontium and magnesium concentrations. Strontium, as Sr^{2+}, is present only in volcanic and evaporite rocks. In evaporites, it is always related to Ca^{2+} content; the Sr^{2+}/Ca^{2+} molar ratio characterizes the type of evaporite. In Europe, the Sr^{2+}/Ca^{2+} ratio is about 5 to 8 per milliliter for waters dissolving evaporites of the German Trias; 2 to 5 per milliliter for the Trias of Provence and the Pyrenees; and 1 to 2 per milliliter for Cretaceous and about 1 or less per milliliter for Miocene substrata. Strontium is a very good indicator of sulfate solution when SO_4^{2-} is reduced by oxygen consumption. SO_4^{2-} may disappear from solution, but Sr^{2+} remains in solution, as observed in thermal springs of the French Pyrenees (M. Bakalowicz, unpublished).

Oxygen consumption occurs commonly in groundwaters and in organic carbon-rich sediments of surface streams, because of microbial activity and oxidation of organic matter. Redox reactions modify some mineral ion concentrations, mainly SO_4^{2-} and NO_3^-, and dissolved O_2. Such reactions require up to several weeks to be complete. Waters in confined aquifers and in interstitial waters of fine and poorly permeable sediments are especially prone to redox gradients.

Oxygen may also be consumed by well-oxygenated waters, in the vadose zone or in the phreatic zone, near the piezometric surface, for oxidizing some minerals, mainly sulfides. Pyrite, the most common sulfide mineral known in limestones, dolostones, and shales, is oxidized to SO_4^{2-}; iron remains generally insoluble in the presence of dissolved oxygen. High-sulfate waters are Sr^{2+} free. Oxydization of pyrite requires six months or more to reach equilibrium, according to Stumm and Morgan (1981). At springs or creeks fed by waters from sulfide formations, it is common to observe a slow increase in SO_4^{2-} content during the base flow, from less than 10 mg/l (depending on the SO_4^{2-} content of rain water) up to ca. 100 mg/l.

When dissolved oxygen is consumed during decomposition of organic matter, NO_3^- and SO_4^{2-} are generally reduced. NH_4^+ and NO_2^- are present and SO_4^{2-} is generally replaced by H_2S, which imparts a characteristic "rotten egg" odor if the medium is confined. Such reactions always include a complexation of Ca^{2+} by organics, which is ignored in most analyses. Complexation may be demonstrated by comparing Ca^{2+} content from an EDTA titration, which takes up only Ca^{2+} and Ca–ion pairs, and by atomic absorption spectrometry, which quantifies all Ca^{2+}, Ca–ion pairs, and Ca–organic complexes in the sample. In polluted waters, the two methods may differ by 10% (Kempe, 1975).

Last, in waters percolating at the base of soil horizons or drained from subsurface aquifers, SO_4^{2-} may be totally absent, despite its presence in rainwaters. Bacterial activity in soils may be responsible for a reduction of SO_4^{2-} and a production of CO_2. Annual mass balance for SO_4^{2-} showed a considerable immobilization of sulfur, up to 35% of the annual input from rain and atmospheric dusts, during humid and warm seasons (Bakalowicz, 1980).

b. Chloride as a Natural Tracer   When not originating from pollutions, Cl^- from oceanic aerosols is present in precipitation. Except in igneous rocks, the Na^+ and K^+ content of groundwater is related to Cl^- content. Cl^- is a very interesting natural tracer, because it is a conservative ion, not used by any chemical or biological reaction, whereas Na^+ and K^+ are present in geological formations; except evaporites, Cl^- is not. Hence, Cl quantity at the input of a hydrological system might be the same as that at the output, if the observation time is long enough to allow permanent flow for some long period (*e.g.*, over a hydrological year). In such conditions, evaporation and transpiration processes do not change in any way the total Cl^- quantity in the system. Because of consumption by evapotranspiration, water quantity decreases; therefore, the Cl^- content of the output must be higher than that at the input. The Cl_{input}/Cl_{output} ratio may be then used for estimating evapotranspiration of the system (Schoeller, 1962). The annual water balance is given by

$$RAIN = ET + OUTFLOW.$$

The yearly Cl^- mass balance is

$$(Cl^-)_I \times RAIN = (Cl^-)_O \times OUTFLOW.$$

Then,

$$(Cl^-)_I/(Cl^-)_O = OUTFLOW/RAIN$$

or

$$(Cl^-)_I/(Cl^-)_O = 1 - ET/RAIN,$$

which may be written

$$[(Cl^-)_O - (Cl^-)_I] / (Cl^-)_O = ET/RAIN.$$

Frequently, RAIN and $(Cl^-)_I$ are known regionally. For example, in the Central Pyrenees, RAIN is equal to 1000 mm and $(Cl^-)_I$ is about 1 mg/l^{-1}. $(Cl^-)_O$ is 2.2 mg/l^{-1}, given by averaging concentrations of two-week samples over a year. ET is then 545 mm, i.e., the same value given by Turc's method [1954; in Castany (1967)], calculating ET from yearly RAIN and mean air temperature (11°C).

c. Silica, Sodium, and Potassium as Natural Tracers   Dissolved silica originates mainly from silicate minerals and cherts and not from quartz. Quartz is soluble in a few conditions (*e.g.*, high temperature and pH, existing in tropical areas and in some deep thermal systems). Silica solution proceeds from slow reactions. That is why the seasonal evolution of dissolved SiO_2 observed at springs can be related to water residence time inside the aquifer.

Na^+ and K^+ are related to SiO_2, when silicate dissolution occurs. But, they are at first mostly discarded from water flow by plant and soil activity.

The Na^+ and K^+ content may increase in aquifers and in alluvial sediments of a surface stream due to silicate solution and cation exchange on clay minerals. Cations in solution, like Ca^{2+} and Mg^{2+}, are exchanged with K^+ and mainly Na^+ from clays.

Cation exchanges in waters draining low permeable aquifers are known, but they occur in any aquifer when clays are present and when water flow is slow enough for exchange. Schoeller (1962) proposed the use of a cation exchange index (CEI) for estimating that effect:

$$CEI = [mCl^- - (mNa^+ + mK^+)] / mCl^-.$$

The effect is null when $mCl^- = mNa^+ + mK^+$. Pollution does not generally change the CEI value, because Cl^- is related to Na^+ and K^+. Transpiration increases CEI (CEI > 0), because of $Na+$ and chiefly K^+ consumption by plant and soil activity. However, cation exchange produces a decrease in CEI (*i.e.*, CEI < 0), because the increase in Na^+ and K^+ content is independent of Cl^- content. In the infiltration zone of a carbonate aquifer, it is possible to show the role of clays of glacial origin filling in cracks by comparing CEI values of rain with values in fast infiltration areas in widely open cracks with slow infiltration areas. Slow infiltration waters flow vertically through about 300 m from surface in 18 weeks on average, whereas fast infiltration flows through about 100 m in a few days (Bakalowicz and Jusserand, 1986) (see Table II).

2. Three-Phase Systems: The Example of Carbonate Water Chemistry

Three-phase systems are particularly interesting for the study of water flows and especially groundwaters and their relations with surface waters and with air of the infiltration zone. The CO_2–H_2O–carbonate system is the most common and, therefore, the most useful system for demonstrating water–air relations. It is very informative for carbonate and karstic aquifers.

The CO_2–H_2O–carbonate system was studied in detail by Roques (1962, 1964), who first showed that the relations between the three phases must be taken into account in order to describe the state of the solution

TABLE II Cation Exchange Effect, Shown in Infiltration Waters of the Niaux Cave Area (Central Pyrenees, France)

	TDS	Cl^-	Na^+	K^+	CEI
Rain water	25	1.0	0.6	0.10	−0.02
Fast infiltration	240	1.8	0.8	0.15	0.23
Slow infiltration	285	1.8	5.5	0.35	−3.89

Note. Concentrations are given in mg/l; the cation exchange index (CEI) is given in %.

and to predict its evolution. The reactions between the three phases are all reversible and produce an equilibrium chain (Table III) (Garrels and Christ, 1965 ; Stumm and Morgan, 1981; Bakalowicz, 1980).

When contact among the three phases is long enough, all the reversible reactions are in equilibrium with respect to each other. The HCO_3^- and Ca^{2+} concentrations of the solution are then fully determined by its CO_2 content (remember that the CO_2 content of a solution is given by the CO_2 partial pressure, or pCO_2, of the gaseous phase related to the solution). If time is too short for equilibrium with respect to either the solid phase, by dissolving carbonate rock, or the gaseous phase, by dissolving or degassing CO_2, then HCO_3^- and Ca^{2+} concentrations of the solution depend upon pCO_2 and on deviation of equilibrium. This is the general situation. Deviation from equilibrium conditions is given by the difference between equilibrium pH for calcite and measured pH of the solution (dpH); dpH is the same as the saturation index for calcite (SIc) (Back *et al.*, 1966). When the solution does not retain the highest possible dissolved carbonate, the solution is undersaturated with respect to calcite (dpH < 0). If CO_2 degassing has

TABLE III   Chemical Equilibria of the CO_2–H_2O–Carbonate System

Equilibria	Equilibrium constants
(1) Between liquid and gaseous phases	
CO₂ solution or escape	
(1) $(CO_2)g \leftrightarrow (CO_2)_l$	$K_0 = (CO_2)_l/pCO_2$
(2) Inside the liquid phase	
CO₂ hydration and ionizations	
$CO_2 + nH_2O \leftrightarrow (CO_2, nH_2O)$	
(2) $(CO_2, nH_2O) + pH_2O \leftrightarrow HCO_3^- +$	$K_1 = (HCO_3^-) (H_3O^+)/(CO_2)$
$H_3O^+ + (n\text{-}p\text{-}2)H_2O$	
(3) $(HCO_3^- + H_2O \leftrightarrow CO_3^{2-} + H_3O^+$	$K_2 = (CO_3^{2-}) (H_3O^+)/(HCO_3^-)$
Forming ion pairs	
(4) $HCO_3^- + Me^{2+} \leftrightarrow MeHCO3^+$	$K_3 = (MeHCO3^+)/(HCO_3^-)(Me^{2+})$
(5) $CO_3^{2-} + Me^{2+} \leftrightarrow MeCO_3^0$	$K_4 = (MeCO3^0)/(CO_3^{2-})(Me^{2+})$
Dissociation of water	
(6) $2H_2O \leftrightarrow H_3O^+ + OH^-$	$K_e = (H_3O^+)(OH^-) = 10^{-14}$
Forming ion pair with SO_4^{2-}	
(7) $Me^{2+} + SO_4^{2-} \leftrightarrow MeSO_4^0$	$K_5 = (MeSO4^0)/(Me^{2+})(SO_4^{2-})$
(3) Between liquid and solid phases	
Carbonate dissociation or precipitation	
(8) $MeCO_3 \leftrightarrow CO_3^{2-} + Me^{2+}$	$K_s = (Me^{2+})(CO_3^{2-})/(MeCO_3)$

(4) Electrical neutrality of the solution
(9) $2mCO_3^{2-} + mHCO_3^- + 2mSO_4^{2-} = 2mCa^{2+} + mH_3O^+ + mMeHCO_3^+$ with $mMe_{tot} = mMe^{2+} + mMeHCO_3^+ + mMeCO_3^0 + mMeSO_4^0$, Me^{2+} being Ca^{2+}, Mg^{2+}, or Sr^{2+} $mHCO_{3tot} = mHCO_3^- + mMeHCO_3^+$ $mCO_{3tot} = mCO_3^{2-} + mMeCO_3^0$ $mSO_{4tot} = mSO_4^{2-} + mMeSO_4^0$

Note. The total concentrations (m_{tot}) are given by chemical analysis.

occurred (*e.g.*, caused decreasing pCO_2 in air) and if degassing was not equilibrated by precipitating a part of dissolved carbonate, the solution is supersaturated (dpH > 0).

The two calculated variables, pCO_2 and dpH, control the CO_2–H_2O–carbonate system as a whole. Ca, Mg, and HCO_3 concentrations and pH are totally dependent on pCO_2 and dpH and therefore are significant indicators of hydrogeological conditions in the aquifer (Jacobson and Langmuir, 1970, 1974; Fleyfel and Bakalowicz, 1980).

a. The Fictitious CO_2 Partial Pressure of Waters—Geochemical Significance CO_2 content could be determined by analysis, but it is transient; measured CO_2 concentrations are always lower than calculated values. Therefore, it is better to calculate the CO_2 content of waters from Eqs. (1) and (2) (Table II), knowing pH, temperature, and alkalinity ($HCO_3^- + CO_3^{2-}$ content).

In reality, the CO_2 content of water depends on the partial pressure of CO_2, pCO_2 of the underground atmosphere in the infiltration zone, and water temperature and local atmospheric pressure. It is then more advisable to give the results not as a content in mg/l but as a pCO_2 as a percentage of the total atmospheric pressure. The only interesting hydrological information is related to infiltration zone pCO_2. Then, by calculating pCO_2 of a fictitious gaseous phase, it becomes possible to compare the CO_2 content of aquifers and to explain space or time variations as consequences of geological and/or hydrological processes. Table IV gives pCO_2 values of waters from some carbonate aquifers of southern France. The space variations show four different effects.

(1) An effect of the biophysical nature of soil and cover. CO_2 originating from soil roughly varies from 0.4 to 5%, depending on soil nature and on plant cover. pCO_2 is higher in low-permeable, thick soils than in sandy,

TABLE IV   Calculated pCO_2 of Some Karst Spring Waters in the South of France

Springs	pCO_2 (%)	Elevation (m)	Origin of anomaly
Baget	0.8	500	
Aliou	0.4	500	Open conduit
Chichoué	0.2	1800	Elevation
Lamalou	1.0	200	
Le Lez	5.0 to 10	100	Deep CO_2

Note. Baget, Aliou, and Chichoué are located in the central Pyrenees; le Lez and Lamalou, near Montpellier.

thin soils. Forests produce much more CO_2 than meadows or cultivated areas.

(2) An elevation effect. The more elevated the aquifer, the lower the pCO_2 emerging from it. This effect is of climatic type, because soil CO_2 production also is a function of temperature. A high mountain spring (at 1800 m in elevation) may have pCO_2 about 0.2%, whereas in the foothill springs pCO_2 is about 0.8 to 1.0%.

(3) An aeration effect. The more aerated the infiltration zone, the lower the pCO_2. Aeration generally results from large openings of caves and karst, which are conduits to the external atmosphere, and pCO_2 sometimes is <0.1%.

(4) An internal production of CO_2. In tectonically active areas, like the French Mediterranean coast, deep CO_2, from decarbonatation and from the Earth's mantle, reaches up to the surface, through aquifers. pCO_2 may be more than 10% and much more in CO_2-enriched (sparkling) waters.

b. The Saturation Index dpH—Geochemical Significance   The dpH is calculated from the set of nine equations given in Table III, based on pH, temperature, and all the ion concentrations. dpH calculation is more complicated than that of pCO_2, because it requires iterative computations of each species content from total ion content given by analysis. For instance, when calculating K_3 of expression (4) in Table III,

$$K_3 = (CaHCO3^+)/(HCO_3^-)(Ca^{2+}),$$

(Ca^{2+}) is calculated from mCa^{2+}, in a function of ionic strength. But, mCa^{2+} is calculated from mCa_{tot} and from all Ca species:

$$mCa_{tot} = mCa^{2+} + mCaHCO_3^+ + mCaCO_3^0 + mCaSO_4^0.$$

All the species molalities, like mCa^{2+}, are a function of ionic strength and of equilibrium constants. Therefore the calculation must be iterative and requires a computer.

Negative dpH values are related to the short residence time of water in contact with carbonate rock only if the pH of the solution causes carbonate solution. Sometimes, pH mainly results from other reactions, involving inorganic or organic acids. That situation prevails in some surface waters, like peat bog waters or polluted streams, and dpH is of little significance, as is calculated pCO_2. Values between -1.0 and -0.1 are common for karstic springs, generally during flood season. Values below -1.0 are relatively rare and have to be related to poorly carbonated aquifers, excepted for noncarbonate reactions.

Equilibrium values of dpH ($-0.05 < dpH < 0.05$) are often significant if a long residence time (more than a few weeks) occurs inside the phreatic zone of the aquifer. For example, all of the dissolved CO_2 entering the system may be used for carbonate solution. Some authors (Stumm and

Morgan, 1981; Dreybrodt, 1988) compared carbonate solution in a closed system (*i.e.*, without any gaseous phase containing CO_2) with that in an open system, (*i.e.*, CO_2 gaseous phase present). From a thermodynamic viewpoint, two-phase and three-phase systems are more appropriate than comparing open and closed systems, the first one being related to phreatic conditions and the second one, to vadose or the infiltration zone.

Field data show that waters percolating through soils are nearly always supersaturated with respect to calcite (dpH > 0). The infiltration zone is a CO_2 reservoir, fed by CO_2 produced in soils and carried downward by solution and gas diffusion. CO_2 distribution is heterogeneous, particularly in karst aquifers, where open conduits are common and responsible for aeration with external air. Waters infiltrate through areas with different pCO_2. Then, calcite supersaturation is due to CO_2 exchanges between underground air and percolating waters when pCO_2 varies. But, positive values of dpH may also be produced by bad sampling conditions or by too long a delay before analyzing alkalinity and pH.

C. Degree of Confidence of Data and Methods for Interpretation of Water Chemistry Data

A rough calculation of the ionic balance may show most of the errors from analyses or monitoring and from data transcription. In such an approach, one assumes that the ion concentrations are given directly by the analysis; i.e., ion pairs are neglected. For example, the analysis gives the total calcium content Ca_{tot}, which is

$$Ca_{tot} = Ca^{2+} + CaHCO_3^+ + CaCO_3^0 + CaSO_4^0,$$

and, in the rough calculation of ionic balance, Ca^{2+} is assumed to be equal to Ca_{tot} and $CaHCO_3^+$ is assumed to be zero; and so on for all the ions. The approximate error on the ionic balance is then given in a percentage by

$$\text{ionic balance (IB)} = 100 \times [S(\text{anions}) - S(\text{cations})]/[S(\text{anions}) + S(\text{cations})],$$

where

$$S(\text{cations}) = Ca_{tot} + Mg_{tot} + Na_{tot} + K_{tot} + Sr_{tot}$$

and

$$S(\text{anions}) = HCO_{3tot} + Cl_{tot} + SO_{4tot} + NO_{3tot}.$$

If IB is lower than -5% or higher than $+5\%$, the sample must be discarded; the present analytical methods are accurate enough to give an IB value lower than 5%. For carbonate waters, whose mineralization is calcium and carbonate for more than 80%, IB should be lower than 1% if alkalinity and total hardness are measured by a volumetric method, soon after sampling. But all the errors and mistakes cannot be shown by this way. For instance, transposition of samples from different groups or systematic errors due to an ion measurement can be shown by statistical methods, such as factor analysis.

Finally, errors of pH may be shown only after computing chemical equilibria. They appear particularly because of irrealistic values of pCO_2 or from use of saturation indexes. As shown previously, rough data cannot be used directly for any interpretation or modeling. We can recommend some methods to assist interpretation of water chemistry data.

1. Equilibrium Computer Programs

For example, for carbonate equilibria, Tillmans's (Muxart and Birot, 1977) and Trombe's (1952) diagrams provide rough estimates of species content assuming that a pure solution of $CaCO_3$ exists. But in natural waters results are too far from the reality. In the same way, approximate calculations were proposed by Roques (1972), using an abacus and by Bakalowicz (1979, 1980), using empirical relationships for correcting effects of Mg^{2+} and SO_4^{2-} ion pairs. These methods work well only for low Mg^{2+} and SO_4^{2-} concentrations, but not for waters with an Mg^{2+} content higher than 4 mmol/l (50 mg/l) and an SO_4^{2-} content higher than 1.5 mmol/l (150 mg/l). Computer programs are available to make calculations (*e.g.*, WATSPEC and WATEQ). We use SOLUTEQ, developed in Laboratoire Souterrain du CNRS in France on WATSPEC basis (Wigley, 1977). These models are accurate enough for studying most of groundwaters, because generally only the primary ions are of interest. Other programs provide more accurate and more complete results, but also require more detailed empirical data as drivers.

2. Graphic Description

Description may be performed by means of several graphic methods. Some of them classify water samples as a function of their chemical composition in diagrams. Others describe the variations of a parameter using histograms and derived techniques.

a. Diagrams Are Used for Comparing the Entire Chemical Content of Different Waters Piper's diagram is the most common. Many program packages refer to it for sorting the relative importance of ions in water samples. The principle is a triangular diagram, one for main anions (HCO_3^-, SO_4^{2-}, and Cl^-) and one for main cations (Ca^{2+}, Mg^{2+}, and [$Na^+ + K^+$]). Each side is measured from 0 to 100% of the related ion content. This method is very well suited to rock geochemistry, because ion content is given as a percentage of the unity mass of rock. But Piper's diagram is not very useful for describing water chemical concentrations. Moreover, Piper's diagram does not take into account the water mineralization, which depends on the residence time of the water–rock contact, among other factors.

The most efficient graphic method is Schoeller's diagram (Schoeller, 1962). The y axis represents concentrations, in milliequivalents per liter, following a log scale. Along the x axis, anions and cations are distributed in the following order: Ca^{2+}, Mg^{2+}, Na^+, K^+, Cl^-, SO_4^{2-}, alkalinity (CO_3^{2-} + HCO_3^-), and NO_3^-. It is a good method when only a few samples are available, but it is useless for large data sets. Figure 1 gives examples from a karst spring and a thermal spring.

b. Histogramms Are Used for Describing Frequency Distribution   The mineralization time series of a spring or of a river may be considered the random function of the regionalized variable "mineralization" (Matheron, 1965). The frequency distribution of such a variable represents the probability law defining the random function (Bakalowicz, 1976, 1977), if sampling is representative of the seasonal variations. The frequency distribution is then representative of the functioning of the aquifer, with a one- or two-week sampling period. A catalog of reference distributions for karst aquifers was done with data from well-known limestone aquifers of France (Fig.2). It

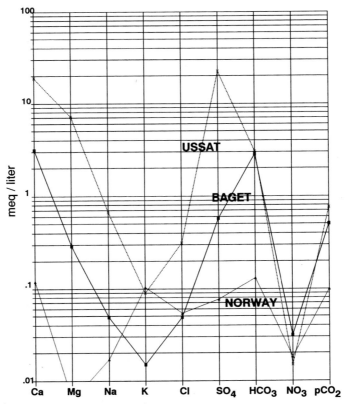

FIGURE 1   Schoeller's diagram example. Water samples are a karstic spring (le Baget investigation area, Central Pyrenees), a thermal spring (Ussat-les-Bains, Central Pyrenees), and a surface stream in a karstic environment in Norway.

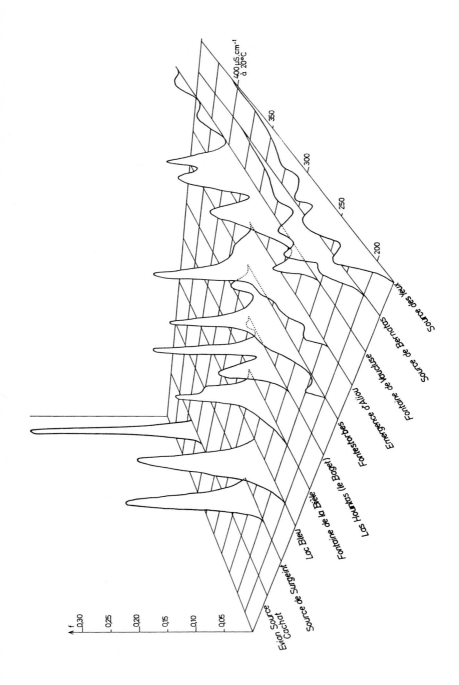

clearly shows that (1) porous and homogeneous (Evian Cachat type) or fissured carbonate (Surgeint type) aquifers show frequency distributions that are weakly spread and unimodal, as a consequence of the homogenization of chemical characteristics, in a phreatic zone without any functional karst drainage structure; (2) porous and inhomogenous karstic aquifers show a wide dispersion of the distribution and the appearance of several modes, as the consequence of a nonhomogeneization of chemical characteristics of infiltration waters, the karst drainage extension being then related to the range and the number of modes; and (3) surface streams show characteristics similar to those of karst aquifers. The shape of frequency distribution is then related to the geochemical heterogeneity of the whole catchment, to the extension of groundwaters, and to their function relations.

At least 25 samples are needed for this approach; they must be regularly distributed during the same hydrological year. A minimum of 7 classes is required to give a good representation of the distribution; 10 to 15 classes are generally enough. The frequency of the main class must be higher than 12–15%.

3. Factor Analysis

Many hydrological systems, such as karst aquifers and surface streams, are so complicated that the flow observed at their outlet, as well as anywhere inside them, is generally highly variable in quantity and in quality. The number of variables is usually too large for easy synthesis of the data. Relationships between variables (*i.e.*, temperature, pH, ion concentrations, pCO_2, dpH, I.B., and some ion ratios) and objects or samples are best done by means of factor analysis. Two different methods are very well suited to chemical data: principal component analysis (PCA) (Cailliez and Pagès, 1976; Bouroche and Saporta, 1980, Zhou *et al.*, 1983; Greenacre, 1984) and discriminating factor analysis (DCA) (Bishop *et al.*, 1975; Seber, 1984; Lebart *et al.*, 1984).

III. CASE STUDIES OF WATER GEOCHEMISTRY

Groundwater chemical concentrations are known to be variable in space and in time, depending on environmental and flow conditions. Three exam-

---

FIGURE 2   Reference catalog of frequency distributions of electric conductivity for carbonate aquifers. The increasing complexity and spreading of the distributions are related to karst conduit development and functioning, from porous (Evian aquifer) and fissured carbonate aquifers to highly developed conduit aquifers.

ples studied with factor analysis will show how space and time variations in
water chemistry may be interpreted in terms of groundwater hydrogeology.

A. Water Geochemistry in a Karst System

Water samples were collected from karst environments (surface streams
at sinkholes, springs, and cave seepage waters) from the Bovertun area in
the central part of southern Norway to the Fauske area in the southern part
of northern Norway. Summer flood conditions prevailed because of heavy
rains (Bakalowicz, 1984). Field data included water temperature, pH, alkalinity, and total hardness; other analyses produced Mg^{2+}, Na^+, K^+, Cl^-,
and SO_4^{2-} concentrations. Ca^{2+} was calculated from total hardness and
Mg^{2+} content. pCO_2, dpH, and the IB were computed by SOLUTEQ (cited
above) and from speciation and equilibria. Several successive factor analyses
were necessary to select the appropriate samples and variables for extracting
the hydrogeological information.

The final PCA considers 18 samples with 10 variables (Table VI), 8
measured (pH, Cl^-, SO_4^{2-}, HCO_3^-, K^+, Na^+, Mg^{2+}, and Ca^{2+}) and 2
computed (pCO_2 and dpH).Results are given in Table V and Fig. 3. Four
factors (F) explain 86.6% of the total variance. The structure is strong
because F1 explains 51.1% of the variance; F2 (18.1%) and F3 (10.4%)

TABLE V   Water Chemistry Data from a Karst Environment in Norway

	Principal components							
	Sample space				Variable space			
Sample	Axis 1	Axis 2	Axis 3	Variable	Axis 1	Axis 2	Axis 3
1	−2.9992	2.1466	−0.6475	pH	0.7899	−0.5679	−0.0453
2	−3.7476	0.89	−1.2657	Cl	−0.6122	−0.0696	−0.5067
3	−0.3245	−0.2518	1.1419	SO$_4$	0.373	0.3025	−0.6983
4	−1.6387	−1.6918	0.2882	HCO$_3$	0.9257	0.261	0.0113
5	−2.9642	0.421	−1.0242	K	−0.4201	0.4654	−0.3207
6	3.5273	−0.4421	−1.7773	Na	0.8178	0.0479	−0.2672
7	−0.6867	−1.0063	−1.221	Mg	0.691	0.3953	0.1566
8	−0.7549	−1.692	0.5027	Ca	0.9447	0.1806	−0.0574
9	0.113	−0.9522	0.7298	pCO_2	−0.0371	0.9291	0.2813
10	0.2779	−0.4326	0.4349	dpH	0.9579	−0.2103	0.0167
11	0.3324	−0.3957	0.2416				
12	0.3907	−1.6755	−0.553				
13	3.9341	2.0295	1.2275				
14	2.9276	−0.7866	0.666				
15	−0.0551	−0.64	0.655				
16	−2.4837	2.2402	1.9496				
17	0.4896	−0.1322	0.0103				
18	3.6618	2.3715	−1.3788				

TABLE VI   Main Results of PCA on Water Chemistry Data from a Karst Environment in Norway

Sample	t (°C)	pH	Cl (mg/l)	SO$_4$ (mg/l)	HCO$_3$ (mg/l)	K (mg/l)	Na (mg/l)	Mg (mg/l)	Ca (mg/l)	pCO$_2$ (%)	dpH
1	8.5	6.85	0.7	4.2	14.5	5.5	0.5	0.1	4.08	0.17	-2.93
2	9	6.82	1.9	3.7	7.9	4	0.38	0.07	2.4	0.1	-3.44
3	8.5	7.45	0.25	3	36	0.5	0.6	0.95	12.08	0.11	-1.49
4	7	7.6	0.95	3	12.2	0.28	0.3	0.1	4.78	0.03	-2.2
5	7.5	6.7	1.8	5.5	6.7	0.2	0.3	0.1	4.9	0.11	-3.35
6	7.05	8.09	0.25	7	94.6	0.5	4.63	0.7	26.34	0.06	-0.15
7	7.2	7.6	1.9	4	23.2	0.7	2.42	0.95	8.48	0.05	-1.69
8	6.2	7.7	0.9	1.8	17.7	0.3	2.52	0.95	3.98	0.03	-2.04
9	6.3	7.71	0.1	3.5	38.4	0.62	0.83	0.9	11.18	0.06	-1.27
10	6.7	7.64	0.1	4.2	42.7	0.88	1	1.05	12.18	0.08	-1.25
11	8.7	7.65	0.1	4.8	43.3	0.76	0.8	1.05	12.18	0.08	-1.21
12	7.3	8.04	0.1	5.5	17.7	0.58	1.5	0.7	9.28	0.01	-1.33
13	7.3	7.76	0.1	3.5	127	0.88	2.95	5.3	28.24	0.18	-0.33
14	7.5	8.03	0.1	2.5	108	0.43	3.24	0.6	28.44	0.08	-0.12
15	7.5	7.58	0.1	3.5	35.4	0.68	1.52	0.7	9.58	0.08	-1.48
16	7.5	6.55	0.1	3	11.6	0.35	1	0.25	3.7	0.28	-3.37
17	6.5	7.5	0.1	4.5	36	0.78	2.9	1.1	10.58	0.09	-1.52
18	6.8	7.65	0.1	8.5	105	0.97	3.84	3.1	26.94	0.19	-0.55

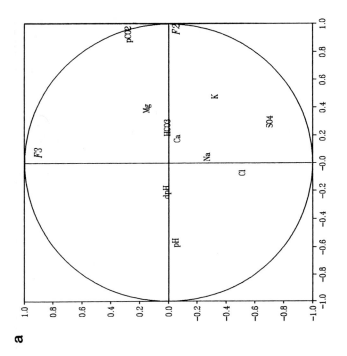

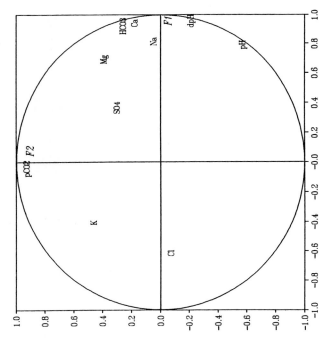

114

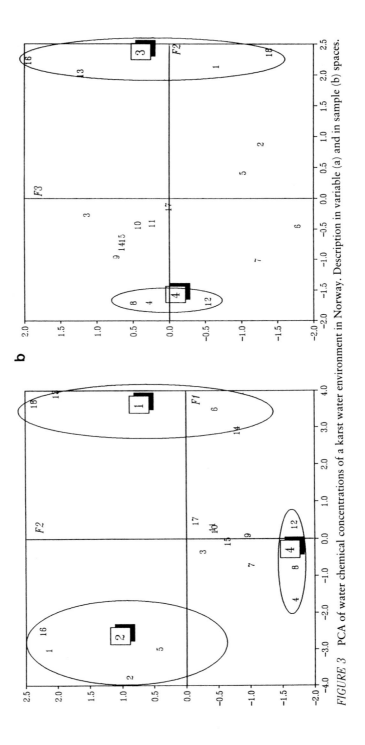

FIGURE 3 PCA of water chemical concentrations of a karst water environment in Norway. Description in variable (a) and in sample (b) spaces.

115

are of little importance. In the variable space (Fig. 3a), the best described variables are close to the circle; that is to say, K^+, SO_4^{2-}, and Cl^- are badly represented by the main plan, F1–F2 (69.2%).

F1 is related to HCO_3^-, Ca^{2+}, dpH, and, at a lower level, pH and Na^+, opposed to Cl^- (Fig. 3). F1 can be interpreted as the time required for water to obtain relatively high carbonate mineralization (*i.e.*, the residence time of water in contact with limestone). It opposes waters that have a long residence time in contact with carbonate rocks and relates to short-residence-time waters. F2 is related to pCO$_2$ and opposed to pH. It may be interpreted as the CO_2 factor opposing CO_2-rich waters, or groundwaters, and relating to CO_2-poor waters, or surface waters. F3, related to Cl^- and SO_4^{2-}, and F4, related to K^+ and opposed to SO_4^{2-}, are difficult to explain without additional environmental information (Cl^- content of rain waters, distance from the sea, distance from large cities, etc.).

Thus, the two most interesting diagrams are given by F1–F2 and F2–F3 spaces, describing, respectively, 69.2 and 28.4% of the total variance. Four clusters were identified (Fig. 3b). (1) Cluster 1, linking samples 6, 13, 14, and 18, with the highest HCO_3^- and Ca^{2+} concentrations, which were weakly undersaturated. These waters of relatively long residence time are seepage waters (samples 6, 14, and 18). The only karst water appears in sample 13 and represents an aquifer fed only by infiltration water and not by surface streams through sinkholes. (2) Cluster 2 grouped samples 1, 2, 5, and 16 of low Ca^{2+} and HCO_3^- concentrations, which were strongly undersaturated. These waters had a very short residence time in contact with carbonate rocks. They were from surface streams (samples 1, 2, and 5) and from the karstic outlet of a lake (sample 16). (3) Cluster 3 represented CO_2-rich waters, from seepage (samples 13 and 18) and surface waters (samples 1 and 16), possibly rich in dissolved organic matter, partly responsible for the pH value. (4) Cluster 4 grouped CO_2-poor waters from surface stream waters (samples 4, 8, and 12), where the catchment area extends above the tree line. The elevation cannot be taken into account, because waters of the same origin (sample 2, 5, or 9, for instance) are richer in CO_2; other characteristics, such as soil cover, may explain CO_2 content. For karst groundwater geochemistry, this example shows the importance of (1) the residence time of water in the aquifer, which depends on karstic drainage development, and (2) the CO_2 content, which is determined by recharge characteristics of the aquifer (*i.e.*, infiltration through soils and surface flow through sinkholes).

B. Variations of Water Chemistry of a Karstic Spring during a Flood

The study of a flood (November 29 to December 8, 1973) was conducted during recharge of the Baget karst system, an experimental area located in the Central Pyrenees. This flood was fully representative of short-term variations in water chemical content (Bakalowicz, 1979). Thirty-four samples (1 to 16, increasing stage of the flood; 17 to 21, first maximum; 22 to 24, second maximum; 25 to 34, falling stage) were taken and nine variables, measured [Ca^{2+}, Mg^{2+}, Cl^-, SO_4^{2-}, and HCO_3^-; temperature (t); and electric conductivity (C)]. dpH and pCO_2 were computed.

The structure of the variations was very strong: only three factors corresponded to more than 92% of the total variance (42.3% for F1, 34.3% for F2, and 15.5% for F3.). In the variable space (Fig. 4a), F1 was defined by Cl^-, pCO_2 opposed to SO_4^{2-}, temperature, and dpH. F2 was characterized by Ca^{2+}, HCO_3^-, and C and partly by pCO_2. F3 was dominated by dpH. The opposition of pCO_2 and Cl^- to both temperature and SO_4^{2-} for F1 shows very well the contrasting variations, which are due to the different origins of two water groups: (1) waters stored in the epikarst and subject to evapotranspiration, which concentrates Cl^- ions, and to CO_2 production in the soil and (2) waters stored in the phreatic zone (saturated karst) in the deep part of the aquifer and subject to higher temperature and to slow chemical processes, such as sulfide oxydization producing SO_4^{2-} ions. The dpH occurred for the same factor, suggesting that SO_4^{2-}-marked waters were close to the equilibrium and Cl^--marked waters were slightly undersaturated. Therefore, F1 may be defined as the spatial origin of waters. The residence time is longer for "deep" saturated waters than for subsurface, undersaturated waters. In sample space, F1 opposes samples of the increasing stage and relates to those of the second flood peak; that is to say, waters stored close to the spring flow out before waters stored close to the surface. F2 could be the factor responsible for the heterogeneity of water chemistry, especially carbonate content, in the saturated karst. It supports Ca^{2+}, HCO_3^-, and less pCO_2 and Mg^{2+}. In the sample space (Fig. 4b), waters marked by SO_4^{2-} ions develop along F2, showing the increase in dissolved carbonate at the beginning of the flood.

The relationship between opposing dpH and discharge is characteristic of F3. It means that the higher the discharge (*i.e.,* increasing flow velocities), the more undersaturated waters became. Outflow of waters undersaturated with respect to calcite reflected fast transit from the surface to the spring. Therefore, F3 may be interpreted as water velocity between its storage place and the spring.

In sample space (Fig. 4b), samples are grouped in three (main plan, F1−F2) or four clusters (plan F1−F3), which clearly correspond to the four parts of the hydrograph: increasing stage, first and second maxima, and falling stage. Clearly, the function of the karst aquifer does not permit homogenization of waters, owing to the fast transit of some waters to the spring.

C. Spatial and Seasonal Variations in Chemistry of the Hyporheic Zone of a Creek

The following example is taken from a comprehensive ecological study of a creek and its hyporheic zone, in the Central Pyrenees (Rouch, 1988;

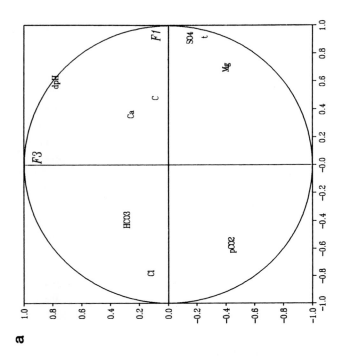

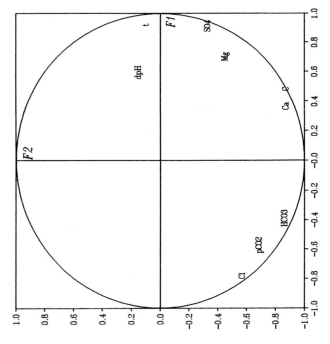

118

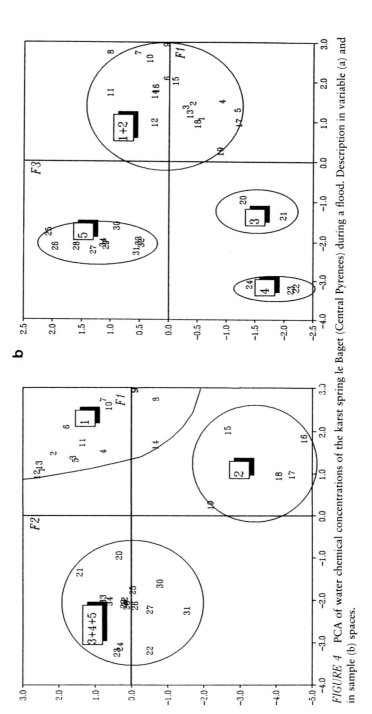

FIGURE 4 PCA of water chemical concentrations of the karst spring le Baget (Central Pyrenees) during a flood. Description in variable (a) and in sample (b) spaces.

119

Rouch *et al.*, 1989). Every two months during a hydrologic year, biological, physical, hydrodynamic, and chemical data were collected at 20 points at a 75-m^2 site to a depth of 0.60 m in the bed sediments and in the water flowing in the creek. The aim of the study was to define the connections among the surface stream, its underflow inside the bed sediments, and possible groundwaters in calcareous alluvia. Such connections were suggested by other studies, using various methods (Bencala, 1993; Triska *et al.*, 1989).

A total of 12 variables were used: 10 were measured (pH, Ca^{2+}, Mg^{2+}, Na^+, K^+, SiO_2, NO_3^-, Cl^-, SO_4^{2-}, and HCO_3^-) and 2 were computed (dpH and pCO$_2$). Table VII summarizes the mean chemical characteristics observed in the sediments of the creek. Chemical concentrations varied spatially during the year. Do the observed space variations characterize each site of underflow, despite seasonal changes? Are they imposed by the chemical content of the surface water?

The observed variations obviously depend on time. Seasonal variations of surface water chemistry showed low TDS during flood time and high TDS during the low stage. In the same way, waters from the hyporheic zone showed identical variations. However, a strong geochemical heterogeneity was also observed, suggesting spatial distribution of water chemistry. The first is called "time structure" and the second, "space structure."

Therefore, in order to define the causes of such structures, two qualitative variables (time and space) were defined and added to the initial data file in the shape of a number (site A, 1; site B, 2; etc.) (first month, 1; second month, 2; etc.). Discriminating factor analysis (DFA) was used, because it works with both qualitative and quantitative variables, unlike PCA.

In the first analysis, the time variable was tested. Surface stream samples were introduced as supplementary data. The three main factors explained only 71.5% of the total variance (F1, 26.7%; F2, 24.9%; and F3, 19.9%). This relatively low percentage means that the discrimination between the samples is mostly controlled by variables other than time. Clearly the space heterogeneity was important (see data in Table VII).

The second analysis, with space as a qualitative variable, was studied in more detail. The three main factors explained 85.7% of the total variance, (F1, 52.2%; F2, 23.8%; and F3, 9.7%). Therefore, in this case spatial structure plays a much more important part in geochemical discrimination of waters than time structure. Spatially (F1–F2 plan; Fig.5), waters from area B were distinct from other areas; area D was intermediate. We used the F2 axis to compare waters from areas E and G with those from areas A, J, H, and F; areas C, I, and D were then intermediate in relation. The percentage of samples, well sorted in their appropriate group, was low (41.5%), because areas were arbitrarily designed as geometrical. The groups were related to areas of complicated geomorphology.

The F1 axis was characterized by waters with a high HCO_3^-, Ca^{2+}, and CO_2 content, opposed to high-SO_4^{2-}-content waters. Area B was known

TABLE VII Elemental Statistics on the Water Chemistry of a Creek and Its Underflow

Area		pH	Ca (mg/l)	Mg (mg/l)	Na (mg/l)	K (mg/l)	SiO_2 (mg/l)	NO_3 (mg/l)	Cl (mg/l)	SO_4 (mg/l)	HCO_3 (mg/l)	dpH	pCO_2 (%)
A	a	7.83	61.1	5.1	1.2	0.53	2.5	1.2	1.8	27	168	0.19	0.22
	s.d.	0.15	3.3	0.9	0.2	0.09	1.5	1.2	0.2	12	9	0.15	0.08
B	a	7.58	64	5	1.2	0.71	3.3	1	1.8	24	183	0.02	0.41
	s.d.	0.12	3.7	0.7	0.2	0.07	1	0.8	0.2	11	8	0.12	0.12
C	a	7.79	61.1	5.1	1.2	0.52	2.9	1.1	1.9	27	171	0.16	0.24
	s.d.	0.12	3.6	0.9	0.2	0.1	0.8	0.9	0.4	11	8	0.12	0.07
D	a	7.7	62.9	5.2	1.2	0.61	3.1	1.2	1.8	26	178	0.1	0.32
	s.d.	0.16	3.8	0.8	0.1	0.08	0.7	1.2	0.3	12	11	0.13	0.16
E	a	7.76	61.3	5.3	1.4	0.54	3.8	1.2	1.8	27	171	0.13	0.25
	s.d.	0.11	3.4	0.7	0.7	0.08	0.7	1	0.3	11	7	0.11	0.06
F	a	7.83	61.2	5.3	1.2	0.59	3.1	1.4	1.8	26	170	0.2	0.21
	s.d.	0.08	3.6	0.9	0.1	0.08	0.8	0.9	0.3	12	8	0.08	0.04
G	a	7.74	61.6	5.1	1.3	0.45	3.2	1	1.8	26	171	0.11	0.27
	s.d.	0.09	3.4	0.6	0.1	0.12	0.8	0.9	0.3	12	6	0.1	0.06
H	a	7.86	60.5	5.1	1.3	0.55	2.2	1.2	1.8	27	168	0.22	0.19
	s.d.	0.07	3.3	0.8	0.1	0.1	1.2	1.2	0.3	12	8	0.08	0.04
I	a	7.78	60.3	5.3	1.3	0.45	2.5	1.1	1.7	27	168	0.14	0.24
	s.d.	0.13	4	1.2	0.2	0.15	0.9	1	0.3	12	7	0.13	0.08
J	a	7.88	60.3	5	1.3	0.51	2.6	1.2	1.8	26	167	0.23	0.19
	s.d.	0.13	3.3	0.9	0.2	0.13	0.7	1.1	0.3	12	7	0.14	0.06
Creek	a	8.19	60.4	5.3	1.3	0.57	2.7	1.7	1.8	29	165	0.53	0.09
	s.d.	0.06	3.6	0.8	0.1	0.12	0.6	1.4	0.3	13	6	0.08	0.01

Note. a, average; s.d., standard deviation.

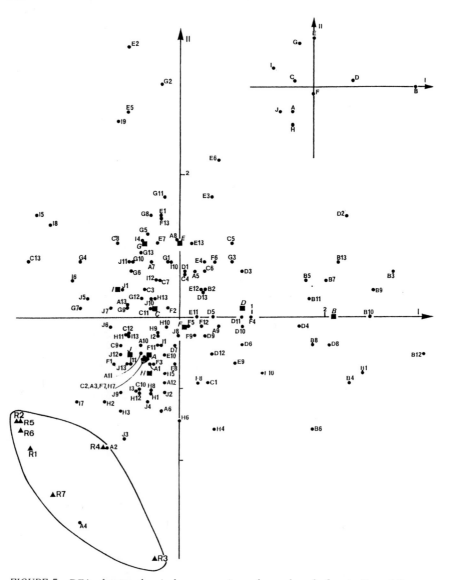

FIGURE 5   DFA of water chemical concentrations of a creek underflow in Central Pyrenees, related to space variations. The creek samples, labeled R, with triangles, are inserted as supplementary samples. Underflow samples are represented by circles and the centers of gravity by squares.

as a reducing zone in which dissolved O_2 concentrations were very low and in which SO_4^{2-} was low or absent because of reduction processes. Therefore, the F1 axis characterized chemical processes controlled by O_2, of which area B is the pole, without any connection with an oxygenated medium such as surface stream waters.

The F2 axis is defined on its negative end by a slight NO_3^- enrichment, higher pH, and positive dpH, showing a weak supersaturation (areas A, J, H, and F). Its positive end is defined by SiO_2 and Na^+ enrichment (areas E and G). The position of creek samples in the neighborhood of areas A, J, H, and F may be interpreted as following: the water residence time in sediments of those areas is not long enough to noticeably increase SiO_2 and Na^+ concentrations. The CO_2 content is slightly lower in those areas than in the others, which means a mixing of surface water with groundwater, as shown also by higher dpH values. But surface stream waters do not considerably feed the underflow because the CO_2 content is closer to the underground value than to the surface value. Underflow waters are essentially of groundwater origin in areas A, J, H, and F. At the opposite end, waters from areas E and G show a long residence time in contact with silicate sediments.

Thus, in this situation, the water chemistry of the hyporheic zone was influenced more by spatial than seasonal structure. The water geochemical characteristics were distributed in groups corresponding to different hydrogeological conditions.

(1) Areas in which the groundwater composition was similar to that of surface water, suggesting a moderate dilution or inflow by surface water and a short residence time in a highly permeable zone (areas A, J, H, and F of upstream gravel banks).

(2) An area in which water chemistry was marked by a high pCO_2, representing a clear undersaturation (dpH < 0); high SiO_2 concentrations; and low SO_4^{2-}, due to reduction processes. These characteristics imply the absence of surface water inflow and a long residence time in a confined medium with low permeability (area B).

(3) Areas in which waters were marked by an increase in SiO_2 and Na^+ and characterizing groundwaters flowing in a medium of low permeability, but not confined (areas E and G). From these results, it was possible to define spatial changes in stream geomorphology that influence groundwater chemistry. DFA is used for sorting the samples into three groups, corresponding to those defined here: 1 for the reduction area (area B), 2 for areas with high permeability, and 3 for areas with low permeability.

A new DFA was then computed. Samples that were not very well sorted in the previous analysis were considered supplementary samples. The main plan F1–F2, explains 100% of the total variance (F1, 65.9%; F2, 34.1%). The two axes have the same definition as that given previously. Then it is possible to draw a map of the chemical characteristics of underflow waters using the sample coordinates, indicating their adherence to the most appropriate group (Fig.6).

Hyporheic water geochemistry was essentially under the control of groundwaters, as shown by high CO_2 content, for instance, and not much influenced by surface waters. Exchanges between the hyporheic zone and stream flow were weak at the study site, but increased upstream. Water

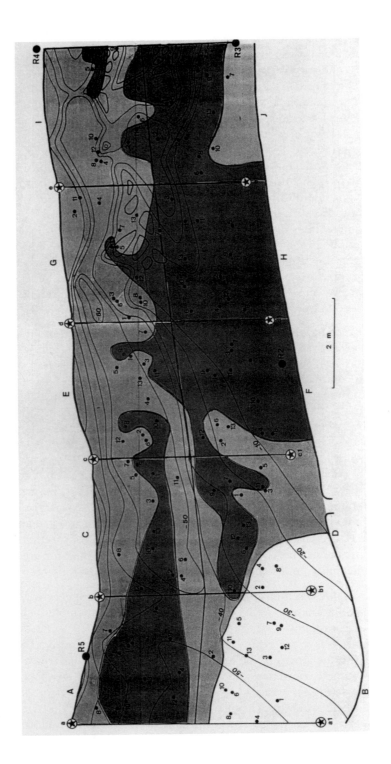

geochemistry of the hyporheic zone showed variations related to space and to seasons. The main factors responsible for such variations depend on flow conditions in the creek channel (flow velocity) and in the sediments (residence time). Among the conclusions, it is important to note that the interpretation of water chemical data may be corroborated by hydrogeological data (piezometry and permeability), crustacean distribution, topography, sediment size, and the organization of the creek bed. Logical relations were demonstrated in relation to the geochemical results.

IV. CONCLUSION

Most ecological and environmental studies include continuous monitoring of several physical and chemical parameters (*e.g.*, pH, temperature, electric conductivity, dissolved oxygen, redox potential, and some ion concentrations, such as Na^+ or Li^+). By means of a theoretical approach and with the help of examples, the relationship between groundwater chemistry and hydrogeological conditions can be shown. Geochemical characteristics and their space and seasonal variations are defined by (1) the chemical composition of aquiferous rocks, (2) the residence time of groundwater in the aquifer (depending especially on flow rate and permeability), and (3) the flow conditions, such as two-phase flow in the infiltration zone, involving a gas phase, or a single one-phase flow, in a confined aquifer, over a long time period, allowing oxygen consumption in reduction reactions or mixing between surface water and groundwater. The examples detailed above clearly show the utility of geochemical and water chemistry analyses in demonstrating the influences of biological and physical processes on groundwater quality and in demonstrating the relative changes that can occur as groundwater and surface water interact.

REFERENCES

Back, W., Cherry, R. N., and Hanshaw, B. B. (1966). Chemical equilibrium between the water and minerals of a carbonate aquifer. *NSS Bull.* **28**(3), 119–126.

Bakalowicz, M. (1976). Géochimie des eaux karstiques. Une méthode d'étude de l'organisation des écoulements souterrains. *Actes Colloq. Hydrol., 2nd,* Pays Calcaire, Besançon, pp. 49–58.

Bakalowicz, M. (1977). Etude du degré d'organisation des écoulements souterrains dans les aquifères carbonatés par une méthode hydrogéochimique nouvelle. *C. R. Acad. Sci. Paris Seances* **284**, 2463–2466.

FIGURE 6 Map of hydrogeochemical characteristics of the underflow. The reduction area, SO_4^{2-} poor and enriched in CO_2 and HCO_3^-, is represented in white; the channel area, enriched in SiO_2 and in Na^+, is represented in light gray; and the gravel banks, CO_2 and HCO_3^- poor, are represented in dark gray.

126   *M. Bakalowicz*

Bakalowicz, M. (1979). Contribution de la géochimie des eaux à la connaissance de l'aquifère karstique et de la karstification. Thèse, Univ. P. et M. Curie, Paris-6.

Bakalowicz, M. (1980). Un précieux informateur hydrogéologique: Le système chimique CO_2-H_2O-carbonate. *Proc. Colloq. Cristallisation, Déformation, Dissolution Carbonates*, Bordeaux, pp. 11–23.

Bakalowicz, M. (1984). Water chemistry of some karst environments in Norway. *Nor. Geogr. Tidsskr.* 38(3/4), 209–214.

Bakalowicz, M., and Jusserand, C. (1986). Etude de l'infiltration en milieu karstique par les méthodes géochimiques et isotopiques. Cas de la grotte de Niaux (Ariège, France). *Bull. Cent. Hydrogéol. (Neuchâtel)* 7, 265–283.

Bencala, K. E. (1993). A perspective on stream-catchment connections. *J. North Am. Benthol. Soc.* 12(1), 44–47.

Bishop, Y., Fienberg, S., and Holland, P. (1975). "Discrete Multivariate Analysis." MIT Press, Cambridge, MA.

Bouroche, J. M., and Saporta, G. (1980). "L'analyse des données," P.U.F. Collection Que saisje? Presses Univ. de France.

Cailliez, F., and Pagès, J. (1976). "Introduction à l'analyse des données." S.M.A.S.H., Paris.

Castany, G. (1967). "Traité pratique des eaux souterraines." Dunod, Paris.

Castany, G. (1982). "Principes et méthodes de l'hydrologie." Dunod, Paris.

Chapelle, F. H. (1993). "Groundwater Microbiology and Geochemistry." Wiley, New York.

Drever, J. I. (1982). "The Geochemistry of Natural Waters." Prentice-Hall, Englewood Cliffs, NJ.

Dreybrodt, W. (1988). "Processes in karst systems. Physics, chemistry and geology." Springer-Verlag, Berlin.

Fleyfel, M., and Bakalowicz, M. (1980). Etude géochimique et isotopique du carbone minéral dans un aquifère karstique. *Proc. Colloq. Cristallisation, Déformation, Dissolution Carbonates*, Bordeaux, pp. 231–245.

Fontes, J. C., Louvat, D., and Michelot, J. L. (1989). Some constraints on geochemistry and environmental isotopes for the study of low fracture flows in crystalline rocks. The Stripa case. *Isot. Tech. Study Hydrol. Fractured Fissured Rocks, Proc. Advis. Group Meet.*, IAEA, Vienna, pp. 29–67.

Garrels, R. M., and Christ, C. L. (1965). "Solutions, Minerals and Equilibria." Harper & Row, New York.

Greenacre, M. (1984). "Theory and Applications of Correspondence Analysis." Academic Press, London.

Hakenkamp, C. C., Valett, H. M., and Boulton, A. J. (1993). Perspectives on the hyporheic zone: Integrating hydrology and biology. Concluding remarks. *J. North Am. Benthol. Soc.* 12(1), 94–99.

Hubert, P. (1984). "Eaupuscule. Une introduction à la gestion de l'eau." Ellipses, Paris.

Jacobson, R. L., and Langmuir, D. (1970). The chemical history of some spring waters in carbonate rocks. *Ground Water* 8(3), 5–9.

Jacobson, R. L., and Langmuir, D. (1974). Controls on the quality variations of some carbonate spring waters. *J. Hydrol.* 23, 247–265.

Kempe, S. (1975). Ca and Mg organic complexes in the water. *Ann. Spéléol.* 30(4), 695–698.

Lebart, L., Morineau, L., and Warwick, K. M. (1984). "Multivariate Descriptive Analysis: Correspondence and Related Techniques for Large Matrices." Wiley, New York.

Margrita, R., Guizerix, J., Corompt, P., Gaillard, B., Calmels, P., Mangin, A., and Bakalowicz, M. (1984). Réflexions sur la théorie des traceurs. Applications en hydrologie isotopique. *Proc. Colloq. Int. Hydrol. Isot. Mise Valeur Ressour. Eau*, Vienne, pp. 653–678.

Matheron, G. (1965). "Les variables régionalisées et leur estimation." Masson, Paris.

Muxart, T., and Birot, P. (1977). L'altération météorique des roches. *Publ. Dép. Géogr. Phys. Univ. Paris-Sorbonne* 4, 1–279.

Palmer, M. A. (1993). Experimentation in the hyporheic zone: Challenges and propectus. *J. North Am. Benthol. Soc.* **12**(1), 84–93.

Roques, H. (1962). Considérations théoriques sur la chimie des carbonates. *Ann. Spéléol.* **17**, 11–41, 241–284, 463–467.

Roques, H. (1964). Contribution à l'étude statique et cinétique des systèmes gaz carbonique-eau-carbonate. *Ann. Spéléol.* **19**(2), 255–484.

Roques, H. (1972). Sur une nouvelle méthode graphique d'étude des eaux naturelles. *Ann. Spéléol.* **27**(1), 79–92.

Roques, H., and Ek, C. (1973). Etude expérimentale de la dissolution des calcaires par une eau chargée de CO_2. *Ann. Spéléol.* **28**(4), 549–563.

Rouch, R. (1988). Sur la répartition spatiale des Crustacés dans le sous-écoulement d'un ruisseau des Pyrénées. *Ann. Limnol.* **24**, 213–234.

Rouch, R., Bakalowicz, M., Mangin, A., and D'Hulst, D. (1989). Sur les caractéristiques chimiques du sous-écoulement d'un ruisseau des Pyrénées. *Ann. Limnol.* **25**(1), 3–16.

Schoeller, H. (1962). "Les eaux souterraines." Masson, Paris.

Seber, G. A. F. (1984). "Multivariate Observations." Wiley, New York.

Stumm, W., and Morgan, J. J. (1981). "Aquatic Chemistry," 2nd ed. Wiley, New York.

Triska, F. J., Kennedy, V. C., Avanzino, R. J., Zellweger, G. W., and Bencala, K. E. (1989). Retention and transport of nutrients in a third-order stream in northwestern California: Hyporheic processes. *Ecology* **70**(6), 1893–1905.

Trombe, F. (1952). "Traité de spéléologie." Payot, Paris.

Turc, L. (1954). Le bilan d'eau des sols: Relations entre les précipitations, l'évaporation et l'écoulement. *3èmes Journées de l'Hydraulique, Soc. Hydrotechnique de France, "Pluie, évaporation, filtration et écoulement,"* Alger, 12–14 avril 1954. p. 36–44.

Wigley, T. M. L. (1977). WATSPEC: A computer programm for determining the equilibrium speciation of aqueous solution. *Tech. Bull.—Br. Geomorphol. Res. Group* **20**, 1–48.

Zhou, D., Chang, T., and Davis, J. C. (1983). Dual extraction of R-mode and Q-mode factor solutions. *J. Int. Assoc. Math. Geol.* **15**(5), 581–606.

Plate 2   *Upper:* Example of tropical karst: The cone karst of Hefeng country in Hubei province (central China). Photograph courtesy of J.P. Barbary. *Lower:* Example of high mountain karst: The proglacial karst of Wildhora, altitude 2800 m (Valais, Switzerland). Photograph courtesy of R. Maire.

5

Karst Geomorphology and Environment

R. Maire and S. Pomel

Laboratoire Environnement
Centre d'Etudes de Géographie Tropicale, CNRS
Domaine Universitaire de Bordeaux
33405 Talence Cedex, France
and
Laboratoire CIBAMAR, Géologie-Recherche
Université Bordeaux I
Avenue des Facultés
33405 Talence Cedex, France

I. INTRODUCTION

Karstic milieus differ from other physical environments because of their subterranean dimension. The karstic milieu is both a record and a sensitive indicator of the state of the environment as it traps part of the turbid elements (rocks, soils, organic matter, etc.) from the surface environment. A large geographical variety of karstic landscapes exists (Sweeting, 1973; Jennings, 1985; Ford and Williams, 1989). Karstic resources, especially drinking water reserves, lime, phosphates, ore, and oil deposits, are essential for economic development. Carbonated rocks compose the substratum of many densely populated regions, and, for example, a third of the surface of France is covered in either limestone (Jura, Alps, Pyrenees, Provence, Languedoc, Aquitaine) or chalk (Paris basin). Half of the population of Austria is supplied with drinking water from karstic sources (Trimmel, 1980). Moreover, we now know that karst milieus are populated by subterranean organisms (Ginet and Decou, 1977), and different studies have shown the role of geomophologic factors in the distribution of both terrestrial and aquatic fauna (Turquin *et al.*, 1974; Reygrobellet *et al.*,1975; Gibert *et al.*, 1983; Gibert, 1986). Thus the geomorphologic context of the karst aquifer provides a variety of habitats and helps to determine the structure of subterranean communities.

Taking into account the multiplicity of morphotectonic and bioclimatic conditions of karst areas (sedimentary basins, folded ranges, uplifted carbonated platforms, etc.), the objectives of this chapter are to show the physicochemical behavior of surficial karst (exokarst) and underground karst (endokarst) in connection with the passage through the karst of suspended matter (*e.g.*, sedimentary deposits) or matter in solution. Finally this chapter aims to demonstrate that the karstic environment varies according to the history of the population involved and the level of economic development.

II. CONCEPTS AND METHODOLOGY

A. Karst: A Record of Environmental Change

A karst forms a sensitive and vulnerable natural environment that filters suspended solids, but allows almost all soluble and bacterian contaminations through. Hence, karst acts as an environmental record and an indicator *sensu lato* of the way in which it functions. Because of their great permeability (fissures and drains) and the regime of emergences, karstic environments provide information on hydrodynamic action: runoff transfer and calculation of water supplies (Mangin, 1974, 1975; Bögli, 1980; White, 1988). At the same time, it filters the suspended particulates, the endokarstic system acts as a trap for sediments, which record previous hydroclimatic activity, and, thus, paleoclimatology is often based on karst deposits (Maire and Quinif, 1988; Quinif, 1994).

The karstic geosystem is subdivided into compartments (Table I), which comprise, from upstream to downstream, the absorption zone (exokarst and epikarst), the subterranean transfer zone (endokarst with free runoff and saturated zone), the outlet (exsurgence or resurgence), and the piedmont zone (tufa, travertine, crust, and detritic formations). The behavior of each of these compartments as far as runoff, filtering capacities, and contamination are concerned is very diversified; however, endokarst behavior remains preponderant.

The filtering capacities of exokarst and endokarst are conditioned by structural geomorphologic, pedologic, and bioclimatic parameters. For example, the filtering and trapping capacities of the karstic environment with regard to natural and human contamination are variable and depend on permeability. In calcareous massifs, permeability is governed by fissuring and the joints, which ensure water transit and the formation of ducts. In thin, highly porous limestone, the water transit may also take place through the porous system of the rock. This is the case with the slighty diagenetized bioclastic Oligo–Miocene limestones in the Aquitaine region and in some chalky facies.

Karstic environments also react directly and rapidly to edaphic factors (vegetation and soils) and climatic factors and especially to any change

TABLE I   Functional Compartments in the Karstic Geosystem

Karst system	Geodynamic, hydrologic, and bioclimatic parameters	Processes
Input	Tectonics	Mechanical erosion
Exokarst (surficial karst)	Lithology	(gelifraction, glacial
	Geomorphology	erosion, etc.)
	Climate	Dissolution
	Hydrology	Weak filtration
	Soils	
	Vegetation	
Throughput	Fracturation	Dissolution
Epikarst (infiltration zone)	Folding	Mechanical erosion
Endokarst (underground karst)	Limestone/impervious	Detrital sedimentation
	contact	(infillings)
	Percolation	Chemical precipitation
	Ponors and sinkholes	(speleothems)
		Weak filtration
Output	Uplift and base level	Chemical precipitation
Springs	Fracturation	(tufa and travertine)
	Limestone/impervious	
	contact	

in these. Consequently the process is essentially subject to the rhythm of bioclimatic phases of stability (biostasy) and instability (rhexistasy), in accordance with the concept established by Erhart (1967). Thus, contamination in the karst records series of events, showing sequences (functioning) and discontinuity (breaks), which are indicators of the present environment, of paleoenvironments and of periods of destabilization. The endokarstic milieu thus constitutes a "box," in which imprisoned records (deposits and contamination) are direct indicators of functional processes.

Human-mediated environmental change (forest clearing, fires, agriculture, and overgrazing) causes soil erosion, increases turbidity and delays formation of concretions (speleothems). Human activities tend to accelerate the destabilization of processes—a phenomenon that also appears without human intervention when there is a climatic change. Consequently, in order to evaluate the part played by human contamination (agriculture, fires, industries, and waste dumps) in a given milieu, it is necessary to compare the action of natural processes with and without humans, in order to compare the effects of natural and human destabilization (Goudie, 1990).

B. The Variety of Karstic Environments and Constraints on Economic Development

On a global scale, the variety of karstic landscapes and milieus is exceptional; carbonated rocks can be found in all climatic zones, from the polar circle to the equator, and in all geologic structures, from volcanized island arcs to recently folded mountain ranges and from platforms to sedimentary basins. The importance of climate in karstogenesis has been emphasized especially in Germany since Lehmann's pioneer works (1936) on karsts in Java and subsequently in France since Corbel's study (1957) of karsts in northwest Europe and Nicod's study (1967) in Provence. Current karstologic research is very diversified and is also investigating more and more subterranean environments (Salomon and Maire, 1992; Julian, 1992). Considerable progress has been made in applied research due to the drilling for oil (Dubois *et al.*, 1993) and mining (Nicolini, 1990).

To the natural karst varieties studied by the karstologists should be added the economic history and development constraints that geographers have to study. This double approach has resulted in a geography of karstic environments that takes development into account. The mutual impact of constraints and developmental factors affects both nonrenewable and quasi-nonrenewable resources (soils, fossil aquifers, and ore deposits) and renewable resources (water and agricultural produce).

The increasing accumulation of waste throughout the globe and the need to get rid of toxic products and wastewater have led to diverse projects: underground storage, depth injection, sometimes in karstified terrain, etc. Impact studies do not always make it possible to foresee the long-term evolution of pollutant migrations and of the relations among various aquiferous systems. This was stressed by Aus and Kreysing (1977) in their study of burying and drilling projects in Germany, in the karstic Malm situated under the mollassic basin between the Alps and the Danube. The general problem raised is that of storing pollutants in a deep extensive karstic aquifer that is currently not in use, but that might one day become essential as it contains extensive reserves. Demands for groundwater, and especially karstic water, must be forecast in the long term, taking into account reserves that are nonrenewable on a human scale.

C. Methodology and Study Techniques

Besides classic and fundamental hydrogeology and karst hydrobiology (hydrochemistry, chemical tracing, and bacteriology) techniques, other complementary methods are used for endokarst sedimentary deposits as follows.

1. Micromorphology

Micromorphology uses an optical microscope to study microstructure organization in rock, infill, and soil samples on thin sections or polished surfaces. Closely related to petrography, micromorphology is based on micropedology, pointing out, by form analysis, the significance of overlapping structures and microsequences (Bullock *et al.*, 1985). These results are mainly qualitative but are indispensable, because they can be interpreted in terms of environment, paleoenvironment, and contamination intensity (Genty, 1993a,b).

2. Digital Imaging

Digital imaging complements micromorphology and is performed using images of thin sections photographed by a videocamera mounted on an optical or electron microscope. It enables quantified observation of pretreated samples (colored or impregnated with colored resin or metal) to enhance porous relief, carbonate types, etc. This analysis works on binary images (black and white pixels) or in color. Dimension (area and pore perimeter) and shape (elongation, contour, circularity, and shape types) parameters are measured using customized software, such as BIOLAB. Coupled with nondestructive analyses (fiber optic microscopy), these techniques also lend themselves to filter studies of solid contaminants (turbids and dust) that cross through the karst system.

It is necessary to analyze both the solid phase (crystals, clasts, microsequences, and contaminants) of the samples and the porous space. Once the coherence and significance of the various parameters measured have been validated, the advantages are speed of measurement, reliability (accurate and easily reproduced measurements), and measurement of parameters impossible to quantify by simple observation.

3. Radio Dating

Some of the karst deposits, such as calcite concretions, act as geologic chronometers and thermometers. Volcanic products (lava and ash), not envisaged here, are also well known in radio dating: the potassium/argon method (K/Ar) and tephrochronology (mineral signature of an eruption dated beforehand by K/Ar). Thus it is possible to accurately date intervolcanic soils or various deposits sealed in lava flows or contaminated by ash. This enables the identification, in the case of cinerites, of distal contributions compared with atmospheric circulation, and a regional stratigraphy of karst infillings can be established.

In the karst cavities, calcite concretions are dated by the U/Th method (up to 350,000 years), which can also be used for travertines and recent lacustrine limestone. This method is reliable for samples presenting a closed geochemical system (no internal corrosion) and without detritic contamination (Maire and Quinif, 1988; Quinif, 1989, 1994). Thermoluminescence (TL) and electronic paramagnetic resonance (EPR), with which it possible to date even older calcite, are used, but the results of these methods are not yet sufficiently reliable. All the radio dating methods require stratigraphic and geomorphologic tests and, if possible, a series of analyses to discount deviating results.

III. GEODYNAMIC CONDITIONS AND FILTERING CAPACITIES OF EXOKARST

The filtering capacities of surface karst (exokarst) vary in relation to geologic and geomorphologic conditions, the absence or presence of pedologic or alluvial covering, etc. But, in general, surface filtration is not very effective. The essential role played by tectonics (uplift) in the vertical development of karstification, by creating a hydraulic potential and by changing the regional base level during morphogenesis (valley deepening), is immediately apparent (Ford and Williams, 1989; Maire et al., 1991).

A. Influence of Morphotectonic Conditions

On a regional scale, the most important discriminating factor for karstogenesis is the morphotectonic pattern. Karsts occur in several types of large morphotectonic units: sedimentary basins, slightly raised platforms, extremely raised platforms, and folded mountain ranges. In detail, geologic structures and paleostructures may overlap, as in central and southern China, and on all scales they condition the present functioning and paleofunctioning of the karsts (Barbary et al., 1991, 1994). Valley formation and the sinking of karstic systems, for example, are directly linked to uplift, resulting in a more or less complex terracing of the drain systems, depending on the tectonic phases.

1. Karsts in Sedimentary Basins

The karsts of the Paris and London basins, in the Swabian and Franconian Jura, in southern Poland, and of the Mississipi basin are characterized by very similar compartments and poor hydraulic potentials due to the weak slopes of the structural layers. However, the resultant hydrogeologic valleys can be very extensive. In the Paris basin water gaps of between 10 and 15 km are numerous, as in the chalky basin of La Vanne, southeast of Paris (Castany, 1967). In the region of Lake Baïkal, in Siberia, a water gap of 200 km has been reported (p.c. Marian Pulina). The filtering of pollutants depends on vegetation and on the nature and thickness of the soil profile or alluvial covering. As these sedimentary basins are farmed, fertilizers and pesticides are often found in variable quantities in karstic exsurgences; *e.g.,* nitrates are the most widespread type of soluble pollution (the European standard for drinking water is ≤ 50 mg/l, with 10 mg/l for the United States).

2. Karsts in the Slightly Raised Carbonated Platforms

The evolution of karstic landscapes depends mainly on the proximity of the general base level (low altitude) and on the permeability of the detritic covering if it exists. The most remarkable example of this is the calcareous platform in Florida, where the surface is situated at an altitude of between 0 and 30 m. The carbonated series dating from the Cretaceous period to the present day is more than 2000 m thick. Moreover, it is covered with an Appalachian mantle of siliceous sands that can rise to several tens of meters in the central and southern peninsulas. In and between sands and limestone are the well-known phosphate deposits, the origin of which has not yet been established. In northern Florida, where the karst occurs in outcrops (Suwannee and Talahassee), karstic filtration is closely connected to vegetal covering and takes place particularly in epikarstic zones. In the south, the arenaceous covering is thick enough to contain aquifers of good quality. Karstification is still developing under the sands and regularly results in the formation of spectacular redoubled funnels (sinkholes) in urbanized regions such as Orlando (Beck, 1984). There are also subsidal coastal platforms such as the famous pitching-peak karsts in southeast Asia: the Bay of Halong (North Vietnam), the coasts of Thailand, and Palawan (Philippines).

3. Karsts of Extremely Uplifted Carbonated Platforms

The karsts of platforms in the Yangtze basin in southern China, the Grand Causses in France, the Pamir–Alai Plateau in central Asia, and the Darai Plateau in Papua New Guinea have a hydrodynamic behavior pattern that is completely different from that of platform karst at very low altitudes. In south and central China (Guangxi, Guizhou, Yunnan, and Hubei), the large Paleozoic and Triassic carbonated platforms have undergone very extensive evolution since their emersion during the Jurassic period. During the Tertiary, they were rejuvenated by the Himalayan surrection and sometimes raised to altitudes of more than 1000–1500 m. At present, exokarstic filtration has little efficiency because of intense deterioration of the forest and because of the multiplicity of interrupted streams and tunnel caves (Maire *et al.*, 1991). Nonetheless, in Guizhou and Hubei, the pollution of springs by nitrates is moderate (6–16 mg/l) because of the moderate exclusive use of farmyard manure (Collignon and Jin, 1991).

In the arcs of volcanic islands, the limestone platforms may have undergone a broad surrection. In the eastern Mediterranean, the Aegean arc has karsts that have been raised to an altitude of more than 2400 m (Mt. Ida and Levka Ori in Crete) since the end of the Miocene (Maire and Pomel, 1992). In New Guinea, in contact with the Australian and Pacific plates, the karst of the Huon Peninsula has been raised to more than 4000 m since the end of the Pliocene (Maire, 1990).

4. Karsts of Folded Mountain Ranges

The Jura, Alps, and Dinaric mountain ranges and the Carpathian, Taurus, Zagros, Andes, and Rocky Mountains all exemplify standard types, according to latitude and climate. Altitude determines a strong hydraulic potential and usually an accentuation of superficial karstic forms (sinkholes, ouvalas, and peaks) that impede filtration. In Croatia, the Velebit karst, at an altitude of between 1400 and 1700 m, is one of the most highly developed in the world, because of the very high rainfall (>5 m/year): the nival hollows can exceed 100 m in depth.

In the forest zones of these folded mountain ranges, part of the filtration takes place at the level of the pedologic covering. However, agricultural contamination by nitrates—the most widespread chemical pollution in Europe—is not prevented by the soils covering the karst. More serious still is the fact that there is no filtration when there is a concentration of interrupted streams. It is precisely in these middle-mountain interrupted streams (*e.g.*, the Jura/ France) that the standard types of pollution take place: waste tips, sewers, pig farms, and various kinds of chemical pollution.

B. Influence of Bioclimatic and Pedologic Conditions

Geomorphologic evolution is closely connected to pedogenetic conditions and consequently to climate and vegetation (Trudgill, 1985). These pedoclimatic and edaphic factors, basic to morphopedology, also determine the nature of the absorption zone (epikarst) and the characteristics of water and contaminant transfer to the endokarst.

1. Modification of Bioclimatic Parameters

Throughout geologic periods, the modification of bioclimatic parameters raises the question of morphopedologic inheritance. Some karsts have evolved under a siliceous weathered cover (karsts of Perigord in France, Silesia in Poland and Florida) and some after the clearing of their permeable blanket (Dinaric karsts) (Salomon and Astruc, 1992). Climatic breaks and tectonic movements have made it possible for geologic covers and weathered and decayed rocks to erode. Under acid and wet pedologic compresses, the karst surface undergoes an intense and relatively homogeneous alteration of a cryptokarstic nature. In humid intertropical regions, characterized by the absence of major climatic changes, this action has been long lasting and homogeneous enough to make it possible for typical forms in patterned ground karsts (Jamaica, Java, New Guinea, etc.) to develop. On the contrary, in temperate zones, the marked succession of climatic cycles during the Plio–Quaternary has precluded the development of this type of morphology.

2. Altitude and Exposure

In all latitudes, topographic features influence local climate patterns and condition edaphic factors, runoff, and infiltration. The vertical bioclimatic zonation of Alpine mountains is a typical example. But there are more complex steppings, *e.g.*, on the eastern side of the Argentine Andes, where the semiarid piedmont slope, under the influence of the foehn, gives way between 1000 and 2000 m to the cloudy rain forest ("Nebelwald").

3. The Succession of Vegetal and Pedologic Covering

Edaphic and vegetal changes modify the characteristics of subterranean transfers of water and suspended solids. In a situation of bioclimatic equilibrium (biostasy), the water is enriched in CO_2 in the clay–humus complex of soils, increasing dissolving capacity, and enters the endokarst with a very reduced suspended load of turbids (Fig. 1A). The vegetal and pedologic covering, very often of the forest type, provides filtration if the water does not accumulate at one particular point. As infiltration is slowed down, the epikarstic dissolution is high and endokarstic concretioning increases. This is the case in temperate forest karsts (*e.g.*, Vercors in France and karst in Slovenia) and humid tropical karsts (Borneo and New Guinea).

In situations of bioclimatic change (rhexistasy; Fig. 1B), the vegetal and pedologic covering becomes discontinuous: it does not act as a filter and the load of turbid particles increases. As the water flows quickly into epikarst and endokarst, it is less rich in CO_2. Its overall dissolving capacity is lower, but it subsists at deep levels: this is the case in general in semiuncovered mediterranean karsts and in subalpine karsts. Subterranean concretion diminishes and is contaminated by products of soil erosion; detritic infillings increase. This is true of contrasting climatic regions and of destabilizations

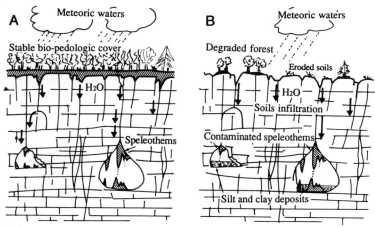

FIGURE 1   (A) Karstic milieu functioning in bioclimatic balance (biostasy). (B) Karstic milieu functioning in bioclimatic imbalance (rhexistasy).

of human origin (soil erosion resulting from deforestation, fires, and shifting cultivation) (Quinif, 1994).

4. Filtration

Filtration is the least effective process of the whole karstic geosystem in comparison with global transit. Soils in karstic regions are heterogeneous, sometimes discontinuous, and relatively young, withdrawals occurring constantly. Moreover, as the soils are often less thick on limestone substrata than on others, they play an important part only in the concentration of water. The range of soils on different calcareous substrata can be explained by karst filtration capacity, underground drainage characteristics, and hydrodynamic properties (Verheye, 1989). In chalky terrain, where the porosity is greater than that in calcareous rock, the filtration is slight, *e.g.*, in the chalky karsts of Normandy (Rodet, 1991). In the forest-covered karsts of temperate humid mountains, soils are not very deep on account of their young age, linked to the sequence of glaciations during the Quaternary. As a result, filtration is good at the level of fissures covered by soil, but bad at the level of deep cracks and pit entrances.

In very humid tropical zones, hyperkarstification under the vegetal compress results in thin soil and extensive development of the epikarstic section. Where there is very high monsoon rainfall, even under rain forest cover, the water absorbed by the interrupted streams is always turbid during the wet season—proof of relatively ineffective karstic filtration. On the other hand, when there is regular convection rainfall, the decayed rock and soils absorb a large part or all of the flow, so that interrupted streams do not function and the karstic reservoir plays its part completely. Phenomena of this type have been thoroughly documented in the central mountain ranges in New Guinea and New Britain (Maire, 1990).

Last, only cryptokarsts developing under a thick arenaceous cover provide very effective filtration.

C. Climatic Change and Role of Inheritance

Identifying climatic change by studying sediments trapped in karstic depressions and cavities enables us to evaluate the natural functioning of the environment. In the natural environment, there are regions that are propitious to the recording of climatic and environmental changes, because they form traps for sediments and chemical deposits. The most notable are karstic depressions (poljes, ouvalas, and sinks) and the basin zones that form receptacles for sediments. By examining natural sections or core drillings, the regional paleoclimatic mechanism can be grasped. Some morphologies are also the result of climatic inheritance.

Many indicators of climatic and bioclimatic operation are constant features of all environments: vegetation, soils, and detritic deposits. The state of vegetation coverings conditions the pedogenesis and the denudation of soils. Morphologic and micromorphologic studies of the soils and the weathered rocks enable us to trace the history of the karstic bioclimatic environment. The surficial pellicular organization (SPO) (*e.g.*, crusts), which record rapid modifications of the environment, is an important feature of the study of the pedologic covering of karsts. Each destabilization or climatic change diminishes protection of the soil and facilitates the lateral processes. Thus calcareous or salted crusts may appear, as well as silcretes, iron and manganese duricrusts, and various hardpans. These SPOs condition the surface hydrology, because they concentrate infiltration and detritic withdrawal.

The study of iron crusts has been used for a long time to differentiate between and to characterize the main families of soils (ferrisols group). They are known to be chemical indicators of soil functioning: the dynamics of iron, the oxidoreduction process, the iron/humus relation, etc. Their use as indicators of environmental processes, however, and particularly of the present or inherited tropical karstic zones is a broad field that has yet to be fully investigated. The resistance to erosion of iron crusts makes them noteworthy "black boxes" of hydric functioning and of surface pedology. The paleocrusts have complex facies that have recorded cycles or breaks in functioning: condensed crusts record fluctuations in the saturation groundwater level, nodular crusts show improvement in land drainage, massive hardpans of truncated alterites and microlayered hardpans show surface reorganization after soil clearing, gravelly crusts bear witness to limb stripping, etc.

During the current period, crusts in karstic zones enable us to observe pellicular organizations and elements other than iron (carbonates, silica, dolomites, and manganese) and to progress in functional hypotheses. In Vietnam, following the massive use of herbicides and the scouring of ferrallitic soils, metric gravelly crusts affect the deep argillic horizons. Specific crusted SPOs and gravelly SPOs appear, and also a number of relatively irreversible processes: condensation of horizons, amorphization of iron humates, rubefaction of ferruginous clays, truncation of profiles and oxidation, loss of fine particles, depression of the landscapes, channeling of the soil catenas, and reversal of weathering dynamics (rising alteration and modified oxidoreduction mechanisms, etc.). All these mechanisms influence and mark out the dynamics of karst.

In Alpine karsts, quaternary glaciations have modified the limits of forests and glaciers and, consequently, the conditions governing flow, infiltration, dissolution, and concretioning. During the interglacial periods, the forest zone gains altitude, slowing infiltration slightly and leading to the development of subterranean concretioning. In uncovered karsts above the forest line, infiltration is rapid and diffuse and therefore carries a low detritic charge. Deep down, the waters are aggressive and concretioning is reduced. In the glacial periods, these same karsts (in and above forest zones) can progress under ice cover. Flow is concentrated, the turbid charge increases, and endokarstic concretioning does not occur. This also holds for the proglacial karsts. Last, the solid or dissolved load in running water leads to three kinds of surface deposit: lacustrine detritric laminated deposits, chemical deposits in springs or riverbanks (tufa and travertines), and lacustrine limestone.

D. Cryptokarsts and Filtration

Cryptokarsts develop under a generally nonimpermeable coverage: ice, weathered and decayed rocks, volcanic ash, alluvial deposits, moraines, etc. A porous aquifer may form above the karstic aquifer if an alluvial cover exists. The two aquifers may be related or not. This type is found in alpine karsts with overdeepened glaciokarstic basins that have been clogged with fluvioglacial sediments (Alps and Jura in France). Volcanic flows can form a quasiimpermeable blanket covering a karstic zone; they may subsequently collapse (*e.g.,* Middle-Atlas, Morocco; Martin, 1981) and are sometimes related to hydrovolcanic or hydrothermal phenomena (Pomel, 1987): *e.g.,* Swabian Alps in Germany and the Antrim Plateau in Ireland. The volcanic material is an important aquifer in contact with the epikarst (*e.g.,* Coirons massif in Ardèche, France).

Usually the cover is thin and the filtration is weak. Ice, high-altitude forest soils, and moraines provide no protection. In the case of subglacial karsts in the Alps and the Rocky Mountains, snowmelt waters quickly flow through the glacier, using subglacial channels in which water flows torrentially, and reach the calcareous floor in less than 24 hr. The duration of the water transit between a loss (mill) and the glacial portal does not exceed a few days. In all cases there is no filtration; however, the snowmelt waters may collect dissolved atmospheric pollution trapped in the ice (radioactive or chemical fallout, etc.). As for morainic deposits, because of their heterometric composition, filtration is virtually absent. Under volcanic coverage (ignimbrites, cinerites, pyroclastites, and glowing ash clouds) there is strong interaction with karstification (West Indies, Indonesia, and New Guinea). There is maximum influence of the humid compress, and filtration is very slight.

Forest clearing on thick alterite covers leads to soil erosion and subsequently to the withdrawal of soil from the endokarst when cryptolapies heads begin to appear. This phenomenon is well known in the Chinese karsts belonging to the "stone forest" (or "shilin") type in the provinces of central and southern China. In the Yunnan Province (Lunan), it takes the form of red tertiary deposits originating in the weathering of Permian basalts. Because the sediment is at least partly permeable, a cryptokarstic uniform dissolution takes place, through the action of acid compresses in contact with rock and with the deposit (Maire *et al.,* 1991). In the Franken Alps (Germany) and in Quercy–Périgord (southwest France), the composition of bocage farmland facilitates the withdrawal of cryptokarst and the transit of arenaceous coverages and of weathered and decayed rocks (Salomon and Astruc, 1992). The formation of podzols with differentiated profiles is thus accelerated, and the pedologic covers no longer play a part in filtration and purifying, particularly as regards turbid pollutants, nitrates, metals, and polysaccharides.

E. Filtration and Withdrawal of Pollution in the Epikarst

In rural areas, there are multiple sources of pollution: waste tips, sewers, pig farms, dairies, cheese factories, sugar refineries, paper factories, slaughterhouses, various types of workshops, fertilizers and manures (nitrates and phosphates), pesticides, and herbicides. In France there are thousands of free or public trash dumps in calcareous regions (dell bottoms, chasms, unused quarries). The slow decomposition of the garbage creates a progressive absorbtion of chemical pollutants (sulfates and nitrates) and organic ones (bacteria) through the screes and the karst fissures. The pollutants are withdrawn directly and contaminate the karstic springs.

The concentrated absorbtion of pollutants takes place when there is a deliberate or accidental discharge (a break in an oleoduct, pipe, tank, or cistern), in sinks, small subterranean streams, or streams with subfluvial discharges. In Würtemberg–Hohenzollern in Germany, for example, the employees of an electroplating factory cleaned out a cyanide tank in a depression and contaminated a spring supplying a population of 14,000;

fortunately the water supply was cut off in time (Stiegele and Klee, 1974, p. 47). In calcareous mountains, the development of mass tourism (ski resorts and mountain refuges) is the cause of massive direct concentrated bacterial pollution. Two cases of direct discharges into karst fissures without treatment of water and discharges following a malfunction in a filtration plant (saturation of decantation basins and inadequate temperature) have been recorded.

In karstic regions, the rarity and poor distribution of water resources, their concentration at particular points (springs), and their extreme vulnerability make control of resources problematic. It has become necessary to diversify sources of supply. The Angoulême area in Charente (France) is a typical example. The large emergence of the Touvre River supplies a population of more than 100,000 with drinking water. Any sizeable pollution of the supply basin can deprive a large part of the population of water. It has therefore become necessary to find a complementary source, independant of the Touvre karstic system and capable of meeting at least a third of the demand (Chamayou, 1979).

IV. STRUCTURE AND FILTRATION CAPACITY OF ENDOKARST

The filtration capacity of endokarst is generally low because groundwater flows freely and rapidly in the karstic drains. Suspended solids in the groundwater may be trapped in karst voids, if the geometry of walls and slope allows sedimentation.

A. Formation of Groundwater Flowpaths

The heterogeneity and anisotropy of karstic environments are their principal hydrodynamic characteristics (Chapter 2 Mangin). Water flows simultaneously through fissures, depositing concretions in the galleries by releasing CO_2, and through galleries (organized drains), depositing alluvials and various infillings that may harden (Renault, 1967, 1968).

The transfer of contaminants through a fissured and karstified aquifer depends on drainage capacity and on transmissivity (Castany and Margat, 1977); it is not the same when the pollutants are directly injected by discharge toward a karstic drain and when they enter a system of small karstified fissures. In karstic drains the propagation speed is high and usually does not exceed a few days or even one or two weeks (in cases of high transmissivity). In cracked zones, on the other hand, widened only a little by dissolution, dispersivity is the rule and the speeds are lower—between 1 and 1000 times that on the aquifer scale (weak transmissivity) (Drogue, 1977).

Last but not least, tectonics play an essential part in the karstification of limestone. The deepening of valleys and of subterranean networks takes place at the same time as the uplift. When uplift is very rapid, karstification may lag behind the deepening of the valleys, resulting in springs perched in the flanks of canyons (*e.g.*, Mexico and New Britain) (Maire, 1984).

B. Endokarstic Infillings: Records of the Natural Environment

Protected from external erosion and from meteoric weathering, the endokarstic milieu is a remarkable preserver and ensures a good tranfer function. Subterranean deposits are subjected to weak or buffered diagenesis (no bioturbation) and, with the exception of antetertiary paleokarstic infillings, show only slight alteration. There are two types of endokarstic infillings (Fig. 2): detritic deposits (loams and clays with laminated deposits and conglomerates, etc.) (Quinif, 1994) and chemical deposits (speleothems) of calcite, aragonite, or gypsum (Hill and Forti, 1986).

1. Sequences

Detritic deposits are organized in sequences with sometimes interstratified stalagmitic concretions: floors, flows, and stalagmites. These calcite deposits enable us to date infillings and therefore paleoclimatic phases. In mountain zones, speleothems increase during periods of warming, because there is a direct relation between the formation of concretions and the

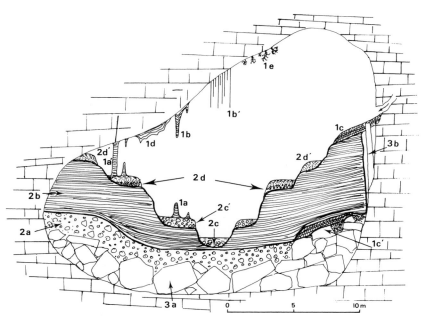

FIGURE 2   The different types of endokarstic infillings. 1. Speleothems (chemical deposits). 1a, stalagmite; 1b, stalactite; 1b', straw stalactite; 1c, flowstone; 1d, drapery; 1e, helictite. 2–3. Fluviatile and breakdown sediments (clastic deposits). 2a, pebbles with clayed matrix; 2b, laminated loam (varves); 2c, lower fluviatile terrace; 2c', pebbly terrace; 2d, terrace with graded bedding; 2d', sandy-loamy terrace; 3a, block (roof collapse); 3b, wall collapse.

production of pedogenic CO_2. Thus, under the denuded high alpine karsts, there is practically no concretion because of the rarity of biogenic CO_2. In humid tropical environments, where vegetation and pedologic cover seem stable (when humans are absent), phases of destabilization are due to climatic breaks. For example, the subterranean concretions of the subtropical karsts in southern China often show alternating sequences of white native calcite and red–brown calcite with soil particles. Their paleoenvironmental significance is noteworthy. The white, pure calcite indicates the stability of edaphic conditions (biostasy): the water is filtered and carries few rock and soil particles. In contrast, the red–brown calcite indicates a break in bioclimatic conditions, because the infiltrating water is loaded with soil elements and iron, as a result of the withdrawal of fersiallitic and ferruginous red soils (Maire, 1991).

2. Microstructures

In the speleothems, successions of clear and dark laminæ can have an annual significance, as has been demonstrated in the case of recent concretions in artificial tunnels (Genty, 1993a,b). The microsequential analysis of speleothems (*i.e.*, the study of growth phases, discontinuity, and contamination) is thus a remarkable tool for the study of recent, historic, prehistoric, and even older environments. Although the calcite layers found in the concretions correspond to crystallization sequences, the textures of calcitic deposits are also valuable indicators of hydrologic and chemical processes. For example, different textures, related to patterning, size, crystal transparency, and the presence of oxides, metals, and organic matter depend on the hydrologic and bioclimatic environment. That is why it is important to be able to quantify accurately the appearance and texture of these sequences, as well as the nature and the abundance of solid contaminants (Figs. 3–5).

Different types of crystal are indicators of the kinetics of crystallization and therefore of the speed and the percolation flow rate linked to hydroclimatic conditions. Precipitation can be rapid (columnar calcite with palissadic structure) or slow (grainmeshed microcrystalline, microsparitic, or micritic calcite). Solid contaminants can influence the nature and type of crystallization. For example, in Mediterranean karsts, during cooling periods, condensed levels with micritic calcite finely polluted by the withdrawal of rubefacted soils are to be found.

3. Influence of Environmental Change

During the Quaternary, the cycle of climatic breaks brought about, at fairly regular periods, a destabilization of the vegetal cover and of the soils, even leading to their disappearance in mountains and at extreme latitudes (direct influence of glaciation). Soils, pollen, macroremains, phytolites, and ash resulting from natural fires are all to some extent trapped in karstified and fissured calcareous environments. The same can be said of volcanic ash, which provides evidence of former eruptions, and of concretion shearings, testifying to paleoseisms and well preserved in karstic galleries. This study of "natural contamination" is essential and must be carried out together with the study of contamination of human origin (*i.e.*, pollution). Speleothems and detritic infillings are also indicators of the local base level. In nonsaturated regimes, concretion increases. In embedded regimes, laminated sediments or varves settle; the preexisting floors, previously shaped in the atmosphere of the ducts, may be subjected to superficial and internal dissolution.

C. Behavior of the Endokarst in the Presence of Pollution

1. Self-Purification

In karstic systems, self-purification is possible only if the bacteria present have enough oxygen to metabolize the organic elements. This implies water rich in dissolved oxygen (aerobic environment), nonmassive organic pollution, the absence of inhibiting chemical substances, and a relatively long residence time (at least 30 days). In karstic systems, these conditions are seldom fulfilled. Oxygenation is adequate in the mountain networks, where flow is free, but the transit time of the water is generally short (one to eight days), especially during flood periods (12 to 36 hr). In fissured and karstic systems, in which the flow rate is slow, self-purification can take place only in anaerobiosis: fermentation produces methane and sulfides. These anaerobic processes tend to infect the environment and endanger the survival of subterranean communities.

An excess of organic pollution (*e.g.*, pig manure and dairy and cheese factory washing water) requires a biochemical demand for oxygen (BDO) which often exceeds the capacity of subterranean streams. Chemical pollution of water (detergents and hydrocarbons) perturbs self-purification, as a film is formed on the surface of the water, reducing and hindering the oxidation of microorganisms.

Direct ecological effects are the result of mutual interactions between the behavior of the endokarst in the presence of pollutants and that of subterranean organisms. These effects affect survival as well as the distribution and dynamics of populations. The consequences for the viability of organisms or for their ecological interactions can be very different according to the diverse types of karst.

2. Temporary Trapping

The temporary trapping of pollutants in the endokarst is a source of problems because it can correspond to a higher concentration of polluting material than that in the original discharge. Human contamination is trapped in the same way as natural contamination. The trapping of turbid particles by gravity, in particular in saturated or lacustrine subterranean environments, depends on sedimentation structure. The trapping of light objects by flotation takes place mainly in episaturated zones (*i.e.*, within the range of flood fluctuations) and in zones of free flow during periods of low water (Maire,

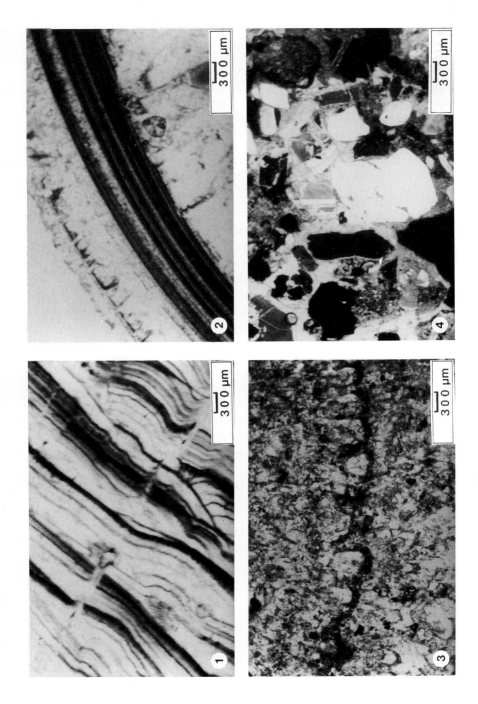

1979). Trapping by absorption, especially at the level of cave clays, leads to a concentration of toxic substances such as heavy metals, pesticides, and radioisotopes. Aquatic underground fauna are contaminated by way of concentration, via the food chain (Gibert, 1986).

The deferred and concentrated restitution process is a threat in fissured and karstic systems, on account of numerous temporary traps (siphons, saturated roofs, etc.) and of delayed water restitution by fissured aquifers with low transmissivity. Sedimented or adsorbed polluting materials can also be freed during rises and enter-tapped exsurgences.

3. Buffer Effect

The buffering action of karst water in the presence of acid pollution is well known. For example, in the southern Chinese karsts (Guizhou) the water attains a remarkable level of acidity (pH 2.8–3.2) when it is in contact with coal seams rich in iron sulfides (pyrite). These deposits have been worked by small mining operations since 1980. After flowing over 500 m of limestone, the water attains a pH of 6.5, and iron is precipitated as oxides, hydroxides, and carbonates (Collignon and Jin, 1991). The abundance of nitrates and phosphates in some karst waters leads to spring eutrophication.

V. IMPACT OF HUMAN ACTIVITIES ON THE KARST MILIEU

The impact of human activities on the karstic milieus varies with ecogeographic parameters: relief, climate, vegetation, population history, wars, and sociocultural and politicoeconomic environment (Demangeot, 1990). The most sensitive zones are the bioclimatic margins, such as semiarid zones, coastal regions, and mountains, which introduce, generally over a short catena, considerable modification of natural conditions. These rural areas experience strong, ancient human pressures. For thousands of years, wars have had an effect on the forest environment, in antiquity (deforestation) and the Middle Ages (neglect and reworking of agricultural land during the Hundred Years War) as well as in the twentieth century, with the colonial and postcolonial wars (use of napalm and dioxine-based defoliants, inducing rapid soil crusting by iron migration). Finally, human pollution intensity in

---

FIGURE 3   Solid contamination examples in endokarst (Papua New Guinea). (1) Rhythmic distribution of fine organic matter withdrawn into a stalactite—Mount Bangeta, Huon peninsula (sample BANG 2-22); natural light; lens, 2.5 × 3.2. (2) Fine organic matter laminae in a pearl—Nare Cave, New Britain (sample NARE 3-26); natural light; lens, 2.5 × 3.2. (3) Turbid contamination (organic matter and soil particles) in a pearl—Nare Cave, New Britain (sample NARE 2-30); natural light, lens, 2.5 × 3.2. (4) Coarse harpan sand sequence—underground terrace of Kavakuna II Cave, New Britain (sample KA II-29); polarized light; lens, 2.5 × 3.2.

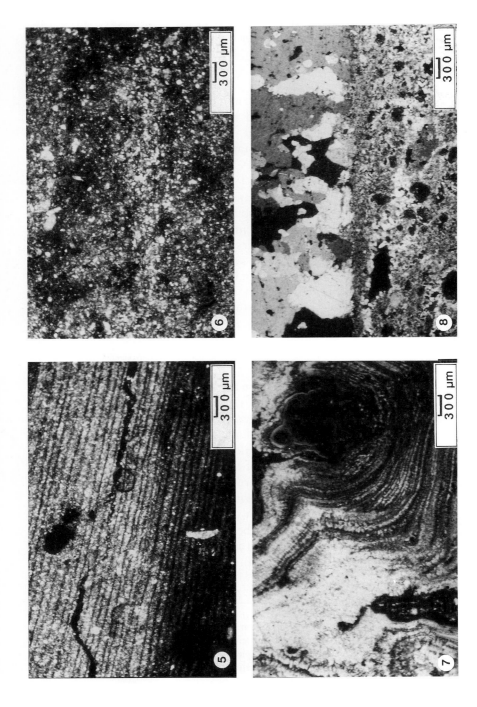

calcareous zones varies with socioeconomic and political systems and their succession (Delannoy, 1985).

A. Destabilization of the Rural Environment

The different indicators of human pressure, dependent on the growing need for space and the development of production systems (drainage, irrigation, land clearing, and installations of various sorts), lead to modifications in the makeup of the landscape and, sometimes, to dysfunction. The study of destabilization processes and the regulation of human-influenced systems lead to historical geography of landscapes, with emphasis on the crisis elements.

Anthropogenic soil (anthropized and anthropic soils) have provided good records of human activity since the deforestation and land burning that preceded neolithization. Anthropized soil reflects a fairly minor level of human activity (*e.g.*, ancient cultivated horizons reclaimed in pedogenesis, certain isohumic ashy soils, acid soil and podzols linked with massive presence of resinous trees, and hardpans). On the other hand, anthropic soil reflects a high degree of human activity: this is cultivated and overgrazed soil (*e.g.*, compacted soil in overgrazed zones, overcrusted soil, or salted soil in irrigated regions). Even in humid equatorial mountain regions, deforestation is often ancient, as the prehistoric "gardens" of the Papuan Highlands, clearly visible from the air, illustrate (Paijmans, 1976).

The concept of mountain and cave as refuges enables the history of anthropization in karstic mountains to be envisaged. This notion is different from that of a survival medium because humans settle in mountain regions in search of more favorable land and a more favorable climate. The notion of the mountain as a refuge also covers the idea of protection during wartime. This is well known in karstic regions, where pinnacles, canyons, and tunnel caves have served as shelters and fortresses, as in Slovenia, Greece, Cuba, or southern China (during the civil war).

The human destabilization threshold is reached at varying population

---

FIGURE 4 Solid contamination examples in southern Chinese endokarst (Hubei Province). (5) Sequence of fine laminated quartz deposits rich in organic matter and iron oxides (soil erosion) at the base of the varve cross-section (−20.5 m, contact with base blocks)—Dadong Cave, Wufeng (sample WU 491-2-29, slide 20,740); polarized light; lens, 2.5 × 3.2. (6) Detail of sequences of gray–white varves, rich in quartz, and red–brown varves, rich in soil particles (clays, iron oxides, and organic matter)—Dadong varve cross-section (−7 m) (sample WU 491-11 bis-15); polarized light; lens, 2.5 × 3.2. (7) Budding stalactite contaminated by coarse pedologic products and fine particles of organic matter and fine quartz—Dadong fossil gallery, Wufeng (sample DAD 518-A-32); natural light; lens, 2.5 × 3.2. (8) Noncontaminated stalagmitic floor on a sequence of microbreccia contaminated by pedologic matter—Longdong fossil network, Wufeng (sample DAD 513-A-38); polarized light; lens, 2.5 × 3.2.

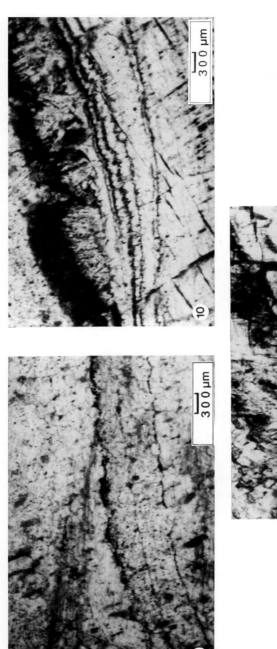

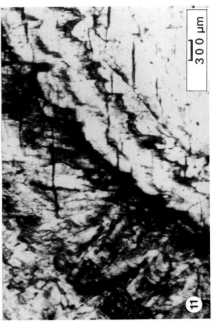

densities, depending on topographic and edaphic characteristics, such as the human occupation time span, the evolution of cultivation and breeding systems, and agrarian installation modes and their upkeep. If humid mountain deforestation favors soil erosion and rainy season flooding, systematic organization of cultivated terraces, on the other hand, provides soil stability and is therefore a regulating factor.

B. Human Impact as a Function of Economic Level and Geographic Situation

Karsts in temperate zone, industrialized countries are of a large area and are subject to major human impact (cities, industries, agriculture, and tourism). Carbonated zone urban development is particularly present in sedimentary basins (Paris and London basins and southern Germany) and large valleys (Rhône, Rhine, and Danube). In Europe, it is not a lack of financial means but a persistent lack of awareness that has predominated for decades after the postwar reconstruction. Environmental protection polices also vary greatly from one country to another.

Karsts in the former Communist Bloc countries are numerous: the Carpathians, south Silesia, Dobrudja, Crimea, the Caucasus, the Urals, Siberia, Pamir—Alai, etc.) (Jakucs, 1977). In eastern Europe, in particular, the karst areas have been and still are subject to much heavier pollution than in western European countries, because of the inefficiency of the former communist economy, upheld by a planned, centralized system. However, the fall of the communist system between 1989 and 1991 does not present a medium-term solution to the very serious water pollution problems, because, paradoxically, the relative vitality of these countries depended on the very weakness of their economies (Dembinski, 1988).

The karsts of the Mediterranean countries shoulder a heavy historical burden because of deforestation and overgrazing, those scourges dating back several millennia, which are still not under control, due to continuing forest fires and intensive livestock breeding (Demangeot, 1990)! These very calcareous regions are the cradles of several civilizations and have always been situated on invasion routes. The disappearence of a large portion of the soil explains why the term "karstic," when used by many naturalists, is often synonymous with poor, bare land. In addition, because of poor precipitation

---

FIGURE 5 Solid contamination examples in Cretan endokarst (Greece). (9) Fine charcoal particles in stalagmite (U/Th dated between 15,000 and 20,000 years)—Tripatzani Cave (−10 m) (sample TZ 247-20); natural light; lens, 2.5 × 3.2. (10) Charcoal laminae in parietal flow (U/TH dated)—cut off cavity, Nida polje edge (sample CI1/C-16, slide IDNI 208); natural light; lens, 2.5 × 3.2. (11) Stalagmite contaminated by soil debris, fine organic matter, turbid particles, and charcoal—cut off cavity, Anogia region (sample 241 C-18); natural light; lens, 2.5 × 3.2.

distribution, water resources, although potentially very great, are spatially limited. Thus many towns are sited close to karstic springs, such as Aix-en-Provence, Nîmes, and Montpellier in southern France.

In the karsts in humid tropical regions, the degree of anthropization depends not only on the karst evolution level (depressions, peaks, canyons, and poljes) but also on population history. For example, the densely populated southern Chinese karsts differ from those, almost uninhabited, of the "green hell" of New Guinea, where accessibility and population currents are not the same. The Chinese subtropical karsts, to the south of Yangtze are the most extensive in the world (Yuan Daoxian, 1991). Population density and ancient population history make this the most remarkable calcareous region in terms of anthropization history and its relation with the karstic environment. Crops are grown in terraces up to 1500 m in altitude, and the forest has almost disappeared. This agrarian development model impedes erosion, but very widespread deforestation over all substrates—the forest now covers no more than 12% of the territory—remains one of modern China's greatest environmental problems.

Tropical coastlines, atolls, and islands are vulnerable through contact with the sea and, sometimes, the proximity of highly populated zones (Guilcher, 1988; Bourrouilh-le-Jan, 1992). They belong to poor or developing countries (Haiti and the Philippines) as well as rich ones (Australia, Florida, the Bahamas, and the Marianas). Their associated ecosystems are coral reefs and mangrove swamps, both good indicators of environmental processes (Guilcher, 1988). The coral reefs, with a total area of 700,000 km^2, are very rich in different species, due to the age of the ecosystem, in place for millions of years in tropical regions (Salvat, 1990).

VI. CONCLUSION

Development problems and pollution imply an increasing emphasis on both pure and applied karstologic studies, with the aim of reconstructing environmental history and forecasting future processes, such as deforestation, soil erosion, pedologic crusts,water pollution, etc. Thus, the development of scientific speleology, particularly in the sequential study of subterranean in-fillings, provides a new approach to paleoclimatology. The role played by karsts in the evolution of human settlements has been understood for a long time, as excavations into shelters under rocks and in caves have shown. Close study of the relations among external in-fillings (depressions and poljes), soils, and underground infillings is beginning to produce results, leading to the reconstruction of the history of human influence on the natural environment. Karstology complements prehistoric, archeological, and historical studies and constitutes a new tool for historical geography.

New research initiatives are needed. For example, karst hydrology has been well quantified by observing system exits (emergences) (Mangin, 1975;

White, 1988; Ford and Williams, 1989), but the inputs are as yet insufficiently understood with respect to turbid particles and dust. The role of solid contamination in the modification of underground ecology also remains a vast field of investigation. In this context, global karstologic studies are likely to provide answers to the questions asked by planners. Such research is ecogeographical in nature and implies a multidisciplinary approach (Demangeot, 1990).

Finally, karsts and their aquifers, given their vital importance and vulnerability to pollution, represent test environments for groundwater management and protection policies. In this view, attention must be paid to karst areas in eastern and southern countries of the world, because well-balanced environmental management is highly desirable to avoid new economical and ecological catastrophes.

REFERENCES

Aus, H., and Kreysing, K. (1977). Les facteurs géologiques et al pratique des eaux résiduaires industrielles dans des couches profondes du sous-sol en R.F.A. *Actes Colloq. Nat. Prot. Eaux Souterraines Captées Aliment. Hum.* Orléans-la-Source, pp. 11–24.

Barbary, J. P., Maire, R., and Zhang, S., eds. (1991). "Karsts de Chine (Gebihe 89)," Karstol. Mém. 4.

Barbary, J. P., Maire, R., and Zhang, S., eds. (1994). "Karsts de Chine centrale (Donghe 92)," Karstol. Mém. 6.

Beck, B. F., ed. (1984). "Sinkholes: Their Geology: Engineering and Environmental Impact," Proc. 1st Multidiscip. Conf. Sinkholes, Orlando, FL. A.A. Balkema. Rotterdam-Boston.

Bögli, A. (1980). "Karst Hydrology and Physical Speleology." Springer-Verlag, Berlin and New York.

Bourrouilh-le-Jan, F. (1992). Evolution des karsts océaniens (karsts, bauxites et phosphates). *Karstologia* 19, 31–50.

Bullock, P., Fedoroff, N., Jongerius, A., Stoops, G., and Tursina, T. (1985). "Handbook for Thin Section Description." Waine Research Publications, London.

Castany, G. (1967). "Traité pratique des eaux souterraines." Dunod, Paris.

Castany, G., and Margat, J. (1977). "Dictionnaire français d'hydrogéologie." Bur. Rech. Géol. Min., Orléans.

Chamayou, J. (1979). Protection et diversification des ressources en eau potable du grand Angoulême. *Actes Colloq. Nat. Lyon: "Connaître le sous-sol: Un atout pour l'aménagement urbain."* pp. 439–444.

Collignon, B., and Jin, Y. (1991). Etudes hydrochimiques et bastériologiques. *Karstol. Mém.* 4, 118–131.

Corbel, J. (1957). Les karsts du NW de l'Europe. *Mém. Doc. Inst. Et. Rhodaniennes, Univ. Lyon* 12, 1–531.

Delannoy, J. J. (1985). Le karst: Un témoin des mutations socio-économiques dans la Sierra de Zongolica (Mexique). *Ann. Soc. Géol. Belg.* 108, 77–83.

Demangeot, J. (1990). "Les milieux naturels du globe," Collect. Géog., 3rd rev. édMasson, Paris.

Dembinski, P. H. (1988). "Les économies planifiées," Collect. Points (économie). Seuil, Paris.

Drogue, C. (1977). Propagation d'une solution vers un forage dans un aquifère fissuré et karstique. Rôle de la structure du milieu. *Actes Colloq. Nat. Prot. Eaux Souterraines Captées Aliment. Hum.*, Orléans-la-Source, pp. 151–162.

Dubois, P., Sorriaux, P., and Soudet, H. J. (1993). Rospo Mare (Adriatique): Un paléokarst pétrolier du domaine méditerranéen. *Karstologia* 21, 31–42.

Erhart, H. (1967). "La genèse des sols en tant que phénomène géologique," Collect. Evol. des sols. Masson, Paris.

Ford, D., and Williams, P. (1989). "Karst Geomorphology and Hydrology." Unwin Hyman, London.

Genty, D. (1993a). Mise en évidence d'alternances saisonnières dans la stratigraphie interne des stalagmites. Intérêt pour la reconstitution des paléoenvironnements continentaux. *C.R. Acad. Sci.* 317(Pt. II), 1229–1236.

Genty, D. (1993b). Les spéléothèmes du tunnel de Godarville (Belgique): Un exemple exceptionnel de concrétionnement moderne, intérêt pour l'étude de la cinétique de la précipitation de la calcite et de sa relation avec l'environnement. *Speleochronos, Fac. Polytech. Mons* 4, 3–29.

Gibert, J. (1986). Ecologie d'un système karstique jurassien; hydrogéologie, dérive animale, transits de matières, dynamique de la population *de Niphargus* (Crustacés Amphipode). *Mém. Biospéol.* 13(40), 1–380.

Gibert, J., Laurent, R., and Maire, R. (1983). Carte hydrogéomorphologique au 1/10000 du karst de Dorvan (Jura méridional, Ain, France). Présentation et principales données sur l'hydrogéologie et l'hydrochimie de ce karst. *Karstologia* 2, 33–40.

Ginet, R., and Decou, V. (1977). "Initiation à la biologie et à l'écologie souterraine." Delarge, Paris.

Goudie, A. (1990). "The Human Impact on the Natural Environment." Blackwell, Oxford.

Guilcher, A. (1988). "Coral Reef Geomorphology." Wiley, Chichester and New York.

Hill, C., and Forti, P. (1986). "Cave Minerals of the World." National Speleological Society, Huntsville.

Jakucs, L. (1977). "Morphogenetics of Karst Regions (Variants of Karst Evolution)." Hilger, Bristol.

Jennings, J. N. (1985). "Karst Geomorphology." Basil Blackwell, Oxford and New York.

Julian, M. (1992). Quelques réflexions théoriques sur le karst. *In* "Karsts et évolutions climatiques: Hommage à Jean Nicod" (J. N. Salomon and R. Maire, eds.), pp. 31–42. Presses Univ. de Bordeaux.

Lehmann, H. (1936). "Morphologische Studien auf Java," Geog Abh. 3. Stuttgart.

Maire, R. (1979). Comportement du karst vis-à-vis des substances polluantes. *Ann. Soc. Géol. Belg.* 102, 101–108.

Maire, R. (1984). Séismes néotectoniques et évolution des versants en Papouasie-Nouvelle Guinée. Effets des séismes sur les reliefs de forte énergie. *Méditerranée* 1/2, 57–66.

Maire, R. (1990). La Haute Montagne Calcaire (karsts, cavités, remplissages, Quaternaire, paléoclimats). *Karstol. Mém.* 3, 1–731.

Maire, R. (1991). Les remplissages karstiques. *Karstol. Mém.* 4, 132–149.

Maire, R., and Pomel, S. (1992). Les Levka Ori (Crète-Grèce): Un jalon miocéne dans l'évolution des karsts méditerranéens. *In* "Karsts et évolutions climatiques: Hommage à Jean Nicod" (J. N. Salomon and R. Maire, eds.), pp. 237–246. Presses Univ. de Bordeaux.

Maire, R., and Quinif, Y. (1988). Chronostratigraphie et évolution sédimentaire en milieu alpin dans la galerie Aranzadi (Gouffre de la Pierre Saint-Martin, Pyrénées, France). *Ann. Soc. Géol. Belg.* 111(1), 61–77.

Maire, R., Zhang, S., and Song, S. (1991). Genèse des karsts subtropicaux de Chine du Sud (Guizhou, Sicuan, Hubei). *Karstol. Mém.* 4, 162–186.

Mangin, A. (1974). Contribution à l'étude hydrodynamique des aquifères karstiques. *Ann. Spéléol.* 29(3), 283–332; (4), 495–601.

Mangin, A. (1975). Contribution à l'étude hydrodynamique des aquifères karstiques. *Ann. Speleol.* 30(1), 21–124.

Martin, J. (1981). Le Moyen-Atlas Central (Maroc). Etude géomorphologique. Thèse, Univ. Paris.

Nicod, J. (1967). Recherches morphologiques en basse-Provence calcaire. Thèse, Univ. Aix-en-Provence (Ophrys).

Nicolini, P. (1990). "Gitologie et exploration minière." Lavoisier, Paris.

Paijmans, K. (1976). "New Guinea Vegetation." CSIRO, Canberra.

Pomel, S., ed. (1987). "Volcans et karsts," Spéc. Trav. URA 903, XVI. CNRS, Aix-en-Provence.

Quinif, Y. (1989). La datation uranium-thorium. *Speleochronos, Fac. Polytech. Mons* 1, 3–22.

Quinif, Y. (1994). Les remplissages karstiques: Concepts et méthodologie. *In* "Enregistreurs et indicateurs de l'environnement en zone tropicale." Presses Univ. de Bordeaux.

Renault, P. (1967). Contribution à l'étude des actions mécaniques et sédimentologiques dans la spéléogenèse. *Ann. Spéléol.* 22(1), 5–21.

Renault, P. (1968). Contribution à l'étude des actions mécaniques et sédimentologiques dans la spéléogenèse. *Ann. Speleol.* 22(2), 209–267; 23(1), 259–308; (3), 529–593.

Reygrobellet, J. L., Mathieu, J., Laurent, R., Gibert, J., and Renault, P. (1975). Répartition du peuplement par rapport à la géomorphologie de la grotte de la cascade de Glandieu (Ain). *Spelunca Mém.* 8, 195–204.

Rodet, J. (1991). La craie et ses karsts. Thèse, Univ. Paris-Sorbonne.

Salomon, J. N., and Astruc, J. G. (1992). Exemple en zone tempérée d'un paléo-cryptokarst tropical exhumé: La cuvette du Sarladais (Périgord, Fr.). *In* "Karst et évolutions climatiques: Hommage à Jean Nicod" (J. N. Salomon and R. Maire, eds.), pp. 431–447. Presses Univ. de Bordeaux.

Salomon, J. N., and Maire, R., eds. (1992). "Karsts et évolutions climatiques: Hommage à Jean Nicod." Presses Univ. de Bordeaux.

Salvat, B. (1990). Menace et sauvegarde des espèces des récifs coralliens. *Cah. Outre-Mer* 172, 489–501.

Stiegele, P., and Klee, O. (1974). "Plus d'eau potable pour demain?" Laffont, Collection Réponses/Ecologie, Paris.

Sweeting, M. M. (1973). "Karst Landforms." Columbia Univ. Press, New York.

Trimmel, H. (1980). Möglichkeiten und probleme der karstlandschaftsnutzung und raumordnung-Insbesondere in Österreich. *Acta Symp. Int. Util. Aree Carsiche,* Trieste, pp. 125–133.

Trudgill, S. T. (1985). "Limestone Geomorphology." Longman, London.

Turquin, M. J., Bouvet, Y., Renault, P., and Pattée, E. (1974). Essai de corrélation entre la géomorphologie d'une cavité et la répartition spatiale de son peuplement actuel. *Actes Congr. Suisse Spéléol.,* 5th, Interlaken, pp. 46–60.

Verheye, W. (1989). Impact des propriétés hydrodynamiques du substrat karstique sur la nature du sol en milieu méditerranéen. *Karstologia* 14, 1–8.

White, W. B. (1988). "Geomorphology and Hydrology of Karst Terrains." Oxford Univ. Press, Oxford.

Yuan D. (1991). "Karst of China." Geological Publishing House, Beijing, China.

Geomorphology of Alluvial Groundwater Ecosystems

M. Creuzé des Châtelliers,* D. Poinsart,†
and J.-P. Bravard†

*Université Lyon I
U.R.A. CNRS 1451, "Ecologie des Eaux Douces et des Grands
Fleuves"
Laboratoire d'Hydrobiologie et Ecologie Souterraines
69622 Villeurbanne Cedex, France

†Université Lyon III
Département de Géographie, L.A. CNRS 260
69239 Lyon, France

I. INTRODUCTION

A river valley may be considered a trap for groundwaters insofar as the alluvial filling allows the storage of a large volume of water inside the interstices (Fig. 1). This groundwater network arises from a complex interaction of morphodynamic processes linked to the water cycle.

The interaction of surface water and groundwater in the alluvium provides a variety of habitats for many aquatic animals. Early studies, such as those of Vejdovsky (1822), suggested the regular occurrence of organisms in the groundwater of alluvial plains. Leruth (1938) probably first suspected the role played by this groundwater environment in diversifying the riverine biotope (Chappuis, 1942; Karaman, 1954; Pennak, 1939). Control of interstitial communities by underground fluxes of water was examined as far back as 1953 by Angelier while studying the coastal river Têt (Pyrénées Orientales, France). Groundwater movements through alluvial aquifers depend largely on the landforms and fluvial structure of the substrata (Freeze and Cherry, 1979; Thibodeaux and Boyle, 1987; White *et al.*, 1987). This suggests a high involvement of geomorphologic processes in the distribution and diversity of interstitial habitats. In the surface water area, the role of geomorphology as a prevailing factor in shaping the structure of biotic communities, either animal (Amoros and Bravard, 1985; Castella *et al.*, 1984; Huryn and Wallace, 1987; Hynes, 1970; Johnston and Naiman, 1987; White, 1990) or vegetal (Fortner and White, 1988; Hendricks and White, 1988; Lodge *et al.*, 1989; Petit and Schumacker, 1985; Swanson and Lienkaemper, 1982), has been recognized. Erosion and sedimentation processes influence not only the dynamics of river channel features, such as bars or pools, but also the exchanges between surface water and groundwater flowing through the substratum.

The purpose of this chapter is to summarize the geomorphologic context of the alluvial aquifer, which determined the structure of groundwater communities. Structure and lithography of alluvial deposits determine the hydrogeologic characteristics of aquifers; also we examine the alluvial filling, discontinuities, and geomorphologic typology of valleys. Then we study the complex sedimentary units of the floodplain, which reflect the chronosequences of valley geomorphology. Finally we envisage human activity affecting river dynamics and alluvial aquifers.

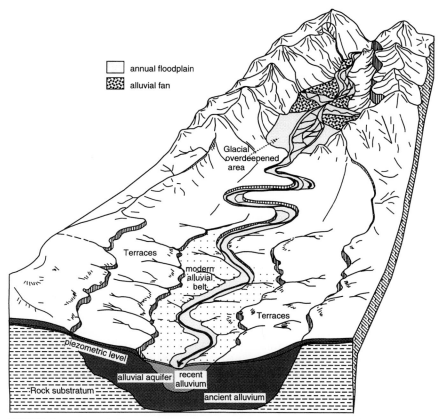

FIGURE 1 Types of alluviation in mountain valleys and plains [rearranged after shetches from Peiry (1988, Figs. 1 and 2, p. 8) and from U.S. Geological Survey (1987, Fig. 48, p. 73)].

II. ALLUVIAL FILLING

The composition of quaternary layers is the result of various processes that can succeed one another or coexist in a same valley. This makes it possible to define aquifers of different permeability. For an interstitial dweller, the age of the formation is unimportant but the structure of the alluvium allowing the circulation of water is certainly a vital factor for the animals (Henry, 1976).

In the regions of the world influenced by the Quaternary glaciations, thick layers of glacial sediments were often deposited. Glacial tills are highly heterogeneous, ranging from glacial clay, which forms a very fine matrix, to stones and blocks of different sizes, sometimes huge. These characteristics and the frequently occurring consolidation of the ancient deposits often limit their permeability (from 10^{-4} to 10^{-5} m/sec^{-1}) and also limit the possibility for a diverse animal assemblages or large populations. In the lower areas of glacial influence, especially in piedmont plains, glacial tills are poor in fine materials; they have been washed away by the proglacial flow. These deposits present highly favorable hydrologic conditions for the settlement of fauna. For example, in the southeastern part of New York, broad areas were covered by glaciers of the Wisconsin age. In this area, the large grain size and the heterogeneous sizes of the deposits cause high porosity and the interstitial habitats host a rich and diversified fauna (Strayer, 1988). Stocker and Williams (1972) give a porosity value of 20 to 35% for similar glacial alluvium of the Speed River (Ontario). These high values can explain the great density of organisms founded by Coleman and Hynes (1970) inside the interstitial water of this river.

In the area of the glacial overdeepening, the deposits of glaciolacustrine material, formed by clays and silts, can be thick but give only weakly active aquifers. In the Rhône Valley (France), such accumulations can reach a thickness of 90 m and constitute aquifers with very low permeability, when compared with the overlaying fluvial deposits of the Holocene.

The glaciofluvial deposits are materials eroded by melting water and shaped by fluvial transport; the coarse load, washed of most fine particles, allows circulation of the interstitial water. They are aquifers of the alluvial piedmont plain, which have high porosity and permeability. This kind of deposit forms the major aquifers of river valleys in Switzerland (Huggenber and Zobrist, 1990). The aquifer of the Ouche river, a tributary of the Saône river (France), is formed by the gravel deposits of Würmian age (Wisconsin); washing of the fine sediments opened large interstices, allowing the development of large populations of stygobite organisms (Henry, 1976; Magniez, 1976). In the same way, the Würmian glaciofluvial corridors near Lyon (France) are characterized by steep slopes, close to 3‰, with high values of permeability (from 4×10^{-3} to 7×10^{-2} m/sec^{-1}). The thickness of these highly dynamic aquifers is up to 10 m, and they directly influence the interstitial communities of the underflow of the Rhône River. In this system, the vertical exchanges between surface water and groundwater are strong. In the Flathead River (Montana), where the Pleistocene glaciation also determined the geomorphology of the valley, a layer of coarse fluvioglacial alluvium overlies an impermeable clay formation. This aquifer system is connected to the river, and large stone fly larvae (*Isocapnia* and *Paraperla*) together with amphipods (*Stygobromus*), which are strictly limited to the groundwater environment, are commonly found as far down as 10 meters deep and up to 3 km from the river channel (Stanford and Ward, 1988).

Fluvial deposits constitute the principal part of the alluvial filling and form aquifers, which are most favorable for the development of abundant groundwater communities. Generally, gravels and sands dominate in these alluvia, arranged by a horizontal and vertical grain-size gradation. Orghidan (1959) used the term "hyporheic biotope" for the interstitial habitat formed by the upper layer of alluvium, which is in "contact" with the running water. This biotope may be considered a zone in which two distinct aquatic systems (epigean and hypogean) mix and that is characterized by the transfer of energy (*e.g.*, temperature, oxygen, and carbon) and organic matter (dead or live).

The nature and size of alluvial deposits vary considerably in space, according to the source of the sediments, the lithology of the watershed, and the transport capacity of the flow. Hence, most alluvial rivers have substrata with a highly variable grain-size distribution (Fig. 2). The porosity and the permeability may also be considered a major factor in the activity of organisms. Schwoerbel [1961; cited in Hynes (1970)] has shown that the interstitial fauna is richest within deposits that originate from disaggregation of metamorphic rocks, characterized by large open spaces between the grains. Alluvial deposits located downstream of soft sedimentary rocks, such as limestones, offer less favorable conditions because of the clogging of interstices by the fine materials of the clay fraction. Indeed, clay layers are not easily penetrable by organisms and provide a more homogeneous milieu, usually without contrasting microforms. Sandy layers, very porous but with smaller interstitial voids, allow only the development of fauna consisting of minute organisms. However, many organisms are able to burrow inside very fine loose sediments, which locally fill the interstices, so the interstratified beds of silt and sand do not always hinder the migration of animals, at least the youngest ones.

III. ALLUVIAL DISCONTINUITIES

Grain-size distribution varies in space, particularly in an upstream–downstream direction in relation to the flow strength, and in time, linked

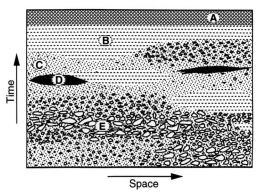

FIGURE 2   Vertical and lateral heterogeneity of alluvium deposits. Stratified units: A, surficial silts; B, fine sands and silts (low permeability); C, medium and fine sands (permeable); D, lenses of clay (impermeable); E, gravels and cobbles (very permeable) [after Castany (1985, Fig. 7, p. 8)]

to the climatic modifications that influence the hydrology, the sediment load, and consequently the morphodynamic behavior of the fluvial system. The response of the stream to the hydroclimatic changes is complex because the watershed acts as a mediator: it can be more or less sloping, be more or less wooded, and have a lithology that is more or less prone to erosion. For that reason, successive climatic phases and, consequently, the variability of the fluvial pattern lead to the heterogeneity of alluviation, which causes the morphodynamics of landscapes in alluvial valleys (Stanford and Ward, 1993).

A. Longitudinal Discontinuities along the Valley

The pattern of the alluvial filling may be modified along a river that passes from one plain to another. The tributaries play a great role in this spatial heterogeneity. They may inject a coarse solid load, inducing an aggradation of the downstream reach; if the lithology of the watershed of the tributary is different from that of the main channel, the dual origin of the deposit can be easily demonstrated. A tributary can also bring an abundant solid load of variable grain size. Downstream of such a confluence, the alluvial filling should be more disturbed in its composition that in its volume. This is the case downstream of the Isère–Rhône confluence (France), where the alluvial filling of the Rhône valley loses its permeability (Henry, 1963): the alluvia of the Isère river, rich in silt and clay, of a glacial origin, clog the interstitial voids of the sandy and gravelly alluvium of the Rhône river.

B. Vertical Discontinuities by Fossilization of Geomorphologic Units

An active alluvial filling is generally characterized by progressive aggradation. This is controlled either from upstream, through, for example, a strong supply of materials, or from downstream, through, for example, eustatic fluctuations or subsidence movements.

The aggradation can fossilize various forms [e.g., "paleopotamons," in Roux, (1982); "fossile channels," in Bravard *et al.* (1986); or "paleochannels," in Stanford and Ward (1993)]: bars, raised banks, abandoned arms, silted depressions, and anastomosing or braiding meanders (Petts and Foster, 1985; Reineck and Singh, 1980; Smith, 1981). On another scale, it can fossilize alluvial fans or terraces (Allen, 1965; Schumm, 1977). According to fluvial patterns, these different units are present or absent (Fig. 3).

The fossilized alluvial fans can be of very large dimensions. Several alluvial fans can be superimposed at different stages of the filling (aggradation); these alluvial fans can also overlap great areas of the plain and push away the main channel, with its own filling occupying a large part of the valley (Fig. 4A); Schumm (1977, p.182) gives the example of the upper

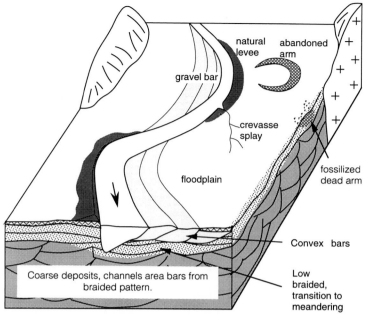

FIGURE 3   Active and buried unit landforms of an alluvial plain.

Mississippi, between St. Louis and St. Paul. In some extreme cases, torrential tributaries can transport considerable quantities of a solid load very progressively, to form a natural dam across the valley and thus create a lake [for an example, see the case of the Piceance Basin; U.S. Geological Survey (1987)]. In the Rhône valley the alluvial fans originate from the active Holocene stages, when the Durance river and, to a lesser extent, the Ardèche river were much bigger than today. All along the stream, these particular areas act as node points, at which groundwater flowing inside the alluvial fan comes into contact with the groundwater of the plain. The arrangement of two diversely stratified formations, one belonging to the fluvial fan and the other belonging to the alluvial plain, may induce many local irregularities that will influence the global permeability of the field. These irregularities create hydraulic discontinuities that may divide the more permeable layers.

Another origin of vertical discontinuities is associated with the changes of the alluvial pattern in time. At a millenary scale, the geometry of the cross-sectional profile of the valley and the grain size of the sediments can be modified when, for example, the fluvial pattern changes from a straight stream with a gravelly solid load to a sinuous stream, where the transported material consists mainly of fine particles in suspension (Schumm, 1963) (Fig. 5).

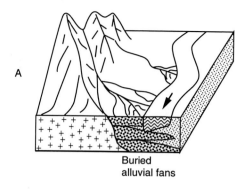

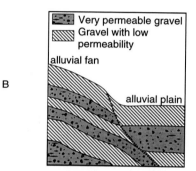

FIGURE 4   (A) The alluvial fan pushes away the channel, which carries away part of the deposits [after Allen (1965, Fig. 36, p.165)]. (B) Cross-section of the alluvial fan. Considering similar permeabilities, the velocity of subterraneous flow depends on the sloping of the strate.

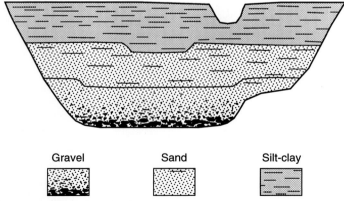

FIGURE 5   Cross-section of a valley fill showing the change of a channel cross-section as the sediment load becomes progressively finer and as the straight bed-load channel at the base of the fill adjusts to become a sinuous suspended load at the top of the fill [from Schumm and Brakenridge (1987, Fig. 5, p. 225), with permission from the Geological Society of America, Colorado].

C. Discontinuities Linked to the Structure of Deposits

The inclination of the substratum varies according to the conditions of deposition. Generally, sudden accumulations linked to the breaking of loads, associated with high flows, induce strong inclinations. This is the case of alluvial fans (Fig. 4B), fluvioglacial transition fans, dissipating areas, and scroll bars. Braided channels are characterized by a crisscross bedding, which represents figures of scour and deposit taking place during the filling. The variability of the flow strength of running waters contributes strongly to the formation of inner discontinuities, and the aquifers associated with the rivers show a high vertical heterogeneity. The intercalations of sandy beds or clayey beds between gravelly layers induce discontinuities that can influence the distribution of organisms. A good example is given by the distribution of the isopod *Proasellus walteri,* which resides inside the stratified aquifer located between two tributaries of the Saône river (France), the Tille and the Norge (Henry, 1976). The recent Quaternary deposits contain several groundwater horizons inside beds characterized by different grain sizes. The upper sheet runs freely inside a sandy layer 1 to 3 m thick; it is here that the hypogean isopods are most abundant, in association with *Bathynella,* minute phreatic water crustaceans. The lack of large-sized predators, which are limited in their movements by the very small size of the voids, may be a favorable factor in the development of these communities. In the layers containing less sand, a larger isopod, *Proasellus strouhali puteanus,* is present (Henry, 1976).

Only rarely do alluvial deposits have a subhorizontal structure, a typical feature of sedimentation basins. Usually the layers are irregular because they have been deposited as ripples on the bottom surface. In the alluvial floodplain, the structure of these deposits is also influenced by the selective action of the vegetation. But in all cases, the inclination of the sedimentary units is an important element in the permeability of the deposits and the groundwater flow velocity.

IV. GEOMORPHOLOGIC TYPOLOGY OF ALLUVIAL VALLEYS

The nature and the structure of the deposits depend on their relation to some large geomorphologic units. The position in an upstream–downstream gradient within the watershed and the interference of external factors, such as tectonics, eustatic movements, or hydroclimatic fluctuations, determine the basic typology.

A. High-Altitude Valleys with Limited Storage and Discontinuous Processes

Low stream orders are generally characterized by steep slopes, rocky stream bottoms, and proximity to the valley sides. Such valleys are better fitted to alluvial transit that to alluvial storage. Sedimentary deposits are, therefore, ephemeral and never very thick. Inputs of inorganic sediments are derived from side processes that are discontinuous in space and time. Examples include landslides, slumps, mud flows, torrential lava, or ravining. They are sources of materials with specific characteristics, especially a weak degree of smoothing and a broad grain-size distribution.

On valley walls formed of impervious formations with a strong slope, groundwaters can form a particular biotope, named the "hypothelminorheic milieu" by Mestrov (1962). Rouch (1968) underlines the fact that this interstitial environment is inhabited by abundant hypogean communities. Among the species found here, some are ubiquitous, widely distributed in all groundwater types, whereas other species, such as *Antrocamptus chappuisi, A. longifurcatus,* and *Elaphoidella pyrenaica,* seem confined to this particular biotope. For the groundwater flows of the valley side many authors have reported diversified assemblages. To define hydraulic connections between this shallow subsurface runoff and the subsurface environment it is necessary to have an global analysis of the whole system (Rouch, 1968).

The aggradational shapes that characterize the valley bottoms of fast running streams have been studied extensively in the northwest coastal chain of North America, especially the western slopes of the Cascade Chain (Oregon). Perched steps consist of boulder berms 1 to 2 m wide, which were deposited by the debris flows inside the reach where the valley broadens and the slope diminishes (4%). In this case, it is the coexistence and the alternation of fluvial and torrential processes that explain the characteristics of the bed and of the alluvial filling. Within a few years, the river again forms its basic geomorphologic units (pools and riffles) that had been disturbed by torrential lava (debris flow), but the alluvial flow remains influenced by these processes. The interstitial milieu of a small stream that runs in this type of mountain valley may sometimes harbor a dense fauna. In a brook in the French Pyrénées, contained by abrupt banks and flowing over metamorphic calcite, Rouch (1968, 1988) found large communities of Crustacea; the groundwater flows inside very heterogeneous deposits consisting of several layers of different permeability. The movement of the animals in this environment is possible thanks to the large size of the interstitial voids. Magniez (1976) points out that large individuals of the isopod *Stenasellus virei* can live only inside the coarsest aquifer beds, but young individuals, of smaller size, can easily migrate within the more sandy areas. So, it appears that the morphodynamic processes that occur in the stream valley play an important role in the diversity of underground habitats.

B. Gorges and Canyons

The fluvial bed of canyons, and more generally of all V-shaped valleys, is an original shape adapted to side processes. In deep gorges, the surface water may flow in direct contact with the bedrock. Nevertheless, Henry (1976) emphasizes that interstitial communities can inhabit gravelly bars. In those areas where the stream is encased inside the metamorphic rocks, contact with the groundwater environment is very low. Delamare-Deboutteville and Paulian (1953) emphasized that interstitial life is possible only within the continuous structure, which is both porous and pervious. Nevertheless, the hydraulic continuity between distinct systems does not necessarily imply a biological continuity between the assemblages. In 1968, Bou made such an observation in a study of the aquifer associated with the Albi River (France). In the contact areas between the karstic and the phreatic systems, the fissures of the encasing massif created a hydraulic continuity, which could connect the hypogean communities, but no mixing of populations was observed (Rouch, 1968). Henry (1976) and Magniez (1976) have described similar biological discontinuities: the karstic fauna, which consists of *Stenasellus virei hussoni* or *Proasellus cavaticus*, may be living close to the phreatic fauna, consisting of *Stenasellus virei* or *Proasellus valdensis*, without any real mixing of the two communities.

In the bed of a stream encased in bedrock canyons, the riffles consist of a large colluvium from the valley sides. In the channel canyon of the Colorado River, the materials of most riffles originate from the period of high postglacial discharges and are not set in movement by the present discharges; this makes them very stable (Graf, 1979). In the Ardèche canyon (France), in the Massif Central plateau, the riffles consist of angular blocks of metric size, originating from landslides, and gravels 20 to 30 cm in diameter. These gravels are derived from paleoflow, which filled the valley bottom with a 20-m-thick sediment layer during the Würm. The variability of the grain-size distribution of riffles and their strong break-off slope produce high permeability of bed sediments. In this type of milieu, permeability may be high enough for the deep migration of epigean organisms to take place. In Catalonia (Spain) the river Ter has a floor consisting of slates and granitic boulders; the interstitial milieu is highly pervious with well-oxygenated water (Sabater, 1987). The conjunction of aerated water and a heterogeneous deposit appears optimal for the development of interstitial communities.

C. Mountain Valleys with Glacial Overdeepening and Complex Upbuilding

Quaternary outwash processes often explain complex sedimentation processes that vary spatially along valley trains in association with deglaciation. The Isère valley, upstream of the town of Grenoble (France), between the crystalline massif of Belledonne and the calcareous massif of the Chartreuse, is a good example of such changes (Fig. 6). Released from the ice of the last glacial period, the valley was filled by several hundred meters of

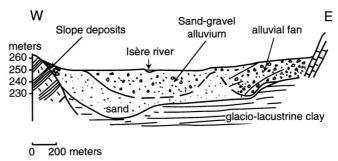

FIGURE 6   The sedimentary deposits of Grésivaudan Valley, France [after Dubus and Fourneaux (1968, Fig.4, p. 509)].

lacustrine clay and then by 20 to 30 m of fluvial sand; the sand was progressively replaced by gravels and pebbles, with a simple sandy matrix from the
torrential Holocene river, the Isère. On the edges of the valley, the coarse
material of the alluvial fan penetrates inside the fluvial deposits and forms
preferential pathways for the groundwater flowing from the valley margins.

In alpine valleys (France) with high relief and steep slopes, alluvial filling
has taken place within a narrow thalweg, at the same time as rock slides
and debris flows from the valley sides. These highly permeable formations
are overlaid by the alluvia consisting of the mud and clogging waters from
torrential rivers. Hence, the deep aquifer flows actively and is recharged
by the river but is more or less independent of surface fluctuations. The
connections between surface water and groundwater are generally restricted
to springs emerging through cracks in the clay substrata or artesian wells.

D. The Filling of Subsident Alluvial Plains

Subsident faults, active during the Quaternary period, accumulate alluvia in characteristic ways. For example, the Little Plain of Hungary is part
of the pannonian rift and has a sediment layer over 200 m thick, which
formed since the Günz period (300,000 BP), and a curved base of the pebbly
deposits deposited since Riss glaciation (100,000 BP) and particularly since
the beginning of the Holocene period (10,000 BP). It demonstrates the active
character of the subsidence in the area, where the Danube builds up an
extensive alluvial fan (Fig. 7). The sandy and gravelly nature of this filling,
close to alpine sedimentary sources, yields highly permeable alluvium deposits and accounts for the activation of groundwater circulation. The rift of
the Little Plain trapped most of the gravel load issuing from mountains,
and the Danube river of the Holocene period has a relatively fine solid
load downstream of the Transdanubial mountains: a pebbly floor of the
Pleistocene is overlain by a layer consisting mainly of a sandy matrix, which
reduces permeability in the downstream direction. It is for the aquifer of
the depression of Skoplje, a part of the Danube plain (Romania), that Motas

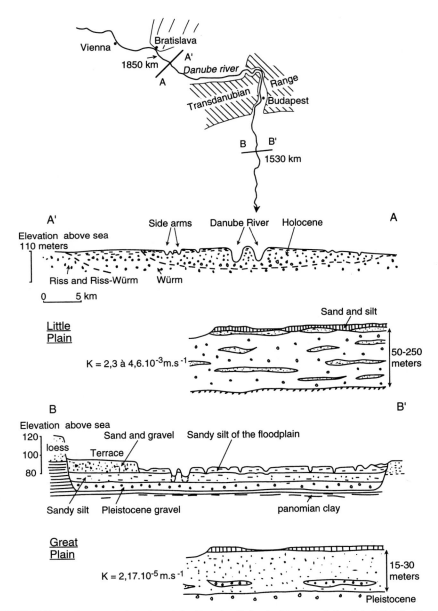

FIGURE 7 Sedimentary deposits of the Danube Plains in Hungary [after Pécsi (1971, Fig. 2, p. 23, and Fig. 7 p. 26), with kind permission from Pergamon Press, Oxford].

(1962) described a true mine of new organisms. In Austria, near Vienna, the alluvial aquifer is formed by recent deposits, consisting of gravel, sand, and silty clay and harboring abundant and diversified interstitial fauna (Danielopol, 1976, 1989).

In the Hercynian massif of Central Europe, the rift of the Rhine River is another subsidence area dating back to the beginning of the Secondary era. Stretching for over 300 km between Basel and Mannheim, the graben is 35 to 40 km wide. The longitudinal profile of the Rhine river seems to be affected by this tectonic feature and its slope declines gradually from 1‰ upstream to 0.5‰ downstream. It seems that the tectonics induced a strong accumulation of gravels and sand, between 10 and 250 m thick. This alluvial filling formed the Rhine aquifer, the most important alluvial aquifer system in Europe (Simler *et al.*, 1979). Phreatobiological studies of wells have shown abundant and diversified fauna (Hertzog, 1936, 1938; Moniez, 1889). In contrast, studies show that portions of the aquifer influenced by polluted water of the Rhine River are very impoverished; the fauna is restricted to the sites recharged by unpolluted groundwater from the infiltration of the Ill River, a French tributary of the Rhine (Creuzé des Châtelliers *et al.*, 1992).

It is particularly difficult for streams to maintain their longitudinal profiles in these subsident areas. A practical example is given by the Saône river, a tributary of the Rhône river (France). Since the Würm glaciation, the main channel has experienced tilting of the long profile due to the subsidence of the plain. Now, the slope, which is close to 0.1‰, does not allow the transport of coarse material, and so this old bottom-bed load is quasiimmobile at present. So, the feeble geomorphologic dynamics of those sectors certainly do not allow the developments of large interstitial biodiversty, but there is here a genuine lack of data.

E. Fluvial and Fluvioglacial Terrace Valleys

Terraces are created when a stream flows through a fluvial or fluvioglacial alluvial fill (including the alluvial fan) (Harvey and Renwick, 1987; Schumm *et al.*, 1984, p. 20; Striedter, 1988). They may appear at the surface or they can be fossilized (through subsequent in-filling); sometimes several generations of terraces are superimposed (Fig. 8). This kind of structure leads to discontinuities in the alluvia, which may then influence the way groundwater flows. Last, terraces can be inset or stepped (Fig. 9); in the latter case, if they correspond to the topography of the surface, then the aquifers are not connected (as in the example of Garonne, France; Bodelle and Margat, 1980). Between Basel and Marckolsheim, the flood plain of the Rhine is inset in the low Würmian terrace, which slopes down from 260 to 180 m; bordered by a 13- to 15-m-high terrace near Mulhouse, it slopes gradually northward under the Holocene alluvia, which fossilise it at Strasbourg. A diagram of the alluvial infill of the valley of the Main and the Regnitz shows how the Late Glacial and Holocene deposits are inset in the groundwater of the Würmian Pleniglacial (Fig. 10); the upper and middle alluvial terraces comprise seven distinct sedimentary sequences, dated by dendrochronology from subfossil trunks and whose surface has subsided

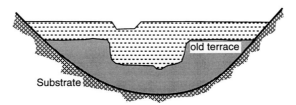

Two alluvial fills; old terraces are buried under recent deposits.

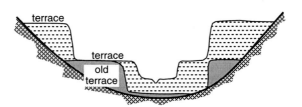

Two successive alluvial fills. Old terraces are buried;
two levels of terraces are shaped into recent deposits.

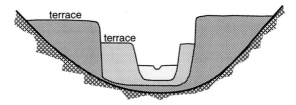

Three successive alluvial fills. Two generations of inset terraces.

FIGURE 8   Heterogeneity of the sedimentary fill, related to successive erosion and accumulation phases [after Schumm (1977, Figs. 6–22, p. 211)]. Groundwaters, totally or to some extent, occupy the alluvial fill.

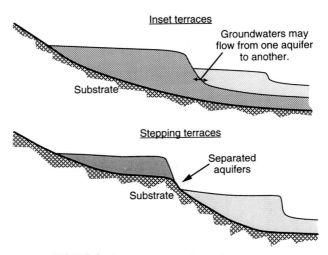

FIGURE 9   Inset terraces and stepping terraces.

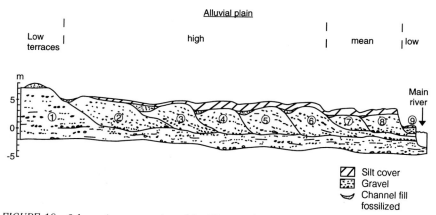

FIGURE 10   Schematic cross-section of the Würm and Holocene fill of the Main and Regnitz Valleys [after Schirmer (1988, Fig. 1, p. 154)].

by a maximum of 2 m. These sandy–pebbly sequences came into being during centennial phases of strong morphodynamic activity separated by periods of rest (Schirmer, 1988). It is probable that such complex activity led to the occurrence of discontinuities within each sequence (grain size variability peculiar to the juxtaposition of synchronous geomorphologic units; material from the bed of the channel and banks) or between discontinuities differentiated by their age and their formation. In the Tobacco Rivers (United States), Stanford and Gaufin (1974) and Stanford and Ward (1988) found many larvae of stone flies (Plecoptera) down to a depth of 4 m.

F. The Valleys of Plains and Plateaus with Holocene Infills and Transformations

In the large valleys situated downstream from the mountain watersheds (in young mountains or ancient massifs) the fluctuations in the Holocene climate resulted in variations of the hydric and mineral fluxes passing through the alluvial plains. The streams reacted by adjusting their geometry, and this led to complex metamorphoses or changes in the channel patterns (from braids to meanders and vice versa) and to variations in the altitude and gradient of the profile, with alternating phases of degradation and aggradation. Thus the alluvial infills are internally highly complex, despite the apparent homogeneity of the floodplains.

In the regions of the plains and plateaus of northwest Europe, the valleys affected by the periglacial processes during the cold periods of the Quaternary underwent powerful phases of infilling with coarse material. The Seine, for example, whose basin was not directly influenced by the glaciary processes, flows over an infill composed of silex pebbles, sand, and carbonated grains from the rock fragments that came down from the low sedimentary plateaus of the Paris basin; in the Seine valley this infill is several kilometers wide. The discharges of the Seine and its tributaries were considerably reduced during the Holocene, and they were not powerful enough to rework the material already deposited.

V. THE SPATIAL COMPLEXITY OF MODERN-DAY FLUVIAL DEPOSITS

We have seen above that the structure and lithology of alluvial deposits determine the hydrogeologic characteristics of the aquifers and their connections with the surface hydrographical network. Fluvial dynamics can create large hydraulic and sedimentary discontinuities within the upper layers of the infill and thereby determine the structure of underground communities. We now distinguish between transport and deposition processes, which act on the active tract of the river bed and are responsible for a repeating pattern of cut and fill alluviation, and on the other hand the complex sedimentary units of the alluvial flood plain, which reflect the chronosequence of valley geomorphology.

A. The Channel Pattern Is the Geomorphologic Expression of Processes on the Scale of the Decade or the Century

Water and sediments in transit are independent variables of the first order that determine the morphology of the channel (Schumm and Brakenridge, 1987, p. 223). The geometry of the channel is controlled by first-order variables such as grain size, channel load, and the gradient of the floodplain, and the channel pattern adjusts to the temporal variations in water and sediment flux (Schumm and Brakenridge, 1987, p. 227; Starkel, 1983, p. 220). Adjustment to these variables is limited by the degree of stability of the geometry of the channel (*e.g.,* bedrock control or simply the cohesion of bank sediments or their size, which may be due to paleoprocesses). However, in alluvial plains, the channel pattern usually adjusts to variables of the first order.

Apart from the particular case of a straight channel, which generally appears over short distances, or through human influence, the three main types of river pattern are braided, meandering, and anastomosed. Any one of these morphodynamic types will affect the connections among a stream, the associated alluvial aquifers, and the dynamics of the biocenoses.

1. Braided Patterns

Braiding is characteristic of streams with a high gradient and high energy levels, which have an abundant gravel load and a high transport capacity that nevertheless fails to evacuate the bulk of the alluvia. That is why braiding is generally accompanied by a raised alluvial level and a large number of weakly sinuous channels that become very unstable during flooding. For most of the year, surface water seeps into the alluvia and feeds groundwater. The substrata of braided streams are usually very porous and permeable; the finest particles are washed either downstream by declining discharge or in the depths, through seepage. However, there are intermediary forms of braided patterns, in which a main channel is just beginning to form. In such instances, the coarse load and the energy of the water tend to decrease and the sandy fraction can thus be deposited in abundance. As the stream bed becomes clogged to a greater or lesser degree, the biological diversity of the interstitial subfluvial environment is reduced, as vertical exchange between the river and the deep groundwater environments becomes limited or clogged. In European mountains today sources of coarse sediments tend to decrease (Bravard, 1991; Peiry, 1988), and this reduction may lead to this kind of phenomenon. For example reaches of the Rhône used to be braided (*e.g.*, on the Brégnier–Brangues site); the bed is degraded, whereas at the same time sand is deposited either in long stretches or in the alluvial interstices. Hence, very little interstitial fauna exists.

The lithology of the watershed plays a considerable role in that it determines the kind of load in transit. The braiding of the tributaries of the Rhône that flow from the crystalline Massif Central deposits bars of pebbles that have been brought down from the coarse periglacial structures; their sandy matrix comes from the breaking down of granites. The Drôme, on the other hand, flowing as it does from the sedimentary prealps made up of limestone and Jurassic marls, forms bars in which coarse alluvia are embedded in a silty–clayey matrix deposited during falling discharge, by which the first few decimeters are infiltrated, thus decreasing the permeability of the deposits.

2. Sinuosities and Meandering

When there is a deficit in the bed load, the energy of the stream can be dissipated by a lengthening of the river's course. The force of the flowing water is stronger than the resistance of the material emerging at the banks, and sinuosities develop through lateral erosion. The material wrenched from the concave banks and deepening pools is mainly deposited on the convex side of the channel. Hence, bars grow at their apex and downstream, whereas the concave banks recede. Classically, the grain size and porosity of alluvia decrease downstream. The internal edge of the point bar is slowly aggraded by the deposition of fine sediments, which form a natural levee.

3. Anastomosing Channels

As in the braided pattern, the anastomosed river have several channels. That is why these two geomorphologic models are so often confused. In contrast to braided streams, the anastomosed pattern characterizes low-energy streams, which transport sandy–silty alluvia. Anastomosed channels may also form on the edges of braided channels that are actively aggrading.

The weak transport capacity of the water can be seen in the sinuous shape of the channels, which are relatively stable thanks to the cohesion of the fine material at the banks and the development of the riparian zone. This pattern is generally associated with marshy zones that form on the edges of the plain. During flooding, the seepage of surface waters is very slow and its volume diminishes, because the marshy zones show that the piezometric level is very close to the surface. Smith (1983) remarks that an anastomosed river system can be distinguished from braiding or meandering river systems by the presence of wetlands.

The very nature of the deposits in anastomosed systems is probably unfavorable to the existence of powerful underground fluxes and a richly populated interstitial environment. A good example has been described in the Rocky Mountains of Colorado, where the South Platte River has a well-developed aquifer system. The river rises 3000 m above sea level and flows down into the great plains to the east, where it follows an anastomosing pattern. Sand and gravel make up 90% of the deposits that form an aquifer with a high level of hydraulic conductivity (Ward and Voelz, 1990). In this system, the interstitial fauna appears to develop best in the boulders, pebbles, and gravel that form the deposits of the upper part of the valley and in the more dynamic confluence zones (Ward and Voelz, 1990). Hydraulic conditions and the properties of the substratum, which depend on the geomorphological processes found in the valley, are key factors in the way the underground biocenoses are organized.

B. Floods Are Responsible for Short-Term Adjustments

Bed forms are shaped by high-volume flows. We must distinguish between exceptional floods and ordinary floods. The former are likely to modify the channel, which then gradually returns to its equilibrium. If such exceptional floods take place repeatedly, this is a sign of an imbalance in the system, and the geometry of the channel and even the river pattern may be affected or permanently modified. Small and medium-sized floods, with an annual recurrence (Q1) to quinquennial recurrence (Q5), reshape the elementary forms that already exist within the flow channel; these are the only ones to be mentioned here. In particular, a bank-full discharge (Q1.2 to Q1.5 flood recurrence) is the discharge that maintains the bed forms.

Bed-load transport is the fundamental element that shapes the bed of the stream. Alluvia transported in suspension are usually deposited along the edges of the channel and reinforce the patterns of the floodplain, which are much more stable than those of the channel.

1. The Different Patterns That Develop in the Channel

A stream with a sand or gravel bed, whatever its pattern, is made up of three basic elements: (1) pools, (2) riffles, and (3) lateral or median bars that are formed by erosion, transport, and deposit of bed load during floods

(Keller and Melhorn, 1978). Pools are degradation forms shaped by the convergence of the fluxes in the concavity, whereas riffles are formed by the deposition of the coarser material or by slight erosion in those reaches of the channel where the fluxes are divergent and therefore have a lower transport capacity. In the case of straight, sinuous, or meandering channels, these shapes are found at regular intervals. If deposited material is abundant, the elements are not sorted well in the channel, and as a result most of the shapes migrate downstream; this is characteristic of braided systems. If there is less material, the coarser elements form riffles, generally more stable shapes that contribute to the relative stability of the bars that are situated laterally in sinuous streams with one channel only (Church and Jones, 1982).

In gravelly streams with contrasted regimes, the riffles may reach the surface of the water when the flow is at its lowest. The longitudinal profile of a thalweg is thus made up of a sequence of partial slopes that are more or less steep and that form between the upper and the lower points of a bed. Each riffle creates a break in the slope, and at low-water levels this leads to a break in the waterline; the water seeps down, upstream from the break in the slope, and reemerges downstream (Fig. 11). At the surface, current velocity increases at the riffle. The water infiltrating upstream of the riffles is rougher and better oxygenated than that at other points. The coarse grain size means that water circulates more easily in the interstices, which help to maintain the porosity of the environment by leaching finer particles. In any stream, the presence of riffles encourages exchanges between the surface and the subterranean environments and increases the heterogeneity of the interstitial environment. In Salem Creek near Elmira (Ontario, Canada), Godbout and Hynes (1982) found that the structure of the interstitial communities is directly linked to the hydrogeologic characteristics induced by the break in the slope. This phenomenon has been noted by other authors (Bretschko, 1983; Marmonier and Dole, 1986; Marmonier, 1988) for riffles or gravel bars at which the surface water tends to infiltrate upstream from the break and then flow underground to reemerge downstream. As it flows through the substrata the physicochemical characteristics of the water are

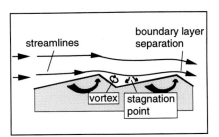

FIGURE 11   Influence of gradient breaks along the channel [after Thibodeaux and Boyle (1987), reprinted with permission from Macmillan Magazines].

modified, and it gradually loses its "surface signature." The oxygenation of interstitial water thus tends to fluctuate in the infiltration zones upstream of the break in the slope, whereas there is usually considerable deoxygenation in the zones downstream. These conditions, brought about by the morphodynamics of the channel, shape the structure of the interstitial communities. The work of Rouch (1968, 1988) on the Lachein stream in the French Pyrénées is one of the most thorough studies of this subject. The dynamics of surface flows have a direct influence on the zones of deposition and deepening (riffles, pools, and bars), each with its specific porosity and grain size, and determine the distribution of oxygenated and anaerobic zones below the channel. This spatial heterogeneity of the environment leads to considerable diversity in the distribution of interstitial organisms (Rouch, 1988).

2. Shapes That Develop on the Borders of the Channel and the Banks, at the Edge of the Floodplain

Several types of deposit are active in the floodplain when overspill occurs. Low zones retain water and decantation deposits occur. Elsewhere on the surface of the plain, even a very weak flow can carry with it the finest particles. Last, the combing effect of vegetation on the water may lead in places to a break in the load and the haphazard deposition of suspended particles of all sizes.

In certain units sedimentation is complex. In abandoned arms there are often two kinds of deposit: first of all, when a meander is cut off, a sandy alluvial plug is formed at either end; next, floods bring with them decantation deposits, which are usually smaller, ranging from sand to silt and clay. The meander is then filled up with large quantities of organic matter (Fig. 12).

Floods may also affect the movement of bedload outside the channel. The bedload is spread over the floodplain when a breach opens suddenly in the concave bank because of current pressure; these deposits may be made up of the bedload, and when the floodwaters recede they are often covered by alluvia that have been transported in suspension.

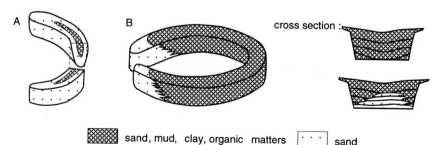

FIGURE 12   Complex sedimentation in abandoned arms [various types: chute cutoff (A) and neck cutoff (B) [after Schumm (1960) in Allen (1965, Fig. 32, p. 156)].

When the water is rising or the flood is at its peak level, the upper layer of deposits is more or less unstable, but at a variable depth there is always a zone of physical stability. In an alluvial plain the composition of the interstitial biocenoses of the channel that undergoes the full force of the floods is different from that of the ones that are to be found in the sediments at the edge of the plain. In the abandoned arms of the river, as in, for example, the Eberschuttwasser on the Danube (Danielopol, 1983, 1984) or in the Grand Gravier on the Rhône (Dole, 1983; Dole and Chessel, 1986), the hypogean fauna is more abundant and diversified than that in the alluvia of the active channel, in which the alluvia are frequently restructured. The perennial nature of the environmental conditions that characterize the surface of the fringe zones of the plain encourages the development of underground populations (Danielopol, 1989), provided the fluxes of groundwater in the aquifer unclog and oxygenate the interstitial milieu (Dole, 1985; Marmonier, 1988).

3. The Influence of Flooding on Exchanges between the River and Groundwater

The aquifers of an alluvial plain have in common the fact that they all converge on a major drainage axis, the river, which divides them into global aquifer/river systems. In such systems Castany (1985) considers three interacting components: (1) the reservoir, heterogeneous and formed by sediments, (2) the groundwater, and (3) the surface water. The flow direction of the groundwater depends on the position of the piezometric surface compared with the level of water in the river (Castany, 1982, 1985). The stream tends to drain the aquifer when the waterline is low compared with the piezometric surface (draining system). On the other hand surface water supplies groundwater if the piezometric surface is lower. Drainage and infiltration zones may alternate along a stream, but usually there is alternance in time following the surface hydrology. The periodic fluctuations in the piezometric level caused by rising and falling water levels in the channel stimulate bacterial activity in the unsaturated interstitial zone (Danielopol, 1989) and facilitate the transit of organic matter transported by overspill water in the floodplain.

VI. HUMAN ACTIVITY AFFECTING RIVER DYNAMICS AND ALLUVIAL AQUIFERS

Humans interfere directly with stream channels and floodplains, either on the surface or in the first few meters of the aquifer (embankments, dams, impermeabilization of different surfaces, and dredging or quarrying for gravel and sand). Such works also have a direct influence on the aquifers and the connections between groundwater and river. Human interference, whether it affects water and alluvial quantities or modifies discharge rates, inevitably gives rise to geomorphologic readjustments.

A. Degradation and Aggradation of Longitudinal Profiles

When a river is embanked its banks are narrowed, the water flows more swiftly between them, and the channel becomes more and more incised. Very often, at the same time as this erosion occurs, there is redeposition downstream, and aggradation takes place. This combination of degradation and aggradation is called hydraulic tilting of the longitudinal profile. In incised sites, groundwater is drained more in low-water periods, whereas surface waters supply groundwater in the raised reaches.

The Rhône to the east of Lyon (France) was channelized from 1847 to 1858, over an 18-km-long stretch (the Miribel canal). Hydraulic instability followed rapidly, and the upstream incision was estimated at 4 m for a downstream aggradation of 5 m (Winghart and Chabert, 1965). Since the beginning of the twentieth century, part of the water in the river has been diverted into a second canal that leads to a hydroelectric power station, and the Miribel canal receives a minimum discharge of 30 m^3/sec^{-1} for 245 days per year on average, whereas the interannual mean is 460 m^3/sec^{-1} (for the hydrologic period 1960–1990). During low water the hydraulic tilting had already reinforced the drainage of groundwater in the incised part and intensified infiltration in the aggraded part; the artificially induced decrease in discharge again intensified this tendency. Interstitial fauna (sampled at a depth of 50 cm in the alluvia) reflects these geomorphologic and hydrogeologic conditions. In the zones in which lateral groundwater is drained, there is a hypogean fauna. In the upstream part, the permanent infiltration of surface water is reinforced by pumping at the Crépieux–Charmy station, which supplies most of Lyon with its water; the animals living in the interstitial environment are mainly epigean and as such adapted to living conditions in surface water (Creuzé des Châtelliers and Reygrobellet, 1990; Creuzé des Châtelliers, 1991).

After the construction of a dam and flow regulation, the longitudinal profile of the Tennessee River underwent hydraulic instability with modification of aquifer/river connections. In the aggraded part, an increase in flooded areas led to aggradation of the floodplain due to sand and silt deposits. These deposits were about 10 cm deep for the first two years and then 7 cm deep the following year (Hupp, 1986). The infiltration and restitution of flood waters slowed down. Indirectly incision concentrates the flow in the floodplain, limits submersion of the floodplain, and contributes indirectly to a reduction of the recharge phase of the groundwater from the flooded surface. If the groundwater is recharged for a shorter period and thus supplied with less water, drainage by the river when the floodwaters subside is also decreased.

Incision of the channel can reach down to older alluvia that are too coarse for the present competency of the stream; these deposits form a pavement and the bed is stabilized. Such alluvial pavements are reshaped only in the case of exceptional floods (Bray and Church, 1980). Such deposits have two components of very different grain size: large cobbles or even boulders on the one hand and on the other hand a matrix of fine sediments (sands, silts, and even clays) that are more or less cemented and that limit the infiltration of surface water. As their floodwaters fall some streams transport and deposit fine sediments (clays or marls) and may contain recently reshaped gravel bars with clogged interstices. This clogging reduces the habitability of the environment for interstitial fauna.

B. Dams

Dams cause aggradation of the bed upstream as the water flow is slowed down and the level of the water is kept at an artificially high altitude; groundwater drainage is totally absent but the infiltration of surface water through the alluvia is intensified by the pressure exerted on the bed (Fig. 13). Whatever the type of dam, downstream discharge is inevitably reduced. Depending on the reason for the dam's contruction, either the flow is regulated throughout the year or it decreases when part of the water is diverted or used for water supplies or irrigation. In the former case, the disappearance of extreme discharges diminishes the extent and duration of groundwater/river exchanges. In the latter case, the quantity of water removed can vary considerably. Since the Serre-Ponçon dam (1961) was built on the Durance (France), the minimum discharge in the channel is 2 m^3/sec^{-1}, and at such times groundwater is very much drained. The mean discharge went from 193 m^3/sec^{-1} (during the period 1892–1950) to 11 m^3/sec^{-1} (during the period 1961–1980; Vivian, 1989).

C. Dredging

In unstable streams, in which the transport capacity is above the available load, dredging accentuates the sedimentary deficit and encourages inci-

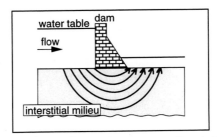

FIGURE 13   Flow pattern of groundwaters in a porous milieu, influenced by a dam [after Freeze and Cherry (1979, Fig. 10a, p. 478)].

sion. Dredging disturbs the equilibrium of the channel and leads to readjustment of the longitudinal profile; modifications take place both upstream and downstream of the dredging area.

In reaches where the slope is low and where the present competency of the water no longer makes it possible for the coarse pebbly load to be carried, fossil sediments form a particularly fragile habitat in that these types of structure and grain size will not be renewed if there is a disturbance or human alteration. In this type of river, the trenches from which alluvial material is extracted are zones at which the current is slowed and thus fine sediments are trapped; this is detrimental to the permeability of the upper part of the aquifer and thus to its habitability.

VII. CONCLUSION

Most of the world's rivers are alluvial, and the geomorphology of many, if no most, was substancially determined by Quaternary cut and fill morphogenesis. Substrata of alluvial rivers are heterogeneous in space and time, and discontinuities in local and longitudinal profiles create complex interstitial habitats for groundwater biota in associated aquifers. Variations in porosity and permeability create discontinuities in the connections between aquifers and river channels and lead to complex distribution pattern of biota.

Phreatobiological studies of alluvial rivers too often disregard the implications of hypogean geomorphic complexity, because it is so difficult to demonstrate. However, groundwater ecosystems formed in alluvia saturated by river flow are inherently patchy environments as a function of ancient and present fluvial processes. Hence, coupling of geomorphic and phreatobiologic study is needed to advance a complete understanding of groundwater ecology of alluvial rivers.

ACKNOWLEDGMENTS

Financial support by CNRS (Program for Interdisciplinary Research on the Environment) is acknowledged. We thank Jack Stanford (Montana), for reviewing this chapter and for his helpful remarks. Vladimir Vanek (Sweden) kindly made criticisms of an early manuscript and Glynn Thoiron (Lyon, France) edited the English text.

REFERENCES

Allen, J. R. L. (1965). A review of the origin and characteristics of recent alluvial sediments. *Sedimentology* 5(2), 89–191.

Amoros, C., and Bravard, J. P. (1985). L'intégration du temps dans les recherches méthodologiques appliquées à la gestion écologique des vallées fluviales: Exemples des écosystèmes aquatiques abandonnés par les fleuves. *Rev. Fr. Sci. Eau* 104, 349–364.

Angelier, B. (1953). Recherches écologiques et biogéographiques sur la faune des sables submergés. *Arch. Zool. Exp. Gen.* **90**, 37–167.

Bodelle, J., and Margat, J. (1980). "L'eau souterraine en France," Colloq. 'Les objectifs de demain.' Masson, Paris.

Bou, C. (1968). Faune souterraine du Sud. Overt du Marrif Central. II. Contribution á la connaissance de la faune des eaux souterraines de l'Albigeois. *Ann. Spéléol.* **23**(2), 441–473.

Bravard, J.-P. (1991). La dynamique fluviale à l'épreuve des changements environnementaux: Quels enseignements applicables à l'aménagement des rivières? *Houille Blanche* **7**(8), 515–521.

Bravard, J.-P., Amoros, C., and Pautou, G. (1986). Impact of civil engineering works on the successions of communities in a fluvial system. *Oikos* **47**, 92–111.

Bray, D. I., and Church, M. (1980). Armured versus paved gravel beds. *J. Hydraul. Div., Am. Soc. Civ. Eng.* **106**, 1916–1937.

Bretschko, G. (1983). Die Biozönen der Bettsedimente von Fliessgevassern-ein Beitrag der Limnologie zur naturnahen Gewässerregulierung. *In* "Wasserwirtschaft, wasservorsorge, Forschungsarbeiten, Bundesministerium fûr Land und Forestwirtschaft." Wien. pp. 1–161.

Castany, G. (1982). "Principes et méthodes de l'hydrogéologie." Dunod, Paris.

Castany, G. (1985). Liaisons hydrauliques entre les aquifères et les cours d'eau. Systèmes global aquifère/rivière. *Stygologia* **1**(1), 1–125.

Castella, E., Richardot-Coulet, M., Roux, C., and Richoux, P. (1984). Macroinvertebrates as describers of morphological and hydrogeologic type of aquatic ecosystems abandoned by the Rhône River. *Hydrobiologia* **119**, 219–225.

Chappuis, P. A. (1942). Eine neue Methode zur Untersuchung der Grundwasser-fauna. *Acta Sci. Math. Nat. Kolozsvar* **6**, 1–7.

Church, M., and Jones, D. (1982). Channels bars in gravel-bed rivers. *In* "Gravel Bed Rivers, Fluvial Processes, Engineering and Management" (R. D. Hey, J. C. Bathurst, and C. R. Thorne, eds.), pp. 291–338. Wiley, Chichester.

Coleman, M. J., and Hynes, H. B. N. (1970). The vertical distribution of the invertebrate fauna in the bed of a stream. *Limnol. Oceanogr.* **15**, 31–40.

Creuzé des Châtelliers, M. (1991). Dynamique de répartition des biocénoses interstitielles du Rhône en relation avec des caractéristiques géomorphologiques (secteurs de Brégnier-Cordon, Miribel-Jonage et Donzère-Mondragon). Thèse, Univ. Lyon I.

Creuzé des Châtelliers, M., and Reygrobellet, J. L. (1990). Interactions between geomorphologic processes, benthic and hyporheic communities: First results on a bypassed canal of the french upper Rhône River. *Regulated Rivers* **5**, 139–158.

Creuzé des Châtelliers, M., Marmonier, P., Dole-Olivier, M. J., and Castella, E. (1992). Structure of interstitial assemblages in a regulated channel of the river Rhine (France). *Regulated Rivers* **7**, 23–30.

Danielopol, D. L. (1976). The distribution of the fauna in the interstitial habitats of Riverine Sediments of the Danube and the Piesting (Austria). *Int. J. Speleol.* **8**, 23–51.

Danielopol, D. L. (1983). Der Einfluss organischer Verschmutzung auf das Grundwasser-Ökosystem der Donau in Raum Wien and Niederösterreich. *Forschungsber., Bundesminist. Gesund. Umweltschutz* **5**, 5–159.

Danielopol, D. L. (1984). Ecological investigations on the alluvial sediments of the Danube in the Vienna area—a phreatobiological project. *Verh. Int. Ver. Limnol.* **22**, 1755–1761.

Danielopol, D. L. (1989). Groundwater fauna associated with riverine aquifers. *J. North Am. Benthol. Soc.* **8**(1), 18–35.

Delamare-Deboutteville, C., and Paulian, R. (1953). Recherche sur la faune interstitielle des sédiments marins et d'eau douce à Madagascar. *Mém. Inst. Sci. Madagascar, Sér. A-A.* **VIII**, 8, 1–10.

Dole, M. J. (1983). Le domaine aquatique souterrain de la plaine alluviale du Rhône à l'Est de Lyon. Ecologie des niveaux supérieurs de la nappe. Thèse, Univ. Lyon I.

Dole, M. J. (1985). Le domaine aquatique souterrain de la plaine alluviale du Rhône à l'Est de Lyon, 2. Structure verticale des peuplements des niveaux supérieurs de la nappe. *Stygologia* **1**(3), 270–290.

Dole, M. J., and Chessel, D. (1986). Stabilité physique et biologique des milieux interstitiels. Cas de deux stations du Haut-Rhône. *Ann. Limnol.* **22**(1), 69–81.

Dubus, J., and Fourneaux, J. C. (1968). Les ressources en eau souterraine de la plaine du Grésivaudan. *Rev. Géogr. Alpine* **3**(4), 497–516.

Fortner, S. L., and White, D. S. (1988). Interstitial water patterns: A factor influencing the distributions of some lotic aquatic vascular macrophytes. *Aquat. Bot.* **31**, 1–12.

Freeze, R. A., and Cherry, J. A. (1979). "Groundwater." Prentice-Hall, Englewood Cliffs, NJ.

Godbout, L., and Hynes, H. B. N. (1982). The three dimensional distribution of the fauna in a single in a stream in Ontario. *Hydrobiologia* **97**, 87–97.

Graf, W. L. (1979). The development of montane arroyos and gullies. *Earth Surf. Processes Landforms* **4**, 1–14.

Harvey, A. M., and Renwick, W. H. (1987). Holocene alluvial fan and terrace formation in the bowland fells, northwest England. *Earth Surf. Processes Landforms* **12**, 225–249.

Hendricks, S. P., and White, D. S. (1988). Hummocking by lotic *Chara*: Observations on alterations of hyporheic temperature patterns. *Aquat. Bot.* **31**, 13–22.

Henry, J. P. (1976). Recherches sur les Asellidae hypogées de la lignée *Cavaticus*. Thèse, Univ. Dijon.

Henry, M. (1963). Contribution à l'étude des nappes phréatiques. *Ann. Ponts Chaussées* **5**, 1–50.

Hertzog, L. (1936). Crustacés des biotopes hypogés de la vallée du Rhin d'Alsace. *Bull. Soc. Zool. Fr.* **61**, 350–372.

Hertzog, L. (1938). Grustaceen aus unterindischen Biotopen des Rheintales bei Strazurg. III. Mitteilung. *Zool. Anz.* **123**(3), 45–56.

Huggenber, P., and Zobrist, J. (1990). La recherche hydrogéologique à l'EAWAG. *Nouv. EAWAG* **29**, 10–14.

Hupp, C. R. (1986). Determination of bank widening and accretion rates and vegetation recovery along modified west Tennessee streams. *Proc. Int. Symp. Ecol. Aspects Tree-Ring Anal.*, Palisades, New York.

Huryn, A. D., and Wallace, J. B. (1987). Local geomorphology as a determinant of macrofaunal production in a mountain stream. *Ecology* **68**(6), 1932–1942.

Hynes, H. B. N. (1970). "The Ecology of Running Waters." University of Toronto, Ontario.

Johnston, C. A., and Naiman, R. J. (1987). Boundary dynamics at the aquatic-terrestrial interface: The influence of beaver and geomorphology. *Landscape Ecol.* **1**(1), 47–57.

Karaman, S. (1954). Über unsere unterrirdische Fauna. *Acta Mus. Macedonici Sci. Nat.* **1**, 195–216.

Keller, E. A., and Melhorn, W. N. (1978). Rhythmic spacing and origin of pools and riffles. *Geol. Soc. Am. Bull.* **89**, 723–730.

Leruth, R. (1938). La faune de la nappe phréatique du gravier de la Meuse à Hermalle-sous-Argenteau. *Bull. Mus. R. Hist. Nat. Belg.* **14**, 1–37.

Lodge, D. M., Krabbenhoft, D. P., and Striegl, R. G. (1989). A positive relationship between groundwater velocity and submersed macrophyte biomass in Sparkling Lake, Wisconsin. *Limnol. Oceanogr.* **34**(1), 235–239.

Magniez, G. (1976). Contribution à la connaissance de la biologie des Stenasellidae. Thèse, Univ. Dijon.

Marmonier, P. (1988). Biocénoses interstitielles et circulation des eaux dans le sous-écoulement d'un chenal aménagé du Haut-Rhône français. Thèse, Univ. Lyon I.

Marmonier, P., and Dole, M. J. (1986). Les Amphipodes des sédiments d'un bras court-circuité du Rhône. Logique de répartition et réaction aux crues. *Rev. Fr. Sci. Eau* **5**, 461–486.

Mestrov, M. (1962). Un nouveau milieu aquatique souterrain, le biotope hypothelminorhéique. *C. R. Acad. Sci. Paris* **254**(14), 2677–2679.

Moniez, R. (1889). Faune des eaux souterraines du département du Nord et en particulier de la ville de Lille. *Rev. Biol. Nord Fr.* **89**(Fol. 1), 175–315.

Motas, C. (1962). Procédés des sondages phréatobiologiques. Division du domaine souterrain. Classification écologique des animaux souterrains. Le psammon. *Acta Mus. Macedonici Sci. Nat.* **8,** 135–173.

Orghidan, T. (1959). Ein neuer Lebensraum des unterirdischen Wassers, das hyporheische Biotop. *Arch. Hydrobiol.* **55,** 392–414.

Peiry, J. L. (1988). Approche géographique de la dynamique spatio-temporelle des sédiments de la plaine alluviale de l'Arve (Haute-Savoie, France). Thèse, Univ. Lyon III.

Pennak, R. W. (1939). The microscopic fauna of the sandy beaches. *In* "Problems of Lake Biology," Publ. No. 10, pp. 94–106. Am. Assoc. Adv. Sci., Washington, DC.

Pécsi, M. (1971). The development of the Hungarian section of the Danube valley. *Geoforum* **6,** 21–32.

Petit, F., and Schumacker, R. (1985). L'utilisation des plantes aquatiques comme indicateur d'activité géomorphologique d'une rivière Ardennaise (Belgique). *Colloq. Phytosociol. Vég. Géomorphol., 13th,* Bailleul, 1985, pp. 691–710.

Petts, G. E., and Foster, I. (1985). "Rivers and Landscape." Edward Arnold, London.

Reineck, H. E., and Singh, I. B. (1980). "Depositional Sedimentary Environments-Fluvial Environment," 2nd ed. Springer-Verlag, Berlin and New York.

Rouch, R. (1968). Contribution à la connaissance des Harpacticides hypogés (Crustacés Copépodes). *Ann. Spéléol.* **23,** 5–165.

Rouch, R. (1988). Sur la répartition spatiale des Crustacés dans le sous-écoulement d'un ruisseau des Pyrénées. *Ann. Limnol.* **24**(3), 213–234.

Roux, A. L. (ouvrage collectif publié sous la direction de) (1982). "Cartographie polythématique appliquée à la gestion écologique des eaux. Etude d'un hydrosystème fluvial: Le Haut-Rhône Français." CNRS Centre Région, Lyon.

Sabater, F. (1987). On the interstitial Cladocera of the River Ter (Catalonia, NE Spain) with a description of the male of *Alona phreatica. Hydrobiologia* **144,** 51–62.

Schirmer, W. (1988). Holocene valley development on the Upper Rhine and Main. *In* "Lake, Mire and River Environments during the Last 15000 Years" (G. Lang and C. Schluchter, eds.), pp. 153–160. Balkema, Rotterdam, The Netherlands.

Schumm, S. A. (1960). The effect of sediment type on the shape and stratification of some modern fluvial deposits. *Am. J. Sci.* **258,** 177–184.

Schumm, S. A. (1963). Sinuosity of alluvial rivers on the Great Plains. *Geol. Soc. Am. Bull.* **74,** 1089–1100.

Schumm, S. A. (1977). "The Fluvial System." Wiley, Chichester.

Schumm, S. A., and Brakenridge, G. R. (1987). River responses. *In* "The Geology of North America; North America and Adjacent Ocean During the Last Deglaciation" (W. F. Ruddiman and H. E. Wright, Jr., eds.), Vol. 3, pp. 221–240. Boulder, Colorado, Geological Society of America.

Schumm, S. A., Harvey, M. D., and Watson, C. C. (1984). "Incised channels, morphology, dynamics and control." Water Resour. Publ., Littleton, Colorado.

Schwoerbel, J. (1961). Uber die Lebensbedingungen und die Besiedlung des hyporheischen Lebensraumes. *Arch. Hydrobiol., Suppl.* **25,** 182–214.

Simler, L., Valentin L., and Duprat, A. (1979). La nappe phréatique du Rhin en Alsace. *Sci. Géol.* **60,** 1–266.

Smith, D. G. (1981). Aggradation of the Alexandra-north saskatchewan river, Banff partk, Alberta. *In* "Fluvial Geomorphology" (M. Morisawa, ed.), pp. 201–219. Allen & Unwin, London.

Smith, D. G. (1983). Anastomosed fluvial deposits: Modern examples from Western Canada. *Spec. Publ. Int. Assoc. Sedimentol.* **6,** 155–168.

Stanford, J. A., and Gaufin, A. R. (1974). Hyporheic communities of two Montana rivers. *Science* **185,** 700–702.

Stanford, J. A., and Ward, J. V. (1988). The hyporheic habitat of river ecosystems. *Nature (London)* **335**, 64–66.

Stanford, J. A., and Ward, J. V. (1993). An ecosystem perspective of alluvial rivers: Connectivity and the hyporheic corridor. *J. North Am. Benthol. Soc.* **12**(1), 48–60.

Starkel, L. (1983). The reflection of hydrologic changes in the fluvial environment of the temperate zone during the last 15000 years. *In* "Background to Paleohydrology" (K. J. Gregory, ed.), pp. 213–235. Wiley, Chichester.

Stocker, Z. S. J., and Williams, D. D. (1972). A freezing core method for describing the vertical distribution of sediments in streambed. *Limnol. Oceanogr.* **17**(1), 136–138.

Strayer, D. (1988). Crustaceans and mites (Acari) from hyporheic and other underground waters in southeastern New York. *Stygologia* **4**(2), 192–206.

Striedter, K. (1988). Le Rhin en Alsace du Nord au subboréal. Genèse d'une terrasse fluviatile holocène et son importance pour la mise en valeur de la vallée. *Bull. Assoc. Fr. Etude Quat.* **25**(33), 5–10.

Swanson, F. J., and Lienkaemper, G. W. (1982). Interactions among fluvial processes, Forest vegetation, and aquatic ecosystems, South Fork Hoh River, Olympic National Park. *In* Ecological research in National Parks of the Pacific Northwest. *Proc. Conf. Sci. Res. Natl. Parks, 2nd,* San Francisco-Corvallis, 1979, pp. 30–34.

Thibodeaux, L. J., and Boyle, J. D. (1987). Bedform-generated convective transport in bottom sediment. *Nature (London)* **325**, 341–343.

U.S. Geological Survey (1987). Oil shale, water resources, and valuable minerals of the Piceance Basin, Colorado: The challenge and choices of development (compiled by O. J. Taylor). *Geol. Surv., Prof. Pap. (U.S.)* **1310**, 1–143.

Vejdovsky, F. (1822). Thierische organismen der Brunnenwässer von prag. *Selbstverlag, Prag.*, pp. 1–70.

Vivian, H. (1989). Hydrological Changes of the Rhône River. *In* "Historical Change of Large Alluvial Rivers: Western Europe" (G. E. Petts, ed.), pp. 57–77. Wiley, Chichester.

Ward, J. V., and Voelz, N. J. (1990). Gradient analysis of interstitial meiofauna along a longitudinal stream profile. *Stygologia* **5**(2), 93–99.

White, D. S. (1990). Biological relationships to convective flow patterns within stream beds. *Hydrobiologia* **196**, 149–158.

White, D. S., Elzinga, C., and Hendricks, S. (1987). Temperature patterns within the hyporheic zone of northern Michigan river. *J. North Am. Benthol. Soc.* **6**(2), 85–91.

Winghart, J., and Chabert, J. (1965). Haut-Rhône à l'amont de Lyon: Étude hydraulique de l'île de Miribel-Jonage. *Houille Blanche* **7**, 1–21.

BIOLOGICAL ORGANIZATIONS AND CONSTRAINTS IN GROUNDWATER

7

Microbial Ecology
of Groundwaters

A. M. Gounot

Université Lyon I
Ecologie Microbienne
U.R.A. CNRS 1450
69622 Villeurbanne Cedex, France

I. INTRODUCTION

The microbiology of terrestrial subsurface environments is an area that has received little attention from microbiologists. Until recently, most people thought that subsurface environments were virtually sterile. Terrestrial subsurface environments are largely inaccessible and are not easy to reach and sample aseptically. Early bacteriological analyses of spring and well water samples detected only pathogenic bacteria or pollution indicators originating from the surface. More basic studies on cave microbiology have been conducted since 1930. However, caves are contamined by microorganisms, native to the surface, which are brought in by runoff waters, airstreams, animals, and humans, and it is difficult to know if microorganisms indigenous to the subsurface are present.

The presence of a rich and fairly abundant microflora in the aquifer environment was first demonstrated by Wolters and Schwartz (1956). In the 1970s, basic and ecological research began on shallow aquifers, especially in the United States and Germany. Interest in groundwater microbiology increased in the 1980s (Ghiorse and Balkwill, 1983; Hirsch and Rades-Rohkohl, 1983). Groundwater, often from fossil aquifers, is used at an ever-increasing rate, and pollution of this groundwater has become a serious problem. Aquifer systems have been investigated with specific attention to the presence, abundance, and metabolic potential of resident microbial populations and with regard to the behavior of organic pollutants in groundwater. Deep aquifer contamination and risks from deep waste disposal, including radionuclides, stimulated development of subsurface microbiology research programs, especially in the United States (U.S. Department of Energy, U.S. Geological Survey, and U.S. Environmental Protection Agency), in Great Britain (Institute of Geological Science and Water Research Center), and in Germany (Umweltbundesamt). These programs focused on detrimental and beneficial microbial activities and the prospect of bioremediation and promoted basic exploratory research to characterize the microbial populations and activities in sediment and groundwater systems, especially in the deep subsurface. Research was conducted by interdisciplinary teams of microbiologists, geohydrologists, and geochemists. Innovative drilling and sampling technologies were developed. The results were published in a special issue of *Microbial Ecology,* Vol. 16 (1988), a special issue of the *Geomicrobiology Journal,* Vol. 7 (1989), and the "Proceedings of the First International Symposium on Microbiology of the Deep Subsurface" (1990). In France some studies were conducted on deep formations in anticipation of future nuclear waste storage (Daumas *et al.,* 1986) and geothermal development (Camus *et al.,* 1987).

For microbiologists, "groundwater" refers to all subsurface water found beneath soil horizons, because for microorganisms living in microhabitats, all available forms of water may be important. Consequently the groundwater habitat is implicitly synonymous with the terrestrial subsurface (Madsen and Ghiorse, 1993). This chapter is devoted to the autochthonous microorganisms of pristine subsurface systems. It mainly focuses on porous aquifers because, in most cases, karsts are colonized by surface microorganisms. In alluvial aquifers, the ecotone habitats, such as river banks or lake sediments, which are transitional zones to the subsurface, are inhabited by a specific hyporheic fauna. However, these habitats are colonized by surface microorganisms that migrate much farther than surface fauna. The hyporheic zone is not a true groundwater system for microbiological investigations. Special subsurface environments, such as oil and gas, sulfur, and coal deposits, which have concentrated only on select microorganisms, such as the sulfate-reducing and the sulfur-oxidizing bacteria fouling and corroding subsurface structures, are not dealt with. Supplementary information can be found in the reviews of Ghiorse and Wilson (1988) and Madsen and Ghiorse (1993), which deal with the microbial ecology of the terrestrial subsurface; Chapter 5, "Microbiology," in "Progress in Hydrogeochemistry" (Matthess *et al.,* 1992); and the book "Ground-Water Microbiology and Geochemistry" (Chapelle, 1993).

II. METHODS

Many more microorganisms, especially autochthonous microorganisms, are attached to sediment particles, and only a few are found in free water samples. Sediment samples must be taken because analyses of free groundwater cannot document true microbial populations of aquifer systems. Basic microbiological research on subsurface sediment is slowed by the complexity of aseptic sampling and by the high cost of adequate drilling. We can question the validity of the results of many earlier studies in which the samples were not collected aseptically or were collected from long-standing wells rather than sediment core samples or in which the cultural techniques did not suit the autochthonous microorganisms. The development of successful microbiological studies of subsurface environments relied first on the availability of aseptic sampling techniques and second on the adaptation of existing microbiological methods to the study of microbial populations and activities in subsurface material.

A. Sampling Methods

The subsurface aquifer system is not homogenous: free groundwater, interstitial water, and well water are quite different as far as chemistry and biota are concerned. It is difficult, tedious, and expensive to obtain uncontamined, undisturbed, and representative samples of sediments, especially deep sediments, for microbiological investigations. Sophisticated drilling and sampling procedures that prevent contamination of samples by surface soil and water, as well as drilling tools and fluid, have been described (McNabb and Mallard, 1984; Leach, 1990; Phelps *et al.,* 1989a; Harvey and George, 1987; Smith and Harvey, 1990). Disturbance and invasion of the formation surrounding the borehole must be avoided.

Sampling methods for well water are described by Cullimore (1993). Hirsch and Rades-Rohkohl (1988) considered that wells should be at least one year old because young wells contain higher numbers of microorganisms due to underground disturbances from the drilling. Prepumping of a stagnant well is necessary before pumping the actual aquifer water.

B. Methods for Microbiological Analyses

Ghiorse and Wilson (1988) gave an overview of the various methods that have been adapted from soil and aquatic microbiology. However, a number of them are difficult to adapt because their low sensitivity does not permit detection of low population density and low levels of activity in the subsurface. Furthermore, the validity of all the methods involving release of bacteria from sediment particles is to be questioned due to the difficulty of desorbing cells from surfaces. Balkwill *et al.* (1977) have shown that at most about 30–40% of the attached organims come off. Cullimore (1993) described methods for well water analyses and biofouling studies.

1. Microscopic Methods

The presence of microorganisms in environmental samples and some of their typical morphological features (*e.g.,* size, shape, eukaryotic or pro-karyotic cell organization, cell wall type, internal storage bodies, and exter-nal polymer structures) were established by direct light and electronic micros-copy (Bone and Balkwill, 1988). For water samples, concentrating cells on membrane filters and staining with a fluorescent dye improved visibility using the fluorescence microscope. This was less successful for sediment samples due to particle interference (and due to the difficulty of desorb-ing cells from surfaces). Bacterial counts in subsurface water or sediment samples have been achieved by staining with acridine orange (Wilson *et al.,* 1983; Webster *et al.,* 1985; Balkwill *et al.,* 1988) or DAPI (4',6'-diamino-2-phenylindole)(King and Parker, 1988). In addition, the proportion of living bacteria were obtained by using a tetrazolium dye reduction (Beloin *et al.,*1988; Marxsen, 1988), but the red spots characterizing respiring cells under the microscope were often recognized only with difficulty due to the very small size of cells in subsurface sediments and water. King and Parker (1988) improved the method of counting viable cells by the use of nalixidic acid, which resulted in an increase of length of living cells. However, some groundwater bacteria were resistant to this inhibitor.

New technologies, such as confocal laser microscopy (Caldwell *et al.,*1992), immunology and immunocytochemistry, flow cytometry, and im-age analysis, have provided new insights into surface aquatic microbiology (Hobbie and Ford, 1993). They offer promise for future investigations of subsurface microbial ecology.

2. Cultural Methods

Standard cultural procedures (*e.g.,* plate counts, most probable number counts, and enrichment culture procedures) have been employed in many subsurface studies (Ghiorse and Wilson, 1988; Hirsch and Rades-Rohkohl, 1988; Hirsch *et al.*, 1992b; Cullimore, 1993), including cave studies (Mason-Williams and Benson-Evans, 1958; Gounot, 1967). Cultural methods are more sensitive than most other methods of detection and provide information on physiology, potential metabolic activities, and microbial diversity because they allow the isolation and characterization of bacteria. However, most classical methods of identification are adapted to bacteria of medical interest and are of little use when attempting to identify soil and water bacteria, especially those from the subsurface (Amy *et al.*, 1992). After isolation, morphological and physiological characterization of subsurface bacteria based on numerical taxonomy using a miniaturized test system was performed by Kölbel-Boelke *et al.* (1988a) and Hirsch and Rades-Rohkohl (1992). Molecular analysis of nucleic acids provides information about diversity and offers promise (Jimenez, 1990).

3. Biochemical Indicators

Biochemical indicators of metabolic activity and detection of specific molecular markers in natural samples (*e.g.*, ATP, GTP, phospholipids, and muramic acid) provide estimations of microbial biomass and activity (White *et al.*, 1983, 1990; Balkwill *et al.*, 1988).

4. Radioisotope Methods

Bacterial growth rates and activities can be measured directly in environmental samples by measuring radioisotope-labeled substrate transformation and uptake (Thorn and Ventullo, 1988). These procedures do not require the removal of microorganisms attached to sediment particles. However, a great deal of caution must be exercised when applying these methods to estimate bacterial growth rates in aquifer samples because in some nutrient-depleted groundwater only a small fraction of assimilated thymidine may be incorporated in new DNA (Harvey and George, 1987).

5. In Situ Incubation

In situ incubations in open or closed systems are used to observe microbial colonization under environmental conditions (Hirsch and Rades-Rohkohl, 1990; Cullimore, 1993).

III. TYPES, NUMBERS, AND DISTRIBUTION OF MICROORGANISMS

A. Karstic Systems

Since the first observations of Schreiber (1929) and Dudich (1930), microbiological investigations in cave environments have scarcely developed with the classical methods of soil microbiology. No studies have been performed utilizing the modern methods of microbial ecology. Cave biotopes are so diverse that it is difficult to provide an overview of their microbiota.

Water and sediment samples collected in caves contain microorganism densities of about 10^2 to 10^4 bacteria per milliliter of water or 10^4 to 10^8 bacteria per gram of dry sediment. Functional activities (such as the oxidation of sulfur compounds, iron, or manganese; nitrification; denitrification; nitrogen fixation; cellulolysis; etc.) were suspected from chemical analyses or shown by cultural methods (Magdeburg, 1933; Birstein and Borutzky, 1950; Mason-Williams and Benson-Evans, 1958; Fischer, 1959; Caumartin, 1959, 1963, 1964; Varga and Takats, 1960; Gounot, 1967, 1969a, 1970, 1973, 1974; Mason-Williams, 1969; Brock *et al.,*1973; Perrier, 1973a,b; Dickson, 1975; Pasqualini *et al.,* 1978; Kilbertus and Schwartz, 1981). Sarbu (1990) found sulfur-oxidizing bacteria in a cave without direct connection to the surface and receiving thermomineral waters.

Lists of genera and species of microorganisms found in caves were deceptive because classical methods of identification were not suitable, and the strains identified were probably surface bacteria. Most investigated sites were probably contaminated by surface waters, air, animals, and humans. The presence of dead plant fragments and pieces of decaying wood, carried down by rainwater runoff, snow thawing, or subterranean river floods, is evidence of contaminations. Many surface bacteria are brought in with them and can develop on this exogenous organic matter (Kilbertus and Schwartz, 1981; Eichem *et al.*, 1993). Therefore, it is difficult to discern the autochthonous bacteria from the more numerous exogenous microorganisms. Physiological studies are helpful in recognizing resident bacteria because they can be adapted to subsurface conditions (Gounot *et al.*, 1970; Gounot, 1973). The presence of such bacteria in an uncontaminated fossil gallery was clearly demonstrated by Perrier (1973a,b, 1977).

Photosynthetic cyanobacteria and eucaryotic algae can develop in faintly illuminated cave entrances (Palik, 1960; Dobat, 1970). Although some cyanobacteria can also develop in the dark, those found in dark galleries probably originated from surface water. Fungi likewise can develop on decayed wood and leaves and therefore would not be autochthonous.

B. Subsurface Aquifer Systems

Cooperative studies, utilizing more appropriate methods of microbial ecology, allow a preliminary overview on the presence, number, distribution, activities, and diversity of microorganisms in groundwater and sediments. Few studies deal with the vadose zone, below the soil horizons. Bacterial numbers are low and decrease with depth (Beloin *et al.*, 1988; Bone and Balkwill, 1988; Colwell, 1989). Bacteria are much more abundant in the saturated zone. Estimations of bacterial density in shallow aquifers depend on the method used. In general, total microscopic counts range between 105 and 107 bacteria per gram of dry sediment or milliliter of aquifer water. Viable cell counts range from zero to values nearly as high as the total microscopic counts (Chapelle *et al.*, 1987; Hirsch and Rades-Rohkohl, 1988; Kölbel-Boelke *et al.*, 1988b). Numerous works are mentioned by Ghiorse and Wilson (1988) and Madsen and Ghiorse (1993). Aquifer regions polluted with high levels of utilizable carbon sources may show higher plate counts than unpolluted regions. The lower counts given by cultural methods probably originate from the inability of microorganisms to grow on the media and the growth conditions employed.

In deep aquifers, the presence of bacteria was first demonstrated by biochemical markers (White *et al.*, 1983, 1990) and later by total microscopic counts and viable cell counts. Some levels seem sterile; others contain substantial numbers of viable bacteria (up to 10^8/g) (Bianchi *et al.*, 1987; Balkwill, 1989; Sinclair and Ghiorse, 1989; Fredrickson *et al.*, 1990). It is assumed that these bacteria are autochthonous because they differ morphologically and physiologically from surface bacteria or from bacteria introduced by drilling fluids (Bianchi, 1986; Fredrickson *et al.*, 1990). The *in situ* morphology, diversity, and physiological state of bacteria were investigated with light and electron microscopy (Balkwill and Ghiorse, 1985; Balkwill *et al.*, 1988; Bone and Balkwill, 1988; Smed-Hildmann and Filip, 1988; Bengtsson, 1989; Sinclair and Ghiorse, 1989). Some structures were typical of nutrient-deprived cells, but nevertheless dividing cells were also seen.

Biomass measurements, based on membrane lipids, cell wall components, and ATP analyses, correlated with direct microscopic counts (Balkwill *et al.*, 1988; Beloin *et al.*, 1988; Obst and Holzapfel-Pschorn, 1988). Global microbial activities were demonstrated by measurement of respiration or metabolism of radiolabeled substrates (Thorn and Ventullo, 1988; Hicks and Fredrickson, 1989; Phelps *et al.*, 1989b; Pedersen and Ekendahl, 1990). More specific activities such as nitrification, denitrification, and anaerobic metabolic processes were shown by a miniaturized test system (microtiter plates) (Kölbel-Boelke *et al.*, 1988b) or 15 N gas chromatography–mass spectrometry (Bengtsson and Annadotter, 1989), and were even demonstrated in deep sites (Francis *et al.*, 1989; Fredrickson *et al.*, 1989; Jones *et al.*, 1989; Madsen and Bollag, 1989; Pedersen and Ekendahl, 1990).

Morphological and physiological characterization of isolated strains provided information on microbial diversity according to the site or the depth and on metabolic capacities (Balkwill and Ghiorse, 1985; Bianchi *et al.*, 1987; Bone and Balkwill, 1988; Kölbel-Boelke *et al.*, 1988a,b; Balkwill, 1989; Balkwill *et al.*, 1989; Hirsch *et al.*, 1992a). Bone and Balkwill (1988) compared surface soil and subsurface microflora at the site of a shallow aquifer in Oklahoma. The specific types of microorganisms that were numerically predominant in the aquifer sediments were entirely different from those that were predominant in the surface soil. The aquifer microflora is probably less diverse than the surface soil microflora but its diversity does not decrease with increasing depth. Water and sediment communities, as well as communities from different sampling sites and communities from different depths of the same sampling site, differed in their qualitative and quantitative morphotype composition and physiological capabilities. Observations made by Hirsch *et al.* (1992a) resulted in the same conclusions.

The taxonomy of groundwater microorganisms is poorly developed (Bianchi *et al.*, 1987; Kölbel-Boelke *et al.*, 1988b; Balkwill *et al.*, 1989; Breen *et al.*, 1989; Jimenez, 1990). Hirsch *et al.* (1992a) reviewed distributions and taxa of microorganisms observed in groundwater and sediment: 24 different genera were representated, such as *Pseudomonas, Achromobacter, Acineto-bacter, Aeromonas, Alcaligenes, Chromobacterium, Flavobacterium, Mor-axella, Caulobacter, Hyphomicrobium, Sphaerotilus, Gallionella, Arthro-bacter,* and *Bacillus.* Generally, Gram-negative bacteria predominated in sandy horizons of aquifers and the proportion of Gram-positive bacteria was higher in silty clay layers (Thorn and Ventullo, 1988; Fredrickson *et al.*, 1990). Filamentous bacteria and spores have rarely been observed. Even the Gram stain was often doubtful. A high percentage of isolated strains was not identifiable even at the genus level (Bianchi *et al.*, 1987; Balkwill *et al.*, 1989; Amy *et al.*, 1992).

Molecular analysis of DNA or 16 S ribosomal RNA of strains isolated from deep aquifers showed a much greater diversity than that shown in phenotypic tests and confirmed that no strain was identical to a surface strain (Jimenez, 1990; Jimenez *et al.*, 1990; Reeves *et al.*, 1990; Stahl *et al.*, 1990; Stim *et al.*, 1990). This might reflect a long period of adaptation to the environmental conditions of the deep subsurface. Plasmid distribution among bacteria isolated from two pristine groundwater aquifers and a con-taminated groundwater aquifer was studied by Ogunseitan *et al.* (1987). Plasmids (between 3.5 and 202 kb) were found in all three sites but more isolates contained plasmids in the polluted site. The properties usually associ-ated with the presence of plasmids (antibiotic resistance, tolerance to heavy metal salts, and bacteriocin production) occurred at low frequencies and were not restricted to plasmid-bearing strains.

Eucaryotic microorganisms have been detected in groundwaters (Hirsch and Rades-Rohkohl, 1983; Beloin *et al.*, 1988; Sinclair and Ghiorse, 1987, 1989; Sinclair, 1990; Madsen and Ghiorse, 1993). The presence of algae and photosynthetic bacteria must indicate connections to the surface envi-ronment (Sinclair and Ghiorse, 1989). Hirsch and Rades-Rohkohl (1983, 1988) found yeasts, and low numbers of fungi were found in some subsurface samples and even in a deep subsurface aquifer (Sinclair and Ghiorse, 1989; Madsen and Ghiorse, 1993). Protozoa were detected in groundwater samples from various depths (Sinclair and Ghiorse, 1989; Madsen and Ghiorse, 1993). Protozoa enumeration results, obtained using a most probable num-ber technique that relies upon predation of bacterial prey, indicate a moder-ate population density (1–100 cyst-forming protozoa per gram) in shallow aquifers. This was confirmed by analyzing phospholipid fatty acids extracted directly from sediment. Numbers were lowest in clay confining layers and highest in coarse-textured sands in which bacterial abundance was also elevated. Cyst-forming flagellates and amoebae appeared to be the principal types of protozoa in groundwaters. Ciliates were never found in sediment samples from depths greater than 0.5 m (Madsen and Ghiorse, 1993). Ingestion of bacterial prey is the dominant mode of nutrition of free-living protozoa. Their abundance depends on bacterial density and metabolic potential (Madsen and Ghiorse, 1993). After contamination with organic pollutants, the increased availability of carbon sources and bacterial populations allowed populations of protozoa to increase to significant levels (Harvey and George, 1987; Sinclair, 1990).

IV. ADAPTATION OF BACTERIA TO THE SUBSURFACE ENVIRONMENT

The conditions of the subsurface environment seem rather unfavorable to microbial life: the lack of light excludes photosynthetic organisms and their primary production for heterotrophic microorganisms; organic carbon and probably some other elements (*e.g.*, nitrogen or phosphorus) may be scarce; some deep aquifers are poorly aerated; and the temperature is rather low in shallow aquifers in temperate or cold regions. However, it is now well known that bacteria are able to live in several extreme environments. Thus, it is not surprising that they have developed structural and physiological adaptations to the conditions of subsurface sediments.

A. Nutrient Availability

Most pristine aquifers are oligotrophic with respect to sources of energy. All autochthonous subsurface microorganisms must be chemotrophic. In subsurface habitats, the activity of chemolithotrophic bacteria can be restricted by several factors, especially by low temperature in cold areas. In unpolluted subsurface aquifers available nitrogen (ammonium) is usually lacking for nitrifying bacteria. On the contrary, the Fe^{2+} oxidizing *Gallionella* are frequently observed in groundwaters. They are microaerophilic, and therefore the low concentrations of oxygen are particularly beneficial for them. However, they were recently shown to be capable of heterotrophic growth (Hallbeck and Pedersen, 1991). The Fe^{2+}/Mn^{2+} oxidizing bacteria *Leptothrix*, also frequently observed, are probably not autrotrophic. Thiobacteria, which are generally acidophilic, are found in thermal subsurface waters rich in sulfide (Daumas *et al.*, 1986; Sarbu, 1990). Hydrogen oxidizing bacteria were cultured from 35% of the deep subsurface sediments examined by Fredrickson *et al.* (1989).

Most bacteria isolated from subsurface sediments are chemoorganotrophic. Except for peat or lignite layers, the concentration of organic carbon in pristine subsurface water and sediments is very low. Most pristine ground-waters contain less than 1 mg dissolved organic carbon per liter, and aquifer solids contain only trace amounts of organic carbon (Ghiorse and Wilson, 1988). However, some shallow groundwaters contain 9–15 mg of dissolved organic carbon per liter (Matthess *et al.*, 1992). Heterotrophic life in these ecosystems is supported by organic compounds that filter down from the soil above. Most readily metabolized compounds are consumed by the surface microorganisms before they reach the water table aquifer (Alföldi, 1988; Ghiorse and Wilson, 1988). It is suggested that subsurface organic matter consists mostly of humic substances, naphtenic acids, and phenolic compounds derived from either the organic matter in sedimentary rocks or lignin degradation products derived from plant residues from the surface. Many heterotrophic bacteria are able to grow in these oligotrophic conditions. They can develop with the very low concentrations of organic carbon (average, 0.1 to 1 mg) found in groundwaters. They are often able to degrade complex molecules such as humic substances (Grøn *et al.*, 1992), hydrocarbons, or phenolic compounds.

Carbon dioxide is usually abundant in soil and near surface groundwaters. This inorganic carbon source is used by all autotrophic bacteria. Heterotrophic bacteria also require carbon dioxide, but it cannot be their main source of carbon. Heterotrophic activity may be a source of carbon dioxide in some aquifers (Chapelle *et al.*, 1987; McMahon and Chapelle, 1990).

Soluble nitrogen and phosphorus concentrations are often extremely low and probably limiting in groundwater (Ghiorse and Wilson, 1988). Autochthonous subsurface bacteria can develop in these oligotrophic conditions (Hirsch and Rades-Rohkohl, 1983; Ghiorse and Wilson, 1988). Moreover they can mobilize insoluble phosphate, as well as carbon and nitrogen in the form of ligands, from insoluble organic matter such as humic substances. Most of the nonphotosynthetic dinitrogen fixing bacteria, which are autotrophic for nitrogen, are heterotrophic for carbon. Dinitrogen fixation requires much energy and it is limited by low organic carbon availibility and, in cold groundwaters, by low temperature.

In surface environments, several morphological and physiological features enable bacteria to grow in oligotrophic conditions (Hirsch *et al.*, 1979; Morgan and Dow, 1986; Poindexter, 1987; Roszak and Colwell, 1987; Fletcher, 1990). Appendages may increase the cell surface for the sorption of nutrients. Some bacteria produce a sheath and many of them produce extracellular polysaccharides. These structures retain nutrients, especially cations, because they are usually negatively charged. Such adaptive structures are often observed in bacteria from pristine subsurface sediments (Balkwill and Ghiorse, 1985; Ghiorse and Wilson, 1988; Matthess, 1990; Hirsch *et al.*, 1992a; Gounot, 1991) (Fig. 1). They may be lost, however, in laboratory subcultures, even in dilute media. Some appendages and exopolymers allow cell attachment to solid surfaces, resulting in biofilms, which can protect bacteria from abiotic and biotic stresses (*e.g.*, toxic substances, sub-

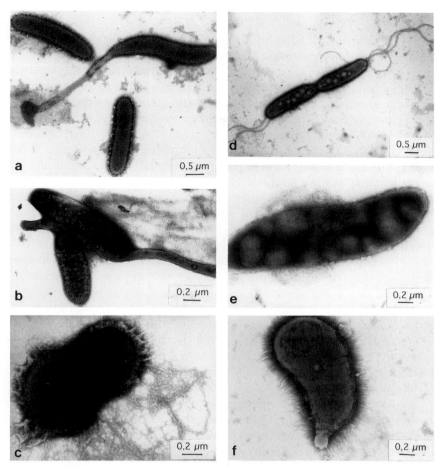

FIGURE 1 Transmission electron micrographs of groundwater bacteria (phosphotungstic acid negatively stained preparations). (a) A stalked Gram-negative cell and rod-shaped Gram-negative cells (direct observation in shallow groundwater). (b) A long, thin, flexible cell, with mesosomes, associated with two rod-shaped Gram-negative cells (direct observation of shallow groundwater). (c) A Gram-negative coccus with exopolymers (pure culture isolate from a subsurface sediment). (d) A Gram-negative dividing cell with polar flagella and light refractile storage granules of poly-β-hydroxybutyrate (pure culture isolate from shallow groundwater). (e) A Gram-negative cell containing large poly-β-hydroxybutyrate storage bodies (pure culture isolate from a deep aquifer). (f) A Gram-negative cell with a short stalk and exopolymers projecting from the surface (fimbriae) ((pure culture isolate from a deep aquifer).

optimal pH, grazing by protozoa, and attack by bacteriophages). Furthermore, attachment prevents pulling out by the water currents and favors cross-feeding. In the groundwater environment, most bacteria are associated with particles and are thus not free living in the pore water (Kölbel-Boelke and Hirsch, 1989; Albrechtsen, 1990; Pedersen and Ekendahl, 1990).

B. Oxygen and Other Electron Acceptors

Although it is difficult to measure oxygen content *in situ,* it seems that most pristine aquifer sediments contain sufficient O_2 or other electron acceptors to support a minimum level of microbial respiratory activity. In stable sedimentary environments, including groundwater systems, the different processes for decomposition of organic matter are segregated in space, time, or both. The reduction sequence is O_2, NO_3^-, MnO_2, $Fe(OH)_3$, SO_4^{2-}, HCO_3^-, and N_2 (Ghiorse and Wilson, 1988; Patrick and Jugsujinda, 1992) and depends on the amount of organic matter. Because of the nutrient limitation and the slow growth rate, oxygen demand is low except in aquifers contaminated by high levels of utilizable organic carbon. Most pristine drinking-water aquifers do contain significant amounts of O_2 (Ghiorse and Wilson, 1988). Aerobic and facultatively anaerobic bacteria are dominant, even in deep aquifers (Bianchi, 1986; Hicks and Fredrickson, 1989; Madsen and Bollag, 1989; Van Beelen and Fleuren-Kemilä, 1989). More than 10% of the bacteria from deep subsurface sediments were found to be microaerophilic, growing optimally under low oxygen partial pressures (Benoit and Phelps, 1990).

Most of the heterotrophic bacteria isolated from groundwaters and sediments are potentially able to reduce nitrate (Bianchi *et al.,* 1987; Francis *et al.,* 1989). Because O_2 is the preferred electron acceptor, this anaerobic respiration takes place when oxygen tension is low, but not zero. However, nitrate concentrations are usually low in groundwaters, unless there is pollution from agricultural fertilizers (Mariotti, 1986; Francis *et al.,* 1989; Fustec *et al.,* 1991). Therefore, in pristine aquifers denitrification can be limited by low concentrations of nitrate and organic carbon.

Manganese oxides and iron oxyhydroxides, which may act as electron acceptors in bacterial respiration (Ehrlich, 1987, 1990; Ghiorse, 1988; Lovley, 1991), are abundant in many subsurface zones. When anaerobic conditions develop, bacterial reduction of these solid-phase minerals may account for the oxidation of organic matter and mobilization of Fe and Mn in groundwaters (Gounot and Haroux, 1986; Gounot *et al.,* 1988; Chapelle and Lovley, 1990; Lovley, 1991; Gounot, 1991).

The other common alternate electron acceptors, SO_4^{2-} and HCO_3^-, can be found in large amounts in shallow subsurface zones (Ghiorse and Wilson, 1988). However, sulfate-reducing bacteria and methanogenic bacteria, which are obligate anaerobes, are not found in oligotrophic pristine sediments. They are present in sediments containing significant amounts of organic matter (Bianchi, 1986; Jones *et al.,* 1989), particularly in oil-containing sediments.

C. pH and Dissolved Salts

In shallow aquifers, the pH value is usually buffered by bicarbonate. Most bacteria are neutrophilic but they tolerate a pH range up to three pH units. Therefore, this factor does not seem to be limiting. Acidophilic bacteria (*e.g.*, *Thiobacillus*) and alcalophilic bacteria are known in surface acidic or alkaline waters, respectively. One can assume that such bacteria can live in acidic or alkaline groundwaters, provided that there is not another limiting factor.

Microbial activities can modify the pH value close to the cells, particularly in biofilms, without a significant change in the free water. This can greatly influence salt solubility, oxidation or reduction processes, and corrosion in the vicinity of bacteria.

Unlike shallow aquifers, old water in deeper aquifers may contain higher salt concentrations and toxic metal concentrations (McNabb and Dunlap, 1975) and are expected to be colonized by bacteria resistant to these conditions. The effect of salinity on subsurface bacterial activities has been investigated in Australia, where the high salinity of some groundwaters is a worrisome problem (Bauld *et al.*, 1990).

D. Temperature

Most bacteria can develop within a range of 30 to 40°C. They are eurytherms. The temperature range of most mesophilic bacteria is 10–45°C, with an optimum between 30 and 37°C. Bacteria which can grow at 0°C are called psychrophiles if they do not grow above 20°C, or psychrotrophs (psychrotolerants) if they do grow above 20°C. True psychrophilic bacteria are restricted to permanently cold habitats. They are found in cold caves, with temperatures of 5°C or below (Gounot, 1970, 1973). Psychrotrophic bacteria are widespread in superficial soils and waters in cold and temperate regions. In caves situated in temperate regions, all autochthonous bacteria are psychrotrophic, growing well at 20°C and above; none are obligate psychrophiles (Gounot, 1973; Perrier, 1973a,b; Brock *et al.*, 1973). However, in a cold cave in Romania stenothermic bacteria were found: they grew at 10 and 20°C but not at 0 and 28°C (Gounot, 1969b).

Little is known about the thermal adaptations of bacteria in aquifer sediments. Generally geothermal temperature gradients show an average increase in ordinary formations of about 2°C per 100 m of depth (Cullimore, 1993). In the shallow aquifers of cold or temperate regions, temperature must have selected psychrotolerant bacteria (Alföldi, 1988; Seppänen, 1988; Hirsch and Rades-Rohkohl, 1992). Low temperature can be a limiting factor for autotrophic activities and nitrogen fixation in cold groundwaters (Gounot, 1970). The deeper formations are favorable to mesophilic bacteria (Balkwill, 1989). Thermal subsurface waters contain thermophilic bacteria (Daumas *et al.*, 1986). Because the temperature of subsurface sediments is stable, the presence of bacteria not adapted to the *in situ* temperature can be used as an indicator of contamination by the suface water (Gounot *et al.*, 1970; Gounot, 1973).

E. Hydrostatic Pressure

Hydrostatic pressure increases with depth. Assuming that in most cases the average rate of increase is approximately 1 atm per 10 m, the hydrostatic pressure reaches about 100 atm at 1000 m, which is not inhibiting for bacteria (Ghiorse and Wilson, 1988; Cullimore, 1993).

F. Darkness

Lack of light prevents the development of organisms with photosynthetic pigments. However, a large number of nonphotosynthetic bacteria are pigmented. Such bacteria have been found in shallow groundwaters (Hirsch *et al.*, 1992a). Many bacteria produce carotenoid pigments, which protect them against harmful photooxidation reactions. In some cases, *e.g.*, in *Corynebacteriacae*, the synthesis of carotenoids is induced by light (Mulder and Antheunisse, 1963). Having isolated many coryneform bacteria from caves, we wondered whether the lack of light resulted in lack of pigmentation, as in the cave fauna. Among 515 *Arthrobacter* strains isolated from cave sediments, less than 1% were pigmented. In surface soils, 19% of *Arthrobacter* strains produced pigments. The percentage was higher in ice field sediments submitted to intense light. However, the absence of pigments was not observed in other groups of bacteria for which the synthesis is not photoinducible (Gounot, 1973).

G. Sediment Texture and Porosity

We have already mentioned the beneficial effects of solid surfaces on bacterial growth in oligotrophic conditions, resulting in higher numbers of attached bacteria in the biofilm than in the bulk water phase. The morphological and physiological types are different in the aquifer water and in particulate sediment (Kölbel-Boelke and Hirsch, 1989; Hazen *et al.*, 1990; Krzanowski *et al.*, 1990). Gram-negative straight or curved rods are largely predominant in free waters, whereas Gram-positive bacteria can be found in large numbers in sediment samples. Motility is common among water isolates, whereas sediment organisms were less often motile (Hirsch *et al.*, 1992a). Attached bacteria are more active (Hazen *et al.*, 1990). In addition, the porosity of sediment, which varies greatly according to its texture, is an important parameter in the vadose zone (Perrier, 1973a,b; Colwell, 1989) and in the saturated zone (Fredrickson *et al.*, 1989; Levine and Ghiorse, 1990; Tonso and Klein, 1990). In aquifers, bacteria are generally present in larger numbers and more active in sandy sediments than in sediments rich in clay minerals (Beloin *et al.*, 1988; Thorn and Ventullo, 1988; Phelps *et al.*, 1989b). The low porosity of clay sediments may result in low oxygen, nutrients, and even water availability. Gram-positive bacteria are more frequently isolated from clay sediments, particularly from unsaturated cave silts (Gounot, 1967). In the vadose zone, microorganisms are submitted to water potential stress and are able to survive under dessication conditions (Perrier, 1973a,b; Kieft *et al.*, 1990). Furthermore, each sample of sediment consists of many different microhabitats, and the distribution of groundwater microorganisms was found to be patchy (Hirsch and Rades-Rohkohl, 1992).

V. ACTIVITY AND FUNCTION OF MICROORGANISMS IN GROUNDWATERS

A. *In Situ* Activity

The question of *in situ* activity of subsurface bacteria is still largely unanswered. There is no doubt that viable microorganisms are present there in large numbers and with a great diversity. Indeed, all of the subsurface isolates from pristine aquifers are capable of growth on very diluted media. Laboratory experiments (sample incubation and studies with microcosms) showed that these microorganisms are active and that they can manage to divide in the unfavorable conditions of their natural environment. Cave silt samples sterilized by gamma irradiation were inoculated either with a diluted silt suspension (containing native microbiota) or with a mixed culture of *Pseudomonas* sp. or *Arthrobacter* sp. isolated from this silt and were incubated at the cave temperature (10°C). *Arthrobacter* sp., *Pseudomonas* sp., and other Gram-negative bacteria did multiply; *Bacillus* and actinomycetes, suspected of being allochthonous in this cave, did not develop under cave conditions (Perrier, 1977).

In-well incubation techniques demonstrated attachment of native bacteria on glass microscope slides and growth of native microorganisms in an incubator device (Cullimore, 1993). The evidence of *in situ* growth within undisturbed sediment by incubation of sterilized sediment for 12 weeks in well water was reported by Hirsch and Rades-Rohkohl (1990). These experimental sediments were entirely colonized and the groundwater had supplied enough carbon and energy to support their growth.

Direct microscopic examinations of subsurface samples revealed the presence of dividing cells, mesosome-like structures, and storage granules (Balkwill and Ghiorse, 1985; Bone and Balkwill, 1988; Balkwill, 1989; Gounot, 1991), suggesting that the cells are metabolically active *in situ*.

B. Geochemical Role

In superficial waters and soils, the geochemical role of bacteria is well known. Bacteria are directly responsible for processes such as mineralization of organic matter, nitrification, denitrification, dinitrogen fixation, oxidation, and reduction of sulfur compounds. Furthermore, by altering physico-chemical conditions, bacteria are indirectly involved in many processes of oxidation or reduction, precipitation or solubilization, adsorption or desorption, and complexation. All these functions can take place in groundwaters (Seppänen, 1988), and they have been well described in pristine and polluted aquifers (Ghiorse and Wilson, 1988). In pristine aquifers, they are slowed down by oligotrophic conditions.

Organic carbon is usually rather rare in pristine subsurface sediments, except in some organic layers. Subsurface bacteria are potentially able to degrade many natural or xenobiotic compounds. This activity can become significant in contaminated areas (Ghiorse and Wilson, 1988). At a noninhibitory concentration, the contaminant is degraded after a lag period, accounting for bacterial adaptation. The long time of residence of contaminants in aquifers allows ample opportunity for microbial adaptation. The rate of biodegradation would be limited by the supply of an appropriate electron acceptor (oxygen or nitrate); nevertheless anaerobic biodegradation can be effective as well. Many bacteria are able to degrade complex organic molecules such as phenolic compounds and hydrocarbons (Arvin et al., 1988; Hicks and Fredrickson, 1989). The oxidation of aromatic compounds can be linked to the reduction of ferric oxides (Lovley, 1991).

The food chain in aquifers is primarily heterotrophic, dependent either on the influx of dissolved organic carbon or on organic materials of sedimentary origin that subsequently may have been transformed (Madsen and Ghiorse, 1993). In hyporheic zones, the amount of organic matter is sufficient for a significant microbial productivity (Hendricks, 1993), which may contribute to the nutrition of invertebrates, as in surface waters (Decho and Castenholz, 1986; Perlmutter and Meyer, 1991). With the exception of karstic systems, all life-forms larger than microorganisms are effectively excluded from most groundwater habitats. On the contrary, microbial life is found even in deep subsurface layers. In these cases the food chain must be restricted to bacterial assimilation of the low amount of dissolved organic carbon or sedimentary organic materials and to bacterial predation by protozoa. In spite of the fact that the potential for biomass production by chemolithotrophic bacteria has been demonstrated in some subsurface sediments (Fredrickson et al., 1989), aerobic chemosynthetic food chains have not been described in pristine groundwater (Madsen and Ghiorse, 1993).

In unpolluted subsurface habitats, the activities of the nitrogen cycle are limited by the low availability of nitrogen compounds and energy sources. Organic as well as inorganic nitrogen compounds are not abundant in subsurface pristine sediments. Nitrifying bacteria are very few or absent. Nitrogen fixation by heterotrophic bacteria is limited by low concentrations of organic matter and, in cold sediments, by the low temperature. We have already mentioned that most bacteria isolated from groundwaters and sediments are able to reduce nitrate when organic carbon and nitrate are present. In polluted aquifers, these bacteria can carry out both degradation of organic compounds and denitrification.

Iron and manganese are usually abundant in subsurface sediments; the geochemical role of bacteria in iron and manganese cycles is, therefore, particularly important (Ghiorse, 1984; Nealson *et al.*, 1988; Chapelle and Lovley, 1990; Ehrlich, 1990; Lovley, 1991; Gounot, 1994). In aerobic conditions, soluble Fe(II) and Mn(II) are oxidized, and the oxides or hydroxides are precipitated. At lower redox conditions, bacterial reduction of these solid-phase minerals, by direct electron transfer or production of reducing metabolites, results in solubilization and migration of Fe(II) and Mn(II) (Gounot and Haroux, 1986; Haroux, 1987; Gounot *et al.*, 1988; Di Ruggiero, 1989; Di Ruggiero and Gounot, 1990; Gounot and Di Ruggiero, 1991; Gounot, 1994). Furthermore, Fe(II) and Mn(IV) oxides strongly adsorb a wide variety of toxic trace metals in the sediment (Kepkay, 1985; Balikungeri and Haerdi, 1988). The remobilization of heavy metals is a potential hazard when Fe(II) and Mn(IV) are reduced (von Gunkel and Sztraka, 1986; Bourg *et al.*, 1989; Lovley, 1991).

In caves, bacteria are involved in the formation of concretions (Billy and Chalvignac, 1976) or "mondmilch" genesis (Caumartin and Renault, 1958; Pochon *et al.*, 1964; Mason-Williams, 1959), in saltpeter formation (Fliermans et al., 1974), and in the formation of sulfur deposits (Sarbu, 1990). According to Coman (1979), biological processes play an important part in cave genesis, not only in karst but also in other formations. Silicate weathering was demonstrated in a shallow petroleum-contaminated aquifer (Hiebert and Bennett, 1992).

Microbially mediated salt ion precipitation can have detrimental effects by plugging or corrosion in extraction and piping of groundwater; in extraction, piping, and storage of petroleum and natural gas; or in metal solubilization. It is beneficial for bioremediation of contaminated groundwater systems. Biological treatments are already used *in situ* for the elimination of iron and manganese (Göttfreund *et al.*, 1985; Breaster and Martinell, 1988; Jaudon *et al.*, 1989; Dumousseau *et al.*, 1990) or nitrate (Collin *et al.*, 1987; Boussaid *et al.*, 1988; Breaster and Martinell, 1988; Janda *et al.*, 1988; Mercado *et al.*, 1988; Hamon and Fustec, 1991).

C. Origin of Groundwater Microorganisms

Microorganisms may colonize groundwater environments by active or passive migration via percolation from the surface above, lateral migration from the recharge areas, or introduction into the sediments during deposition (Ghiorse and Wilson, 1988; Sargent and Fliermans, 1989; Madsen and Ghiorse, 1993).

Migration of microorganisms is controlled by flow-length-dependent transport processes (advection–dispersion and adsorption–desorption) and predominantly by filtration (Matthess *et al.*, 1989). Extended propagation of microbial populations occurs only in large fractures, fissures, and solution holes, whereas filtration controls the transport behavior of microorganisms in porous aquifers. Allochthonous bacteria and viruses may move only a short distance in nonfissured aquifers (Gerba and Bitton, 1984). They are usually eliminated by time-dependant biological, chemical, and physical processes, and their number decreases exponentially with time (Matthess *et al.*, 1989; Matthess, 1990). They cannot reach confined zones and deep aquifers.

In pristine cave silts, bacteria found in a fossil gallery were well adapted to physical and chemical conditions: low, stable temperature; oligotrophy; and matrix potential (Perrier, 1977). They were perhaps brought in with sediments when the gallery was active and flooded by groundwater.

The presence of sterile layers between layers containing numerous microorganisms proves that deep subsurface microorganisms cannot originate from vertical migration in these aquifers (Bianchi *et al.*,1987). All bacteria isolated from pristine aquifers were found to be physiologically and genetically different from surface bacteria. Recent colonization by surface bacteria seems unlikely.

Lateral migration through aquifer sediments is poorly understood. It may be rapid in some porous alluvial aquifers and very slow in other aquifers, particularly in deep and confined aquifers. In the vegetative state, bacterial cells have a diameter of between 0.5 and 5 μm. However, many starved bacterial cells shrink in diameter size to 0.1 to 0.5 μm and become nonattachable. Such stressed cells are able to pass through porous structures and colonize small pores even in rock formations (Morita, 1985; Cullimore, 1993). However, many observations suggest that, in pristine sediments, the bacteria are nearly all attached (Kölbel-Boelke and Hirsch, 1989; Albrechtsen, 1990). Thus, the movement of these microorganisms over a long distance would appear difficult. Many more experiments are needed to answer fundamental questions about the sources of subsurface microorganisms (Madsen and Ghiorse, 1993).

VI. CONCLUSION

Even though groundwaters are often oligotrophic, bacteria are present in subsurface sediments extending into deep zones in which no other organisms are able to survive. Bacteria can be active in these environments and are adapted to low-nutrient and microaerophilic conditions. Most bacteria in groundwaters are epilithic. In pristine aquifers, some bacterial activities can be limited by the low availability of nutrients and energy. In polluted zones, these activities can be spontaneously enhanced, or they can be stimulated by acting on environmental conditions. Groundwater bacteria are potentially able to degrade many natural or xenobiotic compounds such as pesticides, fertilizers, and industrial chemicals, which can reach aquifers.

Basic research is needed to better characterize the microbial populations and activities in subsurface sediments in uncontamined areas. Very little is known about the movement of bacteria through aquatic sediments, about the stratification of activity zones, or about the grazing pressure of subsurface protozoa on bacterial population density. It is necessary to study how hydrogeological and geochemical factors can affect microbial abundance, their distribution, and their activities. Such information is essential for *in situ* bioremediation attempts.

ACKNOWLEDGMENTS

I thank Professor P. Hirsch for his constructive criticism of the first draft of the manuscript. Professor Stanford also reviewed the manuscript and made many interesting suggestions.

REFERENCES

Albrechtsen, H. J. (1990). Bacteria and surfaces in the groundwater environment. *In* "Microbiology of the Deep Subsurface" (C. B. Fliermans and T. C. Hazen, eds.), S4, pp. 85–86. WSRC Information Services, Aiken, South Carolina.

Alföldi, L. (1988). Groundwater microbiology: Problems and biological treatment. State-of-the art report. *Water Sci. Technol.* 20, 1–31.

Amy, P. S., Haldeman, D. L., Ringelberg, D., Hall, D. H., and Russell, C. (1992). Comparison of identification systems for classification of bacteria isolated from water and endolithic habitats within the deep subsurface. *Appl. Environ. Microbiol.* 58, 3367–3373.

Arvin, E., Jensen, B., Aamand, J., and Jørgensen, C. (1988). The potential of free-living ground water bacteria to degrade aromatic hydrocarbons and heterocyclic compounds. *Water Sci. Technol.* 20, 109–118.

Balikungeri, A., and Haerdi, W. (1988). Complexing abilities of hydrous manganese oxide surfaces and their role in the speciation of heavy metals. Intern. *J. Environ. Anal. Chem.* 34, 215–225.

Balkwill, D. L. (1989). Numbers, diversity and morphological characteristics of aerobic, chemoheterotrophic bacteria in deep subsurface sediments from a site in South Carolina. *Geomicrobiol. J.* 7, 33–52.

Balkwill, D. L., and Ghiorse, W. C. (1985). Characterization of subsurface bacteria associated with two shallow aquifers in Oklahoma. *Appl. Environ. Microbiol.* 50, 580–588.

Balkwill, D. L., Rucinsky, T. E., and Casida, L. E., Jr. (1977). Release of microorganisms from soil with respect to electron microscopy viewing and plate counts. *Antonie van Leeuwenhoek* 16, 73–81.

Balkwill, D. L., Leach, F. R., Wilson, J. T., McNabb, J. F., and White, D. C. (1988). Equivalence of microbial biomass measures based on membrane lipid and cell wall components, adenosine triphosphate, and direct counts in subsurface aquifer sediments. *Microb. Ecol.* 16, 73–84.

Balkwill, D. L., Fredrickson, J. F., and Thomas, J. M. (1989). Vertical and horizontal variations in the physiological diversity of the aerobic chemoheterotrophic bacterial microflora in deep southeast coastal plain subsurface sediments. *Appl. Environ. Microbiol.* 55, 1058–1065.

Bauld, J., Evens, W. R., and Kellett, J. R. (1990). Groundwater systems or the Murray Basin, Southeastern Australia. In "Microbiology of the Deep Subsurface" (C. B. Fliermans and T. C. Hazen, eds.), S2, pp. 83–96. WSRC Information Services, Aiken, South Carolina.

Beloin, R. M., Sinclair, J. L., and Ghiorse, W. C. (1988). Distribution and activity of microorganisms in subsurface sediments of a pristine study site in Oklahoma. *Microb. Ecol.* **16**, 65–97.

Bengtsson, G. (1989). Growth and metabolic flexibility in groundwater bacteria. *Microb. Ecol.* **18**, 235–248.

Bengtsson, G., and Annadotter, H. (1989). Nitrate reduction in a groundwater microcosm determined by ^{15}N gas chromatography-mass spectrometry. *Appl. Environ. Microbiol.* **55**, 2861–2870.

Benoit, R. E., and Phelps, T. J. (1990). Microaerophilic bacteria from subsurface sediments. In "Microbiology of the Deep Subsurface" (C. B. Fliermans and T. C. Hazen, eds.), S4, pp. 87–96. WSRC Information Services, Aiken, South Carolina.

Bianchi, A. (1986). Les types bactériens hétérotrophes survivant dans les sédiments quaternaires et pliocène supérieur du delta de la Mahakam. *C. R. Seances Acad. Sci., Ser. 3* **303**, 449–451.

Bianchi, A., Hinojosa, M., Garcin, J., Delebassee, M., Normand, M., Ralijoana, C., Sohier, L., Vianna Doria, E., and Villata, M. (1987). Etude bactériologique des sédiments quaternaires et pliocène supérieur du delta de la Mahakam (Kalimantan, Indonésie). In "Le sondage Misedor," pp. 206–224. Technip, Paris.

Billy, C., and Chalvignac, M. A. (1976). Rôle des facteurs biologiques dans la calcification des grottes de Lascaux et de Font-de Gaume. *C. R. Hebd. Seances Acad. Sci., Ser. D* **283**, 207–209.

Birstein, J. A., and Borutzky, E. V. (1950). En russe (La vie dans les eaux souterraines). *Sizn Priesnych vod SSSR* **3**, 683–706.

Bone, T. L., and Balkwill, D. L. (1988). Morphological and cultural comparison of microorganisms in surface soil and subsurface sediments at a pristine study site in Oklahoma. *Microb. Ecol.* **16**, 49–64.

Bourg, A. C., Darmendrail, E., and Ricour, J. (1989). Geochemical filtration of riverbank and migration of heavy metals between the Deule River and the Ansereuilles Alluvion-Chalk Aquifer (Nord, France). *Geoderma* **44**, 229–244.

Boussaid, F., Martin, G., Morvan, J., Collin, J. J., Landreau, A., and Talbo, H. (1988). Denitrification in situ of groundwater with solid carbon matter. *Environ. Technol. Lett.* **9**, 803–816.

Breaster, C., and Martinell, R. (1988). The VYREDOX and NITREDOX methods of in situ treatment of groundwater. *Water Sci. Technol.* **20**, 149–163.

Breen, A., Stahl, D. A., Flesher, B., and Sayler, G. (1989). Characterization of *Pseudomonas geomorphus*: A novel groundwater bacterium. *Microb. Ecol.* **18**, 221–233.

Brock, T. D., Passman, F., and Voder, I. (1973). Absence of obligately psychrophilic bacteria in constantly cold springs associated with caves in Southern Indiana. *Am. Midl. Nat.* **90**, 240–246.

Caldwell, D. E., Korber, D. R., and Lawrence, J. R. (1992). Confocal laser microscopy and digital image analysis in microbial ecology. *Adv. Microb. Ecol.* **12**, 1–67.

Camus, H., Lion, R., Berthelin, J., Desjardin, T., Bianchi, A., and Garcin, J. (1987). Etude des microorganismes présents dans les couches géologiques profondes. *Rapp. CEE Sci. Tech. Nucl.* **EUR 11141 FR.**

Caumartin, V. (1959). Quelques aspects nouveaux de la microflore des cavernes. *Ann. Spéléol.* **14**, 147–157.

Caumartin, V. (1963). Review of the microbiology of underground environments. *Bull. Natl. Speleol. Soc.* **25**, 1–14.

Caumartin, V. (1964). Essai sur une étude au microscope électronique de la microflore des sédiments argileux de cavernes. *Int. J. Speleol.* **1**, 1–17.

Caumartin, V., and Renault, P. (1958). La corrosion biochimique dans un réseau karstique et la genèse du mondmilch. *Notes Biospéol.* **13**, 87–109.

Chapelle, F. H. (1993). "Ground-Water Microbiology and Geochemistry." Wiley, New York.

Chapelle, F. H., and Lovley, D. R. (1990). Fe^{3+}-reducing bacteria in deep coastel plain aquifers: A mechanism for the origin of high iron concentration in groundwater. *In* "Microbiology of the Deep Subsurface" (C. B. Fliermans and T. C. Hazen, eds.), S5, pp. 19–24. WSRC Information Services, Aiken, South Carolina.

Chapelle, F. H., Zelibor, J. L. Jr., Grimes, D. J., and Knobel, L. L. (1987). Bacteria in deep coastal plain sediments of Maryland: A possible source of CO_2 to groundwater. *Water Resour. Res.* **23**, 1625–1632.

Collin, J. J., Landreau, A., Talbot, H., Martin, G., and Morvan, J. (1987). Procédé de dénitrification des eaux souterraines en vue de leur potabilisation. Eur. Pat. Office, EP 0.133.405 B1.

Colwell, F. S. (1989). Microbiological comparison of surface soil and unsaturated subsurface soil from a semiarid high desert. *Appl. Environ. Microbiol.* **55**, 2420–2423.

Coman, D. (1979). Essai sur une interprétation écologique de l'origine des grottes. *Trav. Inst. Spéol. Emile Racovitza* **18**, 191–199.

Cullimore, D. R. (1993). "Practical Manual of Groundwater Microbiology." Lewis, Chelsea, MI.

Daumas, S., Lombart, R., and Bianchi, A. (1986). A bacteriological study of geothermal spring waters dating from the dogger and trias period in the Paris Basin. *Geomicrobiol. J.* **4**, 423–433.

Decho, A. W., and Castenholz, R. W. (1986). Spatial patterns and feeding of meiobenthic Harpactocoid copepods in relation to resident microbial flora. *Hydrobiologia* **131**, 87–96.

Dickson, G. W. (1975). A preliminary study of heterotrophic microorganisms as factors in substrate selection of troglobitic invertebrates. *NSS Bull.* **37**, 89–93.

Di Ruggiero, J. (1989). Ecologie et physiologie des bactéries réduisant le manganèse. Exemple de la nappe alluviale du Rhône, Avignon (Vaucluse). Thèse, Univ. Lyon I.

Di Ruggiero, J., and Gounot, A. M. (1990). Microbiol manganese reduction mediated by bacterial strains isolated from aquifer sediments. *Microb. Ecol.* **20**, 53–63.

Dobat, K. (1970). Considération sur la végétation cryptogamique des grottes du Jura Souabe (Sud-Ouest de l'Allemagne). *Ann. Spéléol.* **25**, 871–907.

Dudich, E. (1930). Die Nahrungsquellen der Tierwelt in der Aggteleker Tropfsteinhöhle. *Allattorv. Közl.* **27**, 77–85.

Dumousseau, B., Jaudon, P., Massiani, J., Vacelet, E., and Claire, Y. (1990). Origine du manganèse de la nappe alluviale de Beaucaire (Gard, France). Essai de démanganisation *in situ* (procédé Vyredox). *Rev. Sci. Eau* **3**, 21–36.

Ehrlich, H. L. (1987). Manganese oxide reduction as a form of anaerobic respiration. *Geomicrobiol. J.* **5**, 423–431.

Ehrlich, H. L. (1990). "Geomicrobiology." Dekker, New York.

Eichem, A. C., Dodds, W. K., Tate, C. M., and Edler, C. (1993). Microbial decomposition of elm and oak leaves in a karst aquifer. *Appl. Environ. Microbiol.* **59**, 3592–3596.

Fischer, E. (1959). Bakterie dwoch zbiornikow wodnych jarkin tatrzanskich. *Pol. Arch. Hydrobiol.* **6**, 189–199.

Fletcher, M. (1990). Bacterial colonization of solid surfaces in subsurface environments. *In* "Microbiology of the Deep Subsurface" (C. B. Fliermans and T. C. Hazen, eds.), S7, pp. 3–12. WSRC Information Services, Aiken, South Carolina.

Fliermans, C. B., Bohlool, B. B., and Schmidt, E. L. (1974). Autecological study of the chemoautotroph *Nitrobacter* by immunofluorescence. *Appl. Microbiol.* **27**, 124–129.

Francis, A. J., Slater, J. M., and Dodge, C. J. (1989). Denitrification in deep subsurface sediments. *Geomicrobiol. J.* **7**, 103–106.

Fredrickson, J. K., Garland, T. R., Hicks, R. J., Thomas, J. M., Li, S. M., and McFadden,

K. M. (1989). Lithotrophic and heterotrophic bacteria in deep subsurface sediments and their relation to sediment properties. *Geomicrobiol. J.* 7, 53–66.

Fredrickson, J. K., Balkwill, D. L., Zachara, J., Brockman, F., Griffin, E., and Li, S. M. (1990). Microorganisms in deep cretaceous sediments of the atlantic coastal plain: Vertical variations and sampling considerations. *In* "Microbiology of the Deep Subsurface" (C. B. Fliermans and T. C. Hazen, eds.), S3, pp. 53–63. WSRC Information Services, Aiken, South Carolina.

Fustec, E., Mariotti, A., Grillon, X., and Sajus, J. (1991). Nitrate removal by denitrification in alluvial ground water: role of a former channel. *J. Hydrol.* 123, 337–354.

Gerba, C. P., and Bitton, G. (1984). Microbial pollutants: Their survival and transport pattern to groundwater. *In* "Groundwater Pollution Microbiology" (C. P. Gerba and G. Bitton, eds.), pp. 65–88. Wiley, New York.

Ghiorse, W. C. (1984). Biology of iron and manganese-depositing bacteria. *Annu. Rev. Microbiol.* 38, 515–550.

Ghiorse, W. C. (1988). Microbial reduction of manganese and iron. *In* "Biology of Anaerobic Microorganisms" (A. J. B. Zehnder, ed.), pp. 305–331. Wiley, New York.

Ghiorse, W. C., and Balkwill, D. L. (1983). Enumeration and morphological characterization of bacteria indigenous to subsurface environments. *Dev. Ind. Microbiol.* 24, 213–224.

Ghiorse, W. C., and Wilson, J. T. (1988). Microbial ecology of the terrestrial subsurface. *Adv. Appl. Microbiol.* 33, 107–172.

Göttfreund, E., Göttfreund, J, Gerber, I., Schmitt, G., and Schweisfurth, R. (1985). Occurence and activities of bacteria in the unsaturated and saturated underground in relation to the removal of iron and manganese. *Water Supply* 3, 109–115.

Gounot, A. M. (1967). La microflore des limons argileux souterrains: Son activité productrice dans la biocoenose cavernicole. *Ann. Spéléol.* 22, 23–143.

Gounot, A. M. (1969a). Contribution à l'étude des bactéries des grottes froides. *C. R. Congr. Int. Spéléol. 5th,* Stuttgart 1969, Vol. 4, B23, pp. 1–6.

Gounot, A. M. (1969b). Etude préliminaire du peuplement bactérien du limon de la grotte de Peyort (Ariège). *Ann. Spéléol.* 24, 595–601.

Gounot, A. M. (1970). Quelques observations sur le micropeuplement des limons des grottes arctiques. *Bull. Soc. Linn. Lyon* 39, 226–236.

Gounot, A. M. (1973). Recherches sur les bactéries cavernicoles. *C. R. Cong. Natl. Soc. Savantes, Sect. Sci.* 96(3), 257–265.

Gounot, A. M. (1974). Analyse microbiologique d'un limon souterrain des Montagnes Rocheuses. *Ann. Spéléol.* 29, 333–334.

Gounot, A. M. (1991). Ecologie microbienne des eaux et sédiments souterrains. *Hydrogéologie* 3, 239–241.

Gounot, A. M. (1994). Microbial oxidation and reduction of manganese. Consequences in groundwater and applications. *FEMS Microbiol. Rev.* (in press).

Gounot, A. M., and Di Ruggiero, J. (1991). Rôle géochimique des bactéries dans les eaux souterraines: Exemple du cycle du manganèse dans les nappes aquifères. *Hydrogéologie* 3, 249–256.

Gounot, A. M., and Haroux, C. (1986). Manganese transformation in groundwaters. *In* "Microbial Communities in Soil" (V. J. Jensen, A. Kjoller, and L. H. Sorensen, eds.), pp. 293–304. Elsevier, London.

Gounot, A. M., Breuil, C., Brogère, P., and Siméon, D. (1970). Action sélective de la température sur le micropeuplement des grottes froides. *Spelunca Mém.* 7, 141–144.

Gounot, A. M., Di Ruggiero, J., and Haroux, C. (1988). Bacterial manganese transformation in groundwaters. *In* "Current Perspectives in Environmental Biogeochemistry" (G. Giovannozzi-Sermanni and P. Nannipieri, eds.), pp. 371–382. C.N.R.-I.P.R.A., Roma, Italia.

Grøn, C., Tørsløv, J., Albrechtsen, H. J., and Møller Jensen, H. (1992). Biodegradability of dissolved organic carbon in groundwater from an unconfined aquifer. *Sci. Total Environ.* **117/118**, 241–251.

Hallbeck, L., and Pedersen, K. (1991). Autotrophic and mixotrophic growth of Gallionella ferruginea. *J. Gen. Microbiol.* **137**, 2657–2661.

Hamon, M., and Fustec, E. (1991). Laboratory and field study of an *in situ* groundwater denitrification reactor. *Res. J. Water Pollut. Control Fed.* **63**, 942–949.

Haroux, C. (1987). Biogéochimie du manganèse dans les nappes aquifères libres du domaine Rhodanien. Thèse, Univ. Lyon I.

Harvey, R. W., and George, L. H. (1987). Growth determinations for unattached bacteria in a contaminated aquifer. *Appl. Environ. Microbiol.* **53**, 2992–2996.

Hazen, T. C., Jimenez, L., Fliermans, C. B., and Lopez De Victoria, G. (1990). Comparison of bacteria from deep subsurface sediment and adjacent groundwater. *In* "Microbiology of the Deep Subsurface" (C. B. Fliermans and T. C. Hazen, eds.), S2, pp. 141–158. WSRC Information Services, Aiken, South Carolina.

Hendricks, S. P. (1993). Microbial ecology of the hyporheic zone: A perspective integrating hydrology and biology. *J. North Am. Benthol. Soc.* **12**, 70–78.

Hicks, R. J., and Fredrickson, J. K. (1989). Aerobic metabolic potential of microbial populations indigenous to deep subsurface environments. *Geomicrobiol. J.* **7**, 67–77.

Hiebert, F. K., and Bennett, P. C. (1992). Microbial control of silicate weathering in organic-rich ground water. *Science* **258**, 278–281.

Hirsch, P., and Rades-Rohkohl, E. (1983). Microbial diversity in a groundwater aquifer in Northern Germany. *Dev. Ind. Microbiol.* **24**, 183–200.

Hirsch, P., and Rades-Rohkohl, E. (1988). Some special problems in the determination of viable counts of groundwater microorganisms. *Microb. Ecol.* **16**, 99–113.

Hirsch, P., and Rades-Rohkohl, E. (1990). Microbial colonization of aquifer sediment exposed in a groundwater well in northern Germany. *Appl. Environ. Microbiol.* **56**, 2963–2966.

Hirsch, P., and Rades-Rohkohl, E. (1992). The natural microflora of the Segeberger Forst aquifer system. *In* "Progress in Hydrochemistry" (G. Matthess, F. Frimmel, P. Hirsch, H. D. Schulz, and H. E. Usdowski, eds.), pp. 390–412. Springer-Verlag, Heidelberg.

Hirsch, P., Bernhard, M., Cohen, S. S., Ensign, J. C., Jannasch, H. W., Koch, A. L., Marshall, K. C., Matin, A., Poindexter, J. S., Rittenberg, S. C., Smith, D. C., and Veldkamp, H. (1979). Life under conditions of low nutrient concentrations. *In* "Strategies of Microbial Life in Extreme Environments" (M. Shilo, ed.), pp. 357–372. Verlag Chemie, Mannheim.

Hirsch, P., Rades-Rohkohl, E., Kölbel-Boelke, J., and Nehrkorn, A. (1992a). Morphological and taxonomic diversity of groundwater microorganisms. *In* "Progress in Hydrochemistry" (G. Matthess, F. Frimmel, P. Hirsch, H. D. Schulz, and H. E. Usdowski, eds.), pp. 311–325. Springer-Verlag, Heidelberg.

Hirsch, P., Rades-Rohkohl, E., Kölbel-Boelke, J., Nehrkorn, A., Schweisfurth, R., Selenka, F., and Hack, A. (1992b). Methods of studying ground water microbiology: Critical evaluation and method suggestions. *In* "Progress in Hydrochemistry" (G. Matthess, F. Frimmel, P. Hirsch, H. D. Schulz, and H. E. Usdowski, eds.), pp. 325–333. Springer-Verlag, Heidelberg.

Hobbie, J. E., and Ford, T. E. (1993). A perspective on the ecology of aquatic microbes. *In* "Aquatic Microbiology: An Ecological Approach" (T. E. Ford, ed.), pp. 1–14. Blackwell, Boston.

Janda, V., Rudovsky, J., Wanner, J., and Marha, K. (1988). *In situ* denitrification of drinking water. *Water Sci. Technol.* **20**, 215–219.

Jaudon, P., Massiani, C., Galea, J., and Rey, J. (1989). Groundwater pollution by manganese, manganese speciation: Application to the selection and discussion of an *in situ* groundwater treatment. *Sci. Total Environ.* **84**, 169–183.

Jimenez, L. (1990). Molecular analysis of deep-subsurface bacteria. *Appl. Environ. Microbiol.* **56**, 2108–2113.

Jimenez, L., Lopez De Victoria G., Wear, J., Fliermans, C. B., and Hazen, T. C. (1990). Molecular analysis of deep subsurface bacteria. *In* "Microbiology of the Deep Subsurface" (C. B. Fliermans and T. C. Hazen, eds.), S2, pp. 97–113. WSRC Information Services, Aiken, South Carolina.

Jones, R. E., Beeman, R. E., and Suflita, J. M. (1989). Anaerobic metabolic processes in the deep terrestrial subsurface. *Geomicrobiol. J.* 7, 117–130.

Kepkay, P. E. (1985). Kinetics of microbial manganese oxidation and trace metal binding in sediments: Results from an *in situ* dialysis technique. *Limnol. Oceanogr.* 30, 713–726.

Kieft, T. L., Rosacker, L. L., Wilcox, D., and Franklin, A. J. (1990). Water potential and starvation stress in deep subsurface microorganisms. *In* "Microbiology of the Deep Subsurface" (C. B. Fliermans and T. C. Hazen, eds.), S4, pp. 99–111. WSRC Information Services, Aiken, South Carolina.

Kilbertus, G., and Schwartz, R. (1981). Relations microflore-microfaune dans la grotte de Sainte-Catherine (Pyrénées ariégeoises). I. Recherche des sources trophiques. *Rev. Ecol. Biol. Sol* 18, 305–317.

King, L. K., and Parker, B. C. (1988). A simple, rapid method for enumerating total viable and metabolically active bacteria in groundwater. *Appl. Environ. Microbiol.* 54, 1630–1631.

Kölbel-Boelke, J., and Hirsch, P. (1989). Comparative physiology of biofilm and suspended organisms in the groundwater environment. *In* "Structure and Function of Biofilms" (W. G. Characklis and P. A. Wilderer, eds.), pp. 221–238. Wiley, New York.

Kölbel-Boelke, J., Tienken, B., and Nehrkorn, A. (1988a). Microbial communities in the saturated groundwater environment. I. Methods of isolation and characterization of heterotrophic bacteria. *Microb. Ecol.* 16, 17–29.

Kölbel-Boelke, J., Anders, E. M., and Nehrkorn, A. (1988b). Microbial communities in the saturated groundwater environment. II. Diversity of bacterial communities in a pleistocene sand aquifer and their in vitro activities. *Microb. Ecol.* 16, 31–48.

Krzanowski, K. M., Sinn, C. A., and Balkwill, D. L. (1990). Attached and unattached bacterial populations in deep aquifer sediments from a site in South Carolina. *In* "Microbiology of the Deep Subsurface" (C. B. Fliermans and T. C. Hazen, eds.), S5, pp. 25–30. WSRC Information Services, Aiken, South Carolina.

Leach, L. E. (1990). An aseptic procedure for soil sampling in heaving sands using special hollow-stem auger coring. *In* "Microbiology of the Deep Subsurface" (C. B. Fliermans and T. C. Hazen, eds.), S2, pp. 3–18. WSRC Information Services, Aiken, South Carolina.

Levine, S. N., and Ghiorse, W. C. (1990). Analysis of environmental factors affecting abundance and distribution of bacteria, fungi and protozoa in subsurface sediments of the upper atlantic coastal plain. *In* "Microbiology of the Deep Subsurface" (C. B. Fliermans and T. C. Hazen, eds.), S5, pp. 31–45. WSRC Information Services, Aiken, South Carolina.

Lovley, D. R. (1991). Dissimilatory Fe(III) and Mn(IV) reduction. *Microbiol. Rev.* 55, 259–287.

Madsen, E. L., and Bollag, J. M. (1989). Aerobic and anaerobic microbial activity in deep subsurface sediments from the Savannah River plant. *Geomicrobiol. J.* 7, 93–101.

Madsen, E. L., and Ghiorse, W. C. (1993). Groundwater microbiology: Subsurface ecosystem processes. *In* "Aquatic Microbiology: An Ecological Approach" (T. E. Ford, ed.), pp. 167–213. Blackwell, Boston.

Magdeburg, P. (1933). Organogene Kalkkonkretionen in Höhlen. Beiträge zur Biologie der in Höhlen vorkommenden Algen. *Sitzungsber. Naturforsch. Ges. Leipzig* 56–59, 14–36.

Mariotti, A. (1986). La dénitrification dans les eaux souterraines, principes et méthodes de son identification: Une revue. *J. Hydrol.* 88, 1–23.

Marxsen, J. (1988). Investigation into the number of respiring bacteria in groundwater from sandy and gravelly deposits. *Microb. Ecol.* 16, 65–72.

Mason-Williams, A. (1959). The formation and deposition of moon-milk. *Trans. Cave Res. Group. G. B.* 5, 133–138.

Mason-Williams, A. (1969). Comments on the bacterial populations of small pools in caves. *Actes Congr. Int. Spéléol., 4th,* Yougoslavie, 1965, Vol. 4/5, pp. 162–166.

Mason-Williams, A., and Benson-Evans, K. (1958). A preliminary investigation into the bacterial and botanical flora of caves in South Wales. *Publ. Cave Res. Group. G. B.* 8, 1–70.

Matthess, G. (1990). Hydrogeological controls of bacterial and virus migration in subsurface environments. *In* "Microbiology of the Deep Subsurface" (C. B. Fliermans and T. C. Hazen, eds.), S7, pp. 33–46. WSRC Information Services, Aiken, South Carolina.

Matthess, G., Pekdeger, A., and Schroeter, J. (1989). Persistence and transport of bacteria and viruses in groundwater. A conceptual evaluation. *J. Contam. Hydrol.* 2, 171–188.

Matthess, G., Frimmel, F., Hirsch, P., Schulz, H. D., and Usdowski, T. E., eds. (1992). "Progress in Hydrogeochemistry." Springer-Verlag, Heidelberg.

McMahon, P. B., and Chapelle, F. H. (1990). A model of sulfate diffusion and bacterial production of CO_2 in the black creek aquifer, South Carolina. *In* Microbiology of the Deep Subsurface" (C. B. Fliermans and T. C. Hazen, eds.), S2, pp. 137–140. WSRC Information Services, Aiken, South Carolina.

McNabb, J. F., and Dunlap, W. J. (1975). Subsurface biological activity in relation to ground water pollution. *Ground Water* 13, 33–44.

McNabb, J. F., and Mallard, G. E. (1984). Microbiological sampling in assessment of groundwater pollution. *In* "Groundwater Pollution Microbiology" (C. P. Gerba and G. Bitton, eds.), pp. 235–260. Wiley, New York.

Mercado, A., Libhaber, M., and Soares, M. I. M. (1988). *In situ* biological groundwater denitrification: Laboratory studies. *Water Sci. Technol.* 20, 197–209.

Morgan, P., and Dow, C. S. (1986). Bacterial adaptations for growth in low nutrient environments. *In* "Microbes in Extreme Environments" (R. A. Herbert and G. A. Codd, eds.), pp. 187–214. Academic Press, Orlando, FL.

Morita, R. Y. (1985). Starvation and miniaturization of heterotrophs, with special emphasis on maintenance of the starved viable state. *In* "Bacteria in their Natural Environment" (M. Fletcher and G. D. Floodgate, eds.), pp. 11–130. Academic Press, Orlando, FL.

Mulder, E. G., and Antheunisse, R. I. (1963). Morphologie, physiologie et écologie des *Arthrobacter. Ann. Inst. Pasteur, Paris* 105, 46–74.

Nealson, K. H., Tebo, B. M., and Rosson, R. A. (1988). Occurrence and mechanisms of microbial oxidation of manganese. *Adv. Appl. Microbiol.* 33, 279–318.

Obst, U., and Holzapfel-Pschorn, A. (1988). Biochemical testing of groundwater. *Water Sci. Technol.* 20, 101–107.

Ogunseitan, O. A., Tedford, E. T., Pacia, D., Sirotkin, K. M., and Sayler, G. S. (1987). Distribution of plasmids in groundwater bacteria. *J. Ind. Microbiol.* 1, 311–317.

Palik, P. (1960). A new blue-green alga from the cave Baradla near Aggtelek (Biospeologica Hungarica XII). *Ann. Univ. Sci. Budap. Rolando Eotvos Nominatae, Sect. Biol.* 3, 275–286.

Pasqualini, A., Fumanti, B., and Visona, L. (1978). Microflore et activité de groupements fonctionnels dans les sédiments de trois grottes de l'Italie Centrale. *Int. J. Speleol.* 10, 73–105.

Patrick, W. H., and Jugsujinda, A. (1992). Sequential reduction and oxidation of inorganic nitrogen, manganese, and iron in flooded soil. *Soil Sci. Soc. Am. J.* 56, 1071–7073.

Pedersen, K., and Ekendahl, S. (1990). Distribution and activity of bacteria in deep granitic groundwaters of Southeastern Sweden. *Microb. Ecol.* 20, 37–52.

Perlmutter, D. G., and Meyer, J. L. (1991). The impact of a stream-dwelling harpacticoid copepod upon detritally associated bacteria. *Ecology* 72, 2170–2180.

Perrier, J. (1973a). Répartition des bactéries dans les limons argileux souterrains en fonction des facteurs écologiques. *C. R. Congr. Natl. Soc. Savantes, Sect. Sci.* 96(3), pp. 211–225.

Perrier, J. (1973b). L'établissement et l'évolution de la microflore dans les limons souterrains en fonction des conditions de milieu. Thèse, Univ. Lyon I.

Perrier, J. (1977). Colonisation bactérienne de limons souterrains après stérilisation par irradiation. *Rev. Inst. Pasteur Lyon* 10, 175–184.

Phelps, T. J., Fliermans, C. B., Garland, T. R., Pfiffner, S. M., and White, D. C. (1989a). Methods for recovery of deep terrestrial subsurface sediments for microbiological studies. *J. Microbiol. Methods* 9, 267–279.

Phelps, T. J., Raione, E. G., and White, D. C. (1989b). Microbial activities in deep subsurface environments. *Geomicrobiol. J.* 7, 79–91.

Pochon, J., Chalvignac, M. A., and Krumbein, W. (1964). Recherches biologiques sur le mondmilch. *C. R. Hebd. Seances Acad. Sci., Ser. D* 258, 5113–5115.

Poindexter, J. S. (1987). Bacterial responses to nutrient limitation. *In* "Ecology of Microbial Communities" (M. Fletcher, T. R. G. Gray, and J. G. Jones, eds.), pp. 283–317. Cambridge Univ. Press, Cambridge, UK.

Reeves, J. Y., Reeves, R. H., and Balkwill, D. L. (1990). Restriction endonuclease analysis of deep subsurface bacterial isolates. *In* "Microbiology of the Deep Subsurface" (C. B. Fliermans and T. C. Hazen, eds.), S2, p. 115. WSRC Information Services, Aiken, South Carolina.

Roszak, D. B., and Colwell, R. R. (1987). Survival strategies of bacteria in the natural environment. *Microbiol. Rev.* 51, 365–379.

Sarbu, S. (1990). The unusual fauna of a cave with thermomineral waters containing H_2S from Southern Dobrogea, Romania. *Mém. Biospéol.* 17, 191–195.

Sargent, K. A., and Fliermans, C. B. (1989). Geology and hydrology of the deep subsurface microbiology sampling sites at the Savannah River Plant, South Carolina. *Geomicrobiol. J.* 7, 3–13.

Schreiber, G. (1929). Il contenuto di sostanza organica nel fango delle grotte di Postumia. *Atti Accad. Sci. Veneto-Trentino-Istriana* 20, 51–53.

Seppänen, H. (1988). Groundwater: A living ecosystem. *Water Sci. Technol.* 25, 95–100.

Sinclair, J. L. (1990). Eukaryotic microorganisms in subsurface environments. *In* "Microbiology of the Deep Subsurface" (C. B. Fliermans and T. C. Hazen, eds.), S3, pp. 39–51. WSRC Information Services, Aiken, South Carolina.

Sinclair, J. L., and Ghiorse, W. C. (1987). Distribution of protozoa in subsurface sediments of a pristine groundwater study site in Oklahoma. *Appl. Environ. Microbiol.* 53, 1157–1163.

Sinclair, J. L., and Ghiorse, W. C. (1989). Distribution of aerobic bacteria, protozoa, algae and fungi in deep subsurface sediments. *Geomicrobiol. J.* 7, 15–31.

Smed-Hildmann, R., and Filip, Z. (1988). Microorganisms of groundwater and decomposing refuses as viewed by the transmission electron microscope. *Water Sci. Technol.* 20, 233–235.

Smith, R. L., and Harvey, R. W. (1990). Development of sampling techniques to measure *in situ* rates of microbial processes in a contaminated sand and gravel aquifer. *In* "Microbiology of the Deep Subsurface" (C. B. Fliermans and T. C. Hazen, eds.), S2, pp. 19–33. WSRC Information Services, Aiken, South Carolina.

Stahl, D. A., Key, R., and Balkwill, D. L. (1990). Phylogenetic diversity among subsurface microorganisms. *In* "Microbiology of the Deep Subsurface" (C. B. Fliermans and T. C. Hazen, eds.), S2, p. 69. WSRC Information Services, Aiken, South Carolina.

Stim, K. P., Drake, G. R., Padgett, S. E., and Balkwill, D. L. (1990). 16S Ribosomal RNA sequencing analysis of phylogenetic relatedness among aerobic chemoheterotrophic bacteria in deep aquifer sediments from a site in South Carolina. *In* "Microbiology of the Deep Subsurface" (C. B. Fliermans and T. C. Hazen, eds.), S2, pp. 117. WSRC Information Services, Aiken, South Carolina.

Thorn, P. M., and Ventullo, R. M. (1988). Measurement of bacterial growth rates in subsurface sediments using the incorporation of tritiated thymidine into DNA. *Microb. Ecol.* 6, 3–16.

Tonso, N. L., and Klein, D. A. (1990). Particle-size relationship to heterotrophic microbial community characteristics of two deep subsurface samples. *In* "Microbiology of the Deep

Subsurface" (C. B. Fliermans and T. C. Hazen, eds.), S2, pp. 71–79. WSRC Information Services, Aiken, South Carolina.

Van Beelen, P., and Fleuren-Kemilä, A. K. (1989). Enumeration of anaerobic and oligotrophic bacteria in subsoils and sediments. *J. Contam. Hydrol.* **4,** 275–284.

Varga, L., and Takats, T. (1960). Mikrobiologische Untersuchungen des Schlammes eines wasserlosen Teiches der Aggteleker Baradla-Höhle. *Acta Zool. Acad. Sci. Hung.* **6,** 429–437.

von Gunkel, G., and Sztraka, A. (1986). Untersuchungen zum Verhalten von Schwermetallen in Gewässern. II. Die Bedeutung für die Eisen- und Mangan-Remobilisierung für die hypolimnische Anreicherung von Schwermetallen. *Arch. Hydrobiol.* **106,** 91–117.

Webster, J. A., Hampton, G. J., Wilson, J. T., Ghiorse, W. C., and Leach, F. R. (1985). Determination of microbial cell numbers in subsurface samples. *Ground Water* **23,** 17–25.

White, D. C., Smith, G. A., Gehron, M. J., Parker, J. H., Findlay, R. H., Martz, R. F., and Fredrickson, H. L. (1983). The groundwater aquifer microbiota: Biomass, community structure, and nutritional status. *Dev. Ind. Microbiol.* **24,** 201–211.

White, D. C., Ringelberg, D. B., Guckert, J. B., and Phelps, T. J. (1990). Biochemical markers for *in situ* microbial community structure. *In* "Microbiology of the Deep Subsurface" (C. B. Fliermans and T. C. Hazen, ed.), S4, pp. 45–56. WSRC Information Services, Aiken, South Carolina.

Wilson, J. T., McNabb, J. F., Balkwill, D. L., and Ghiorse, W. C. (1983). Enumeration and characterization of bacteria indigenous to a shallow water-table aquifer. *Ground Water* **21,** 134–142.

Wolters, N., and Schwartz, W. (1956). Untersuchungen über Vorkommen and Verhalten von Mikroorganismen in reinum Grundwasser. *Arch. Hydrobiol.* **51,** 500–541.

Adaptation of Crustacea to Interstitial Habitats: A Practical Agenda for Ecological Studies

D. L. Danielopol,* M. Creuzé des Châtelliers,†
F. Moeszlacher,* P. Pospisil,‡ and R. Popa§

*Limnological Institute, Austrian Academy of Sciences
Gaisberg, 116
5310, Mondsee, Austria

†Université Lyon I
U.R.A. CNRS 1451, "Ecologie des Eaux Douces et des Grands Fleuves"
Laboratoire d'Hydrobiologie et Ecologie Souterraines
69622 Villeurbanne Cedex, France

‡Institute of Zoology, University of Vienna
1090 Vienna, Austria

§Institute of Speology "E.G. Racovitza," Rumanian Academy
11 RO-78109 Bucuresti, Rumania

I. INTRODUCTION

Adaptation is a central problem in evolutionary biology and a perennial topic for theoretical and empirical research in subterranean biology (Culver, 1982; Christiansen, 1992). Because one of the aims of ecology is to understand the causes of distribution and abundances of the organisms (Krebs, 1972), studies on organismal adaptation are of paramount interest.

Ecologists dealing with groundwater aspects seldom approach the complex problems of organismal adaptations for animals living permanently in the subsurface environment [see, for a review, Culver (1982) and, for later data, Kane and Culver (1992), Mathieu and Turquin (1992), and Parzefall (1992)].

Danielopol and Rouch (1991) proposed an agenda for ecological research dealing with this topic. The framework proposed by these authors is here reviewed and illustrated with data from an ongoing project on groundwater ecology carried out in Austria.

The Crustacea were chosen as the animal group and the interstitial habitats of unconsolidated sediments, as the paradigmatic environment to demonstrate how one can implement an evolutionary ecological project centered around the adaptation topic. Crustacea are one of the most diversified groups inhabiting many types of groundwaters (Botosaneanu, 1986), and unconsolidated sediments offer these animals a wide variety of habitats. Hence one of the major aims of biologists is to describe both the patterns of adaptations displayed by groundwater Crustacea and the processes by which they originated. An expanded agenda of research is here proposed in order to fulfil this latter aim.

The success of ecological programs centered around the adaptation topic depends not only on the development of new research projects and the accumulation of data but also on the way in which students will be able to form sound and testable hypotheses about specific adaptations of groundwater animals. For this purpose cooperation with evolutionary biologists, with behavioral ecologists, and with geneticists will be an advantage.

In this chapter, we examine adaptations as a major topic in subterranean biology. Then we show that it is possible to implement an ecological program related to adaptation of groundwater Crustacea. Finally we propose evolutionary scenarios and discuss preadaptation and natural selection.

II. ADAPTATION: A MAJOR TOPIC IN SUBTERRANEAN BIOLOGY

Subterranean dwelling organisms are open systems that have to steadily process matter in order to get energy for their maintainance and reproduction. To achieve this goal subsurface organisms have to put their structural and functional features at all levels of internal organization in concordance with the subterranean environment within which they exist (they fit or adjust their biological organization to the surrounding environment). This peculiarity is, in its widest sense, called *adaptation* (the state of being adapted is called *adaptedness;* Burian, 1992), and it is one of the major aspects of living beings. Subterranean dwelling organisms change over time, they evolve adaptations that allow them to fit their environment and hence they are able to continue to play the "existential game." A common process by which adaptation occurs is natural selection (Williams, 1966; Brandon, 1990).

The subterranean aquatic environment is, generally, perceived by natural scientists [for a recent review, see Marmonier *et al.,* (1993)] as less complex than the epigean one, with habitats having longer persistence and lower environmental fluctuations. Such habitats are energetically poor and have simple (short) trophic links (Culver, 1982; Hüppop, 1985). Groundwaters are one of the most oligotrophic systems of the biosphere (Thurman, 1985); because of the total darkness there are only a few primary producers, the chemolitotrophic bacteria. Such environmental parameters as temperature and oxygen are considered to fluctuate within narrow ranges (Illies, 1971).

Groundwater, filling the interstitial space in unconsolidated rocks, forms narrow labyrinthine channels that are interconnected. The meio- and macro-organisms living in such interstitial habitats (Fig. 1) display remarkable convergent morphological and physiological characters, as well as similar ecological strategies. The animals are generally of minute size and elongated shape compared to their surface dwelling relatives, they are blind and unpigmented, and they have reduced limbs and/or elongated sensorial structures, which compensate for the lack of vision. These morphological traits, called by Christiansen (1962) troglomorphic characters, stimulated many biologists to investigate and to speculate on their origin (Delamare-Deboutteville, 1960; Culver and Fong, 1986). Many interstitial animals produce a low number of eggs and develop slowly; *e.g.,* Danielopol (1980) gave such examples for limnic ostracods. These specializations are considered adaptations to the characteristics of the subterranean environment (Culver, 1982).

For a long time it was widely accepted that exclusively subterranean-dwelling animals are one of the best examples of narrowly specialized organisms adapted to stable and persistent environments (Vandel, 1965; Ginet and Decou, 1977); the other classic example being the deep-sea benthic fauna (Margalef, 1983). The close fit of troglobite species to their environment suggested to many biologists that these animals became maladaptive to other environments and that they are "out-of-the-way" organisms (Fage, 1931; Vandel, 1965; Müller, 1974; Botosaneanu and Holsinger, 1991) that survived in the subsurface habitats as in a "closed shop" [this latter simile, used by Westheide (1987), characterizes the interstitial habitats from which epigean-dwelling species, especially large predators, are excluded]. Ecological work on both karstic and interstitial limnic fauna demonstrates that the subterranean environment is much more diverse, the environmental

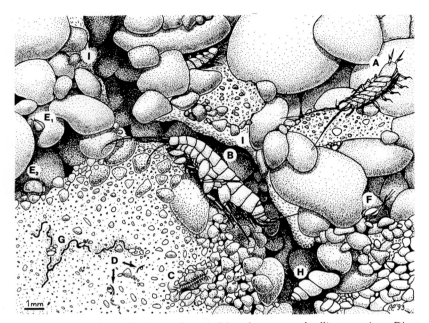

FIGURE 1   The interstitial habitat and some of the subterranean-dwelling organisms. Diagram composed, mainly, after video pictures taken in the Lobau miniaquifer. A, *Proasellus slavus* (Isopoda); B, *Niphargus* sp. (Amphipoda); C, *Bathynella* sp. (Syncarida); D, *Parastenocaris* sp. (Copepoda, Harpacticoida); E$_1$,E$_2$, *Cryptocandona kieferi* and *Kovalevskiella* sp. (Ostracoda); F, *Acanthocyclops gmeineri* (Copepoda, Cyclopoida); G, Oligochaeta; H, *Bythiospeum* sp. (Gastropoda); I, bacterial biofilm.

parameters are more fluctuating than earlier accepted, and subterranean animals display a wider range of adaptations to the ecological space within which they live [Rouch, 1986 (review); Camacho *et al.*, 1992; Mathieu and Turquin, 1992; Danielopol *et al.*, 1992]. Besides species with narrow ecological tolerances to subsurface environmental parameters and restricted spatial distributions (stenaptic species, to use a term coined by E. G. Racovitza; Jeannel, 1950), there are others that are widely tolerant (euryaptic species).

Rouch and Danielopol (1987) showed that many exclusively groundwater-dwelling organisms stem from epigean and ecologically euryvalent species. Within this context it became interesting to study the adaptations of organisms related to the true complexity of groundwater environments, *i.e.*, where the systems are heterogenous and variable on all scales of perception. There is unfortunately little ecological information dealing with this problem [see, for a review, Mathieu and Turquin (1992)].

Danielopol and Rouch (1991, p. 139), noting that "one of the main aims of subterranean ecology is the description of the diversity and complexity of subsurface world and of the adaptive phenomena related to it," proposed an agenda of research centered around these problems.

The need for good ecological data in order to understand various evolutionary processes has been stressed repeatedly (Endler, 1986; Thomson, 1988). Here we show how we implemented an ecological program related to adaptation of groundwater Crustacea.

III. THE SEARCH FOR PATTERN: A PRACTICAL AGENDA

A. The Danielopol and Rouch (1991) Proposals

1. A Historical Definition of Adaptation

Danielopol and Rouch (1991) were attracted by the sequential (hierarchical) way of defining adaptation, depending on the purposes and the level of information available (Brandon, 1990, p. 194; Reeve and Sherman, 1993), *i.e.*, from the descriptive and static concept of adaptedness to the more advanced one, which includes the historical process (*e.g.*, natural selection) by which adaptation originates. Adaptation for Danielopol and Rouch (1991, p. 131), in its static variant, "refers to the characteristic traits of an organism which reflects its relationships with the environment in which the organism lives and which in principle allows or contributes to its existence in this environment." This definition is similar to those of Bock (1979, p. 39), and it was felt that it better fits the needs of groundwater ecologists, who in most cases deal with field and laboratory work without quantifing the efficiency of natural selection [the work of Jones *et al.* (1992) is an exception].

2. The Diversity of the Subterranean Environment

Danielopol and Rouch (1991) emphasized that in the subterranean environment, which is heterogenous and complex on all perceptual scales, a high diversity of adaptive organismal responses would also be expected. Therefore the major task of ecologists should be to describe, as thoroughly as possible, both the subterranean environment and the organisms that inhabit it.

3. The Level of Description: A Holistic and Comparative Approach

Generally, adaptation is seen through the properties of organismal traits. However, evolutionary processes, like natural selection, operate on the whole individual. Danielopol and Rouch (1991) consider that ecologists should study the ways in which whole organisms interact with their environment.

Ecological experience shows that the diversity of adaptations can be hierarchically structured. There are general adaptations and specializations.

The former are those that represent basic solutions to the existential problems of the organisms (Brown, 1958, p. 166). Special adaptations (specializations) are the original solutions produced by different groundwater organisms depending on historical contingencies and the organismal boundary conditions.

In order to better understand the diversity of adaptations, their originality, their functional meaning, and their efficiency, the usage of comparative methods based on careful phylogenetic analysis is an absolute necessity (Harvey and Pagel, 1991). The study of the organismal constraints of other animals that fail to colonize the subterranean environment is also of much interest.

4. The Historical and Functionalist Approach

From the static approach used for the description of adaptations one should expand the investigations toward the problems of evolutionary origins and their processes. We should ask why organisms fit their environment and how well they fit it. Danielopol and Rouch (1991) consider that groundwater organisms live successfully in the subsurface due not only to the natural selection process acting in this environment but also to the acquisition of traits prior to the colonization of subterranean habitats. Many animals succeed because they are preadapted to life in the subsurface.

Most groundwater organisms actively perceive their environment; they are able to discover and to adopt their preferential habitat. Therefore Danielopol and Rouch (1991) propose to differentiate between the process of adaptation and that of adoption (habitat selection, *sensu lato*). This is very little understood by ecologists.

B. Organisms and Their Environment: The Case History of *Proasellus slavus*

We took as the object of our investigations an isopod species belonging to a widely distributed crustacean group in subterrânean waters, the Asellota. *Proasellus slavus* (subspecies *P. slavus vindobonensis;* Fig. 2) belongs to an old phylogenetic lineage of the familly Asellidae, which inhabits both interstitial and karstic waters (Henry, 1976). Several subspecies were described from subterranean waters (Henry *et al.*, 1986), *P. slavus serbiae* and *P. slavus zeii* from Serbia, *P. slavus slavus* from Slovakia, and *P. slavus histriae* and *P. slavus styriacus* from Slovenia. The species also occurs in the alluvial plain of the Drava in Croatia (Lattinger-Penko, 1976). Two subspecies from Austria are known, in Vienna, *P. slavus vindobonensis,* and in Salzburg, *P. slavus salisburgensis.* During the last 15 years the nominate species were collected along the Danube and one of its tributaries (the Seebach) in Lower Austria (Danielopol, 1976a,b, 1983; Marmonier, 1985).

P. slavus, compared with other freshwater asellid species, is a minute

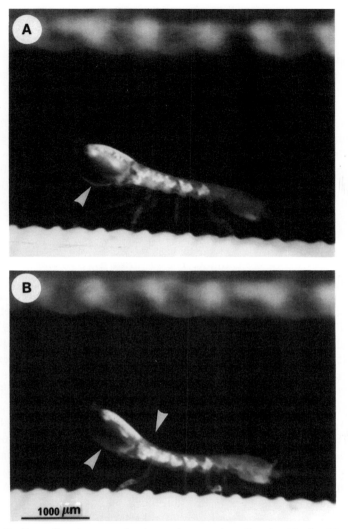

FIGURE 2 *Proasellus slavus* moving in a microaquarium (length of body, 2.8 mm). (A) Position of the operculum, opened wider (see arrow), and (B) dorsal arched position of the body (see arrow).

and delicate isopod. It is blind and unpigmented, and its body is elongate and flexible. Using the video technique that we developed to observe groundwater habitats (Niederreiter and Danielopol, 1991; Pospisil, 1992) as well as on the basis of laboratory observations, we can say that the animal is very active, it moves through the interstitial space rapidly, and it steadily explores its environment with its long antennae. The adult specimens from the Lobau are between 2 and 3.5 mm in length, and their dry weight varies between

0.1 and 0.2 mg. *P. slavus zeii,* which lives in a cave in Serbia, is not larger than the closely located subspecies *P. slavus serbiae* from alluvial sediments of the river Vrana, *i.e.,* 5.5–5.7 mm in length (Henry *et al.,* 1986).

Proasellus and *Asellus* species are detritivores and sediment feeders. Their mouth limbs and digestive system are well developed to extract energy from the ingestion of microorganisms that are attached to the sediment or to organic debris.

The Asellidae, in general, have a well-developed respiratory system (Magniez, 1976). The exopodite of the third pleopod forms an operculum, which, being rigid, efficiently moves the water, which flows toward the respiratory areas of the pleopods (Fig. 2), *i.e.,* the endopodite of the third pair and the endo- and exopodites of the fourth and fifth pleopod pairs. These are very flexible and are in permanent movement too. The operculum also has the role of protecting the respiratory pleopods by closing them in a respiratory cavity.

The interstitial habitats in which we observed *P. slavus* are formed by alluvial sediments of the Danube. The substrate is very heterogenous. Fine sediment accumulates between cobbles and gravel, mainly sand and silt, on which microorganisms develop (Fig. 1). There is, in many cases, a layering of this sediment; sandy layers alternate with silty ones or with well-sorted gravel (Pospisil, 1993; see also Chapter 13). The Lobau area, in its present state (Fig. 3), reflects the old Danube system of meandering channels that were cut off at the end of the last century (Pospisil and Danielopol, 1990). From a hydrogeologic point of view the Lobau groundwaters belong to the Marchfeld aquifer (Danielopol *et al.,* 1994; see also Chapter 13). Figure 4 represents the field area of an intensive research program carried out over a three-year period (Danielopol *et al.,* 1992; Gunatilaka *et al.,* 1994; Pospisil, 1992, 1993, and unpublished). The area forms a well-defined miniaquifer system, of about 1600 m^2, in which the surface water strongly infiltrates the system from the Eberschuttwasser and exfiltrates the system in the next abandoned arm, the Mittelwasser (Dreher *et al.,* 1994). Danielopol (1983) showed that *P. slavus* lives in the superficial sediments, *i.e.,* 50 cm below the bottom of an abandoned arm of the Danube, the Eberschuttwasser (Fig. 3, arrow) (Chapter 13, site A), where the interstitial waters were well oxygenated (more than 1 mg/l^{-1}). We found the species in the wetland Lobau around the Eberschuttwasser (Fig. 3) in areas with varying oxygen conditions (Fig. 5), 7 to 9 m below the ground. Some of the wells are located in well-oxygenated sediments (*e.g.,* well 81) during the whole year; others are located in areas that are better oxygenated during the winter and the spring seasons (more than 1 mg/l^{-1} O$_2$) and remain hypoxic (less than 0.5 mg/l^{-1} O$_2$) the rest of the time (*e.g.,* wells 84 and 96). Other sites remain poorly oxygenated during the whole year (wells 63, 74, and 90).

The oxygen situation is complex and depends on the dynamics of the surface water that infiltrates the miniaquifer and on the biological and

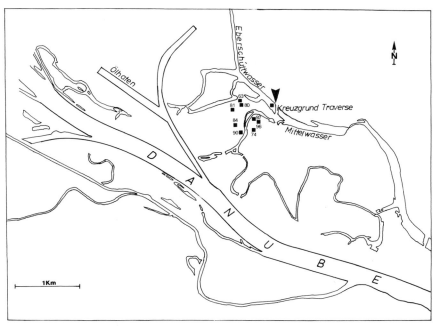

FIGURE 3   The Danube wetland in the Vienna area. Squares and arrow indicate the sites where *P. slavus* was found; the arrow locates the position of the miniaquifer represented in Fig. 4.

chemical oxygen consumption within the subsurface system. Generally, one can speak about well-oxygenated groundwaters during winter and spring and hypoxic waters during summer and a part of autumn (Fig. 5). *P. slavus* forms within this area patchy aggregates of individuals (Fig. 4) that, apparently, move over time (Fig. 6). We found the species in sandy sediments that accumulate between gravel and cobbles. In well T3, which is a metal piezometer, we regularly observed isopods moving on the well's inner wall. This well is covered with a thin bacterial biofilm. Figure 6 shows the quantitative distribution of *P. slavus* around wells T3 and D3. We extracted the isopods from the interstitial space using a Bou–Rouch pump connected to a double packer sampler (Danielopol and Niederreiter, 1987; Pospisil, 1992). One notices that the species occurs at all depths, and the highest abundances are related to the periods when the groundwater is well oxygenated (Fig. 5). At site T3, which is located far from the Eberschuttwasser (Fig. 4), the oxygen level in the layers closely located to the groundwater table (*i.e.*, 1–5 m below the ground) is higher than that in the deeper layers (Fig. 5); the isopods are more abundant in this upper zone (Fig. 6). Using the oxygen data from the Eberschuttwasser area, where *P. slavus* occurs, we constructed a graph of the physiological quality of the interstitial habitat

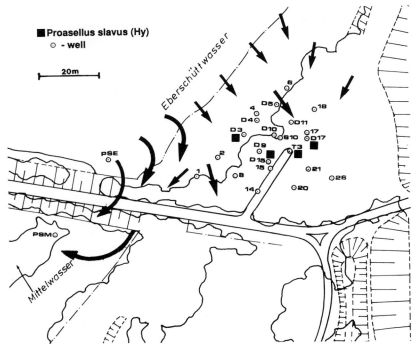

FIGURE 4   The Eberschuttwasser miniaquifer with the position of the multilevel wells (D + number), of normal piezometers (simple numbers), and of the sites where *P. slavus* was found. Diagrammatic representation; arrows indicate the direction of the water circulation.

(Fig. 7), following Huey's (1991) method. When the oxygen concentrations are grouped under three quality classes and the isopod abundances for each class, as mean values for the samples of each quality class are plotted, one can see that the highest values are in the well-oxygenated class (Fig. 7). Therefore the physiological optimum of this euryoxic species lies at higher oxygen concentrations.

We investigated the concentration of the dissolved organic carbon and the bacterial densities and activity. Within the Eberschuttwasser aquifer, at the sites where isopods were found, there is high heterogeneity of the spatial and temporal distribution of these parameters (Danielopol *et al.*, 1992; Gunatilaka *et al.*, 1994; F. Moeszlacher and P. Torreiter, unpublished). We also noticed high patchy distributions of the bacterial biofilms that cover the sediment grains or that accumulate within the interstitial space (Fig. 1).

We used the videocamera technique of Niederreiter and Danielopol (1991) to observe the activity of the isopods in the subsurface space of the alluvial sediments of the Eberschuttwasser. Figure 1 reconstructs diagrammatically a typical situation: *P. slavus* occurs in medium-size pore spaces and does not coexist for long periods of time with the facultative carnivores

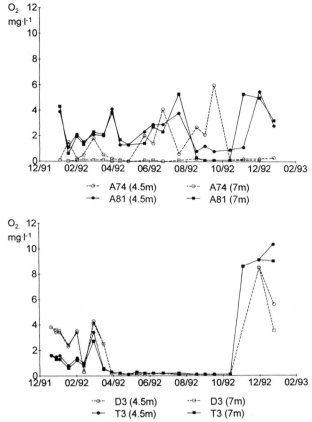

FIGURE 5   Oxygen dynamics in the Lobau; evolution for two different sediment depths below the ground. Oxygen measured with a double packer in multilevel wells. For the location of wells A74 and A81, see Fig. 3; for wells D3 and T3, see Fig. 4.

belonging to the genus *Niphargus* (Amphipoda). In well T3 we noticed how *P. slavus* approaches the *Niphargus* individuals but it recognizes them at a very close distance and avoids their contact (F. Moeszlacher, unpublished). These data are in accordance with those of Marmonier (1985), who reported that, in the deeper layers of the Seebach sediment, the stygobite isopods and amphipods, even if they occur in the same depth layer, select different microhabitats. It was not observed either in the field or in laboratory experiments that *Niphargus* species predate subterranean isopods as described by Henry (1976) or that *P. slavus* (Fig. 1) interacts with other meio- and macroorganisms, such as ostracods, cyclopoids, harpacticoids, gastropods, and oligochaetes. We could not relate the isopod abundances at any sites and depth levels either to the bacterial biofilms or to the accumulation of organic matter. This negative situation could be due to the high mobility

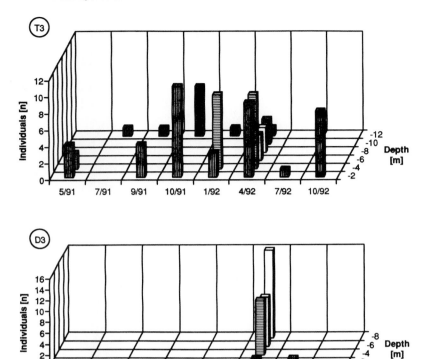

FIGURE 6   Abundances of *P. slavus* in 5-l groundwater samples extracted from various depths using the multilevel wells and a Bou–Rouch pump fixed to a double packer.

of *P. slavus;* the isopods, instead of staying within optimal rich organic patches, move steadily in order to explore the interstitial space, and they stay within a patch and feed only for reduced periods of time. This hypothesis was tested under laboratory conditions. When placed to feed on dead plant leaves under darkness and high oxygen conditions, *P.slavus* alternates long periods of movement, exploring its habitat, with short periods of feeding, where it scrapes the leaf's surface and ingests both plant material and bacteria (Fig. 8); note that replicate observations for every individual were made in infrared light, using a videocamera and recorder (F. Moeszlacher, unpublished). In the field, *P. slavus* feeds on fine sediment, fine organic detritus, and bacteria. Accumulation of fecal pellets, as coarse sediment grains, in some microhabitats was reported by Danielopol (1989) and Danielopol *et al.* (1994).

In other experiments we studied the ventilation activity of the *P. slavus* pleopods as well as the general movement of the animal. For these observations we used miniaquaria, with one individual immersed in water maintained under constant temperature (12°C), normal light conditions, and

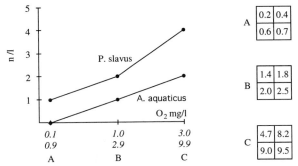

FIGURE 7   The abundances of *P. slavus* and *A. aquaticus* in groundwater samples from the Eberschuttwasser area, ordered following three quality classes of the interstitial habitats, related to the oxygen concentration.

varying oxygen concentrations (9, 1, and 0.1 mg/l^{-1} O_2); we recorded respiratory behavior with video techniques (M. Creuzé des Châtelliers, unpublished). Figure 9 shows that *P. slavus* ventilates actively during long periods of time and maintains a high speed of pleopod movement, independently of the environmental oxygen concentration. The energetic effort to respire under low oxygen conditions (especially at 0.1 mg/l^{-1} O_2) is very high, and the isopods slow down their exploratory movement. During rest periods, most of the individuals adopt a special position of the body, *i.e.*, the pleotelson and the posterior half of the pereion take a dorsal arched position, the operculum is opened wider, and the amplitude of the respiratory pleopod movements is higher. Apparently, through this special behavior, the animal

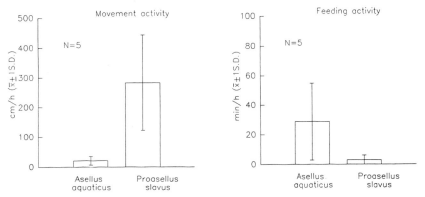

FIGURE 8   Movement and feeding activities of two isopods, under experimental conditions in the laboratory. The food is a dead leaf 1 cm^2 in area placed in a petri dish 6 cm in diameter; *N*, number of individuals successively observed in darkness conditions with an infrared videocamera setup. For the movement, the total length of the track within the 1-hr observation period is represented; for the feeding activity, the percentage of the time spent on the leaf is recorded.

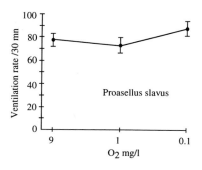

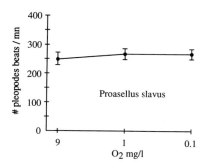

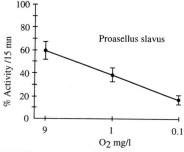

FIGURE 9   Respiration activity and motility of *P. slavus* at three different oxygen concentrations. For each experiment, six individuals were followed. The data are expressed as arithmetic means with ±1 standard error.

can better extract oxygen from its surroundings. The absolute oxygen consumption was measured in a closed minirespirometer designed by I. Bals for the oxymeter "Orbisphere" (F. Moeszlacher and M. Creuzé des Châtelliers, unpublished). *P. slavus,* being a very minute isopod, consumes small quantities of oxygen compared with other larger asellid species (Fig. 10). The respiration rate of *P. slavus* (Fig. 11), after an initial period of 5–6 hr of acclimatization, remains more or less stable for a long time period (at least 36 hr) and is independent of the external oxygen concentration.

C. Comparative Data

In the alluvial sediments of the Eberschuttwasser, at 50 cm depth and in well-oxygenated areas (more than 1 mg/l^{-1} O$_2$), we repeatedly found the epigean- dwelling isopod *Asellus aquaticus* (Danielopol, 1983). The species occurs neither in the deeper sediment layers nor in the groundwater zones located far from the surface water, but rather in habitats with high oxygen conditions, similar to those where we found *P. slavus* (Fig. 7).

A. aquaticus is a large species; the adults attain more than 1 cm in length and its dry weight varies between 1 and 2 mg. The population studied

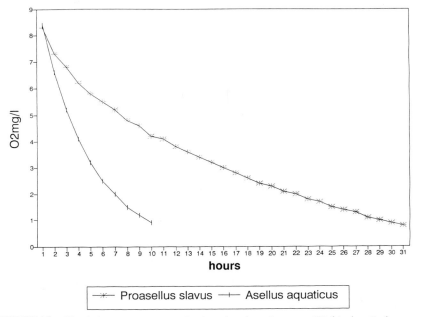

FIGURE 10   Absolute oxygen consumption in a closed respirometer, "Orbisphere"; the curves
follow the mean values of six individuals for the hourly consumption.

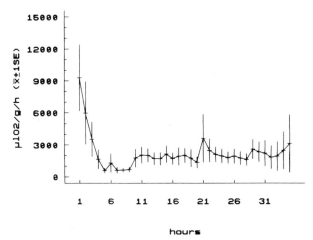

FIGURE 11   Respiration rates of *P. slavus* expressed as oxygen consumption (microliters)
per dry weight and hour. The line follows the mean hourly values of six individuals. Vertical
bars represent ±1 standard error.

originates from a garden pool located in the Vienna area. The specimens investigated have, generally, a strongly pigmented and sclerified body; the eyes are well developed.

The exploration activity of and the search for food by the epigean-dwelling *A. aquaticus* (Fig. 8) are opposite those observed, under the same experimental conditions, for *P. slavus; i.e.,* they spend more time feeding. Also the absolute oxygen consumption (Fig. 10) is high. Under high oxygen concentration the isopods ventilate very little, the opercula remain closed, and the movement of the respiratory pleopods cannot be observed (Figs. 12 and 13). This situation changes dramatically when the animals are placed under low oxygen concentrations (Fig. 12). Both ventilatory activity and pleopod movement (the amplitude and the number of beats) increase and then slightly decrease. The total motility decreases significantly during the hypoxic phase (Fig. 14). It does not bend dorsally, as in the case of *P. slavus,* most probably because of the strong sclerification of the body.

The respiration of the epigean *A. aquaticus* follows the oxyconformer type (Fig. 15); *i.e.,* the respiration rate decreases with the reduction of the oxygen concentration in the external environment. The lethality of this

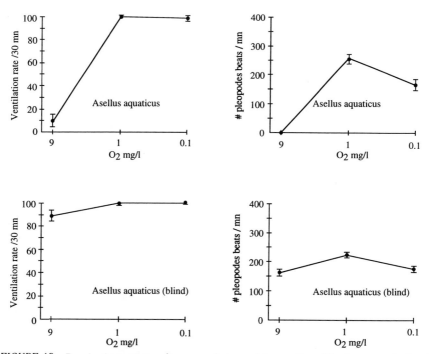

FIGURE 12   Respiration activity of two populations of *A. aquaticus*. The data are calculated as in Fig. 9; $N = 6$.

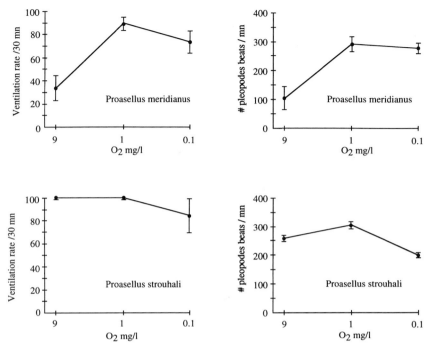

FIGURE 13   Respiration activity of *P. meridianus* and *P. strouhali*. The data are calculated as in Fig. 9; *N* = 6.

isopod population is high under chronic hypoxic conditions and organic load (M. Creuzé des Châtelliers, unpublished). We consider that the low oxygen conditions and the high energetic needs of *A. aquaticus* for both feeding and motility constrain the epigean individuals to actively colonize groundwater habitats located far from surface waters.

We studied a population of *A. aquaticus* from the karstic area of Mangalia, where the Movile cave is located (Sarbu and Popa, 1992). Our material stems from a well in the city of Mangalia as well as from the Movile cave (leg. Radu Popa). In both areas the water contains hydrogen sulfide and the oxygen concentration is low. Excepting the depigmentation and the reduction of the eyes, the individuals resemble those of the epigean-dwelling population studied. Figures 12 and 14 show the respiratory and the motility behavior of these stygobite individuals. They resemble those of the subterranean-dwelling *Proasellus* species, especially those of *P. strouhali* (see below).

P. meridianus is an epigean-dwelling species that occurs also in interstitial habitats (Henry, 1976). We used a population from an abandoned arm of the Rhône (just upstream of Lyon, France), which is cut off from the main channel and supplied by groundwaters. The animals are of small size, the adults are about 4–5 mm long, and the dry weight is about 0.3 mg.

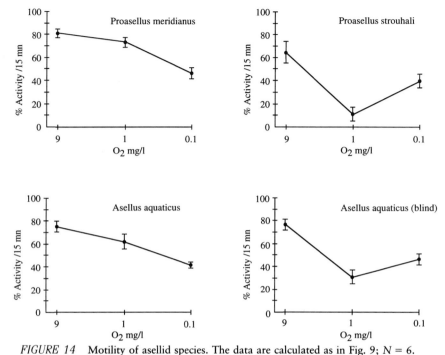

FIGURE 14   Motility of asellid species. The data are calculated as in Fig. 9; N = 6.

The animals are pigmented and oculated. The ventilation activity (Fig. 13) and the total motility (Fig. 14) resemble those of the epigean population of *A. aquaticus.*

Finally we studied a second stygobite species belonging to *Proasellus,* i.e., *P. strouhali* (subspecies *P. strouhali strouhali* from a well, located in Mondsee). The individuals are unpigmented and blind; the adults are about 5.5 mm long and 0.7 mg in dry weight. The ventilation (Fig. 13) activity resembles that of *P. slavus,* whereas the motility (Fig. 14) at low oxygen concentrations follows the pattern of the stygobite population of *A. aquaticus.*

D. Adaptation of *P. slavus* to Interstitial Habitats: A Synthetic View

We investigated simultaneously the peculiarities of the alluvial sediments of a wetland area, with their complex structure and patchy distribution of habitats, as well as the multiple ways in which the isopod *P. slavus* fits this environment.

The minute size of this species allows it to move through the reduced porous space that forms in the sandy and gravelly sediments. Dissolved oxygen in the groundwater of the Lobau is an important electron acceptor

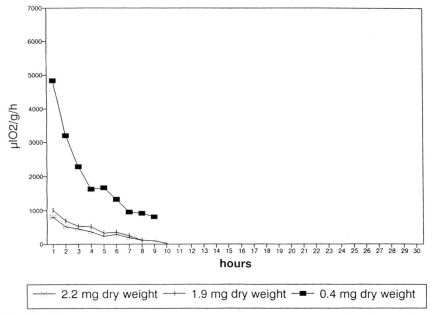

FIGURE 15 Respiration rates of *A. aquaticus* expressed as in Fig. 11. The data are from three different specimens (the largest, the smallest, and the median specimen from $N = 6$).

resource and a key environmental parameter. *P. slavus* developed multiple adaptive solutions to cope with low oxygen concentrations or with variable oxygen situations in space and time.

The oxygen consumption and the ventilatory activity of *P. slavus* are of the oxyregulator type. These adaptations are useful for animals that live in environments with variable oxygen conditions, where normoxic situations in time or space alternate with hypoxic ones, or where the latter dominate throughout the year. Under hypoxic conditions the high energetic expenditure needed to oxyregulate is compensated by a trade-off process; *i.e.*, the motility of the animal decreases significantly. The regulation of ventilatory activity exists in all the hypogean- dwelling isopod species investigated within this project. We consider it a general adaptation, possibly an adaptive pattern, of hypogean Crustacea living under these types of oxygen conditions.

P. slavus developed special adaptations to cope with the low oxygen concentrations in the groundwater, *i.e.*, the adoption of a special position of the body or the increase of the opening of the operculum. The strong exploration activity could also be a special adaptation of this species. The isopod uses the exploration tactic, instead of a "sit and wait" or an "exploitation" tactic, for the acquisition of its food. In patchy environments, with alternations between areas with rich organic resources and/or high oxygen concentrations and areas with a low concentration of organic matter and/

or oxygen resources, the exploration tactic is an excellent solution. For instance, isopods that would sit and aggregate in one place for long periods of time could exhaust the limited resources (*e.g.,* the dissolved oxygen) faster than those animals that permanently change their location and do not form large groups of individuals.

IV. FROM PATTERN TO PROCESS: THE EXPANDED AGENDA

A. Adaptation: Three Evolutionary Scenarios

Theoretically there are three possibilities (called here evolutionary scenarios) by which the integration of the organisms into the subterranean environment is achieved.

(1) Organisms react permanently to the changing environment. This is the solution of the animal "conformers" or "adjusters" (Bruton, 1989). They are limited by the type and degree of environmental change. E.G. Racovitza [in Jeannel (1950)] called this ecological strategy "adaptation limitatrice."

(2) There are animals that develop strong homeostatic capacities, maintaining their activity more or less at the same level, independently of the environmental situation. These are the "regulators." Racovitza, who was inspired by the ideas of Claude Bernard (*i.e.,* the organisms that develop a specialized "milieu interieur" are not subject to the impact of the external environment), called this type of adaptation *seclusion* or "adaptation liberatrice".

(3) Finally, there are animals that are able to construct their environment (Lewontin, 1983). Organisms select their environment and through their activity they change it. There is permanent interactive change that improves both the organism and the environment. Bruton (1989) called these organisms "extraregulators," and Bateson (1984) showed that the adaptive regulation evolves from the conformer type toward the extraregulators.

Organisms that are versatile [in the sense used by Andrews (1991)] (*i.e.,* they can do many things and have a high speed of response to new conditions) have an advantage in heterogenous and variable environments over the conformer specialists (Phillipson, 1981; Huey and Hertz, 1984). The regulators perceive heterogenous and/or widely fluctuating habitats as fine-grained and stable environments (Sheldon, 1993) and are able to compensate for energy allocation (Wieser, 1989).

The adaptation of *P. slavus* to variable oxygen conditions follows the second evolutionary scenario. Danielopol *et al.* (1994) showed that there is no evidence that these isopods within the Lobau area are extraregulators.

The physiological mechanisms and the ecological strategies used by *P. slavus* are excellent (in the sense of efficient) solutions to cope with the heterogenous and variable interstitial environment of the Danube alluvial plain and therefore explain the wide ecological distribution of this species.

B. Preadaptation and Natural Selection

Erikson *et al.* (1983) showed that, before accepting an explanation of an adaptation as the result of the natural selection process, one should test alternative hypotheses. Preadaptation as a process for the origin of adaptations, discussed above, is such an alternative.

In the case of *P. slavus,* we showed that the reduced size of the body is a peculiarity of both karstic and interstitial dwelling subspecies. Therefore we suspect that the small size of the porous space in alluvial sediments did not select for the reduced body size of this isopod species; it could be due to an arrested evolution of the postembryonic development (a pedomorphic feature). Such evidence was presented by Marmonier and Danielopol (1988) and Danielopol and Bonaduce (1990) for various ostracod groups and also discussed by Coineau and Boutin (1992) for Malacostraca crustaceans.

The low absolute oxygen consumption of *P. slavus* from the Lobau (Fig. 10) is similar to the absolute consumption values of other small isopod species, like *Jaera istrii* (F. Moeszlacher, unpublished), which does not occupy subterranean hypoxic habitats. It differs markedly (Fig. 10) when compared with that of the large species *Asellus aquaticus*. This means that the absolute low oxygen consumption of *P. slavus* could be a side effect related to reduced body size, in other words, a preadaptation that could be acquired even outside the subterranean environment. Note that *P. slavus,* according to Henry (1976), is an old, maybe Tertiary, isopod lineage, which possibly lived during those times in epigean low-oxygenated benthic habitats. This is similar to Schiemer's (1987) observations in the case of nematodes, for which low metabolic rates occur frequently in the case of epigean-dwelling species living in oxygen-deficient habitats.

There is a clear convergence between the respiratory behavior (the capacity to regulate respiration rate and the ventilatory activity of the pleopods) of the subterranean-dwelling isopods and those of the epigean species. We have here a strong argument that at least the ventilatory activity, as a general adaptation, was acquired through the natural selection process acting on the stygobite isopods in subterranean habitats with low and/or variable oxygen regimes. Sket (1985), Kane and Richardson (1990), and Kane and Culver (1992), *inter alia,* discussed how natural selection works in the subterranean environment in order to produce troglomorphic characters.

Certainly we need a serious effort in the future to investigate the processes that led to the adaptation of *P. slavus* to interstitial habitats. However, we can now advance the hypothesis that both preadaptation and natural selection played, or are playing, an important role in the evolutionary ecology of this species.

C. Future Research

We predict that adaptation will become a major topic in the evolutionary ecology of groundwater animals. The success of this research direction depends, in our opinion, on the suitability of the hypothesis we shall try to test, as well as on our capacity to cooperate with evolutionary biologists, with behavioral ecologists, and with ecological geneticists, *inter alia*.

In this chapter we showed that oxygen is a major energetical resource to which subterranean organisms adapt. Our experience, as well as those of our colleagues and mentors in Austria and France (see the work of Wieser, Pattee, Schiemer, and their associates), demonstrated that oxygen is an excellent describer (*i.e.*, it can be reliably quantified) for ecological studies. Studying the general and/or special adaptations of groundwater organisms to this parameter is, as it was demonstrated here, a realistic research program with which one can test evolutionary scenarios. Another suitable hypothesis to be tested is the assertion that both preadaptation and natural selection are important processes for the shaping of ecological strategies, physiological specializations, or both.

Rogulj *et al.* (1993) suggested that specialized groundwater organisms, like the ostracod species *Mixtacandona spandli* in the Lobau, with very localized distributions, are adapted to environmental parameters that we do not understand or fail to measure. Such a parameter could be the high concentration of carbon dioxide in the subsurface waters. The importance of this environmental factor for the distribution and evolution of the terrestrial troglobitic fauna was emphasized by Howarth and Stone (1990).

In order to achieve an ecological research program centered around adaptation topics we consider it necessary to reinforce cooperation with evolutionary biologists. It is through their expertise and with their help that ecologists will be able to better understand the history and the phylogenetical relationships of the organisms (within populations, species, and clades) that they study. It is also through cooperation with evolutionary biologists that sound experiments can be organized in order to test the role of natural selection or of other alternative processes.

The role of behavioral ecologists in the study of various aspects of habitat selection (the phenomenon of adoption) related to groundwater organisms is very important as has already been mentioned here. The research of Uiblein *et al.* (1992) on this topic should be intensified, using groundwater invertebrates as experimental organisms [for more thorough theoretical argumentation, see Rosenzweig (1991) and Orian and Wittenberger (1991)].

Finally, a better insight into the ecological genetics of groundwater organisms is necessary. This approach was strongly developed in several laboratories, such as those of Sbordoni in Rome, of Wilkens in Hamburg, of Juberthie in Moulis, and of Culver and Kane in the United States. It is important to understand the genetic adaptation of populations to local habitats. For instance, for *P. slavus* we do not know if the adaptations we observed are specific to the Lobau population or if they occur also in other populations, subspecies, or both. This is a problem that also has a practical aspect (*i.e.,* in order to use such a widely distributed species for ecological monitoring, its precise ecological requirements must be understood).

V. CONCLUSION

This chapter demonstrates that groundwater crustaceans, such as the isopod species *P. slavus,* are well adapted to the subterranean environment, in a way that fits the general principles of evolutionary biology. Studying the adaptations of these animals, following the practical agenda we illustrated here, one can positively contribute not only to the development of groundwater ecology but also to debates that animate general biology topics, a hope that was expressed by Christiansen (1992) and Danielopol (1992).

ACKNOWLEDGMENTS

We are indebted to the "Fonds zur Foerderung der Wissenschaftlichen Forschung" (Vienna), which supported most of this research (*i.e.,* Project P 7881, attributed to D.L.D.). Visits in Mondsee (M.C.d.C. and R.P.) and in Lyon (D.L.D.) were possible due to the scientific exchange program of the Austrian Academy of Sciences with the C.N.R.S., in France, and with the Romanian Academy, in Bucharest. Our friends and colleagues P. Torreiter and C. Griebler (Mondsee); M. Kaiser-Geiger (Salzburg); A. Gunatilaka, J. Dreher, and F. Schiemer (Vienna); B. Rogulj (Zagreb); B. Sket (Ljubljana); J. P. Henry and G. Magniez (Dijon); R. Rouch (Moulis); F. Uiblein (Salzburg); P. Marmonier (Chambéry); J. Gibert, J. Mathieu, and M. J. Turquin (Lyon); D. Culver (Washington, D.C.); T. Kane and S. Sarbu (Cincinnati); and J. Stanford (Polson) discussed various aspects of this chapter with us. I. Bals (Geneva) helped in many ways with the "Orbisphere" products. H. Ployer and I. Gradl (Mondsee) helped to edit the manuscript; H. Benion (Mondsee) and Glynn Thoiron (Lyon) improved the English presentation.

REFERENCES

Andrews, J. H. (1991). "Comparative Ecology of Microorganisms and Macroorganisms." Springer-Verlag, New York.

Bateson, G. (1984). "La nature et la pensée." Seuil, Paris.

Bock, W. J. (1979). A synthetic expanation of macroevolutionary change—a reductionistic approach. *Bull. Carnegie Mus. Nat. Hist.* **13**, 20–69.

Botosaneanu, L., ed. (1986). "Stygofauna mundi." E. J. Brill/Dr. W. Backhuys, Leiden.

Botosaneanu, L., and Holsinger, J. R. (1991). Some aspects concerning the colonisation of the subterranean waters: A response to Rouch & Danielopol. *Stygologia* **6**, 11–39.

Brandon, R. N. (1990). "Adaptation and Environment." Princeton Univ. Press, Princeton, NJ.

Brown, W. L. (1958). General adaptation and evolution. *Syst. Zool.* **7**, 157–168.

Bruton, M. N. (1989). The ecological significance of alternative life-history styles. *In* "Alternative Life-History Styles of Animals" (M. N. Bruton, ed.), pp. 503–553. Kluwer Acad. Publ., Dordrecht, The Netherlands.

Burian, R. M. (1992). Adaptation: Historical perspectives. *In* "Keywords in Evolutionary Biology" (E. Fox Keller and E. A. Lloyd, eds.), pp. 7–12. Harvard Univ. Press, Cambridge, MA.

Camacho, A. I., Bello, E., Becerra, J. M., and Vaticon, I. (1992). A natural history of the subterranean environment and its associated fauna. *In* "The Natural History of Biospeology" (A. I. Camacho, ed.), Monogr. 7, pp. 171–197. Mus. Nac. Cienc. Nat., C.S.I.C., Madrid.

Christiansen, K. (1962). Proposition pour la classification des animaux cavernicoles. *Spelunca* **2**, 76–78.

Christiansen, K. (1992). Biological processes in space and time: Cave life in the light of modern evoltionary theory. *In* "The Natural History of Biospeology" (A. I. Camacho, ed.), Monogr. 7, pp. 453–478. Mus. Nac. Cienc. Nat., C.S.I.C., Madrid.

Coineau, N., and Boutin, C. (1992). Biological processes in space and time: Colonization, evolution and speciation in interstitial stygobionts. *In* "The Natural History of Biospeology" (A. I. Camacho, ed.), Monogr. 7, pp. 423–451. Mus. Nac. Cienc. Nat., C.S.I.C., Madrid.

Culver, D. C. (1982). "Cave Life: Evolution and Ecology." Harvard Univ. Press, Cambridge, MA.

Culver, D. C., and Fong, D. W. (1986). Why all cave animals look alike. *Stygologia* **2**, 208–216.

Danielopol, D. L. (1976a). The distribution of the fauna in the interstitial habitats of riverine sediments of the Danube and the Piesting (Austria). *Int. J. Speleol.* **8**, 23–51.

Danielopol, D. L. (1976b). Sur la distribution géographique de la faune interstitielle du Danube et de certains de ses affluents en Basse-Autriche. *Int. J. Speleol.* **8**, 323–329.

Danielopol, D. L. (1980). Sur la biologie de quelques Ostracodes Candoninae épigés et hypogés d'Europe. *Bull. Mus. Natl. Hist. Nat., Ser. 4*, **2**, 471–506.

Danielopol, D. L. (1983). Der Einfluss organischer Verschmutzung auf das Grundwasser-Oekosystem der Donau im Raum Wien und Niederoesterreich. *Forschungsber. Bundesminist. Gesund. Umweltschutz* **5**, 5–160.

Danielopol, D. L. (1989). Groundwater fauna associated with riverine aquifers. *J. North Am. Benthol. Soc.* **8**, 18–35.

Danielopol, D. L. (1992). New perspectives in ecological research of groundwater organisms. *In* "Ground Water Ecology" (J. A. Stanford and J. J. Simons, eds.), pp. 15–22. Am. Water Resour. Assoc., Bethesda, MD.

Danielopol, D. L., and Bonaduce, G. (1990). The origin and distribution of the interstitial Ostracoda of the species group *Xestoleberisarcturi* Triebel (Crustacea). *CFS, Cour. Forschungsinst. Senckenberg.* **2**, 69–86.

Danielopol, D. L., and Niederreiter, R. (1987). A sampling device for groundwater organisms and oxygen measurements in multi-level monitoring wells. *Stygologia* **3**, 252–263.

Danielopol, D. L., and Rouch, R. (1991). L'adaptation des organismes au milieu souterrain. Réflexions sur l'apport des recherches écologiques récentes. *Stygologia* **6**, 129–142.

Danielopol, D. L., Dreher, J., Gunatilaka, A., Kaiser, M., Niederreiter, R., Pospisil, P., Creuzé des Châtelliers, M., and Richter, A. (1992). Ecology of a organisms living in a hypoxic groundwater environment at Vienna (Austria). *In* "Ground Water Ecology" (J. A. Stanford and J. J. Simons, eds.), pp. 79–90. Am. Water Resour. Assoc., Bethesda, MD.

Danielopol, D. L., Rouch, R., Pospisil, P., Torreiter, P., and Moeszlacher, F. (1994). Ecotonal animal assemblages; their interst for groundwater studies. *In* "Groundwater/Surface Water Ecotones" (J. Gibert, J. Mathieu, and F. Fournier, eds.). Cambridge Univ. Press, Cambridge, UK (in press).

Delamare-Deboutteville, C. (1960). "Biologie des eaux souterraines littorales et continentales." Hermann, Paris.

Dreher, J., Pospisil, P., and Danielopol, D. (1994). The role of hydrology in defining a groundwater ecosystem within the wetland of the Danube at Vienna. *In* "Groundwater/Surface Water Ecotones" (J. Gibert, J. Mathieu, and F. Fournier, eds.). Cambridge Univ. Press, Cambridge (in press).

Endler, J. A. (1986). "Natural Selection in the Wild." Princeton Univ. Press, Princeton, NJ.

Erikson, O., Inghe, O., Tapper, P.-G., Telenius, A., and Torstensson, P. (1983). A note on non-adaptation hypotheses in plant ecology. *Oikos* 41, 155–156.

Fage, L. (1931). Araneae, 5° Serie, précedée d'un essai sur l'évolution souterraine et son determinisme. *Arch. Zool. Exp. Gen.* 71, 99–291.

Ginet, R., and Decou, V. (1977). "Initiation à la biologie et à l'écologie souterraine." Delarge, Paris.

Gunatilaka, A., Dreher, A., and Richter, A. (1994). Dissolved organic carbon dynamics in a groundwater environment. *Verh. Int. Ver. Limnol.* (in press).

Harvey, P. H., and Pagel, M. D. (1991). "The Comparative Method in Evolutionary Biology." Oxford Univ. Press, Oxford.

Henry, J.-P. (1976). Recherches sur les Asellides hypogés de la lignée cavaticus. Thése, Univ. Dijon.

Henry, J.-P., Lewis, J. J., and Magniez, G. (1986). Isopoda Asellota: Aselloidea, Gnathostenetroidoidea, Stenetrioidea. *In* "Stygofauna mundi" (L. Botosaneanu, ed.), pp. 434–464. E. J. Brill/Dr. W. Backhuys, Leiden.

Howarth, F. G., and Stone, F. D. (1990). Elevated carbon dioxide levels in Bayliss cave, Australia: Implications for the evolution of obligate cave species. *Pac. Sci.* 44, 207–218.

Huey, R. B. (1991). Physiological consequences of habitat selection. *Am. Nat.* 137, S91–S115.

Huey, R. B., and Hertz, P. E. (1984). Is a jack-of-all-temperatures a master of none? *Evolution (Lawrence, Kans.)* 38, 441–444.

Hüppop, K. (1985). The role of metabolism in the evolution of cave animals. *NSS Bull.* 47, 137–146.

Illies, J. (1971). "Einfuehrung in die Tiergeographie." Fischer, Stuttgart.

Jeannel, R. (1950). "La marche de l'évolution." Editions Museum, Paris.

Jones, R., Culver, D. C., and Kane, T. C. (1992). Are parallel morphologies of cave organisms the result of similar selection pressures? *Evolution (Lawrence, Kans.)* 46, 353–365.

Kane, T. C., and Culver, D. C. (1992). Biological processes in space and time: Analysis of adaptation. *In* "The Natural History of Biospeology" (A. I. Camacho, ed.), Monogr. 7, pp. 377–399. Mus. Nac. Cienc. Nat., C.S.I.C., Madrid.

Kane, T. C., and Richardson, R. C. (1990). The phenotype as the level of selection: Cave organisms as model systems. *P S A* 1, 151–164.

Krebs, C. J. (1972). "Ecology." Harper & Row, New York.

Lattinger-Penko, R. (1976). Quelques données sur la population de *Proasellus slavus ssp*.n. Sket (Crustacea, Isopoda) dans l'hyporhéique de la rivière Drave près de Legrad. *Int. J. Speleol.* 8, 83–97.

Lewontin, R. C. (1983). Gene, organism and environment. *In* "Evolution from Molecules to Men" (D. S. Bendall, ed.), pp. 273–285. Cambridge Univ. Press, Cambridge, UK.

Magniez, G. (1976). Contribution à la connaissance de la Biologie des Stenasellidae (Crustacea Isopoda Asellota des eaux souterraines). Thèse, Univ. Dijon.

Margalef, R. (1983). "Limnologia." Omega, Barcelona.

Marmonier, P. (1985). Spatial distribution and temporal evolution of *Gammarus fossarum*, *Niphargus sp.* (Amphipoda) and *Proasellus slavus* (Isopoda) in the Seebach sediments (Lunz, Austria). *Jahresber. Biol. Stn. Lunz* 8, 40–54.

Marmonier, P., and Danielopol, D. L. (1988). Découverte de *Nannocandona faba* Ekman (Ostracoda, Candoninae) en Basse- Autriche. Son origine et son adaptation morphologique au milieu interstitiel. *Vie Milieu* 38, 35–48.

Marmonier, P., Vervier, P., Gibert, J., and Dole-Olivier, M.-J. (1993). Biodiversity in ground waters. *TREE* **8**, 392–395.

Mathieu, J., and Turquin, M.-J. (1992). Biological processes at the population level. II. Aquatic populations: *Niphargus* (Stygobiont amphipod) case. *In* "The Natural History of Biospeology" (A. I. Camacho, ed.), Monogr. 7, pp. 263–293. Mus. Nac. Cienc. Nat., C.S.I.C., Madrid.

Müller, P. (1974). "Aspects of Zoogeography." Dr. W. Junk Publ., The Hague.

Niederreiter, R., and Danielopol, D. L. (1991). The use of mini-videocameras for the description of groundwater habitats. *Mitteilungsbl. Hydrogr. Dienstes Oesterr.* **65/66**, 85–89.

Orian, G. H., and Wittenberger, J. F. (1991). Spatial and temporal scales in habitat selection. *Am. Nat.* **137**, S29–S49.

Parzefall, J. (1992). Behavioural aspects in animals living in caves. *In* "The Natural History of Biospeology" (A. I. Camacho, ed.), Monogr. 7, pp. 327–376. Mus. Nac. Cienc. Nat., C.S.I.C., Madrid.

Phillipson, J. (1981). Bionergetic options and phylogeny. *In* "Physiological Ecology" (C. R. Townsend and P. Callow, eds.), pp. 20–45. Blackwell, Oxford.

Pospisil, P. (1992). Sampling methods for groundwater animals of unconsolidated sediments. *In* "The Natural History of Biospeology" (A. I. Camacho, ed.), Monogr. 7, pp. 107–134. Mus. Nac. Cienc. Nat., C.S.I.C., Madrid.

Pospisil, P. (1993). Die Grundwasser Cyclopiden der Lobau in Wien (Oesterreich); faunistische, taxonomische und oekoloische Untersuchungen. Thesis, Univ. Vienna.

Pospisil, P., and Danielopol, D. L. (1990). Vorschlaege fuer den Schutz der Grundwasserfauna in geplanten Nationalpark "Donauauen" oestlich von Wien, Oesterreich. *Stygologia* **5**, 75–85.

Reeve, H. K., and Sherman, P. W. (1993). Adaptation and the goals of evolutionary research. *Q. Rev. Biol.* **68**, 1–32.

Rogulj, B., Danielopol, D. L., Marmonier, P., and Pospisil, P. (1993). Adaptive morphology, biogeographical distribution and ecology of the species group *Mixtacandona hvarensis* (Ostracoda, Candoninae). *Mém. Biospéol.* **20**, 195–207.

Rosenzweig, M. L. (1991). Habitat selection and population interactions: The search for mechanism. *Am. Nat.* **137**, 5–28.

Rouch, R. (1986). Sur l'écologie des eaux souterraines dans le karst. *Stygologia* **2**, 352–398.

Rouch, R., and Danielopol, D. L. (1987). L'origine de la faune aquatique souterraine, entre le paradigme du refuge et le modèle de la colonisation active. *Stygologia* **4**, 345–372.

Sarbu, S., and Popa, R. (1992). A unique chemoautotrophically based cave ecosystem. *In* "The Natural History of Biospeology" (A. I. Camacho, ed.), Monogr. 7, pp. 637–666. Mus. Nac. Cienc. Nat., C.S.I.C., Madrid.

Schiemer, F. (1987). Nematoda. *In* "Animal Energetics" (T. J. Pandian and F. Vernberg, eds.), Vol. 1, pp. 185–215. Academic Press, Orlando, FL.

Sheldon, P. R. (1993). Making sense of microevolutionary patterns. *In* "Evolutionary Patterns and Processes" (D. R. Lees and D. Edwards, eds.), pp. 20–29. Academic Press, London.

Sket, B. (1985). Why all cave animals do not look alike—a discussion on adaptive value of reduction processes. *NSS Bull.* **47**, 78–85.

Uiblein, F., Durand, J. P., Juberthie, C., and Parzefall, J. (1992). Predation in caves: The effects of prey immobility and darkness on the foraging behaviour of two salamanders, Euproctus asper and Proteus anguinus. *Behav. Processes* **28**, 33–44.

Thomson, K. S. (1988). "Morphogenesis and Evolution." Oxford Univ. Press, New York.

Thurman, E. M. (1985). "Organic Geochemistry of Natural Waters." Martinus Nijhoff/Dr. W. Junk Publ., Dordrecht, The Netherlands.

Vandel, A. (1965). "Biospeleology." Pergamon, Oxford.

Westheide, W. (1987). Progenesis as a principle in meiofauna evolution. *J. Nat. Hist.* **21**, 843–854.

Wieser, W. (1989). Energy allocation by addition and by compensation: An old principle revisited. *In* "Energy Transformations in Cells and Organisms" (W. Wieser and E. Gneiger, eds.), pp. 98–105. Thieme, Stuttgart.

Williams, G. C. (1966). "Adaptation and Natural Selection." Princeton Univ. Press, Princeton, NJ.

9

Biotic Fluxes and Gene Flow

T. C. Kane,* D. C. Culver,† and J. Mathieu‡

* Department of Biological Sciences
University of Cincinnati
Cincinnati, OH 45221

† Department of Biology
American University
Washington, D.C. 20016

‡ Université Lyon I
U.R.A. CNRS 1451, "Ecologie des Eaux Douces et des Grands
Fleuves"
Laboratoire d'Hydrobiologie et Ecologie Souterraines
69622 Villeurbanne Cedex, France

I. INTRODUCTION

The concept of the underground ecosystem (Rouch, 1977) with consideration given to animals living in large cavities (*e.g.*, caves) and small cavities (*e.g.*, underflow of streams) has considerable appeal. Such communities share a dependence on external food sources and an absence of sunlight. On the ecological level, they are often united in the same underground drainage basin and share a resource base (Rouch, 1986). On the evolutionary level, however, the unity of the organisms in such communities is not at all apparent. Despite the shared characteristic of eyelessness, the morphology of animals living in large subterranean cavities and small subterranean cavities is not that similar. The isopod *Proasellus albigensis,* from hyporheic habitats, is fusiform with generally small size, whereas its congener *P. vandeli,* from caves, is large with much longer appendages (Henry and Magniez, 1983). The morphology of *P. vandeli* is, however, typical in most respects of other cave-limited forms of both vertebrates and invertebrates. This shared morphological syndrome of cave animals has been termed "troglomorphy" by Christiansen (1962).

The late R. H. MacArthur (1972) said, "To do science is to search for repeated patterns . . ." The troglomorphic pattern is repeated both among disparate taxa of the same cave system (*e.g.,* vertebrates and invertebrates) and among similar taxa of geographically separated cave regions (*e.g.,* trechine carabid beetles of Europe and North America). We will devote our attention to animals living in large cavities (*i.e.,* caves) and focus on evolutionary processes that may produce the repeated troglomorphic pattern. The uniformity of cave environments, particularly with regard to lack of light and constancy of temperature and humidity, may produce similar selection pressures on organisms, leading to parallel or convergent evolution among taxonomically or geographically distinct populations. For aquatic species in particular, migration, gene flow, and the genetic structure of populations are heavily influenced by the hydrology of the system. These in turn affect morphological patterns that are a consequence of shared ancestry. Separating the effects of natural selection from the effects of shared evolutionary history is a fundamental problem in evolutionary biology. Caves are useful in this regard because they are geologically and hydrologically well-defined systems that are often replicated in space. Further, the similar and unusual environmental conditions of caves produce similar selection pressures, whereas the degree of hydrologic separation and distinctness of these systems makes predictions about gene flow and relatedness of populations possible.

Historically (Barr, 1968) biospeleologists have attempted to categorize cave organisms on the basis of their degree of restriction to the cave habitat (troglobites and troglophiles; see Chapter 1). The accuracy and utility of these designations have been questioned (Culver, 1982). Are morphologically unmodified species known only from caves troglobites, or are they troglophiles for which the surface populations have simply not as yet been observed (Hamilton-Smith, 1971)? Conversely, are presumptive troglophiles containing both morphologically unmodified surface-dwelling populations and troglomorphic cave-dwelling populations actually a single species? Finally, is the troglophilic condition merely an intermediate stage in the evolution of troglobites or do troglophiles represent an alternative evolutionary strategy in cave adaptation (Culver, 1982)? In this chapter we consider how the geology and hydrology of caves and the selection pressures of these habitats have influenced adaptation and evolution of aquatic cavernicoles.

II. BACKGROUND

In the concluding section of his lengthy monograph on North American cave fauna, Packard (1888) stated, "The main interest in the foregoing studies on cave life centers in the obvious bearing of the facts upon the theory of descent" (p. 116). Indeed, two central issues in biospeleology are the mode of origin of cave dwelling species and the evolutionary forces producing cave-associated morphological changes (*e.g.*, reduction or loss of eyes and pigmentation and hypertrophy of extraoptic sensory structures). Theories for the origin of cave-isolated species have often postulated major climatic shifts such as the glacial advances and retreats of the Pleistocene (Barr, 1968; Barr and Holsinger, 1985) or lowering of sea levels (Stock, 1980) as the driving forces. Biogeographic information and systematic relationships among taxa, based on morphological data, have been used to support these hypotheses. As Rouch and Danielopol (1987) have pointed out, such theories suggest that invasion of subterranean habitats may often be passive, occurring under "constraint."

Troglomorphy includes both elaborated (*e.g.*, appendages and extraoptic sensory structures) and regressed (*e.g.*, eyes and pigmentation) characters. Adaptationist hypotheses as explanations for the hypertrophied characters have predominated, although these hypotheses have seldom been tested (Jones and Culver, 1989; Jones *et al.*, 1992; Culver *et al.*, 1994). Evolutionary explanations for the reduction or loss of eyes and pigmentation have been more controversial and include both neutralist (fixation of structurally reducing neutral mutations via genetic drift) (Wilkens, 1971, 1988) and selectionist (Jones and Culver, 1989; Jones *et al.*, 1992) theories. Although the precise causes remain uncertain, it is clear that there has been much parallelism and convergence in the evolution of troglomorphy.

The use of morphological data alone to investigate questions of speciation and adaptation in cave animals has proven problematical. Discrete populations may share the same morphological features because of common ancestry or because of independent and parallel evolutionary histories (Christiansen, 1961). Further, distinguishing between selection and genetic drift as causes of evolutionary change requires an understanding of the genetic structure and gene flow patterns among populations (Wright, 1978), yet the genetic basis for morphological differences in cave animals is poorly understood at best. The application of biochemical and, to a lesser extent, molecular (Caccone and Powell, 1987) techniques in studies of cave animals over the past two decades has shed new light on issues of speciation and adaptation. These techniques yield data that are readily amenable to population genetic analysis. In addition, there is emerging a pattern that suggests that the biochemical variation observed in cave animals behaves neutrally in an evolutionary sense and thus reflects evolutionary history rather than adaptation (see Laing *et al.* 1976a,b; Kane *et al.*, 1992b).

III. GENETIC PATTERNS IN CAVE ANIMALS

Review papers by Barr (1968) and by Poulson and White (1969) made two very different predictions about levels of genetic variability in cave dwelling organisms. Barr modified Mayr's (1963) founder effect model for the special case of cave dwelling organisms. Barr suggested that incipient troglobites would undergo genetic bottlenecks initially due to extirpation of surface populations but that levels of genetic variability would subsequently be restored as population sizes increased due to subterranean migration and adaptation to the hypogean environment. Thus the argument for variability levels is largely neutralist, emphasizing the importance of population size. Poulson and White (1969), however, argued that the relatively constant and homogeneous hypogean environment would lead to reduced genetic variability in troglobites over evolutionary time, a distinctly selectionist view.

The accumulated data from allozyme studies of cave animals over the past 20 years are clearly at variance with the Poulson and White (1969) view. Although some species of troglobites with long histories of cave isolation appear to have low levels of genetic variability (Laing *et al.*, 1976a,b; Swofford, 1982), others (Sbordoni *et al.*, 1979; Caccone *et al.*, 1982; Sbordoni, 1982) appear to have levels of variability comparable to those observed in abundant, widely distributed surface dwelling species. Even among related troglobitic taxa with similar ecologies, such as trechine beetles, different patterns of variability are often obtained (Kane *et al.*, 1992a). Further, taxa that show low levels of intrapopulational variability do not necessarily lack genetic variation overall but have the variability partitioned among, as opposed to within, populations (Laing *et al.*, 1976a,b; Kane *et al.*, 1992a). Thus, there appears to be no simple relationship between allozyme variability and cave adaptation, suggesting that the selectionist hypothesis of Poulson and White (1969) is untenable for this type of variation.

One general pattern that does seem to emerge from electrophoretic studies of cave organisms is reduced gene flow among conspecific populations relative to that seen in surface dwelling species (Laing *et al.*, 1976b; Dickson *et al.*, 1979; Kane, 1982; Kane and Brunner, 1986). Restricted gene flow has been reported for both troglobites (Laing *et al.*, 1976a,b) and for cave dwelling populations of troglophiles (Avise and Selander, 1972; Kane *et al.*, 1992b), indicating that disruption of gene flow among local populations may occur fairly early in the evolution of cavernicoles. Furthermore, limitation of gene flow has been observed in limestone regions that are continuous geologically, such as the Mississippian plateaus of the southeastern United States (Barr, 1985; Kane *et al.*, 1992a), which should permit maximum dispersal of cave dwelling organisms through subterranean routes. Often, however, genetic differentiation observed at the biochemical level is not accompanied by corresponding morphological differentiation (Laing *et al.*, 1976a; Kane *et al.*, 1992a).

Clearly patterns of gene flow and genetic differentiation in cave dwelling animals are complicated and not amenable to simple explanations. Not only do the studies summarized above appear to falsify the Poulson and White (1969) hypothesis, but also many are at variance with some of the predictions of the Barr (1968) model. We contend, however, that there is an accumulating database (Christiansen, 1965; Jones and Culver, 1989; Jones *et al.,* 1992) indicating that selection plays a major role in molding the phenotype of both reduced and elaborated morphological features. Thus, genetic structure, and by implication phylogenetic relationships, may be more accurately measured at the biochemical (allozyme) level than at the morphological level. Assessment of biochemical differentiation in a troglobite provides a static picture of the genetic structure of that species. Because troglobites no longer have extant surface dwelling populations, they may be the wrong organisms to study in trying to understand the evolutionary dynamics of cave invasion and cave isolation. Troglophiles, by contrast, do maintain gene flow with conspecific surface dwelling populations and often show intermediate morphological modification with respect to troglomorphic characters. Thus troglophiles may provide an opportunity to observe ongoing troglobite evolution, with the qualification (Culver, 1982) that all troglophiles may not be troglobites *in statu nascendi*. In the following we review four case studies for which detailed genetic (i.e, electrophoretic), biogeographic, and morphologic data have been accumulated. In keeping with the theme of this book, all of the species examined are aquatic, and, consistent with our approach, two studies involve troglophiles and two deal with troglobites.

IV. CASE STUDIES

A. Troglophiles

1. Gammarus minus in West Virginia and Virginia

Gammarus minus (Amphipoda: Gammaridae) is a widespread troglophile that occurs in resurgences (springs), spring runs, and caves throughout the southern Appalachians (Holsinger and Culver, 1970; Holsinger, 1972). In two large karst areas (Fig. 1), one in West Virginia and the other in Virginia, populations of *G. minus* occurring in caves are morphologically distinct from those occurring in resurgences. Individuals from caves have a consistently larger body size, longer antennae, and smaller eyes than individuals living in the resurgences of these drainages (Culver, 1987). Fong (1989) has shown these traits to be highly heritable in both cave and resurgence populations.

The population structure of *G. minus* is largely determined by the local geology and hydrology. Water enters caves from sinking streams at the edge of drainage basins, from vertical percolation of rainfall and runoff through

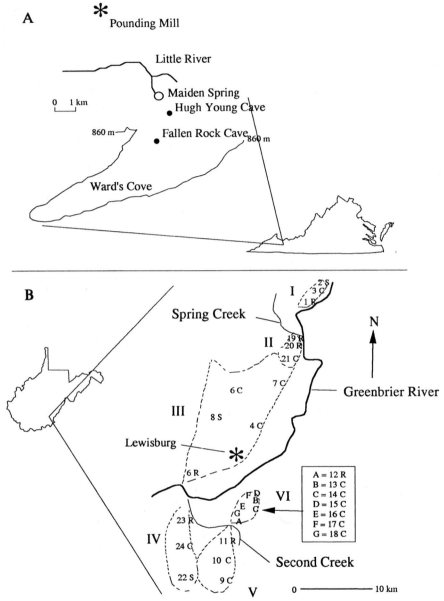

FIGURE 1   Location of Virginia (A) and West Virginia (B) resurgence, cave, and karst window populations of *G. minus* studied. In B the dashed lines indicate the limits of subterranean drainage basins. Roman numerals refer to basins; R, to resurgences; C, caves; and S, to karst windows (nonresurgence springs).

the limestone, and from flow from other connecting caves, and it exits the system at a resurgence. In southeastern West Virginia (Fig. 1) this pattern of cave and resurgence habitats is repeated in a series of closely adjoining basins, which are completely isolated from one another hydrologically (Jones, 1981, 1988). Within each basin troglomorphic populations of *G. minus* in the upstream subterranean portion of the drainage are hydrologically connected with morphologically unmodified populations in the resurgence (Culver, 1987; Culver *et al.*, 1990). The accessible portions of the subterranean drainage in the six West Virginia karst basins and one Virginia (Fig. 1) karst basin examined are characterized by open conduit flow. However, downstream in these systems, passages become water filled and are no longer accessible to the investigator even though dye tracing studies confirm a hydrologic connection with the resurgence (Jones, 1981, 1988). These "phreatic loops" (Ford and Ewers, 1978) continue as closed conduits and may be choked with sediment. These closed conduits are likely to be unsuitable habitats for *G. minus*.

The genetic structure of *G. minus* populations in caves and resurgences has been examined using gel electrophoresis of proteins (Kane and Culver, 1991; Kane *et al.*, 1992b; Sarbu *et al.*, 1993). An analysis of genetic data for 19 loci yields three consistent patterns. (1) In both Virginia and West Virginia resurgence populations are genetically differentiated from upstream cave populations within the same basin (Fig. 2). (2) In West Virginia, where multiple karst basins were examined, clustering of cave populations is generally by basin (compare Figs. 1 and 2). For four of five cases in which multiple cave populations were sampled clusters correspond with basins (Fig. 2). The one exception occurs in the largest and most hydrologically complex of the basins (Basin III of Fig. 1), and the lack of clustering of cave populations appears to be explained by both the current and the historical hydrological characteristics of this basin. (3) There is little genetic differentiation among the seven resurgence populations of the six West Virginia basins (Fig. 2).

Overall differentiation among the 24 West Virginia populations (F_{ST} = 0.486) and among the three Virginia populations (F_{ST} = 0.226) is great. A hierarchical analysis of the West Virginia populations (Table I) indicates that 57% of the among-population genetic variance is due to differentiation among populations within basins. Genetic differences between cave and resurgence habitats within basins (Fig. 2) account for a substantial portion of this differentiation. The remaining 43% of among-population variance can be attributed to differentiation among basins (Table I). Gene flow, measured as N_m, the average number of migrants exchanged per generation, can be estimated from F_{ST} using the relationship (Wright, 1931; Slatkin and Barton, 1989)

$$F_{ST} = 1/(4N_m + 1).$$

For the 24 cave and resurgence populations in West Virginia N_m = 0.264 and for the three populations in Virginia N_m = 0.856. An analysis of the West Virginia populations by habitat type yields F_{ST} = 0.100 and N_m = 2.25 for the seven resurgence populations and F_{ST} = 0.539 and N_m = 0.214 for the 17 cave and karst window populations. Values of

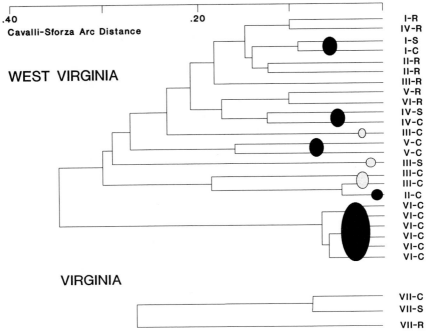

.40   .20   I-R
                                                                             IV-R
Cavalli-Sforza Arc Distance   I-S
                                                                             I-C
                                                                             II-R
WEST VIRGINIA   II-R
                                                                             III-R
                                                                             V-R
                                                                             VI-R
                                                                             IV-S
                                                                             IV-C
                                                                             III-C
                                                                             V-C
                                                                             V-C
                                                                             III-S
                                                                             III-C
                                                                             III-C
                                                                             II-C
                                                                             VI-C
                                                                             VI-C
                                                                             VI-C
                                                                             VI-C
                                                                             VI-C
                                                                             VI-C

VIRGINIA

                                                                             VII-C
                                                                             VII-S
                                                                             VII-R

FIGURE 2   UPGMA dendrograms for West Virginia and Virginia populations of *G. minus* based on Cavalli–Sforza and Edwards arc distance. For The West Virginia populations, solid ellipses indicate the nonresurgence populations for a single karst basin. The three open ellipses indicate the nonresurgence populations for Basin III (see Fig. 1B). [after Kane and Culver (1991)]. The Virginia populations from Ward's Cave are indicated as Basin VII [after Sarbu *et al.* (1993)].

$N_m < 1.0$ should lead to significant divergence among populations, and models suggest that values of $N_m < 0.5$ will cause populations to be unconnected genetically with regard to neutral variation (Trexler, 1988; Slatkin and Barton, 1989). Although gene flow among resurgence populations seems adequate to stem differentiation (compare the N_m value above and Fig. 2), there appears to be little genetic connection between cave and resurgence populations within a drainage (Fig. 2). Cave populations of different drainages in West Virginia also appear to be unconnected (Fig. 2), supporting the hypothesis (Culver *et al.*, 1990) that cave colonization by *G. minus* has occurred independently in each basin. Founder events (Barr, 1968) or periodic genetic bottlenecks (Wilkens, 1988) have often been invoked to account for the divergence of cave stocks from surface ancestors. Such events would be expected to cause a reduction of genetic variation in recent cave colonists such as *G. minus*. However, allozyme data do not indicate that such events have occurred in the course of cave invasion by *G. minus*. The average proportion of polymorphic loci per population (P)

TABLE I Hierarchical F-Statistic Analysis of 24 Populations of *Gammarus minus* in West Virginia (Refer to Fig. 1) Based on 11 Variable Loci

Comparison			
X	Y	Variance component	F_{XY}
Population	Basin	0.83945	0.346
Population	Total	1.47127	0.481
Basin	Total	0.63182	0.206

Note. The two hierarchical levels reflect differentiation among populations within basins and differentiation among the six basins.

and the average proportion of heterozygous loci per individual (H) do not differ between resurgence (8 populations: $P = 0.392$; $H = 0.082$) and cave (19 populations: $P = 0.287$; $H = 0.078$) habitats. Further, these levels of variability are comparable to those reported for many species of abundant surface dwelling invertebrates (Selander, 1976), indicating that cave and resurgence dwelling populations of *G. minus* are not excessively small.

2. Astyanax Fasciatus in Mexico

The Mexican cavefish, *Astyanax fasciatus* (Pisces: Characidae), is undoubtedly the best studied troglophilic vertebrate species. Epigean populations of *A. fasciatus* are hydrologically connected with morphologically modified hypogean populations, showing varying degrees of troglomorphy (Wilkens, 1971,1988; Mitchell *et al.*, 1977). There is direct laboratory evidence (Wilkens, 1971) that cave and surface populations are interfertile. Although opportunities for hybridization are more numerous, field data suggest that successful cases are restricted to caves with abundant food supplies (Mitchell *et al.*, 1977; Romero, 1983; Wilkens, 1988).

The Sierra de El Abra region of Mexico (Fig. 3) has the largest concentration of blind cave populations of *A. fasciatus* (Mitchell *et al.*, 1977; Wilkens, 1988), which are also the best known. Avise and Selander (1972) conducted an electrophoretic survey of 17 loci on six epigean and three cave populations of *A. fasciatus* from this area (Fig. 3). A UPGMA dendrogram (Fig. 4) generated from their data suggests close genetic similarity among epigean populations. Cave populations are differentiated from the epigean populations, and the Pachon cave population is further differentiated from the cave populations of Los Sabinos and Chica (Fig. 4). Surface streams near the Pachon cave belong to a drainage separate from that to the south, where the other two caves occur (Fig. 3; see also Mitchell *et al.*, 1977). A hierarchical F-statistic analysis (Table II) indicates that 41% of the among-population

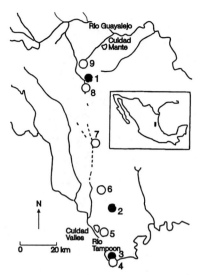

FIGURE 3   Map of the Sierra de El Abra region of Mexico, indicating the locations of the nine *Astyanax fasciatus* populations sampled in the study of Avise and Selander (1972). Cave populations are denoted by solid circles.

genetic variance is due to differentiation between habitat types (caves vs surface streams) and 59% is due to differentiation among populations within a habitat. Overall differentiation among the nine populations is very great ($F_{ST} = 0.401$), and therefore gene flow estimates are low ($N_m = 0.373$), largely due to an apparent lack of gene flow between cave and surface populations and among cave populations (Fig. 4).

The genetic structure of these *A. fasciatus* populations is very similar, both qualitatively and quantitatively, to the structure of *G. minus* popula-

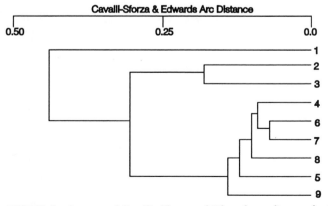

FIGURE 4   UPGMA dendrogram of Cavalli–Sforza and Edwards arc distance for 17 electrophoretic loci in nine populations of *Astyanax fasciatus* [after Avise and Selander (1972)].

TABLE II   Hierarchical F-Statistic Analysis of
Nine Populations of *Astyanax fasciatus* in
Mexico (Refer to Fig. 3) Based on 15
Variable Loci

Comparison			
X	Y	*Variance component*	F_{XY}
Population	Habitat	0.64503	0.283
Population	Total	1.09466	0.401
Habitat	Total	0.44962	0.165

Note. The two hierarchical levels reflect differentiation among populations within habitats (cave and
surface) and differentiation between habitats (cave
vs surface) [data from Avise and Selander (1972)].

tions described previously. In both species, morphologically modified subterranean populations suffer reduced gene flow with conspecific surface dwelling populations. Furthermore, differentiation among cave populations can
be related to drainage divides in each case. Unlike the situation for *G. minus*,
however, cave populations of *A. fasciatus* showed reduced genetic variability
($P = 0.137$; $H = 0.036$) compared with epigean populations ($P = 0.373$;
$H = 0.112$), and the difference in heterozygosity values is statistically significant (Mann–Whitney U test, $P < 0.025$). These data are consistent with
Wilkens' (1988) view that *A. fasciatus* has not actively colonized caves but
rather has done so passively, becoming trapped in caves that capture streams.
Further, there is evidence that epigean fish suffer malnutrition and impaired
gamete production in caves (Wilkens and Huppop, 1986; Wilkens, 1988).
Thus founding populations may undergo further bottlenecks after their
accidental entry into the subterranean drainage.

The above cases suggest that substantial cave-related morphological
change can occur in hypogean populations even when they are in close
proximity to unmodified, conspecific epigean populations. Despite the close
proximity and hydrologic connection between cave and surface populations
for each of these species, the electrophoretic data clearly indicate a disruption
of gene flow between these two habitats in each case. Gene flow between
cave and resurgence populations of *G. minus* may be partially or completely
blocked if the inaccessible downstream subterranean portions of the drainage
are filled conduits (Ford and Ewers, 1978). Such areas may be uninhabitable
by *G. minus* and thus act as ecological barriers to gene flow. Most cave
populations of *A. fasciatus* are physically prevented from swimming back
to the surface because they have been trapped in the subterranean drainage
by stream capture (Mitchell *et al.*, 1977; Wilkens, 1988), precluding hybridization between hypogean and epigean stocks in surface streams. The poten-

tial for hybridization does exist in caves, because epigean forms continue to be washed in and thus come in contact with cave forms. Only in caves with large food supplies, such as Chica Cave (Fig. 3), which has a large bat roost, is there strong evidence that hybridization between the two forms is actually occurring, however (Mitchell *et al.*, 1977; Wilkens and Huppop, 1986; Wilkens, 1988). In food-poor caves epigean forms are apparently nutritionally incapable of reproducing (Huppop, 1987), putting them at a selective disadvantage to the cave modified forms. Further, there is some suggestion (Mitchell *et al.*, 1977) that hypogean forms may be cannibalizing, rather than mating with, epigean individuals.

The large degree of genetic differentiation among cave dwelling populations of *G. minus,* relative to that seen among resurgence populations, appears to reflect multiple invasions of the subterranean drainage by this species. As indicated previously, the clustering of cave populations (Fig. 2) is almost exclusively by drainage basin, suggesting that each represents a separate gene pool. Morphological data (Culver, 1987; Dai, 1989) support the view of an independent invasion of subterranean drainages and, together with the electrophoretic data (Kane and Culver, 1991), suggest that these invasions may have occurred at different times. Multiple isolations of epigean *A. fasciatus* have also undoubtedly occurred in the Sierra de El Abra (Mitchell *et al.*, 1977; Wilkens, 1988). As is the case in many karst areas, surface and subterranean drainage patterns in the Sierra de El Abra do not always correspond (Mitchell *et al.*, 1977). Thus, separate isolates from the surface may subsequently mix and interbreed in this system. Subterranean isolation seems to be influenced by the vertical position of the cave in the limestone. Pachon Cave (Fig. 3) lies high in the limestone, above the hypothesized high-water profile (Mitchell *et al.*, 1977). The absence of allozyme variability in this population (Avise and Selander, 1972) may reflect an absence of gene flow due to lack of hydrologic connection with other cave populations (Mitchell *et al.*, 1977).

B. Troglobites

1. Niphargus in Europe

Ginet and Decou (1977) observed, "La situation présente de la faune aquatique hypogée doit résulter de certains processus du passé" (p. 230). In the case of troglobites, however, past processes can be surmised only through inferences gained from present population patterns. Among troglobitic aquatic invertebrates, the population biology of species of the stygobite genus *Niphargus* (Amphipoda: Gammaridae) perhaps is the best known (Ginet, 1965, 1969; Turquin, 1981; Marmonier and Dole, 1986; Mathieu and Essafi, 1991). Some of the larger species (*e.g., N. virei*), although geographically widespread, are restricted to karst waters and never occur in interstitial habitats. This habitat restriction may limit migration. Populations of *N. virei* are numerous in both the Jura and the Ardeche regions of eastern and southeastern France (Fig. 5). However, the intervening 150 km between these two regions is devoid of *N. virei* despite the fact that the area is partly karstic. Ginet (1952, 1965, 1969) suggested that the absence of *N. virei* from this area could be due to (1) the extirpation of previously existing populations by an alpine glacial advance or (2) present unfavorable ecological conditions that prevent populations from becoming established. To test between these alternatives, Ginet (1952) transplanted a population of *N. virei* into a small cave stream in the intermediate zone (la Balme, Fig. 5). The transplantation was successful (Ginet, 1965, 1969; Barthélémy, 1982), indicating that the ecological conditions in the area are suitable for *N. virei* and leading Ginet to accept the glacial extirpation hypothesis to explain the absence of *N. virei* in the area. The data also suggest that *N. virei* is a poor colonizer as it has not reinvaded the region since the glacial retreat. Individuals of the species have been collected at outflows (Fig. 6) (Turquin, 1976; Vervier, 1988), suggesting that movement in the form of downstream drift does occur, but the above data indicate that these individuals rarely, if ever, become established.

Some species of *Niphargus* are less restricted in their habitat utilization. *Niphargus rhenorhodanensis* occurs not only in karst streams but also in the fluvial sediments of surface rivers (Ginet, 1983; Dole and Chessel, 1986; Marmonier and Dole, 1986) and in the sediments of forest drainage channels (Mathieu *et al.*, 1984, 1987). The small size of *N. rhenorhodanensis* permits it to inhabit the low-porosity-clay, glacial sediments of forest drainage channels. These sediments undergo drying for six months each year, and popula-

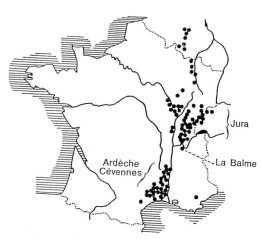

FIGURE 5 Biogeographic distribution of the European amphipod *Niphargus virei* in the groundwater of France. La Balme is the cave site where *N. virei* was successfully transplanted [from Ginet and Decou (1977)].

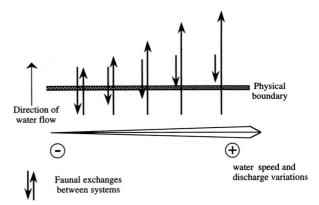

FIGURE 6   Possible washout of *Niphargus virei* at a karst outlet as a function of hydrological fluctuations. In a transition zone the direction of exchanges is directly linked to the water velocity and the intensity of discharge variation [after Vervier (1988)].

tion studies indicate that most movement by *N. rhenorhodanensis* is vertical, following the wetting or drying of the channels (Fig. 7). The low porosity of these sediments and the prolonged period of drying each year probably limit migration severely in this habitat.

Populations of *N. rhenorhodanensis* occurring in the fluvial sediments of river basins and their tributaries have greater opportunities for migration and gene flow. In fact, in these habitats it is not clear whether individuals belong to one continuous population or are subdivided into a series of discrete adjacent populations. Interstitial *N. rhenorhodanensis* occurs in the upper level of the sediments of these systems (Ginet, 1983) to a depth of about 3 m (Dole and Chessel, 1986; Marmonier and Dole, 1986). Maximum densities are found at depths of 50 cm to 1.5 m. There is potential for movement in both the vertical and the horizontal dimensions of this system. There is a significant relationship between the intensity and duration of flooding and changes in the vertical distribution of *N. rhenorhodanensis* (Dole-Olivier and Marmonier, 1992). Horizontal movements during flooding have also been noted for stygobite species of *Niphargopsis* and *Salentinella* and the epigean *Gammarus fossarum* but are less well demonstrated for *N. rhenorhodanensis* (Marmonier and Dole, 1986). They are more difficult to demonstrate in these sediments, because *N. rhenorhodanensis* is found in almost all samples and has the largest local distribution of the species occurring in this habitat.

N. rhenorhodanensis populations in cave systems are subject to the same environmental influences as those described earlier for *N. virei*. High flow results in drift of individuals out of the karst (Gibert, 1984, 1986). The degree of drift is determined in part by the hydrology of the system (Mathieu and Essafi-Chergui, 1990; Mathieu and Essafi, 1991; Essafi *et al.*,

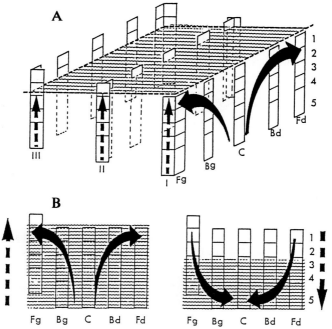

FIGURE 7 Schematic presentation of migration of *Niphargus rhenorhodanensis* within the sediment of drainage canals. Hatch marks indicate water level. Solid arrows indicate colonization as a function of water level. Discontinuous arrows reflect water level changes. I, II, and III are transects, and Fg, Bg, C, Bd, and Fd indicate positions of artificial samplers in the transect. A and B are two different sites (canals) [from Mathieu *et al.* (1987)].

1992). During low-water periods, most *N. rhenorhodanensis* individuals are found in the karst, with very few individuals collected at the outlet or downstream in the sediments. In these situations, individuals that are washed out come into contact with *G. fossarum,* which may prey on them (Essafi, 1990). Washout of *N. rhenorhodanensis* appears to be greater during high-water periods. Individuals swept away from the system by the current found refuge in the sediments of a river bank (Mathieu and Essafi, 1991) and were found at several downstream sampling points.

Genetic data (Sbordoni *et al.,* 1979) further support the view that migration rates in *Niphargus* are low. An electrophoretic survey of 10 populations of *N. longicaudatus* in central Italy (Fig. 8) indicates very great genetic differentiation among populations. A reanalysis of their data for seven variable loci examined in all 10 populations yields $F_{ST} = 0.287$. Sbordoni *et al.* (1979) obtained a significant correlation between Nei's genetic distance (D) and geographic distance between populations. A dendrogram generated from Cavalli–Sforza and Edwards arc distance (Fig. 9) indicates that differentiation among *N. longicaudatus* populations is somewhat greater than

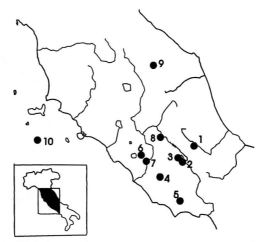

FIGURE 8   Map of central Italy, indicating the location of 10 populations of *Niphargus longicaudatus* [from Sbordoni *et al.* (1979)].

that observed among cave populations of *G. minus* (Fig. 2) and comparable to that observed among cave populations of *A. fasciatus* (Fig. 4). Despite the overall correlation of differentiation with geographic distance, two populations of *N. longicaudatus* separated by only 300 m (2 and 3 in Fig. 9) show substantial genetic differentiation. This suggests that it may be hydrologic relationships, reflecting subterranean connectivity of populations, rather than geographic distance per se, that are the relevant parameter in understanding gene flow. Genetic variability levels in *N. longicaudatus* are extremely high. Sbordoni *et al.* (1979) reported that 15 of 18 loci examined were variable in some or all of the 10 populations sampled. The

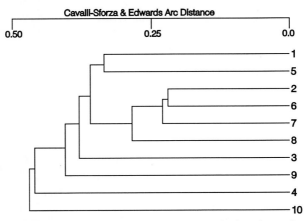

FIGURE 9   UPGMA dendrogram using Cavalli–Sforza and Edwards arc distance for seven loci in 10 populations of *Niphargus longicaudatus* [data from Sbordoni *et al.* (1979)].

average heterozygosity was 0.283, suggesting that these populations are large in size. Thus the genetic structure of these populations is not unlike that observed in cave dwelling populations of *G. minus*, with populations maintaining substantial genetic variation despite having only limited gene flow with other populations.

2. Amblyopsid Fishes in North America

The family Amblyopsidae (Pisces) is represented by four genera and six species in the southeastern United States. Included among these species are an epigean form, *Chologaster cornuta;* a troglophile, *C. agassizi;* and four troglobites, *Typhlichthys subterraneus, Amblyopsis rosae, A. spelaea,* and *Speoplatyrhinus poulsoni.* Poulson (1963, 1964) conducted a series of morphological, physiological, and behavioral studies on five of these species, which indicated an increasing degree of cave adaptation progressing from the *Chologaster spp.* to the extreme troglobites of the genus *Amblyopsis. Speoplatyrhinus,* unknown at the time of Poulson's work, appears on morphological grounds to be even more cave adapted than Amblyopsis (Cooper and Kuehne, 1974), thus continuing the progression.

Swofford (1982) undertook an electrophoretic survey of five of the above species (all but *S. poulsoni*) in an attempt to relate genetic patterns of this phylogenetically cohesive group to ecological differences among its species. Although sample sizes were necessarily small, largely due to the relative rarity of troglobitic species, some consistent patterns emerge. There is a general tendency for cave dwelling populations to be less genetically variable than populations of the epigean *C. cornuta. C. agassizi,* the troglophile, exhibits more intrapopulational variability than the troglobitic species, and, among the troglobites, the highly troglomorphic *Amblyopsis spp.* are the least variable (Table III). When overall levels of polymorphism (P) are considered, however, the values for *C. cornuta, C. agassizi,* and *T. subterraneus* are comparable and relatively large (Table III). Admittedly

TABLE III   Patterns of Genetic Variability in Five Species of Amblyopsid Fishes

Species	H	P	$P_{overall}$
Chologaster cornuta	0.040	0.094	0.58
C. agassizi	0.028	0.068	0.47
Typhlichthys subterraneus	0.019	0.053	0.68
Amblyopsis rosae	0.006	0.026	0.16
A. spelaea	0.000	0.000	0.00

Note. P, average proportion of polymorphic loci per population; $P_{overall}$, proportion of polymorphic loci per species; H, average proportion of heterozygous loci per individual [data from Swofford (1982)].

overall *P* in *A. rosae* and *A. spelaea* remains low (0.16 and 0.0, respectively,;Table III), but the former species has a very restricted range with only two populations sampled, and, in the case of the latter, Swofford (1982) was unable to sample populations in the southern portion of the range, where a greater potential for uncovering variation exists (Hobbs and Barr, 1972). In contrast, *T. subterraneus* has a vast geographic range and in fact is the most widely distributed troglobite in North America (Barr and Holsinger, 1985). Most of the extent of this range was sampled in Swofford's (1982) study (Fig. 10), including portions in which there is overlap with *C. agassizi*. For the purposes of this chapter, a comparison between *T. subterraneus* and the *Chologaster spp.* is the most appropriate.

Genetic differentiation among populations of *C. cornuta* ($F_{ST} = 0.691$) and *C. agassizi* ($F_{ST} = 0.787$) is substantial, but even larger differentiation is observed among populations of *T. subterraneus* ($F_{ST} = 0.941$). A UPGMA dendrogram (Fig. 11), generated from data for *T. subterraneus,* shows a greater divergence among populations than that observed for *N. longicaudatus* or either of the troglophilic species discussed earlier. In areas where *C. agassizi* and *T. subterraneus* overlap, greater differentiation was observed among the *T. subterraneus* populations. Swofford (1982) concluded that this resulted from increased migration among *C. agassizi* populations via surface streams, migration routes unavailable to *T. subterraneus*. Differentiation among *T. subterraneus* populations not only is very great but also occurs on a fairly local scale (compare Figs. 10 and 11). Further, there are cases for which allozyme differentiation and morphological differentiation do not correspond. Poulson (1961) argued that the morphological homogeneity of populations along the eastern Highland Rim and the western edge of the Cumberland Plateau was due to relatively high levels of gene flow along the plateau. Swofford (1982), however, found that two populations

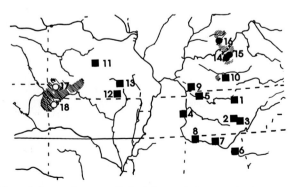

FIGURE 10 Map of the southeastern United States, indicating the location of *Amblyopsis rosae* (open circles), *A. spelaea* (closed circles), and *Typhlichthys subterraneus* (closed squares). Populations sampled by Swofford (1982).

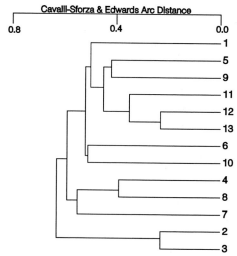

Cavalli-Sforza & Edwards Arc Distance

0.8   0.4   0.0

1
5
9
11
12
13
6
10
4
8
7
2
3

FIGURE 11   UPGMA dendrogram of Cavalli–Sforza and Edwards arc distance for 19 electrophoretic loci in 13 populations of *Typhlichthys subterraneus* [after Swofford (1982)].

toward the center of this region (2 and 3 in Fig. 10) and separated from
each other by only a few kilometers were fixed for alternative alleles at one
locus, indicating an absence of gene flow. Furthermore these populations
are well differentiated electrophoretically from Highland Rim/Cumberland
Plateau populations to the north (1 in Fig. 10) and south (7 in Fig. 10).
Swofford concludes that similar selective pressures, rather than gene flow,
may be maintaining the morphological homogeneity among these populations.

The genetic data also have implications regarding the manner of origin
of *T. subterraneus* from a surface dwelling ancestor to produce its present
distribution. Despite its large distribution, the genetic data indicate that
dispersal by *T. subterraneus* is very limited. These data preclude the possibility that *T. subterraneus* arose from a single isolation of its surface ancestor
and indicate that multiple isolations took place, with some subsequent range
extension through subsurface waters. The morphological resemblance of
populations can result from retention of some ancestral characters and
parallel evolution of troglomorphic features (*e.g.*, eye and pigment degenera-
tion and elaboration of the lateral line system) (Swofford, 1982).

Populations of *Niphargus* and *T. subterraneus* both exhibit reduced
gene flow, with significant differentiation among local populations often
occurring over relatively short geographic distances. The pattern observed
is similar in a qualitative sense to that described previously for cave dwelling
populations of troglophiles. However, the degree of differentiation among
troglobite populations appears to be greater. This may mean that trogloph-

iles (*e.g.*, *C. agassizi*) can disperse through surface streams. In addition, assuming that genetic divergence at allozyme loci results from genetic drift in the absence of gene flow, the degree of differentiation will be a function of effective population size and time since the reduction or elimination of gene flow.

Swofford (1982) makes a strong circumstantial case for the origin of *T. subterraneus* through multiple isolations of the surface dwelling ancestor in caves. He reconciles disparities between morphological and electrophoretic differentiation in *T. subterraneus* populations by arguing for parallel evolution of some morphological features in the absence of gene flow. The mode of origin of subterranean *Niphargus* is less clear. Furthermore, the situation is complicated by the possibility of secondary extinctions of *Niphargus* populations via movement of alpine glaciers (Ginet, 1965, 1969). The occurrence of some *Niphargus* in interstitial habitats (*e.g.*, *N. rhenorhodanensis*) suggests increased dispersal capabilities relative to those species restricted to karst waters. Finally, levels of genetic variability are large in *N. longicaudatus* populations (Sbordoni *et al.*, 1979), indicating that local populations are large and have not gone through recent bottlenecks.

C. Discussion

1. Cave Colonization

The case studies reviewed above provide valuable insight into the process of cave colonization. One of the direct consequences of cave colonization appears to be substantial changes in gene flow levels among populations. This results in significant genetic differentiation, in the case of presumed troglophilic species, between surface and cave dwelling populations as well as among cave dwelling populations. Indeed, there is little evidence of gene flow between cave and surface populations of *G. minus* or those of *A. fasciatus*. This raises the question of whether epigean and morphologically modified hypogean populations should continue to be considered conspecific [see Sarbu *et al.* (1993) with regard to *G. minus*]. Taxonomic issues aside, it is apparent that a considerable amount of evolution in troglomorphic features has occurred in these two cases despite the presence of related surface dwelling populations.

There is evidence for both *A. fasciatus* (Mitchell *et al.*, 1977) and *G. minus* (Kane and Culver, 1991; Sarbu *et al.*, 1993) that subterranean populations are the result of multiple isolations of the surface dwelling ancestor. At least seven independent episodes of cave colonization are indicated for *G. minus* in West Virginia and Virginia, some apparently occurring at different times (Culver, 1987; Kane and Culver, 1991; Sarbu *et al.*, 1993). These isolations, however, appear not to have included genetic bottlenecks for *G. minus*. Cave and spring dwelling populations do not differ in levels of genetic variability, and variability levels are generally large. Assuming genetic equilibrium, H can be related to effective population size (N_E) as

$$H = 4N_E u/(4N_E u + 1),$$

where the rate of mutation to neutral alleles (u) is 10^{-7} (Kimura, 1983). For cave populations of *G. minus* the estimate of N_E is approximately 2×10^5 individuals. It is true that heterozygosity levels in cave populations ($H = 0.036$) of *A. fasciatus* are significantly lower than those observed in populations in surface streams (Avise and Selander, 1972). This still gives an estimate of $N_E = 9 \times 10^4$ for subterranean *A. fasciatus* populations, although the assumption of genetic equilibrium may be less valid for these populations than it is for *G. minus*.

If allozyme variability is primarily a consequence of stochastic factors (*i.e.*, genetic drift) acting on genetically isolated populations, differentiation of populations will be influenced by effective population size and time since reduction or cessation of gene flow. Thus two predictions arise. Troglobites, with a longer history of cave habitation, should be more differentiated than troglophiles. Second, cave dwelling vertebrates, having smaller population sizes (Culver, 1982), should be more differentiated than cave dwelling invertebrates. The four case studies largely support these predictions, with the exception that populations of the troglobite *N. longicaudatus* appear no more differentiated ($F_{ST} = 0.287$) than do those of the troglophile *G. minus* ($F = 0.401$).

2. Troglomorphy and Adaptation

The large amount of biochemical differentiation observed among cave dwelling populations in these studies is in sharp contrast with the similarity in troglomorphic features observed among the same populations. There has been a tradition in cave biology of decoupling the regressive and elaborated troglomorphic characters and developing separate evolutionary hypotheses for their explanation (Culver, 1982; Wilkens, 1988). Because they are less obviously adaptive, evolutionary explanations for eye and pigment loss have often not invoked natural selection (Barr, 1968; Culver, 1982). Wilkens (1971, 1988) has argued that eye and pigment loss in *A. fasciatus* has been a consequence of accumulation of selectively neutral structurally reducing alleles for these characters in small cave populations. This hypothesis, therefore, invokes the same genetic mechanism for the evolution of these reduced morphological features (*i.e.*, genetic drift) that we have used to explain allozyme variation. Elaborated troglomorphic features appear to be lacking in *A. fasciatus*.

Individuals from cave dwelling populations of *G. minus* exhibit both reduced (eyes) and hypertrophied (antennae) troglomorphic features (Culver, 1987). A cross-sectional study on cave populations from two separate drainages in West Virginia (Jones and Culver, 1989; Jones *et al.*, 1992; Culver *et al.*, 1994) demonstrated that selection was operating on both eye size and antennal length. Individuals with smaller eyes and larger antennae enjoyed higher mating success and, in the case of females, higher fecundity (Jones *et al.*, 1992). Jones and Culver (1989) speculated that selection was operating not on the morphology per se, but rather on the function it performs. Processing of chemical and visual sensory information is not independent at the level of the central nervous system of amphipods (Bullock and Horridge, 1965). Thus, sensory compensation in cave dwelling populations may require an evolutionary trade-off, with brain regions devoted to visual inputs reduced, permitting an enhancement of regions devoted to the processing of chemical stimuli. Voneida and Fish (1984) have shown that areas of the optic tectum are innervated by the visual sensory system of fish of surface waters, but the same areas of the central nervous system are innervated by the somatic–sensory system of fish from cave waters. This more holistic view of the troglomorphic pattern suggests that the similar selection pressures of cave groundwater environments provide an adequate explanation for the ubiquity of troglomorphy among both invertebrate and vertebrate species of troglobites.

ACKNOWLEDGMENTS

We greatly appreciate the field assistance that has been provided over the years by numerous students, colleagues, and cavers. We also thank the many landowners who have kindly permitted us access to field sites on their property. Portions of the research summarized here have been supported by National Science Foundation Grants BSR-84154862 and BSR-8818560 to D. C. C. and National Science Foundation Grant BSR-8905220 to T. C. K.

REFERENCES

Avise, J. C., and Selander, R. K. (1972). Genetics of cave-dwelling fishes of the genus *astyanax*. *Evolution (Lawrence, Kans.)* 26, 1–19.
Barr, T. C., Jr. (1968). Cave ecology and the evolution of troglobites. *Evol. Biol.* 2, 35–102.
Barr, T. C., Jr. (1985). Pattern and process in speciation of trechine beetles in eastern North America (Coleoptera: Carabidae: Trechinae). *In* "Taxonomy, Phylogeny and Zoogeography of Beetles and Ants" (G. E. Ball, ed.), pp. 350–407. Dr. W. Junk Publ., Dordrecht, The Netherlands.
Barr, T. C., Jr., and Holsinger, J. R. (1985). Speciation in cave faunas. *Annu. Rev. Ecol. Syst.* 16, 313–337.
Barthélémy, D. (1982). La colonisation artificielle de la rivière souterraine de la Balme (Dépt. de l'Isère) par l'Amphipode *Niphargus virei:* Bilan actuel. *Bull. Soc. Linn. Lyon* 51, 50–256.
Bullock, T. H., and Horridge, G. A. (1965). "Structure and Function in the Nervous System of Invertebrates," Vol. 2. Freeman, San Francisco.
Caccone, A. M., and Powell, J. R. (1987). Molecular evolutionary divergence among North American cave crickets. II. DNA-DNA hybridization. *Evolution (Lawrence, Kans.)* 41, 1215–1238.

Caccone, A. M., Cobolli-Sbordoni, E., De Matthaeis, E., and Sbordoni, V. (1982). Una datazione su base genetico-molecare della divergenza tra specie cavernicole e marine di Sferomidi (gen. *Monolistra e Sphaeroma*, Crustacea, Isopoda). *Lav. Soc. Ital. Biogeogr.* **7**, 853–867.

Christiansen, K. A. (1961). Convergence and parallelism in cave Entomobryinae. *Evolution (Lawrence, Kans.)* **15**, 288–301.

Christiansen, K. A. (1962). Proposition pour la classification des animaux cavernicoles. *Spelunca* **2**, 76–78.

Christiansen, K. A. (1965). Behavior and form in the evolution of cave Collembola. *Evolution (Lawrence, Kans.)* **19**, 529–537.

Cooper, J. E., and Kuehne, R. A. (1974). *Speoplatyrhinus poulsoni*, a new genus and species of subterranean fish from Alabama. *Copeia*, pp. 486–493.

Culver, D. C. (1982). "Cave Life: Evolution and Ecology." Harvard Univ. Press, Cambridge, MA.

Culver, D. C. (1987). Eye morphometrics of cave and spring populations of *Gammarus minus* (Amphipoda: Gammaridae). *J. Crustacean Biol.* **7**, 74–84.

Culver, D. C., Kane, T. C., Fong, D. W., Jones, R., Taylor, M. A., and Sauereisen, S. C. (1990). Morphology of cave organisms—is it adaptive? *Mém. Biospéol.* **17**, 13–26.

Culver, D. C., Jernigan, R. W., O'Connell, J., and Kane, T. C. (1994). The geometry of natural selection in cave and spring populations of the amphipod *Gammarus minus* Say (Crustacea: Amphipoda). *Biol. J. Linn. Soc.* **52**, 49–67.

Dai, N. (1989). The comparative examination of eye anatomy of spring-dwelling and cave-dwelling populations of the amphipod *Gammarus minus*. Thesis, American University, Washington, DC.

Dickson, G. W., Patton, J. C., Holsinger, J. L., and Avise, J. C. (1979). Genetic variability in cave-dwelling and deep-sea organisms with emphasis on *Crangonyx antennatus* (Crustacea: Amphipoda) in Virginia. *Brimleyana* **2**, 119–130.

Dole, M. J., and Chessel, D. (1986). Stabilité physique et biologique des milieux interstitiels. Cas de deux stations du Haut-Rhone. *Ann. Limnol.* **22**, 69–81.

Dole-Olivier, M. J., and Marmonier, P. (1992). Effects of spates on the vertical distribution of the interstitial community. *Hydrobiologia* **230**, 49–61.

Essafi, K. (1990). Structure et transfert des peuplements aquatiques souterrains à l'interface karst-plaine alluviale. Thèse, Univ. Lyon I.

Essafi, K., Mathieu, J., and Beffy, J. L. (1992). Spatial and temporal variations of *Niphargus* populations in interstitial aquatic habitats at the karst/floodplain interface. *Regulated Rivers* **7**, 83–92.

Fong, D. W. (1989). Morphological evolution of the amphipod *Gammarus minus* in caves: Quantitative genetic analysis. *Am. Midl. Nat.* **121**, 361–378.

Ford, D. C., and Ewers, R. O. (1978). The development of limestone cave systems in the dimensions of length and depth. *Can. J. Earth Sci.* **15**, 1783–1798.

Gibert, J. (1984). Répartition spatiale de l'Amphipode souterrain *Niphargus rhenorhodanensis* au sein d'un aquifère karstique. *Verh. Int. Ver. Limnol.* **22**, 1739–1743.

Gibert, J. (1986). Ecologie d'un système karstique jurassien. Hydrogeologie, dérive animale, transit de matieèes, dynamique de la population de *Niphargus* (Crustacé Amphipode). *Mém. Biospéol.* **13**, 380.

Ginet, R. (1952). La grotte de la Balme (Isère); topographie et faune. *Bull. Soc. Linn. Lyon* **1**(2), 4–17, 27–30.

Ginet, R. (1965). Expérience de colonisation souterraine aquatique par *Niphargus;* premier résultats biologiques. *Bull. Soc. Zool. Fr.* **40**, 581–590.

Ginet, R. (1969). Rythme saisonnier des reproductions de *Niphargus* (Crustacés, Amphipodes, hypogés). *Annl. Spéléol.* **24**, 387–397.

Ginet, R. (1983). Les *Niphargus* (Amphipodes souterrain) de la région de Lyon (France). Observations biogoégraphiques, systèmatiques et écologiques. *Mém. Biospéol.* **10**, 179–187.

Ginet, R., and Decou, V. (1977). "Initiation à la biologie et à l'écologie souterraines." Delage, Paris.

Hamilton-Smith, E. (1971). The classification of cavernicoles. *NSS Bull.* **33**, 63–66.

Henry, J. P., and Magniez, G. (1983). Introduction pratique à la systématique des organismes des eaux continentales Françaises. 4. Crustacés Isopodes (principalement Asellotes). *Bull. Soc. Linn. Lyon* **52**, 319–358.

Hobbs, H. H., Jr., and Barr, T. C., Jr. (1972). Origins and affinities of the troglobitic crayfishes of North America (Decapoda: Astacidae). II. Genus *Orconectes. Smithson. Contrib. Zool.* **105**, 84.

Holsinger, J. R. (1972). "The Freshwater Amphipod Crustaceans (Gammaridae) of North America," Biota of Freshwater Ecosystem Identification Manual No. 5. Environmental Protection Agency, Washington, DC.

Holsinger, J. L., and Culver, D. C. (1970). Morphological variation in *Gammarus minus* Say (Amphipoda: Gammaridae) with emphasis on subterranean forms. *Postilla* **146**, 1–24.

Huppop, K. (1987). Food finding ability in cave fish (*Astyanax fasciatus*). *Int. J. Speleol.* **16**, 59–66.

Jones, R. D., and Culver, D. C. (1989). Evidence for selection on sensory structures in a cave population of *Gammarus minus* Say (Amphipoda). *Evolution (Lawrence, Kans.)* **43**, 688–693.

Jones, R. D., Culver, D. C., and Kane, T. C. (1992). Are parallel morphologies of cave organisms the result of similar selection pressures? *Evolution (Lawrence, Kans.)* **46**, 353–365.

Jones, W. K. (1981). A karst hydrology of Monroe County, West Virginia. *Proc. Int. Congr. Spéléol., 8th,* Bowling Green, Vol. 1, pp. 345–347.

Jones, W. K. (1988). Hydrology. *Bull. W. VA. Speleol. Surv.,* No. 9.

Kane, T. C. (1982). Genetic patterns and population structure in cave animals. *In* "Environmental Adaptation and Evolution" (D. Mossakowski and G. Roth, eds.), pp. 131–149. Fischer, Stuttgart.

Kane, T. C., and Brunner, G. D. (1986). Geographic variation in the cave beetle *Neaphaenops tellkampfi* (Coleoptera: Carabidae). *Psyche* **93**, 231–251.

Kane, T. C., and Culver, D. C. (1991). The evolution of troglobites—*Gammarus minus* (Amphipoda: Gammaridae) as a case study. *Mém. Biospéol.* **18**, 3–14.

Kane, T. C., Barr, T. C., Jr., and Badaracca, W. J. (1992a). Cave beetle genetics: Geology and gene flow. *Heredity* **68**, 277–286.

Kane, T. C., Culver, D. C., and Jones, R. T. (1992b). Genetic structure of morphologically differentiated populations of the amphipod *Gammarus minus. Evolution (Lawrence, Kans.)* **46**, 272–278.

Kimura, M. (1983). "The Neutral Theory of Molecular Evolution." Cambridge Univ. Press, New York.

Laing, C. D., Carmody, G. R., and Peck, S. B. (1976a). How common are sibling species in cave inhabiting invertebrates? *Am. Nat.* **110**, 184–189.

Laing, C. D., Carmody, G. R., and Peck, S. B. (1976b). Population genetics and evolutionary biology of the cave beetle *Ptomaphagus hirtus. Evolution (Lawrence, Kans.)* **30**, 484–498.

MacArthur, R. H. (1972). "Geographical Ecology." Harper & Row, New York.

Marmonier, P., and Dole, M. J. (1986). Les Amphipodes des sédiments d'un bras court-circuité du Rhône. Logique de répartition et réaction aux crues. *Rev. Fr. Sci. Eau* **5**, 461–486.

Mathieu, J., and Essafi, K. (1991). Changes in abundance of interstitial populations of *Niphargus* (stygobiont amphipod) at the karst/floodplain interface. *C. R. Seances Acad. Sci.* **312**, 489–494.

Mathieu, J., and Essafi-Chergui, K. (1990). Le peuplement aquatique interstitiel à l'interface karst/plaine alluviale. 1. Cas d'une alimentation en eau essentiellement karstique. *Mém Biospéol.* **17**, 113–122.

Mathieu, J., Martin, D., and Huissoud, P. (1984). Influence des conditions hydrologiques sur l'évolution de la structure spatiale et de la démographie d'une population phréatique de l'Amphipode *Niphargus rhenorhodanensis*. Premier résultats. *Mém. Biospéol.* **11**, 27–36.

Mathieu, J., Debouzie, D., and Martin, D. (1987). Influence des conditions hydrologiques sur la dynamique d'une population phréatique de *Niphargus rhenorhodanensis* (Amphipode souterrain). *Vie Milieu* **37**, 93–200.

Mayr, E. (1963). "Animal Species and Evolution." Harvard Univ. Press, Cambridge, MA.

Mitchell, R. W., Russell, W. H., and Elliott, W. R. (1977). Mexican eyeless characin fishes, genus *Astyanax:* Environment, distribution, and evolution. *Spec. Publ.—Mus., Tex. Tech. Univ.* **12**, 1–89.

Packard, A. S. (1988). Cave fauna of North America. *Mem. Natl. Acad. Sci.* **4**, 1–156.

Poulson, T. L. (1961). Cave adaptation in amblyopsid fishes. Ph.D. Dissertation, University of Michigan, Ann Arbor.

Poulson, T. L. (1963). Cave adaptation in amblyopsid fishes. *Am. Midl. Nat.* **70**, 257–290.

Poulson, T. L. (1964). Animals in aquatic environments: Animals in caves. *In* "Handbook of Physiology" (D. B. Dill, E. F. Adolph, and C. G. Wilber, eds.), Sect. 4, pp. 749–771. Am. Physiol. Soc., Washington, DC.

Poulson, T. L., and White, W. B. (1969). The cave environment. *Science* **165**, 971–981.

Romero, A. (1983). Introgressive hybridization in the *Astyanax fasciatus* (Pisces, Characidae) population at the Cueva Chica. *NSS Bull.* **45**, 81–85.

Rouch, R. (1977). Considérations sur l'écosysteme karstique. *C. R. Hebd. Seances Acad. Sci.* **284**, 1101–1103.

Rouch, R. (1986). Sur l'écologie des eaux souterraines dans le karst. *Stygologia* **3**, 352–398.

Rouch, R., and Danielopol, D. (1987). L'origine de la faune aquatique souterraine, entre le paradigme du refuge et le modèle de la colonisation active. *Stygologia* **3**, 345–372.

Sarbu, S., Kane, T. C., and Culver, D. C. (1993). Genetic structure and morphological differentiation: *Gammarus minus* (Amphipoda: Gammaridae) in Virginia. *Am. Midl. Nat.* **129**, 145–152.

Sbordoni, V. (1982). Advances in speciation of cave animals. *In* "Mechanisms of Speciation" (C. Barigozzi, ed.), pp. 219–240. Liss, New York.

Sbordoni, V., Cobolli-Sbordoni, and De Matthaeis, E. (1979). Divergenza genetica tra popolazioni e specie ipogee ed epigee di *Niphargus* (Crustacea, Amphipoda). *Lav. Soc. Ital. Biogeogr.* **6**, 329–351.

Selander, R. K. (1976). Genic variation in natural populations. *In* "Molecular Evolution" (F. J. Ayala, ed.), pp. 21–45. Sinauer Assoc., Sunderland, MA.

Slatkin, M., and Barton, N. H. (1989). A comparison of three indirect methods for estimating average levels of gene flow. *Evolution (Lawrence, Kans.)* **43**, 1349–1368.

Stock, J. H. (1980). Regression model evolution as exemplified by the genus *Pseudoniphargus* (Amphipoda). *Bijdr. Dierkd.* **50**, 104–144.

Swofford, D. L. (1982). Genetic variability, population differentiation, and biochemical relationships in the family Amblyopsidae. Thesis, Eastern Kentucky University, Richmond.

Trexler, J. C. (1988). Hierarchical organization of genetic variation in the Sailfin Molly, *Poecilia latipinna* (Pisces: Poeciliidae). *Evolution (Lawrence, Kans.)* **42**, 1006–1017.

Turquin, M.-J. (1976). Choix d'un traceur biologique dans un système karstique jurassien. *Ann. Sci. Univ. Besançon* **25**, 423–429.

Turquin, M.-J. (1981). The tactics of dispersal of two species of *Niphargus* (Perennial, Troglobitic Amphipoda). *Proc. Int. Congr. Speleol., Bowling Green, 8th*, Vol. 1, pp. 353–355.

Vervier, P. (1988). Hydrologie et dynamique des peuplements aquatiques de deux systèmes karstiques des gorges de l'Ardèche. Thèse, Univ. Lyon I.

Voneida, T. J., and Fish, S. E. (1984). CNS changes related to the reduction of visual input in a naturally blind fish *Anophtichthys hubbsi*. *Am. Zool.* **24**, 775–782.

Wilkens, H. (1971). Genetic interpretation of regressive evolutionary processes: Studies on hybrid eyes of two *Astyanax* cave populations (Characidae, Pisces). *Evolution (Lawrence, Kans.)* **25**, 530–544.

Wilkens, H. (1988). Evolution and genetics of epigean and cave *Astyanax fasciatus* (Characidae, Pisces). *Evol. Biol.* **23**, 271–367.

Wilkens, H., and Huppop, K. (1986). Sympatric speciation in cave fish? Studies on a mixed population of epi- and hypogean *Astyanax* (Characidae, Pisces). *Z. Zool. Syst. Evolutionsforsch.* **24**, 223–230.

Wright, S. (1931). Evolution in Mendelian populations. *Genetics* **16**, 97–159.

Wright, S. (1978). "Evolution and Genetics of Populations," Vol. 4. Univ. of Chicago Press, Chicago.

10

Species Interactions

D. C. Culver

Department of Biology
American University
Washington, D.C. 20016

I. INTRODUCTION

Because of the simplicity and extremity of the subterranean environment, the relatively small numbers of species, and the highly replicated nature of the habitat, subterranean communities can serve as model systems for the study of interactions among species (Culver, 1982). Two frameworks are especially useful for the consideration of interspecific interactions of groundwater organisms—the food web and associated quantitative and topological descriptors (Pimm, 1982) and the classification of pairwise interactions based on the effect of one species on the population growth rate of the other. The food web allows a close examination of the flow of energy through the ecosystem, whereas pairwise interactions allow a close examination of the myriad ways (trophic and nontrophic) that two species can interact. Thus in this chapter these two aspects will be examined successively.

Groundwater Ecology
Copyright © 1994 by Academic Press, Inc. All rights of reproduction in any form reserved.

271

II. TROPHIC INTERACTIONS

A. Background

Subterranean habitats impose major constraints on trophic interactions. With some exceptions, the major energy input is external to the system (*i.e.*, allochthonous), with photosynthesis missing and chemoautotrophy at least quantitatively unimportant [Ginet and Decou, 1977; see Sarbu (1990) for an example of a macroscopic fauna with a chemoautotrophic food base]. Relative to surface habitats, the variety of resources is reduced. Furthermore, many of the resources available will be detritus of various kinds and sizes, including both aquatic and terrestrial epigean animals washed into the system (Decamps and Rouch, 1973; Turquin, 1975), dissolved organic matter (Delay and Aminot, 1973), and other particulate organic matter. Relative to surface streams, the proportion of larger-size particles of food [e.g., coarse particulate organic matter (CPOM)] may be reduced in hyporheic habitats (Hynes, 1983).

Furthermore, the quantity of detritus is less in many subsurface habitats than that in surface habitats. This is well known for karst waters (Ginet and Decou, 1977), but other interstitial waters have reduced resource levels as well. For example, Mathieu and Gibert provide indirect evidence of this with the finding that *Niphargus rhenorhondanensis,* which inhabits nonkarst subsurface waters, has a metabolic rate intermediate between that of cave and surface amphipods (Mathieu, 1972; Gibert and Mathieu, 1980) and is adapted to intermediate levels of resources. Finally, some hyporheic habitats may actually have elevated levels of some nutrients, especially DOM and FPOM [see Stanford and Ward (1988) for an example].

The impoverishment and monotony of available food resources coupled with the aphotic nature of the habitats have resulted in taxonomic imbalances relative to surface aquatic habitats. Larger aquatic insects, especially visually oriented predators such as Odonata, are underrepresented (Stanford and Ward, 1988; Ward and Voelz, 1990). Most subsurface habitats have a rich crustacean fauna and generally a rich worm fauna (Pennak and Ward, 1986), especially in porous subsurface habitats (see Botosaneanu, 1986).

An additional constraint on subsurface aquatic faunas, with the exception of the fauna of conduits and large cavities, is one of size. Typically, animals are living among gravels and sands in relatively slow moving water, where all organisms are worm like to varying degrees (Botosaneanu, 1986). Thus in many habitats, large predators (*e.g.*, fish and salamanders) will be absent, and predators will be close to the size of their prey.

The above characteristics and constraints of subterranean habitats lead to two conflicting evolutionary pressures and adaptive responses. The first is strong selection for omnivory and a broadening of diet as a result of reduced resource levels. This may explain in part the apparent success of groups that were initially scavengers, omnivores, detritivores, etc., in subterranean habitats and the apparent broadening of diets of subterranean animals. The second evolutionary pressure is strong selection for increased efficiency at food finding, including predation. In the aphorism of Racovitza (1950), that subterranean animals may be "carnassier par prédilection, mais un saprophage par necessité" neatly summarizes one view of the conflicting selective pressures. The conflict between these two selective pressures is that improvement in food-finding ability for a particular kind of food (*e.g.*, live prey and dead insects) may lead to a reduction in food-finding ability for other kinds of food. Likewise, broadening of the diet may result in reduced efficiency in obtaining a particular kind of food. The trade-off of particular interest is that between predation and other kinds of feeding such as saprophagy. These two themes will be emphasized in a review of the feeding habits of two major groups of cave fish—the Mexican cave characin *Astyanax mexicanus* and the North American amblyopsid fishes.

B. Cave Fish: Predators or Scavengers?

River-dwelling *A. mexicanus* populations (Parzefall, 1983) normally school near the bank, down to a depth of 2 m. With rapid swimming movements the fish follow small food particles and prey items and test them by direct contact. Smaller animals, such as insects or young fish, are, according to Parzefall, eaten within seconds. In general, food is obtained visually in open water. In contrast, the cave fish are largely bottom feeders, swimming at an angle of 45° to the bottom and guided to food by chemical signals and water movements. Surface-dwelling fish, in the absence of light, swim at an angle of 80° to the bottom.

Morphologically, cave populations show at least one major elaboration of structure related to food-finding ability. The gustatory region is larger and spread over the ventral region of the head and has a greater density of taste buds (Schemmel, 1974).

In a laboratory study that simulated conditions in the cave, small pieces of food (10 mm^3 of beef heart) were added in darkness to an aquarium containing six cave and six river fish (Huppop, 1987). Cave fish were four times more successful in obtaining food. This is not surprising given their chemical senses and more efficient swimming angle. However, Huppop's experiment mimics scavenging ability rather than predatory ability because it used immobile food particles dropping through the water column. What about predatory behavior? Klimpel and Parzefall (1990) present some evidence that cave fish show more scavenging behavior than do river fish. River fish, whether reared on dry food or living prey (*Artemia* and *Tubifex*), strongly preferred living prey, whereas cave fish were less consistent. Cave fish reared on dry food preferred living prey, but cave fish reared on living prey, actually preferred dry food, although the difference was not statistically significant.

Thus, cave-dwelling individuals of *A. fasciatus* are, by virtue of behavioral and morphological modifications, more efficient bottom foragers and scavengers. However, it seems apparent that they are not more efficient predators in open water than river-dwelling individuals, and are probably less so.

By contrast, cave fish in the family Amblyopsidae have remained predators. Relative to the spring-dwelling *Chologaster agassizi,* cave species (*Amblyopsis rosae, A. spelaea, Speoplatyrhinus poulsoni,* and *Typhlichthys subterraneus*) have a highly elaborated lateral line system (Poulson, 1963; Culver, 1982) that greatly enhances their ability to detect prey. When 1 *Daphnia* was introduced into a 100-l aquarium, *A. spelaea* found it hours before *C. agassizi* did (Poulson and White, 1969). At high food densities, the situation was reversed. *C. agassizi* ate all 10 *Daphnia* introduced into a 5-l aquarium before *A. spelaea* had eaten half. However, the situation at high densities may reflect appetite rather than food-finding differences. The maximum distance of orientation of cave species toward prey is two to five times that of *C. agassizi.* Thus, in contrast to the situation with *A. styanax,* cave Amblyopsidae have become highly efficient predators, but not scavengers. Because the major morphological elaboration in amblyopsids is in the lateral line system and hence in the ability to detect, whereas the major morphological elaboration in the Mexican characins is the gustatory system, this difference may well have its origin in the differences between the surface ancestors of the two groups. That is, surface populations of *Astyanax* may have more elaborated taste systems relative to those of *Chologaster,* whereas *Chologaster* may have a more elaborated lateral line system relative to that of *Astyanax.* The spring-dwelling *C. agassizi* is highly carnivorous and finds prey by tactile organs and neuromast receptors (Hill, 1969). These comments contribute to the continuing controversies concerning preadaptation and general adaptation of cave animals (Rouch, 1986; Danielopol and Rouch, 1991), which are beyond the scope of this discussion.

C. Food Webs

Very few food webs, other than the general webs (Mohr and Poulson, 1966), that do not apply to a particular groundwater community have been published. Pennak and Ward (1986) identify several groundwater community types, but do not present any food webs. Culver (1982) gives a food web for the macroscopic aquatic fauna of Gallohan Cave No. 1, Lee County, Virginia, which is shown in Fig. 1. Several food webs have been published for springs—the ecotone between groundwater and surface water (Gibert *et al.,* 1990). Perhaps the most thorough study is that of Tilly (1968) for Cone Spring, Iowa. A rich theoretical and empirical literature on food web patterns exists (Pimm, 1982; Schoenly *et al.,* 1991). This literature allows comparison of the Cone Spring and Gallohan Cave food webs with food

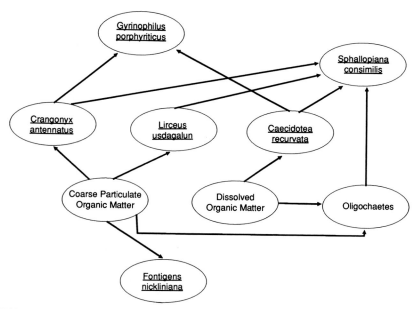

FIGURE 1   Food web for a stream in Gallohan Cave No.1, Lee Co., Virginia. Arrows indicate feeding interactions.

webs in general. Some of these comparisons are shown in Table I. The number of trophic levels in caves and springs is lower, the amount of connectance of the webs is lower, the amount of compartmentation is higher, and, somewhat surprisingly, the amount of omnivory is lower. Pimm (1982) defines omnivory as feeding on more than one trophic level so that the

TABLE I   Food Web Characteristics of Gallohan Cave Stream (see Fig. 1), Cone Spring (Tilly 1968), and a Selection of 12 Food Webs Analyzed by Pimm (1982)

Characteristic	Cone spring	Gallohan cave	Overall
No. of trophic levels	3	2	3–4
Omnivory[a]	0.25	0.0	1.0
Compartmentation[b]	0.224	0.214	0.191
Connectance[c]	2.7	1.8	3.1

[a] Sum of numbers of omnivore links (i.e., feeding at a second trophic level) divided by the number of top predators.
[b] Related to the number of species both species i and j interact with compared with the total number of species that either one interacts with. Consult Pimm (1982) for a technical definition.
[c] Number of predator–prey interactions divided by $N = 1$, where N is the number of species (compartments) in the food web.

reduced number of trophic levels in caves and springs reduces the opportunity for omnivory by this definition. The summary in Table I gives only a hint of the power of the statistical analysis of food web characteristics (Pimm, 1982).

III. INTERACTIONS IN GENERAL

A. Background

Although trophic interactions may predominate in a community, there are a variety of interspecific interactions that a species pair may show. The usual classification of species interactions is shown in Fig. 2. The meaning of the word "effect" of species A on species B can be defined behaviorally, and, otherwise, a population definition allows us the use of differential equation and dynamical systems theory. Species A has a negative effect on species B if an increase in the number of individuals of species A has a negative effect on the growth rate of the population of species B. The pairwise interaction coefficients, first defined by Gause (1934, 1971) for competition (mutual negative effects), can be generalized to all the interactions listed in Fig. 2. If N_i is the population size and dN_i/dt is its rate of increase, then the interaction coefficient is

$$\alpha_{ij} = \frac{\dfrac{(dN_i/dt)}{N_j}}{\dfrac{(dN_i/dt)}{N_i}}$$

(1)

The sign of the interaction coefficient (1) is the same as that of the interactions listed in Fig. 2.

Effect of B on A

	0	+	−
0	None	Commensalism	Amensalism
+	Commensalism	Mutualism	Predation
−	Amensalism	Predation	Competition

Effect of A on B

FIGURE 2   Classification of interspecific interactions. See text for details.

Species interactions among amphipods and isopods in several Appalachian cave streams have been intensively studied (Culver, 1973a, 1976, 1982; Culver *et al.*, 1991). In these gravel-bottomed cave streams, there is an alternation between deeps (pools) and shallows (riffles). The riffles are much shorter than the pools and repeat at a more or less regular interval of five to seven stream widths (Leopold *et al.*, 1964). The amphipods and isopods, which dominate numerically and in terms of biomass, are highly concentrated in riffles (Culver, 1982). This concentration is due to several factors, including a concentration of food, especially detritus; increased O_2; and the absence of salamander predators, which live in pools. Within a riffle, the habitable space itself is highly discontinuous, consisting of the underside of gravels out of the brunt of the current. In this habitat, there are three obvious kinds of interaction:(1) species may compete for food, (2) species may compete for space (the underside of gravels), and (3) species may serve as food for other species. As we shall see, the actual interactions are even more complex than this.

Three cave streams have served as the focus for the study of species interactions among isopods and amphipods in cave streams. Thompson Cedar Cave is a small cave approximately 250 m long in the Powell River drainage in southwestern Virginia (Fig. 3). Organ Cave is a large cave over 60 km long in the Greenbrier River drainage in southern West Virginia (Fig. 3). Alpena Cave is a large cave over 2 km long in the Monongahela River drainage in northern West Virginia (Fig. 3). All are in well-integrated karst basins with extensive cave development (Culver *et al.*, 1992; Stevens, 1988; Medville and Medville, 1971).

B. Species Interactions in Thompson Cedar Cave

Until the severe contamination of the cave by sawmill waste and nearby logging operations in 1987, Thompson Cedar Cave had a small stream with little annual fluctuation in discharge. One amphipod species (*Crangonyx*

FIGURE 3 County map of Virginia and West Virginia, indicating the locations of Thompson Cedar Cave (VA), Organ Cave (WVA), and Alpena Cave (WVA).

antennatus) and two isopod species (*Caecidotea recurvata* and *Lirceus usdagalun*) predominate under gravels in riffles. Larvae of the salamander *Gyrinophilus porphyriticus* occur in pools. Pairs of amphipod and isopod species, along with the appropriate single species controls, were placed in artificial riffles in the laboratory under conditions in which the washout rate could readily be measured. The rate of washout from artificial riffles, in short-term experiments with abundant food, is a measure of the rate of competition (a_{ij}) between the species and can also be used to measure the relative carrying capacity of the environment (K) in terms of space under gravels. The behavioral bases of competition are collisions between individuals on the underside of gravels, which result in washouts (Culver, 1970). Details of the experimental procedure will not be given here, and the interested reader can consult Culver (1973a) and Culver *et al.* (1991). The results can be used to estimate α and K of the standard differential equations of competition,

$$dN_i/dt = r_i N_i (K_i - N_i - \Sigma \alpha_{ij} N_j)/K_i, \tag{2}$$

where N is the population size of species i and j, r is the intrinsic rates of increase, K is the carrying capacity, and α is the competition coefficient. Using the first letter of the species names to signify the species (*e.g.* "a"for *C. antennatus*), Eq. (2) was partially specified for the three amphipod and isopod species as

$$dN_a/dt = r_a N_a (1.4K - N_a - 0.99N_r - 1.32N_u)/1.4k$$
$$dN_r/dt = r_r N_r (1.3K - N_r - 0.32N_a - 1.29N_u)/1.3k$$
$$dN_u/dt = r_u N_u (1.0K - N_u - 1.16N_a - 0.49N_r)/1.0k \tag{3}$$

The intensity of competition between pairs appears to decline over evolutionary time (Culver, 1976). For the three species in Thompson Cedar Cave, *C. antennatus* and *L. usdagalun* compete most strongly, whereas *C. antennatus* and *C. recurvata* interact least.

These laboratory experiments and the resulting partially specified Eq. (3) allow predictions concerning which pairs of species can coexist and whether the triplet of species can stably coexist. Stable coexistence is taken to mean the return to an equilibrium with both (or all three) species present following a small perturbation in the numbers of individuals present. Stability (and instability) can be graphically represented by the phase portraits of the species much utilized in introductory ecology texts. The phase portraits for all three species pairs and the species triplets are shown in Fig. 4. The prediction from this combination of laboratory experiments and differential equations of competition is that, although all three species can coexist stably, only the pair *C. antennatus*–*C. recurvata* can stably coexist. The pairwise interaction between *L. usdagalun* and *C. antennatus* should result in the elimination of *L. usdagalun*. The pairwise interaction between *L. usdagalun* and *C. recurvata* should result in the elimination of *C. recurvata*.

These predictions are consistent with the patterns of species cooccur-

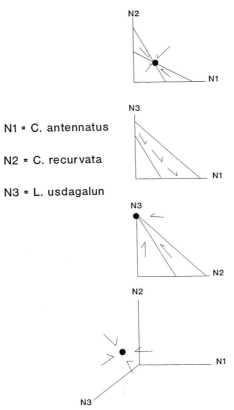

N1 = C. antennatus

N2 = C. recurvata

N3 = L. usdagalun

FIGURE 4   Phase diagrams for competition among *Crangonyx antennatus* (species 1), *Caecidotea recurvata* (species 2), and *Lirceus usdagalun* (species 3). Arrows indicate the direction of change in population size.

rence. Amphipods and isopods occur in three areas of the Thompson Cedar Cave stream (Fig. 5): immediately upstream of the entrance in a gravel-bottomed stream crawl extending for approximately 5 m; immediately downstream of the entrance for a distance of approximately 25 m; and in the final downstream section of gravel bottomed stream, which extends for approximately 20 m. The upstream area has the predicted stable species pair (*C. antennatus* and *C. recurvata*); the middle area has only *L. usdagalun*; and the downstream section has all three species (Culver, 1981a). Microdistribution patterns of these species in other caves near Thompson Cedar Cave are also highly consistent with predictions from laboratory experiments and competition theory (Culver, 1973a, 1981a) as are the temporal correlations of species abundance (Culver, 1981b) and microhabitat niche shifts (Culver, 1982). Although direct field manipulation experiments have not been done in this area (see Culver *et al.*, 1991), there is little doubt that interspecific competition is the dominant interaction and that it has profound

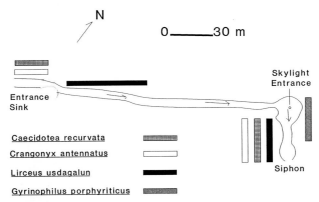

FIGURE 5   Map of distribution of *Crangonyx antennatus, Caecidotea recurvata, Gyrinophilus porphyriticus,* and *Lirceus usdagalun* in Thompson Cedar Cave, Lee Co., Virginia.

and predictable effects on species abundance, niches, and distribution (Culver, 1982).

Finally, there is one set of trophic interactions that is important in Thompson Cedar Cave—predation on the amphipods and isopods by larval *Gyrinophilus porphyriticus.* Although salamanders were not common in Thompson Cedar Cave, salamander predation on *C. antennatus* and especially *C. recurvata* was locally important in the downstream section of the cave (Fig. 5). Predation results in a reduction in both of these species and a proportional increase in the abundance of *C. antennatus* (Culver, 1982). *G. porphyriticus* rarely eats *L. usdagalun* in the field, although it is apparently not toxic (Culver, 1973b).

C. Species Interactions in Organ Cave

The microdistribution and niche separation patterns of amphipods and isopods in the Greenbrier Valley of West Virginia (see Fig. 2) suggested that these too were competitively controlled communities (Culver, 1970, 1973a). However, laboratory riffle experiments and manipulation experiments in Organ Cave (Culver *et al.,* 1991) have demonstrated that the interactions among the amphipods and isopods of Greenbrier Valley caves are a combination of trophic interactions (predation and feeding commensalism) and nontrophic interactions (competition for space on the underside of gravels).

The fauna and hydrology of Organ Cave are described elsewhere in this volume (see Chapter 16), and only the interactions among three pairs of amphipod and isopod species will be considered here. For each species pair, washout rates from artificial riffles over short time spans (hours) and washout rates from natural riffles in Organ Cave streams over longer time spans (months) were measured and compared (Culver *et al.,* 1991).

In laboratory riffles the isopod *Caecidotea holsingeri* had a negative effect on the amphipod *Stygobromus spinatus; i.e.,* it increased its washout rate. This effect is analogous to the competition among the amphipod and isopod species in Thompson Cedar Cave. There was no reciprocal effect, so the interaction can be described as an amensalism (see Fig. 2). In field manipulation experiments, neither species had any effect on the other, so the amensalism disappeared.

In laboratory riffles *C. holsingeri* had a negative effect on the amphipod *Gammarus minus; i.e.,* it increased its washout rate. The reciprocal effect was that of a predator—the total number (the number washed out plus the number remaining in the riffle) of *C. holsingeri* declined in the presence of *G. minus,* indicating that *G. minus* was a predator of the small isopod *C. holsingeri.* Field manipulation experiments confirmed that *G. minus* was a predator of *C. holsingeri.* In the field *C. holsingeri* had a positive effect on the abundance of *G. minus,* presumably by the attraction of *G. minus* to prey concentrations. The negative short-term laboratory effect of *C. holsingeri* on *G. minus* was not detected in the field.

In laboratory riffles the amphipod *Stygobromus emarginatus* had a negative effect on *G. minus; i.e.,* it increased its washout rate. There was no reciprocal effect, so the interaction can be described as an amensalism. In field manipulation experiments, the negative effect of *S. emarginatus* on *G. minus* persisted, but *G. minus* had a hitherto positive effect on *S. emarginatus.* This positive effect, the immediate result of movement of *S. emarginatus* toward the vicinity of *G. minus,* was the consequence of feeding by *S. emarginatus* on feces of *G. minus.* This combination of amenalism and commensalism produced the population dynamics of a predator–prey pair, even though no predation was occurring!

The three pairwise interactions are summarized in Table II. In the laboratory, there were two pairs of amensalists (*S. spinatus–C. holsingeri* and *S. emarginatus–G. minus*) and a predator–prey and competitor pair (*G. minus–C. holsingeri*). In terms of individual population dynamics there were

TABLE II   Comparison of Significant Interspecific Effects in Laboratory Riffle and in Manipulation Experiments in Organ Cave

Effect of	On	Laboratory	Field
C. holsingeri	*S. spinatus*	−	0
S. spinatus	*C. holsingeri*	0	0
C. holsingeri	*G. minus*	−	+
G. minus	*C. holsingeri*	−	0
S. emarginatus	*G. minus*	−	−
G. minus	*S. emarginatus*	0	+

three amensalists, one predator, and two cases of noninteraction. In the field, there were two predator–prey interactions (*G. minus–C. holsingeri* and *S. emarginatus–G. minus*) and one pair of non significant interactions. In terms of individual population dynamics there were: (1) one predator (*G. minus* on *C. holsingeri*); (2) one prey (*C. holsingeri*); (3) one commensalist (*S. emarginatus* of *G. minus*), and (4) one amensalist (*G. minus* of *S. emarginatus*). Thus, short-term laboratory interactions were dominated by negative effects, whereas longer-term field interactions were a mixture of positive, negative, and zero effects.

This difference is echoed in the different patterns of microdistribution that occur at different scales. The distribution of animals under individual rocks support the idea that nontrophic encounters between individuals were an important interaction in the field as were encounters in the laboratory stream. Distributions of animals under individual rocks in three riffles in Organ Cave are summarized in Table III. In all three cases, the observed cooccurrence is less than the expected cooccurrence. In two samples, no case of cooccurrence was found, and in the third, only one case of cooccurrence was found.

At the level of the riffle, however, the effect of antagonistic collisions cannot be easily detected, and, instead, distributions at this scale are concordant with the results from the longer-term field experiments. Although there were no statistically significant differences between observed and expected frequencies, all observed frequencies of cooccurrence in riffles were higher than expected frequencies of cooccurrence (Table III). The weakly positive associations may indicate that the commensalism of *S. emarginatus* on *G. minus* is important and that *G. minus* and *C. holsingeri* are positively correlated at the level of riffles (and negatively correlated at the levels of rocks) as a result of predator–prey interactions (Rose and Leggett, 1990). But other influences, especially differences among streams resulting from migration from other subsurface habitats and historical factors, are likely to be important at the level of riffles (see Chapter 16).

D. Species Interactions in Alpena Cave

A set of laboratory experiments and field manipulations similar to those done in Organ Cave was performed on two coexisting isopod species in Alpena Cave (see Fig. 3): *Caecidotea cannulus* and *C. holsingeri*. Unlike the three species pairs from Organ Cave that were studied, *C. cannulus* and *C. holsingeri* showed no sign of antagonistic interactions in the lab. That is, neither increased the washout rate of the other. In fact, the presence of *C. cannulus* seemed to slightly decrease the washout rate of *C. holsingeri*. In field manipulation experiments in Alpena Cave, there was no evidence of any interaction between the two species. The pattern of cooccurrence under rocks and in riffles (Table III) also reflects the lack of interaction. Cooccur-

TABLE III   Patterns of Observed and Expected Species Cooccurrence under Stones and under Riffles

	Stones				Riffles			
Species pair	N	O	E	P	N	O	E	P
Organ cave								
Gammarus minus	121	0	1.7	.04	12	6	5.8	.80
Stygobromus emarginatus								
Stygobromus spinatus	150	0	0.2	.48	12	1	0.8	.71
Caecidotea holsingeri								
Gammarus minus	168	1	1.8	.45	12	3	3	—
Caecidotea holsingeri								
Alpena cave								
Caecidotea cannulus	184	2	0.8	.20	12	2	1.2	.52
Caecidotea holsingeri								

Note. For each species pair, N is the number of stones (or riffles) observed; O, the observed frequency of cooccurrence; E, the expected frequency; and P, the probability of significant difference between the two. For riffles, P is computed from Fisher's exact test, and, for stones, P is computed from the log-likelihood test.

rence on both scales was slightly but not significantly more frequent than expected. *C. cannula* is much larger than *C. holsingeri,* especially in the area of sympatry, where *C. cannula* is nearly twice as large. This suggested that the species showed character displacement and that the absence of present-day competition was the result of the evolution of character displacement. However, a more careful study of these two species indicated that the size differences between the isopod species reflected differences in the sizes of the rocks in the stream (Culver and Ehlinger, 1980) rather than past competitive interactions.

E. Conclusion

Several generalities emerge from these studies of species interactions among isopods and amphipods. First, the importance of competition (and amensalism) varies from cave to cave. Competitive interactions dominated in Thompson Cedar Cave, were important but not dominant in Organ Cave, and were undetectable in Alpena Cave. The difference may be partly the result of hydrogeologic factors. Alpena Cave shows considerable annual fluctuation in current flow and temperature, which may keep populations low enough in density that interactions are unimportant. By contrast, Thompson Cedar Cave is a highly stable environment. Second, trophic interactions among amphipods and isopods, at least in Organ Cave, are much more complex than previously recognized. Both predation and trophic commensalism were detected. Third, many interactions are not solely of one type. For example, *S. emarginatus* avoided direct contact with *G. minus* (amensalism) but feeds on its fecal material (commensalism). Finally, species interactions can also have major effects on microdistribution patterns. Competition largely controlled the distribution in Thompson Cedar Cave (Fig. 5). Predator–prey interactions can have major effects as well (*e.g., G. minus* and *C. holsingeri* in Organ Cave), a phenomenon noted by others as well (see Leruth, 1938; Chodorowski, 1963; Stock, 1983). Commensalism can be important but remains little studied in general (Hobbs, 1975).

REFERENCES

Botosaneanu, L., ed. (1986). "Stygofauna mundi." E. J. Brill, Dr. W. Backhuys, Leiden.

Chodorowski, A. (1963). Sur la coaction biocenotique chez les cavernicoles aquatiques. *C. R. Hebd. Seances Acad. Sci.* **256**, 2049–2051.

Culver, D. C. (1970). Analysis of simple cave communities: Niche separation and species packing. *Ecology* **51**, 949–958.

Culver, D. C. (1973a). Competition in spatially heterogeneous systems: An analysis of simple cave communities. *Ecology* **54**, 102–110.

Culver, D. C. (1973b). Feeding behavior of the salamander Gyrinophilus porphyriticus in caves. *Int. J. Speleol.* **5**, 369–377.

Culver, D. C. (1976). The evolution of aquatic cave communities. *Am. Nat.* **110**, 945–957.

Culver, D. C. (1981a). The effect of competition on species composition of some cave communities. *Proc. Int. Congr. Speleol., 8th*, Bowling Green, Vol. 1, pp. 207–209.

Culver, D. C. (1981b). Some implications of competition for cave stream communities. *Int. J. Speleol.* **11**, 49–62.

Culver, D. C. (1982). "Cave Life: Evolution and Ecology." Harvard Univ. Press, Cambridge, MA.

Culver, D. C., and Ehlinger, T. J. (1980). The effects of microhabitat size and competitor size on two cave isopods. *Brimleyana* **4**, 103–114.

Culver, D. C., Fong, D. W., and Jernigan, R. W. (1991). Species interactions in cave stream communities: Experimental results and microdistribution effects. *Am. Midl. Nat.* **126**, 364–379.

Culver, D. C., Jones, W. K., and Holsinger, J. R. (1992). Biological and hydrological investigation of the Cedars, Lee County, Virginia, an ecologically significant and threatened karst area. *In* "Groundwater Ecology" (J. A. Stanford and J. J. Simons, eds.), pp. 281–290. Am. Water Res. Assoc., Bethesda, MD.

Danielopol, D. L., and Rouch, R. (1991). L'adaptation des organismes au milieu aquatique souterrain. Réflexions sur l'apport des recherches écologiques récentes. *Stygologia* **6**, 129–142.

Décamps, H., and Rouch, R. (1973). Le système karstique du Baget. I. Premiéres estimations sur la dérive des invertébres aquatiques d'origine épigee. *Ann. Spéléol.* **28**(1), 89–110.

Delay, B., and Aminot, A. (1973). Présence d'acides aminés libres dans les eaux d'infiltration qui circulent dans la zone superficielle des massifs karstiques. *C. R. Hebd. Seances Acad. Sci., Ser. D* **276**, 3289–3292.

Gause, G. F. (1934). "The Struggle for Existence." Williams & Wilkins, Baltimore, MD.

Gause, G. F. (1971). "The Struggle for Existence" (reprint), Dover, NY.

Gibert, J., and Mathieu, J. (1980). Relations entre les teneurs en protéines, glucides et lipides au cours du jeûne experimental, chez deux espèces de *Niphargus* peuplant des biotopes différents. *Crustaceana, Suppl.* **6**, 137–147.

Gibert, J., Dole-Olivier, M. J., Marmonier, P., and Vervier, P. (1990). Surface water groundwater ecotones. *In* "The Ecology and Management of Aquatic-Terrestrial Ecotones" (R. J. Naiman and H. Decamps, eds.), Man & Biosphere Ser., Vol. 4, pp. 199–226. UNESCO, Paris.

Ginet, R., and Decou, V. (1977). "Initiation à la biologie et à l'écologie souterraines." Delarge, Paris.

Hill, L. G. (1969). Feeding and food habits of the spring cavefish, *Chologaster agassizi*. *Am. Midl. Nat.* **82**, 110–116.

Hobbs, H. H., III (1975). Distribution of Indiana cavernicolous crayfishes and their ectocommensal ostracods. *Int. J. Speleol.* **7**, 273–302.

Huppop, K. (1987). Food-finding ability in cave fish (*Astyanax fasciatus*). *Int. J. Speleol.* **16**, 59–66.

Hynes, H. B. N. (1983). Groundwater and stream ecology. *Hydrobiologia* **100**, 93–99.

Klimpel, B., and Parzefall, J. (1990). Comparative study of predatory behavior in cave and river populations of *Astyanax fasciatus* (Characidae, Pisces). *Mém. Biospéol.* **17**, 27–30.

Leopold, L. B., Wolman, M. G., and Miller, J. P. (1964). "Fluvial Processes in Geomorphology." Freeman, San Francisco.

Leruth, R. (1938). La faune de la nappe phréatique du gravier de la Meuse a Hermalle-sous-Argenteau, études biospéologiques. IX. *Bull. Mus. R. Hist. Nat. Belg.* **14**, 1–37.

Mathieu, J. (1972). Métabolisme respiratoire de *Niphargus rhenorhodanensis* (Crustacé, Gammaridé souterrain; premiers résultats. *Ann. Spéléol.* **25**, 179–221.

Medville, D., and Medville, H. (1971). "Caves of Randolph County," Bull. 1. West Virginia Speleol. Surv., Barrackville.

Mohr, C. E., and Poulson, T. L. (1966). "The Life of the Cave." McGraw-Hill, New York.

Parzefall, J. (1983). Field observation in epigean and cave populations of the Mexican characid *Astyanax mexicanus* (Pisces, Characidae). *Mém. Biospéol.* **10**, 171–178.

Pennak, R. W., and Ward, J. V. (1986). Interstitial faunal communities of the hyporheic and adjacent groundwater biotopes of a Colorado mountain stream. *Arch. Hydrobiol., Suppl.* **3**, 356–396.

Pimm, S. L. (1982). "Food Webs." Chapman & Hall, London.

Poulson, T. L. (1963). Cave adaptation in amblyopsid fishes. *Am. Midl. Nat.* **70**, 257–290.

Poulson, T. L., and White, W. B. (1969). The cave environment. *Science* **165**, 971–981.

Racovitza, E. G. (1950). Asellides: 1ère série: *Stenasellus. Arch. Zool. Exp. Gen.* **87**, 1–94.

Rose, G. A., and Leggett, W. C. (1990). The importance of scale to predator-prey spatial correlations, an example of Atlantic fishes. *Ecology* **71**, 33–43.

Rouch, R. (1986). Sur l'écologie des eaux souterraines dans le karst. *Stygologia* **2**, 339–351.

Sarbu, S. (1990). An unusual fauna of a cave with thermo-mineral waters containing H_2S from southern Dobrogea, Romania. *Mém. Biospéol.* **17**, 191–196.

Schemmel, C. (1974). Genetische Untersuchungen zur Evolution des Geschmackapparates bei cavernicolen Fischen. *Z. Zool. Syst. Evolutions forsch.* **12**, 196–215.

Schoenly, K., Beaver, R. A., and Heumier, T. A. (1991). On the trophic relations of insects: A food-web approach. *Am. Nat.* **137**, 597–638.

Stanford, J. A., and Ward, J. V. (1988). The hyporheic habitat of river ecosystems. *Nature (London)* **335**, 64–66.

Stevens, P. J., ed. (1988). "Caves of the Organ Cave Plateau, Greenbrier County, West Virginia," Bull. No. 9. West Virginia Speleol. Surv., Barrackville.

Stock, J. H. (1983). Predation as a factor influencing the occurrence and distribution of small Crustacea in West Indian groundwaters. *Bijdr. Dierk.* **53**, 233–243.

Tilly, L. J. (1968). The structure and dynamics of Cone Spring. *Ecol. Monogr.* **38**, 169–197.

Turquin, M. J. (1975). Incidences des biocénoses terrestres sur le rythme de ponte de l'Amphipode troglobie *Niphargus. Bull. Soc. Zool. Fr.* **100**, 169–176.

Ward, J. V., and Voelz, N. J. (1990). Gradient analysis of interstitial meiofauna along a longitudinal stream profile. *Stygologia* **5**, 93–99.

11

Limits to Biological Distributions in Groundwater

D. L. Strayer

Institute of Ecosystem Studies
Box AB
Millbrook, New York 12545

I. INTRODUCTION

The topic of this chapter may seem overly ambitious, because biological distributions and their controls in groundwater are likely to be as diverse and complex as those in surface waters. Just as in surface waters, the biological communities of groundwaters are highly varied, ranging from rich communities with dense populations of bacteria, fungi, protozoans, and hundreds of species of invertebrates to sparse communities of almost inactive bacteria. The distribution of the groundwater biota is certainly controlled by historical factors, physicochemical variables, biological interactions, and interactions among these broad classes of elements, just as is the case for the surface-water biota. In contrast to our extensive knowledge of surface waters, though, we still know little about the extent of biological distributions in groundwater and the specific factors that control these distributions. Thus, a comprehensive review of biological distributions and their controls in groundwater would require more space and knowledge than is at my disposal. Nonetheless, it may be useful to describe some distributional patterns that are especially characteristic of the groundwater biota, to speculate a little on controlling factors that are either frequently important or peculiar to members of groundwater communities, and to expose the present state of our knowledge about these subjects.

II. COMMON DISTRIBUTIONAL PATTERNS

A. Endemism

Many groundwater invertebrates have strikingly restricted distributions. Even in regions that have been explored fairly intensively, many species of invertebrates are known only from one or two localities (Fig. 1) (Coineau,

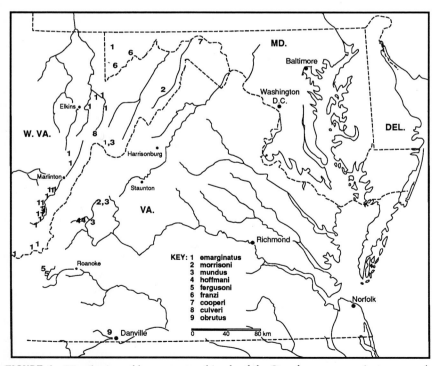

FIGURE 1 Distribution of hypogean amphipods of the *Stygobromus emarginatus* group in eastern North America. Each numeral shows the collection records for a different species. Redrawn from Holsinger (1978).

1971; Lescher- Moutoué 1973). Of course, not all species have small ranges. For example, the hypogean polychaete *Troglochaetus* is known from throughout central Europe and in Colorado, Ohio, and New Hampshire in North America (Husmann, 1962; Pennak, 1971; Tilzer, 1973; D. Strayer, unpublished). Nonetheless, it appears to be generally true that specialized groundwater invertebrates have much smaller distributions than their epigean relatives (Fig. 2). Although a high level of endemicity is characteristic of groundwater invertebrates, we do not yet know whether the microbial communities of groundwaters exhibit a similar degree of endemicity, or whether microbial species are as wide ranging in groundwaters as those in surface waters.

Endemic species of groundwater invertebrates are not distributed randomly around the world, but are concentrated in regions that support very diverse communities. For example, regions such as the Balkan Peninsula (Coineau, 1971; Bole and Velkovrh, 1986; Karaman and Ruffo, 1986; Sket, 1986) and the Edwards Aquifer in Texas (Holsinger and Longley, 1980; Longley, 1981; Herschler and Longley, 1986; Kroschewsky, 1990) support extremely diverse assembleges of specialized, endemic groundwater invertebrates of many taxa: flatworms, snails, amphipods, isopods, syncarids, and so on. These areas are the underground equivalents of the tropical rain forests. Other regions (*e.g.,* Scandinavia and northeastern North America) support only a handful of endemic species (Vandel, 1964; Husmann and Teschner, 1970; Peck, 1988; Strayer, 1988; Williams, 1989).

B. Distance from the Earth's Surface

Biologists who are familiar only with surface waters often are surprised at the great depths to which members of the groundwater biota are found. In karst regions, both microbes and invertebrates (and even vertebrates) are

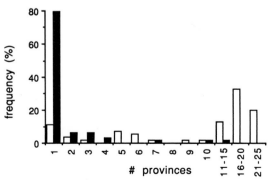

FIGURE 2   Number of zoogeographic provinces (in the sense of Illies) occupied by species of groundwater (black bars) and surface water (white bars) cyclopoid copepods in Europe. Distributional data from Kiefer (1978); list of groundwater species from Lescher-Moutoué (1986). All subspecies of a species are pooled.

found in caves and wells more than 100 m below the Earth's surface. Although few invertebrates penetrate more than 1–10 m down in nonkarst (*i.e.,* interstitial) environments, bacteria frequently are present in nonkarstic aquifers hundreds of meters below the Earth's surface (e.g., Fliermans and Balkwill, 1989; Ghiorse and Wobber, 1989; Fredrickson *et al.,* 1991). Indeed, microbiologists speculate that bacteria may live in groundwaters thousands of meters below our feet, although satisfactorily uncontaminated samples have not yet been recovered from such depths (Ghiorse and Wilson, 1988).

There typically are several differences between biological communities from shallow and deep groundwater. Population densities of invertebrates, protozoans, and bacteria are usually much higher in very superficial groundwaters than in deeper strata (Fig. 3). Although some studies have found a decline in bacterial densities with depth (Chapelle *et al.,* 1987), it appears that bacterial population density in deep groundwaters is a function more of local environmental conditions than of distance from the Earth's surface (Chapelle, 1993). As Fig. 3 suggests, bacteria usually penetrate to much greater depths than do invertebrates and protozoans. The few studies of protozoan distribution (Sinclair and Ghiorse, 1987, 1989; Sinclair *et al.,* 1990; Madsen *et al.,* 1991) suggest that protozoans may also occur to greater depths than do the invertebrates.

The species richness of invertebrate communities usually declines sharply with increasing depth into groundwater. Studies of microbial diversity in groundwater are just beginning, but there are some hints that microbial diversity likewise may be lower in groundwaters than in surficial soils (Bone and Balkwill, 1988).

Paralleling this decline in species richness is a shift in community composition with depth. Invertebrate communities of shallow groundwater are usually a mixture of widespread and generalized surface-water species with a few specialized groundwater species (Bretschko, 1992; Creuzé des Châtelliers *et al.,* 1992). In deeper groundwaters, the specialized groundwater species become relatively more abundant, whereas most of the epigean generalists disappear (Dole, 1983, 1984; Dole and Chessel, 1986; Pennak and Ward, 1986; Danielopol, 1989). This generalization may hold for microbial communities as well; it now appears that bacteria inhabiting deep groundwaters are different from those of surficial soils (Fliermans and Balkwill, 1989) and that there are large differences in bacterial communities among strata in deep aquifers (Fredrickson *et al.,* 1991).

In karst terrain, invertebrates may penetrate to much greater depths than does the interstitial fauna of nonkarst environments, and the saturated zone of karsts sometimes presents high species richness of invertebrate communities (Rouch *et al.,* 1968; Gibert, 1986). Thus the patterns described above are less clearly expressed. This deeper penetration presumably is due to links to surface karstic systems, the frequent exposure of drain waters

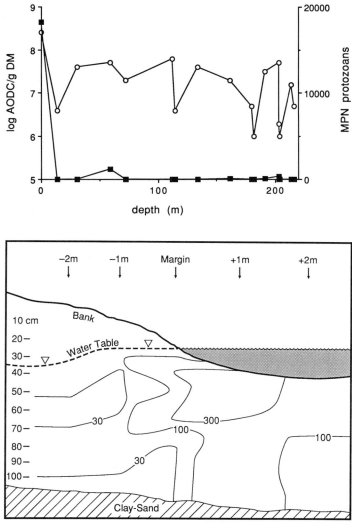

FIGURE 3 (upper) Densities of bacteria (open circles, as acridine orange direct counts) and protozoans (closed squares, as most probable numbers) as a function of depth in aquifers in South Carolina, from data of Sinclair and Ghiorse (1989). (lower) Diagramatic cross-section of the hyporheic zone of a stream in Ontario, showing densities of invertebrates (per 75 cm³) Redrawn from Williams (1989).

to the oxygen-rich atmosphere, and possibly the quicker delivery of water and organic matter to many drains than to most interstitial habitats. Thus, the environmental constraints associated with increasing depth in interstitial habitats may be more loosely (and more variably) imposed on karstic communities.

C. Position along a Flow Path

Few studies have considered the distribution of the biota along a ground-water flow path. Nevertheless, it is clear that community structure typically changes from the recharge zone to the discharge zone, whether the flow path is short or long. Different metabolic groups of bacteria (nitrate reducers, sulfate reducers, methanogens, etc.) often dominate specific zones in kilome-ter-long flow paths (Chapelle, 1993). Hendricks (1992) was able to detect differences in microbial activity along meters-long flow paths between the heads and the tails of riffles. Although studies of groundwater interstitial invertebrates rarely have been placed in a hydrologic context, researchers often find striking differences between the animal communities of upwelling and downwelling zones that are only a few meters apart in the hyporheic zone (Creuzé des Châtelliers, 1991; Stanley and Boulton, 1993). At the scale of karstic systems the distribution of the biota along groundwater flow pathways has been shown [for example, for the Baget system, see Rouch and Carlier (1985), for the Dorvan-Cleyzieu system, see Gibert (1986), and for the Foussoubie system, see Vervier and Gibert (1991); see also Chapter 16].

III. CONTROLS ON THE DISTRIBUTION OF THE GROUNDWATER BIOTA

Because our understanding of the distribution of the groundwater biota is still rudimentary, a discussion of the factors that limit that distribution must necessarily be speculative and incomplete. It is nevertheless possible to identify the factors that ought to exert control over the range and abundance of groundwater organisms, and even an incomplete discussion of these factors, such as the one below, may be helpful in guiding future research.

A. Dispersal

Many groundwater invertebrates disperse so slowly that dispersal abilities set strong limits on their ranges. Perhaps the most remarkable examples of dispersal-limited ranges are the numerous animals that have been essentially unable to leave their places of origin. At least this is the interpretation given for species such as many hypogean thermosbaenaceans, isopods, and amphipods (Cole and Minckley, 1966; Stock, 1977a,b,c, 1980, 1981; Botosaneanu, 1986; Holsinger, 1986; Cals and Monod, 1988), whose present-day ranges closely follow former marine shorelines from which they invaded fresh groundwaters (Fig. 4). The dispersal abilities of these animals apparently are inadequate to reach suitable habitats away from ancient coasts, given even millions of years.

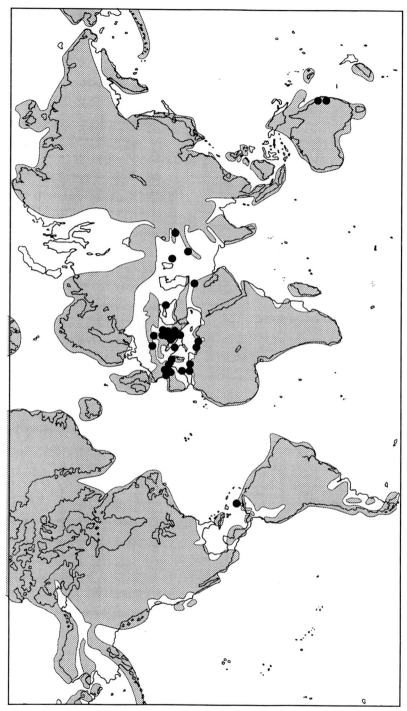

FIGURE 4 Distribution of freshwater microparasellid isopods and Oligocene shorelines. Stippled areas show the extent of land during the Oligocene epoch; closed circles show where freshwater microparasellids have been collected. Modified from Stock (1977a) with additional records from Coineau (1986).

Similarly, many groundwater animals apparently are very slow to recolonize areas that were defaunated by catastrophic disturbances. The best example is due to Pleistocene glaciations, which probably eliminated most or all of the groundwater fauna from glaciated regions [but see Holsinger (1980) for evidence that a few species may have survived glaciation in place beneath the ice]. Even though deglaciation began ca. 18,000 BP and large parts of glaciated North America and Europe were free of ice by 10,000 BP (Flint, 1971), present-day distributions of groundwater animals still reflect patterns of glaciation (Fig. 5).

Such extraordinarily slow dispersal of many groundwater animals suggests that the faunas of isolated aquifers (those on remote islands, deep beneath the Earth's surface, or surrounded by large areas of impermeable sediment, for example) might be relatively depauperate (MacArthur and Wilson, 1967). In addition, the faunas of aquifers affected by anthropogenic pollution might be extremely slow to recover, so that local extinctions of groundwater invertebrates caused by human activities might be essentially irreversible.

Finally, on a still smaller scale, it is possible that dispersal might limit the distributions of invertebrates that require more than one habitat to complete their life cycle. For example, the many insects that have an aerial

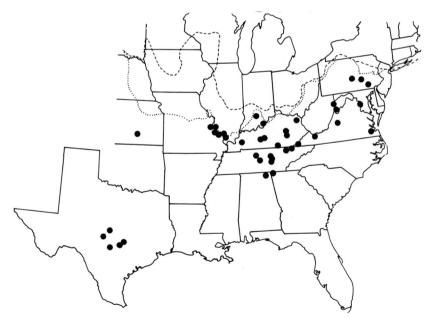

FIGURE 5 Distribution of the groundwater flatworm *Sphalloplana* in North America, from records of Kenk (1977, 1984). The dashed line shows the limit of Wisconsinan glaciers, the dotted line shows the limit of all Pleistocene glaciers, and closed circles show sites at which *Sphalloplana* has been found.

adult stage might not be able to return readily to the surface to emerge and breed. Some insects with aerial adults are nonetheless able to penetrate very far into groundwaters (Stanford and Gaufin, 1974; Stanford and Ward, 1988), so the possession of an aerial adult stage may not actually restrict the distribution of groundwater insects in a significant way.

In summary, the distribution of many groundwater invertebrates is limited by the very poor dispersal abilities of these animals. Although we have little information on the distribution of specific strains of groundwater bacteria, it seems unlikely that such dispersal-limited ranges are common for groundwater bacteria.

B. Redox Status of Groundwater

Redox conditions are related to the availability of electron donors and acceptors in the environment and thus can exert a powerful control on the distribution of the biota. Groundwaters vary from highly oxidized to highly reduced, depending on hydrological and geological conditions and on the activity of the biota. Champ *et al.* (1979) showed how redox conditions can change over large spatial scales along groundwater flow paths. Along flow paths to which oxidants are not added in large amounts (*i.e.*, confined aquifers or even most deep unconfined aquifers), the redox potential will decline as molecular oxygen, nitrate, manganese, iron, and sulfate progressively are reduced. Where groundwater discharges into an oxidized environment [the open oxidant model of Champ *et al.* (1979)], the same substances are oxidized in the reverse order, and redox potential rises. It is almost certain that redox conditions in aquifers vary greatly over small spatial scales (millimeters to meters) as well, due to the local heterogeneity, laminar flows, and anisotropic hydraulic conductivity characteristic of most aquifers. The biological consequences of such fine-scale patterns in redox conditions in aquifers have not been investigated.

Molecular oxygen is the most-well-known redox-active substance in groundwater and might be thought to exert an important control on biological distributions (Danielopol and Niederreiter, 1987). Nonetheless, the evidence that dissolved oxygen controls invertebrate distributions is surprisingly equivocal. Several studies have found strong correlations between concentrations of dissolved oxygen and the density, diversity, and composition of invertebrate communities (Fig. 6) (Husmann, 1975; Rouch, 1988, 1991; Danielopol, 1991; Rouch and Lescher-Moutoué, 1992; Boulton *et al.*, 1992). Nevertheless, it appears that groundwater invertebrates are relatively indifferent to dissolved oxygen, at least compared with surface-water invertebrates. Several authors found no correlation between the abundance of various invertebrates and the concentration of dissolved oxygen in hyporheic zones and shallow groundwaters, despite the frequent occurrence of concentrations of <1 mg/l (Fig. 7; Danielopol, 1976; Pospisil, 1989; Williams,

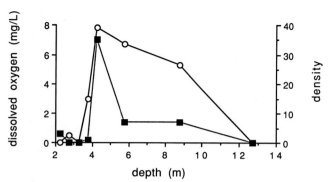

FIGURE 6   Vertical distribution of dissolved oxygen (open circles) and invertebrates (closed squares, as a number per 50-l pumped sample) in an alluvial aquifer in Germany, from data of Husmann (1975).

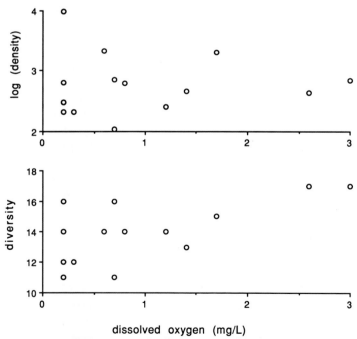

FIGURE 7   Density and number of taxonomic orders of invertebrates found in 3-l Bou–Rouch samples (100-μm mesh) as a function of dissolved oxygen in the alluvium of the North Fork of the Roanoke River in Virginia. Nematodes and turbellarians were assigned one order each. From D. L. Strayer (unpublished data).

1989; Creuzé des Châtelliers and Marmonier, 1993). Whether these results are due to tolerance to low oxygen concentrations by hypogean invertebrates (Danielopol, 1989; Danielopol *et al.*, 1992) or to the existence of microzones of high dissolved oxygen is not yet known. However, different strategies have been proposed by which the integration of the organisms into the groundwater environment is achieved (see Chapter 8). Detailed laboratory studies of the oxygen tolerances of groundwater invertebrates still are lacking.

There is as yet little evidence that the concentrations of redox-active substances other than oxygen affect the distribution of groundwater invertebrates or that groundwater invertebrates are adapted to tolerate anoxia indefinitely. Danielopol (1976) found that hyporheic sediments along the Danube River that contained sulfide supported few or no invertebrates. This is in sharp contrast to marine sediments, in which a rich fauna of specialized invertebrates lives in the anoxic sediments of the "sulfide system" (Fenchel and Riedl, 1970).

The activity of groundwater bacteria is strongly dependent on redox conditions. Naturally, the activities of functional groups such as sulfate reducers, denitrifiers, and methanogens are restricted to areas of suitable redox conditions. For example, Beeman and Suflita (1987) investigated the distribution and activities of sulfate reducers and methanogens in an aquifer beneath a municipal landfill in Oklahoma. Parts of the aquifer (site B) receive very high loadings of organic materials, and intensive anaerobic metabolism upstream depletes or eliminates sulfate [Fig. 8 (upper)]. In other parts of the aquifer (*e.g.*, site A), which receive less organic matter, sulfate concentrations remain high throughout the year. Rates of methanogenesis [Fig. 8 (middle)] and numbers of methanogens were markedly higher in the sulfate-depleted part of the aquifer, whereas numbers of sulfate-reducing bacteria were (naturally) higher where sulfate was present [Fig. 8 (lower)]. It is worth noting that methanogenesis did occur in the presence of sulfate [Fig. 8 (middle)], presumably in sulfate-depleted microhabitats at site A (or by using substrates not immediately available to the sulfate-reducing bacteria). Phenomena similar to those reported by Beeman and Suflita presumably are of very general occurrence in groundwaters.

C. Supply of Organic Matter

Many of the morphological and physiological features of the groundwater biota (*e.g.*, slow metabolic rates of invertebrates, accumulation of storage products such as poly-β-hydroxybutyrate by bacteria, and an apparently high proportion of inactive cells in bacterial communities) suggest that many groundwaters are very food poor. This impression is supported by data from aquifers on dissolved organic carbon, which typically is dilute and dominated by recalcitrant fulvic and humic acids (Leenheer *et al.*, 1974;

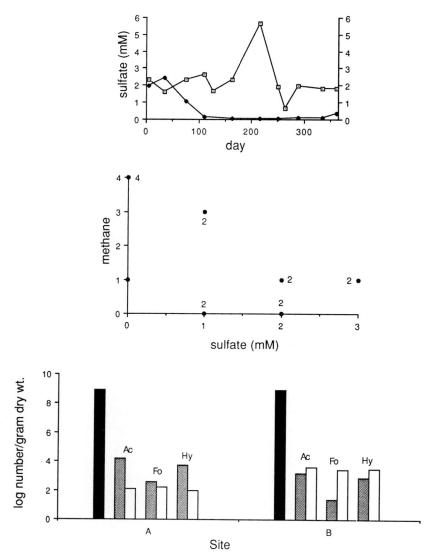

FIGURE 8 (upper) Annual cycle of sulfate concentrations at two sites in an aquifer beneath a landfill in Oklahoma. Site A is represented by open squares; site B, by closed dia-
monds. (middle) Methane concentrations (1, 1–10 mg/l; 2, 10–50 mg/l; 3, 50–100 mg/l;
4, >100 mg/l) as a function of sulfate concentrations in this aquifer in late April. Numbers
next to data points show the number of multiple data points falling at the same point. (lower)
Composition of the bacterial community at sites A and B. Black bars show the total number
of bacteria (acridine orange direct counts), and stippled and white bars show most-probable
number estimates of sulfate-reducing bacteria and methanogens, respectively. Substrates used
are acetate (ac), formate (fo), and hydrogen (hy). From data of Beeman and Suflita (1988).

Thurman, 1985a,b). Also, the metabolic rates of bacterial communities in unpolluted aquifers are very low (Ghiorse and Wilson, 1988; Chapelle and Lovley, 1990). All of these bits of evidence suggest that groundwater communities might frequently be limited by inadequate supplies of suitable food.

As yet, there is little direct evidence that food limits groundwater organisms. Numbers of bacteria and invertebrates sometimes are correlated with the organic matter content of the aquifer (Fig. 9; Fredrickson *et al.*, 1991). Additionally, the hypothesis of food limitation has been tested inadvertently through widespread anthropogenic loading of organic matter to groundwaters. We would expect anthropogenic loading of labile organic matter to stimulate biological communities in groundwaters, unless the organic loading was so great that it exhausted suitable electron acceptors or was accompanied by inputs of toxic pollution. Several studies (Fig. 10) (Smith *et al.*, 1986; Ronen *et al.*, 1987; Ghiorse and Wilson, 1988; Madsen *et al.*, 1991) suggest that anthropogenic loading of organic matter commonly increases the density and activity of bacteria, protozoans, and invertebrates, although the data are hardly definitive.

D. Other Physicochemical Factors

Given the enormous range of physicochemical conditions present in the world's groundwaters, a wide variety of physicochemical factors probably influences biotic distributions at one place or another. Here, I will discuss only three of the most obviously important of these factors: temperature, sediment texture, and salinity.

The temperatures of groundwater typically vary much less than those of surface waters, so it is reasonable to expect groundwater organisms to be stenothermic. Despite early speculations that such stenothermy might

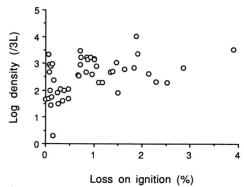

FIGURE 9   Density of invertebrates (number per 3-l Bou–Rouch samples, 100-μm mesh) at various hyporheic sites in eastern North America as a function of the organic matter content of the sediment. $r = 0.42$, $p < 0.01$. From D. L. Strayer (unpublished data).

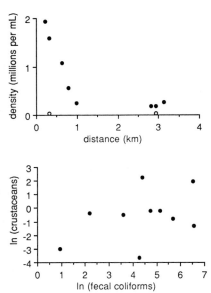

FIGURE 10   (upper) Density of free-living (*i.e.*, unattached) bacteria in a sandy aquifer on Cape Cod, Massachusetts, as a function of the distance downstream from a sewage input. Closed circles represent sites in the sewage plume; open circles represent those outside the plume. Modified from Harvey *et al.* (1984). (lower) Density of crustaceans (number of amphipods plus isopods per liter) as a function of fecal coliform density (number per 100 ml), as an indicator of sewage contamination, in wells in a gravelly aquifer in New Zealand. $r = 0.57, 0.1 > p > 0.05$. From data of Sinton (1984).

exclude groundwater organisms from thermally variable surface waters (Chappuis, 1927), it appears that many groundwater organisms are not especially stenothermic (Ginet, 1960; Vandel, 1964), so a variable temperature regime probably does not generally limit the distribution of the groundwater biota. In contrast, it seems likely that high temperature may set the lower depth limit on the distribution of the groundwater biota (McNabb and Dunlap, 1975; Ghiorse and Wilson, 1988). Typically, temperatures rise about 3°C per 100 m of depth, and known bacteria can tolerate temperatures a little above 100°C, so this lower limit may be a few thousand meters below the surface of the Earth (Ghiorse and Wilson, 1988).

Sediment texture has been shown to affect the distribution and abundance of both microbes (Fig. 11) (Ghiorse and Wilson, 1988; Sinclair and Ghiorse, 1987, 1989; Fredrickson *et al.*, 1991) and invertebrates (Delamare-Deboutteville, 1960; Coineau, 1971; Danielopol, 1989). The importance of sediment texture probably is related chiefly to its influence on hydraulic conductivity (Freeze and Cherry, 1979; Fredrickson *et al.*, 1991), which in turn controls supply rates of dissolved substances such as oxygen, organic carbon, and nitrate. Sediment texture also is correlated with the amount of

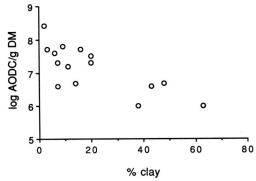

FIGURE 11   Bacterial densities (acridine orange direct counts) as a function of the clay content of subsurface sediments in South Carolina. $r = 0.76$, $p < 0.01$. From Sinclair and Ghiorse (1989).

particle surface area available for biological colonization, the distribution of sizes of pore spaces suitable for use by the biota, and the suitability of the sediment for burrowing (Magniez, 1974), all of which could affect the composition, distribution, and abundance of the biota.

Salinity varies widely in groundwaters, from less than 0.1‰ to more than 100‰, and certainly limits the distribution of the groundwater biota. Very briny groundwaters probably support only a few specialized bacteria (McNabb and Dunlap, 1975). In less extreme environments, salinity has received little attention from groundwater ecologists, but does exert important control on biological distributions. For example, Foraminifera have been reported only from brackish or saline inland groundwaters (Gauthier-Liévre, 1935; Mikhalevich, 1976). Likewise, although many groundwater invertebrates (especially recent invaders from the sea) are euryhaline (Magniez, 1975; Botosaneanu *et al.*, 1986; Botosaneanu, 1986; Karaman and Ruffo, 1986), they are nonetheless restricted to groundwaters within a fairly narrow salinity range (<35‰). Groundwaters more saline than seawater probably generally restrict the biota, and very dilute waters (*i.e.*, <5 mg/l of calcium) may pose problems for invertebrates such as mollusks or crustaceans that have high calcium requirements (Økland, 1983).

Finally, toxic substances, whether naturally derived or the result of human activities, can affect the distribution of the groundwater biota. Notenboom *et al.* (Chapter 18) deal with this topic in detail.

E. Mortality Rate

Many groundwater invertebrates have attributes such as low metabolic rates, long prereproductive periods, small broods, and long intervals between broods, which imply very low rates of natural increase (r) and productivity

(Ginet, 1960; Rouch, 1968; Coineau, 1971; Gourbault, 1972; Lescher-Moutoué, 1973; Magniez, 1975; Danielopol, 1980). Although direct measurements of bacterial productivity in groundwater are scarce, there is likewise some evidence [*e.g.*, thick glycocalyxes and accumulations of storage products such as poly-β-hydroxybutyrate, which suggest unbalanced growth (Ghiorse and Balkwill, 1983; White *et al.*, 1983; Balkwill *et al.*, 1988), and a high proportion of inactive cells in groundwater bacterial communities (Chapelle *et al.*, 1987, 1988; Marxsen, 1988)] that the productivity of groundwater bacterial communities is low. Such low productivity is to be expected of communities in very oligotrophic environments. Because mortality equals productivity (over the long term), populations with low rates of productivity cannot endure high rates of mortality. We can therefore predict that specialized groundwater organisms will either be absent from environments where mortality rates are high, have adaptations to reduce mortality rates, or be able to increase their productivity to meet the demand of high mortality rates where they occur.

There is little hard evidence available to evaluate these ideas. Perhaps the most obvious sources of mortality in groundwater ecosystems are predation and advective losses from spates in near-stream environments or washout from deeper environments. Predation is potentially of great importance. In most surface-water environments, predators are a dominant source of mortality and consume most of the production of bacteria and invertebrates (Pace, 1988; Kemp, 1990; Strayer, 1991). In fact, groundwater habitats often have been seen as providing a refuge for small, interstitial prey from large, epigean predators (Williams, 1984; Strayer, 1991).

We know little about the intensity of predation in groundwater communities or its effects on prey. Stock (1982, 1983) argued that predation by hadziid amphipods affects the distribution of thermosbaenaceans on West Indian islands. The often-observed absence or scarcity of specialized groundwater invertebrates from near-surface environments (Fig. 12; Marmonier, 1986) could be due to excessive mortality from epigean predators. It is equally plausible, however, that mortality from spates eliminates slow-growing groundwater invertebrates from near-surface environments, as has been shown by Marmonier and Creuzé des Châtelliers (1991). If flood-caused mortality exerts a major control on the distribution of groundwater invertebrates, then one would expect to find that the abundance of these animals varies as a function of the physical stability of the substratum, between streams with different hydrological regimes, within a stream between patches of sediment with different resistance to scouring, or within a stream over time as a function of the time since the last catastrophic flood (Marmonier and Creuzé des Châtelliers, 1991).

Likewise, it is possible that the near absence of losses from predation or washout allows slow-growing strains of bacteria to persist in deep groundwater. Data bearing on this idea are not available.

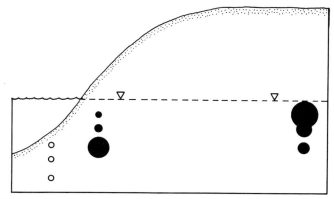

FIGURE 12   Mean annual density of bathynellacean crustaceans in the alluvium of the South Fork of the South Branch of the Platte River, Colorado. The area of the black circles is proportional to bathynellacean density, white circles show sites from which bathynellaceans were absent, and the dashed line shows the approximate position of the water table. From data of Pennak and Ward (1986).

F. Other Biotic Interactions

Just as in epigean habitats, biotic interactions other than predation may influence biological distributions in groundwater. However, these interactions are as yet poorly studied. Culver (1982; see also Chapter 10) has shown that competition can affect the distribution of species of cave-dwelling crustaceans, both on the scale of individual rocks and on the scale of a cave. On a broader scale, Stock (1980) pointed out that the distributions of the amphipods *Niphargus* and *Pseudoniphargus* in Europe are nearly complementary and suggested that *Pseudoniphargus* was unable to invade areas already occupied by *Niphargus*. Furthermore, microbiologists believe that competition for common substrates helps to determine the distributions of groups of bacteria that use different metabolic pathways (*e.g.*, aerobes vs sulfate reducers vs methanogens) (Fig. 8). Nonetheless, we still lack a general understanding of the strength and frequency of competitive interactions in groundwater communities (Chapter 10).

Another factor that may limit the distribution and abundance of groundwater organisms is the availability of suitable prey or hosts for specialized organisms. Obviously, the distribution of a parasite or commensal organism will be limited by the distribution of its host. It has been suggested (Teschner, 1964) that the restriction of most parasitengone mites to superficial groundwaters is caused by a similarly restrictive distribution of most insects, the hosts of many of these mites. This explanation was questioned by Petrova (1984) and Schwoerbel (1986). Possibly better examples of this phenomenon are given by the entocytherid ostracods, commensal on crayfish and isopods

(Danielopol, 1986), and the mermithid nematodes, whose juveniles are parasitic on insects (Kaiser and Schwank, 1985; Poinar, 1991).

Finally, the activities of one species may create an appropriate habitat for another species and thereby extend its distribution. Perhaps the best example is the burrows of some crayfish, which provide habitat for a wide variety of groundwater invertebrates (Horwitz, 1988). Another possibly widespread example was given by Danielopol (1989), who suggested that the feeding activities of some groundwater invertebrates might increase the permeability of aquifers, thereby possibly increasing flow rates of oxygen, organic carbon, and so on. This mechanism has not yet been tested, but has the potential to affect the distribution of many groundwater organisms.

IV. CONCLUSION

Our knowledge of biotic distributions in groundwaters is still in many ways rudimentary. The underground part of the biosphere is extensive, reaching hundreds to thousands of meters below the Earth's surface. We know that the density and composition of groundwater communities vary widely across underground habitats, but we are able to speak only in the most general terms about the causes of this variation. For example, I do not believe that it is now possible to delimit precisely the environmental conditions under which a given group of organisms (*e.g.*, protozoans, copepods) will occur, or to predict the abundance of a particular organism with any precision.

In this chapter, I have listed several factors that might control biotic distributions in groundwaters. One certainly could add additional factors or develop an entirely different conceptual framework of controlling factors. Nonetheless, I do not believe that the study of groundwater ecology will necessarily require consideration of a large set of complex controlling factors peculiar to each field study. Rather, it may be possible to use a small number of integrative variables to develop simple models that can be used to describe biological distributions both within and across habitats (Dole-Olivier and Marmonier, 1992). Two such variables that look especially promising to me are the hydraulic conductivity of the aquifer and the age of the water.

Hydraulic conductivity is a measure of the ease with which water can flow through an aquifer. Along with patterns of hydraulic head, it determines renewal rates of water and dissolved substances (such as oxygen, organic carbon, and nitrate) to an aquifer. Because it is a strong function of sediment grain size, hydraulic conductivity also will reflect the size and connectedness of the pores that serve as the physical habitat of groundwater organisms. Further, hydraulic conductivity is readily estimated from either *in situ* tests (*e.g.*, slug tests) or laboratory analyses of sediment grain size distributions.

The age of the water in an aquifer will be related to the extent to which biogeochemical processes can consume materials supplied from the Earth's surface (*e.g.*, oxygen and labile organic matter) and weather materials from the aquifer. Thus, age will be broadly related to both the food resources and the environment available to the groundwater biota. Age also will be correlated with the remoteness of the aquifer from the Earth's surface. Although age of groundwater can sometimes be estimated by various geochemical methods (Freeze and Cherry, 1979), it often may be sufficient to rank sites according to age (*e.g.*, along a flow path from the recharge zone to the discharge zone) without knowing the absolute age of the water. In other cases, especially in comparing aquifers, knowing the absolute age of the water may be helpful.

Until recently, our understanding of biotic distributions in groundwaters was limited primarily by the paucity of data. As information on groundwater organisms increases, our understanding will be increasingly limited not by the volume of available information, but by our ability to organize and interpret that information. I hope that the material in this chapter provides some useful directions for ways to organize future work on biotic distributions in groundwaters.

ACKNOWLEDGMENTS

Preparation of this chapter was supported by a grant from the National Science Foundation and by a Research Residency from the New York State Library. I thank Judy Bondus, Jon Cole, Stuart Findlay, Chris Hakenkamp, and three anonymous reviewers for their helpful comments on the manuscript and thank Sharon Okada for help in preparing the figures. This is a contribution to the program of the Institute of Ecosystem Studies.

REFERENCES

Balkwill, D. L., Leach, F. R., Wilson J. T., McNabb, J.F., and White, D. C. (1988). Equivalence of microbial biomass measures on membrane lipid and cell wall components, adenosine triphosphate, and direct counts in subsurface aquifer sediments. *Micob. Ecol.* **16**, 73–84.

Beeman, R. E., and Suflita, J. M. (1987). Microbial ecology of a shallow unconfined groundwater aquifer polluted by municipal landfill leachate. *Microb. Ecol.* **14**, 39–54.

Beeman, R. E., and Suflita, J. M. (1988). Evaluation of deep subsurface sampling procedures using serendipitous microbial contaminants as tracer organisms. *Geomicrobiol. J.* **7**, 223–233.

Bole, J., and Velkovrh, F. (1986). Mollusca from continental subterranean aquatic habitats. *In* "Stygofauna mundi" (L. Botosaneanu, ed.), pp. 177–208. E. J. Brill Dr. W. Backhuys, Leiden.

Bone, T. L., and Balkwill, D. L. (1988). Morphological and cultural comparison of microorganisms in surface soil and subsurface sediments at a pristine study site in Oklahoma. *Microb. Ecol.* **16**, 49–64.

Botosaneanu, L. (1986). Isopoda: Anthuridea. *In* Stygofauna mundi" (L. Botosaneanu, ed.), pp. 428–433. E. J. Brill Dr. W. Backhuys, Leiden.

Botosaneanu, L., Bruce, N., and Notenboom, J. (1986). Isopoda: Cirolanidae. *In* "Stygofauna mundi" (L. Botosaneanu, ed.), pp. 412–422. E. J. Brill Dr. W. Backhuys, Leiden.

Boulton, A. J., Valett, H. M., and Fisher, S. G. (1992). Spatial distribution and taxonomic composition of the hyporheos of several Sonoran Desert streams. *Arch. Hydrobiol.* **125,** 37–61.

Bretschko, G. (1992). Differentiation between epigeic and hypogeic fauna in gravel streams. *Regulated Rivers* **7,** 17–22.

Cals, P., and Monod, T. (1988). Evolution et biogéographie des Crustacés Thermosbénacés. *C. R. Seances Acad. Sci.* **307,** 341–348.

Champ, D. R., Gulens, J., and Jackson, R. E. (1979). Oxidation-reduction sequences in groundwater flow systems. *Can. J. Earch Sci.* **16,** 12–23.

Chapelle, F. H. (1993). "Ground-Water Microbiology and Geochemistry." Wiley, New York.

Chapelle, F. H., and Lovley, D. R. (1990). Rates of microbial metabolism in deep coastal plain aquifers. *Appl. Environ. Microbiol.* **56,** 1865–1874.

Chapelle, F. H., Zelibor, J. L., Grimes, D. J., and Knobel, L. L. (1987). Bacteria in deep coastal plain sediments of Maryland: A possible source of CO_2 to groundwater. *Water Resour. Res.* **23,** 1625–1632.

Chapelle, F. H., Morris, J. T., McMahon, P. B., and Zelibor, J. L. (1988). Bacterial metabolism and the delta C_{14} composition of groundwater, Floridian aquifer system, South Carolina. *Geology* **16,** 117–121.

Chappuis, P. A. (1927). "Die Tierwelt der unterirdischen Gewässer." Die Binnengewässer III. E. Schweizerbart'sche Verlagsbuchhandlung, Stuttgart.

Coineau, N. (1971). Les isopodes interstitiels. Documents sur leur écologie et leur biologie. *Mém Mus. Nat. Hist. Nat. Ser. A* **64,** 1–170.

Coineau, N. (1986). Isopoda: Asellota: Janiroidea. *In* "Stygofauna mundi" (L. Botosaneanu, ed.), pp. 465–472. E. J. Brill Dr. W. Backhuys, Leiden.

Cole, G. A., and Minckley, W. L. (1966). *Speocirolana thermydronis,* a new species of cirolanid isopod crustacean from central Coahuila, Mexico. *Tulane Stud. Zool.* **13,** 17–22.

Creuzé des Châtelliers, M. (1991). Dynamique de répartition des biocénoses interstitielles du Rhône en relation avec des caractéristiques géomorphologiques (secteurs de Brégnier-Cordon, Miribel-Jonage et Donzére-Mondragon). Thèse, Univ. Lyon I.

Creuzé des Châtelliers, M., and Marmonier, P. (1993). Ecology of benthic and interstitial ostracods of the Rhône River, France. *J. Crustacean Biol.* **13,** 268–279.

Creuzé des Châtelliers, M., Marmonier, P., Dole-Olivier, M.-J., and Castella, E. (1992). Structure of interstitial assembleges in a regulated channel of the River Rhine (France). *Regulated Rivers* **7,** 23–30.

Culver, D. C. (1982). "Cave Life: Evolution and Ecology." Harvard Univ. Press, Cambridge, MA.

Danielopol, D. L. (1976). The distribution of the fauna in the interstitial habitats of riverine sediments of the Danube and the Piesting (Austria). *Int. J. Speleol.* **8,** 23–51.

Danielopol, D. L. (1980). Sur la biologie de quelques Ostracodes Candoninae épigés et hypogés d'Europe. *Bull. Mus. Natl. Hist. Nat. Sect. A* [4] **2,** 471–506.

Danielopol, D. L. (1986). Ostracoda. Part I: Stygobiont Ostracoda from inland subterranean waters. *In* "Stygofauna mundi" (L. Botosaneanu, ed.), pp. 265–278. E. J. Brill Dr. W. Backhuys, Leiden.

Danielopol, D. L. (1989). Groundwater fauna associated with riverine aquifers. *J. North Am. Benthol. Soc.* **8,** 18–35.

Danielopol, D. L. (1991). Spatial distribution and dispersal of interstitial Crustacea in alluvial sediments of a backwater of the Danube at Vienna. *Stygologia* **6,** 97–110.

Danielopol, D. L., and Niederreiter, R. (1987). A sampling device for groundwater organisms and oxygen measurement in multi-level monitoring wells. *Stygologia* **3,** 252–263.

Danielopol, D. L., Dreher, J., Gunatilaka, A., Kaiser, M., Niederreiter, R., Pospisil, P., Creuzé des Châtelliers, M., and Richter, A. (1992). Ecology of organisms living in a hypoxic groundwater environment at Vienna (Austria); methodological questions and preliminary results. *In* "Ground Water Ecology" (J. A. Stanford and J. J. Simons, eds.), pp. 79–90. Am. Water Resour. Assoc. Bethesda, MD.

Delamare-Deboutteville, C. (1960). "Biologie des eaux souterraines litorales et continentales." Hermann, Paris.

Dole, M.-J. (1983). Le domaine aquatique souterrain de la plaine alluviale de Rhône à l'est de Lyon 1. Diversité hydrologique et biocénotique de trois stations représentatives de la dynamique fluviale. *Vie Milieu* **33**, 219–229.

Dole, M.-J. (1984). Structure biocénotique des niveaux supérieurs de la nappe alluviale du Rhône á l'est de Lyon. *Mém. Biospéol.* **11**, 17–26.

Dole, M.-J., and Chessel, D. (1986). Stabilité physique et biologique des milieux interstitiels. Cas de deux stations du Haut-Rhône. *Ann. Limnol.* **22**, 69–81.

Dole-Olivier, M.-J., and Marmonier, P. (1992). Patch distribution of interstitial communities: Prevailing factors. *Freshwater Biol.* **27**, 177–191.

Fenchel, T., and Riedl, J. M. (1970). The sulfide system: A new biotic community underneath the oxidized layer of marine sand bottoms. *Mar. Biol.* **7**, 255–263.

Fliermans, C. B., and Balkwill, D. L. (1989). Microbial life in deep terrestrial subsurfaces. *BioScience* **39**, 370–377.

Flint, R. F. (1971). "Glacial and Quaternary Geology." Wiley, New York.

Fredrickson, J. K., Balkwill, D. L., Zachara, J. M., Li, S.-M. W., Brockman, F. J., and Simmons, M. A. (1991). Physiological diversity and distributions of heterotrophic bacteria in deep Cretaceous sediments of the Atlantic coastal plain. *Appl. Environ. Microbiol.* **57**, 402–411.

Freeze, R. A., and Cherry, J. A. (1979). "Groundwater." Prentice-Hall, Englewood Cliffs, NJ.

Gauthier-Lièvre, L. (1935). Sur une des singularités de l'Oued Rhir: Des Foraminifères thalassoides vivant dans des eaux sahariennes. *Bull. Soc. Hist. Nat. Afr. Nord* **26**, 142–147.

Ghiorse, W. C., and Balkwill, D. L. (1983). Enumeration and morphological characterization of bacteria indigenous to subsurface environments. *Dev. Ind. Microbiol.* **24**, 213–224.

Ghiorse, W. C., and Wilson, J. T. (1988). Microbial ecology of the terrestrial subsurface. *Adv. Appl. Microbiol.* **33**, 107–172.

Ghiorse, W. C., and Wobber, F. J., eds. (1989). Deep subsurface microbiology. *Geomicrobiol. J.* **7**, 1–131.

Gibert, J. (1986). Ecologie d'un système karstique jurassien: Hydrogéologie, dérive animale, transits de matières, dynamique de la population de *Niphargus* (Crustacé Amphipode). *Mém. Biospéol.* **13**, 1–330.

Ginet, R. (1960). Ecologie, éthologie et biologie de *Niphargus* (Amphipodes Gammaridés hypogés). *Ann. Spéléol.* **15**, 127–376.

Gourbault, N. (1972). Recherches sur les Triclades Paludicoles hypogés. *Mém. Mus. Natl. Hist. Nat. Ser. A* **73**, 1–249.

Harvey, R. W., Smith, R. L., and George, L. (1984). Effect of organic contamination upon microbial distributions and heterotrophic uptake in a Cape Cod, Mass., aquifer. *Appl. Environ. Microbiol.* **48**, 1197–1202.

Hendricks, S. P. (1992). Bacterial dynamics near the groundwater-surface water interface (hyporheic zone) beneath a sandy-bed, third-order stream in northern Michigan. *In* "Ground Water Ecology" (J. A. Stanford and J. J. Simons, eds.), pp. 27–35. Am. Water Resour. Assoc. Bethesda, MD.

Herschler, R., and Longley, G. (1986). Phreatic hydrobiids (Gastropoda: Prosobranchia) from the Edwards (Balcones Fault Zone) Aquifer Region, south-central Texas. *Malacologia* **27**, 127–172.

Holsinger, J. R. (1978). Systematics of the subterranean amphipod genus *Stygobromus* (Crangonyctidae). Part II. Species of the eastern United States. *Smithson. Contrib. Zool.* **266**, 1–144.

Holsinger, J. R. (1980). *Stygobromus canadensis*, a new subterranean amphipod crustacean

(Crangonyctidae) from Canada, with remarks on Wisconsin refugia. *Can. J. Zool.* **58,** 290–297.

Holsinger, J. R. (1986). Zoogeographic patterns of North American subterranean amphipod crustaceans. *In* "Crustacean Biogeography (Crustacean Issues 4)" (R. H. Gore and K. L. Heck, eds.), pp. 85–106. Balkema, Rotterdam.

Holsinger, J. R., and Longley, G. (1980). The subterranean amphipod crustacean fauna of an artesian well in Texas. *Smithson. Contrib. Zool.* **308,** 1–62.

Horwitz, P. (1988). The faunal assemblage (or pholeteros) of some freshwater crayfish burrows in southwest Tasmania. *Bull Aust. Soc. Limnol.* **12,** 29–36.

Husmann, S. (1962). Ökologische und verbreitungsgeschichtliche Studien über den Archianneliden *Troglochaetus beranecki* Delachaux; Mitteilung über Neufunde aus den Grundwasserströmen von Donau, Ybbs, Ötz, Isar, Lahn, Ruhr, Niederrhein und Unterweser. *Zool. Anz.* **168,** 312–325.

Husmann, S. (1975). Versuche zur vertikalen Verteilung von Organismen und chemischen Substanzen im Grundwasser von Talauen und Terrassen; Methoden und erste Befunde. *Int. J. Speleol.* **6,** 271–302.

Husmann, S., and Teschner, D. (1970). Ökologie, Morphologie und Verbreitsgeschichte subterraner Wassermilben (Limnohalacaridae) aus Schweden. *Arch. Hydrobiol.* **67,** 242–267.

Kaiser, H., and Schwank, P. (1985). Mermithidae (Nematoda) aus oberhessischen Fliessgewässern. *Arch. Hydrobiol.* **103,** 347–369.

Karaman, G. S., and Ruffo, S. (1986). Amphipoda: *Niphargus* group (Niphargidae *sensu* Bousfield, 1982). *In* "Stygofauna mundi" (L. Botosaneanu, ed.), pp. 514–534. E. J. Brill Dr. W. Backhuys, Leiden.

Kemp, P. F. (1990). The fate of bacterial production. *Rev. Aquat. Sci.* **2,** 109–124.

Kenk, R. (1977). Freshwater triclads (Turbellaria) of North America. IX. The genus *Sphalloplana*. *Smithson. Contrib. Zool.* **246,** 1–38.

Kenk, R. (1984). Freshwater triclads (Turbellaria) of North America. XV. Two new subterranean species from the Appalachian region. *Proc. Biol. Soc. Washington* **97,** 209–216.

Kiefer, F. (1978). Copepoda non-parasitica. *In* "Limnofauna Europaea" (J. Illies, ed.), 2nd ed., pp. 209–223. Fischer, Stuttgart.

Kroschewsky, J. R. (1990). The Edwards Aquifer Research and Data Center: Objectives and accomplishments. *Stygologia* **5,** 213–220.

Leenheer, J. A., Malcolm, R. L., McKinley, P. W., and Eccles, L. A. (1974). Occurrence of dissolved organic carbon in selected ground-water samples in the United States. *J. Res. U. S. Geol. Surv.* **2,** 361–369.

Lescher-Moutoué, F. (1973). Sur la biologie et l'écologie des copépodes cyclopoïdes hypogés (Crustacés). *Ann. Spéléol.* **28**(3), 429–502, (4), 581–674.

Lescher-Moutoué, F. (1986). Copépoda Cyclopoïda Cyclopidae des eaux douces souterraines continentales. *In* "Stygofauna mundi" (L. Botosaneanu, ed.), pp. 299–312, E. J. Brill Dr. W. Backhuys, Leiden.

Longley, G. (1981). The Edwards Aquifer: The earth's most diverse groundwater ecosystem? *Int. J. Speleol.* **11,** 123–128.

MacArthur, R. H., and Wilson, E. O. (1967). "The Theory of Island Biogeography." Princeton Univ. Press, Princeton, NJ.

Madsen, E. L., Sinclair, J. L., and Ghiorse, W. C. (1991). In situ biodegradation: Microbiological patterns in a contaminated aquifer. *Science* **252,** 830–833.

Magniez, G. (1974). Données faunistiques et écologiques sur les Stenasellidae (Crustacea Isopoda Asellota des eaux souterraines). *Int. J. Speleol.* **6,** 1–80.

Magniez, G. (1975). Observations sur la biologie de *Stenasellus virei* (Crustacea Isopoda Asellota des eaux souterraines). *Int. J. Speleol.* **7,** 79–228.

Marmonier, P. (1986). Spatial distribution and temporal evolution of *Gammarus fossarum, Niphargus sp.* (Amphipoda) and *Proasellus slavus* (Isopoda) in the Seebach sediments (Lunz, Austria). *Jahresber. Biol. Stn. Lunz* **8,** 40–54.

Marmonier, P., and Creuzé des Châtelliers, M. (1991). Effects of spates on interstitial assemblages of the Rhône River. Importance of spatial heterogeneity. *Hydrobiologia* **210**, 243–251.

Marxsen, J. (1988). Investigations into the number of respiring bacteria in groundwater from sandy and gravelly deposits. *Microb. Ecol.* **16**, 65–72.

McNabb, J. F., and Dunlap, W. J. (1975). Subsurface biological activity in relation to groundwater pollution. *Ground Water* **13**, 33–44.

Mikhalevich, V. I. (1976). New data on the Foraminifera of the groundwaters of middle Asia. *Int. J. Speleol* **8**, 167–175.

Økland, J. (1983). Factors regulating the distribution of fresh-water snails (Gastropoda) in Norway. *Malacologia* **24**, 277–288.

Pace, M. L. (1988). Bacterial mortality and the fate of bacterial production. *Hydrobiologia* **159**, 41–49.

Peck, S. B. (1988). A review of the cave fauna of Canada, and the composition and ecology of the invertebrate fauna of caves and mines in Ontario. *Can. J. Zool.* **66**, 1197–1213.

Pennak, R. W. (1971). A fresh-water archiannelid from the Colorado Rocky Mountains. *Trans. Am. Microsc. Soc.* **90**, 372–375.

Pennak, R. W., and Ward, J. V. (1986). Interstitial faunal communities of the hyporheic and adjacent groundwater biotopes of a Colorado mountain stream. *Arch. Hydrobiol. Suppl.* **74**, 356–396.

Petrova, A. (1984). Origine et formation des Acariens stygobiontes. *Khidrobiologiya* **22**, 3–24.

Poinar, G. O. (1991). Nematoda and Nematomorpha. *In* "Ecology and Classification of North American Freshwater Invertebrates" (J. H. Thorp and A. P. Covich, eds.), pp. 249–283. Academic Press, Orlando, FL.

Pospisil, P. (1989). *Acanthocyclops gmeineri n. sp.* (Crustacea, Copepoda) aus dem Grundwasser von Wien (Österreich): Bemerkungen zur Zoogeographie und zur Sauerstoffsituation des Grundwassers am Fundort. *Zool. Anz.* **223**, 220–230.

Ronen, D., Magaritz, M., Almon, E., and Amiel, A. J. (1987). Anthropogenic anoxification (eutrophication) of the water table region of a deep phreatic aquifer. *Water Resour. Res.* **23**, 1554–1560.

Rouch, R. (1968). Contribution à la connaissance des harpacticides hypogés (Crustacés-Copépodes). *Ann. Spéléol.* **23**, 5–167.

Rouch, R. (1988). Sur la répartition spatiale des Crustacés dans le sous-écoulement d'un ruisseau des Pyrénées. *Ann. Limnol.* **24**, 213–234.

Rouch, R. (1991). Structure et peuplement des Harpacticides dans le milieu hyporhéique d'un ruisseau des Pyrénées. *Ann. Limnol.* **27**, 227–241.

Rouch, R., and Carlier, A. (1985). Le systéme karstique du Baget. XIV. La communauté des Harpacticides. Evolution et comparaison des structures du peuplement épigé à l'entrée et à la sortie de l'aquifère. *Stygologia* **1**(1), 71–92.

Rouch, R., and Lescher-Moutoué, F. (1992). Structure et peuplement des Cyclopides (Crustacea: Copepoda) dans le milieu hyporhéïque d'un ruisseau des Pyrénées. *Stygologia* **7**, 197–211.

Rouch, R., Juberthie-Jupeau, L., and Juberthie, C. (1968). Essai d'étude du peuplement de la zone noyée d'un karst. *Ann. Spéléol.* **23**, 717–733.

Schwoerbel, J. (1986). Acari: "Hydrachnellae." *In* "Stygofauna mundi" (L. Botosaneanu, ed.), pp. 652–696. E. J. Brill Dr. W. Backhuys, Leiden.

Sinclair, J. L., and Ghiorse, W. C. (1987). Distribution of protozoa in subsurface sediments of a pristine groundwater study site in Oklahoma. *Appl. Environ. Microbiol.* **53**, 1157–1163.

Sinclair, J. L., and Ghiorse, W. C. (1989). Distribution of aerobic bacteria, protozoa, algae, and fungi in deep subsurface sediments. *Geomicrobiol. J.* **7**, 15–31.

Sinclair, J. L., Randtke, S. J., Denne, J. E., Hathaway, L. R., and Ghiorse, W. C. (1990). Survey of microbial populations in buried-valley aquifer sediments from northeastern Kansas. *Ground Water* **28**, 369–377.

Sinton, L. W. (1984). The macroinvertebrates in a sewage-polluted aquifer. *Hydrobiologia* **119**, 161–169.

Sket, B. (1986). Isopoda: Sphaeromatidae. *In* "Stygofauna mundi" (L. Botosaneanu, ed.), pp. 422–427. E. J. Brill Dr. W. Backhuys, Leiden.

Smith, G. A., Nickels, J. S., Kerger, B. D., Davis, J. D., Collins, S. P., Wilson, J. T., McNabb, J. F., and White, D. C. (1986). Quantitative characterization of microbial biomass and community structure in subsurface material: A prokaryotic consortium responsive to organic contamination. *Can. J. Microbiol.* **32**, 104–111.

Stanford, J. A., and Gaufin, A. R. (1974). Hyporheic communities of two Montana rivers. *Science* **185**, 700–702.

Stanford, J. A., and Ward, J. V. (1988). The hyporheic habitat of river ecosystems. *Nature (London)* **335**, 64–66.

Stanley, E. H., and Boulton, A. J. (1993). Hydrology and the distribution of hyporheos: Perspectives from a mesic river and a desert stream. *J. North Am. Benthol. Soc.* **12**, 79–83.

Stock, J. H. (1977a). Microparasellidae (Isopoda, Asellota) from Bonaire, with notes on the origin of the family. *Stud. Fauna Curaçao Other Caribb. Islands* **51**, 69–91.

Stock, J. H. (1977b). The taxonomy and zoogeography of the hadziid Amphipoda with emphasis on the West Indian taxa. *Stud. Fauna Curaçao Other Caribb. Islands* **55**, 1–130.

Stock, J. H. (1977c). The zoogeography of the crustacean suborder Ingolfiellidea, with descriptions of new West Indian taxa. *Stud Fauna Curaçao Other Caribb. Islands* **55**, 131–146.

Stock, J. H. (1980). Regression model evolution as exemplified by the genus *Pseudoniphargus* (Amphipoda). *Bijdr. Dierkd.* **50**, 105–144.

Stock, J. H. (1981). The taxonomy and zoogeography of the family Bogidiellidae (Crustacea, Amphipoda), with emphasis on the West Indian taxa. *Bijdr. Dierkd.* **51**, 345–374.

Stock, J. H. (1982). The influence of hadziid Amphipoda on the occurrence and distribution of Thermosbaenacea and cyclopoid Copepoda in the West Indies. *Pol. Arch. Hydrobiol.* **29**, 275–282.

Stock, J. H. (1983). Predation as a factor influencing the occurrence and distribution of small Crustacea in West Indian groundwaters. *Bijdr. Dierkd.* **53**, 233–243.

Strayer, D. (1988). Crustaceans and mites (Acari) from hyporheic and other underground waters in southeastern New York. *Stygologia* **4**, 192–207.

Strayer, D. L. (1991). Perspectives on the size structure of lacustrine zoobenthos, its causes, and its consequences. *J. North Am. Benthol. Soc.* **10**, 210–221.

Teschner, D. (1964). Factors limiting mite life in groundwater. *Acarologia* **6**, 357–359.

Thurman, E. M. (1985a). Humic substances in groundwater. *In* "Humic Substances in Soil, Sediment, and Water" (G. R. Aiken, D. M. McKnight, R. L. Weshaw, and P. MacCarthy, eds.), pp. 87–103. Wiley, New York.

Thurman, E. M. (1985b). "Organic Geochemistry of Natural Waters." Martinus Nijhoff/Dr. W. Junk Publ. Dordrecht, The Netherlands.

Tilzer, M. (1973). Zum Problem der Ausbreitungsfähigkeit von limnisch-interstitiellen Grundwassertieren, am Beispiel von *Troglochaetus beranecki* Delachaux (Polychaeta Archiannelida), *Arch. Hydrobiol.* **72**, 263–269.

Vandel, A. (1964). "Biospéologie: La biologie des animaux cavernicoles." Gauthier-Villars, Paris.

Vervier, P., and Gibert, J. (1991). Dynamics of surface water/groundwater ecotones in a karstic aquifer. *Freshwater Biol.* **26**, 241–250.

White, D. C., Smith, G. A., Gehron, M. J., Parker, J. H., Findlay, R. H., Martz, R. F., and Fredrickson, H. L. (1983). The groundwater aquifer microbiota: Biomass, community structure, and nutritional status. *Dev. Ind. Microbiol.* **24**, 201–211.

Williams, D. D. (1984). The hyporheic zone as a habitat for aquatic insects and associated arthropods. *In* "The Ecology of Aquatic Insects" V. H. Resh and D. M. Rosenberg, eds.), pp. 430–455. Praeger, New York.

Williams, D. D. (1989). Towards a biological and chemical definition of the hyporheic zone in two Canadian rivers. *Freshwater Biol.* **22**, 189–208.

THREE

FUNCTIONING OF GROUNDWATER ECOSYSTEMS: CASE STUDIES

Plate 3   *Upper:* The Flathead River in the Kalispell Valley, Montana, U.S.A. The view is southwest toward the town of Kalispell. The river is flowing from right to left to confluence with Flathead Lake, which is just out of the picture (see Fig. 1 in Chapter 14). Photograph courtesy of J. Stanford. *Lower:* The floodplain of the Rhône River, France, upstream of Lyon before the impacts of civil engineering activities, in the Chautagne sector. Photograph courtesy of A.L. Roux.

12

Interstitial Fauna Associated with the Alluvial Floodplains of the Rhône River (France)

M-J. Dole-Olivier,[*] P. Marmonier,[†]
M. Creuzé des Châtelliers,[*]
and D. Martin[*]

*Université Lyon I
U.R.A. CNRS 1451 "Ecologie des Eaux Douces et des Grands Fleuves"
Laboratoire d'Hydrobiologie et Ecologie Souterraines
43 Boulevard du 11 Novembre 1918
69622 Villeurbanne Cedex, France

†Université de Savoie
Département d'Ecologie Fondamental et Appliquée
B.P. 1104
73011 Chambéry, France

I. INTRODUCTION

The alluvial aquifers of the Rhône River are well developed in many areas along its watercourse, particularly near the city of Lyon, where the aquifer has been used for many years as the source of drinking water. In the seminal study Gibert *et al.* (1977) demonstrated the great richness and biodiversity of the interstitial assemblages of the aquifer system. The fauna included benthic, stygophilous, and stygobite forms. Most of these animals were new to science (e.g, *Microcharon reginae, Bathynella* n. sp., and *Niphargus renei*) or had been rarely collected in the past (e.g., *Troglochaetus beranecki* and *Pseudocandona triquetra*). Later studies focused on the complex functioning of the hypogean component of the river (Gibert *et al.,* 1981; Reygrobellet *et al.,* 1981; Seyed-Reihani, 1980; Seyed-Reihani *et al.,* 1982a,b; Ginet, 1982; Mathieu and Amoros, 1982), and Dole (1983a,b, 1984) introduced the notion of the "biological layer," which is the biologically active stratum (upper 2–3 m) in which most biotic and abiotic processes take place (Danielopol, 1980). It is characterized by high densities of interstitial fauna, great quantities of organic matter, and strong interactions between surface waters and deep groundwaters.

Later studies demonstrated that maintenance of interstitial heterogeneity and faunal patterns was a consequence of the hydrologic and geomorphologic properties of the river (Marmonier and Dole, 1986). Water circulation patterns within the sediment create the structure of faunal assemblages both in space (Marmonier, 1988) and in time (*i.e.,* series of floods and low water periods) (Dole-Olivier and Marmonier, 1992b). Dynamic bed sediment degradation–aggradation processes cause discontinuities in the river bed slope, thereby mediating a succession of riffle–pool sequences, and cause evulsion and isolation of secondary channels (arms) in the floodplain. These geomorphic processes also influence the distribution of interstitial fauna (Creuzé des Châtelliers, 1991a,b; Dole, 1984; Dole and Chessel, 1986; Marmonier *et al.,* 1992).

Because it is now recognized that interconnected aquatic, semiaquatic, and terrestrial ecosystems depend on the existence of the alluvial aquifer (Bravard *et al.,* 1986; Amoros *et al.,* 1987), all these studies in groundwater ecology complemented an earlier, multidisciplinary study of the structure and functioning of the Rhône River hydrosystem (Roux, 1976). Moreover, the top part of the aquifer, in contact with the river (often called "hyporheic"), is recognized as the dynamic boundary between the epigean, or benthic, milieu of rivers and the hypogean, or phreatic, milieu of the subterranean biotope of classical studies of groundwater ecology. Epigean and hypogean fauna converge in this zone and fluxes between these two biotopes are fundamental to the ecology of alluvial rivers (Stanford and Ward, 1993). The concept of a dynamic boundary, or "ecotone," in stream ecology was derived in part from studies of groundwater ecology in the Rhône River (Gibert *et al.,* 1990).

In this chapter we review studies on the Rhône River carried out during the last 15 years, which demonstrate the ecological importance of the interstitial fauna and the processes that influence its distribution and abundance. In the first section, we describe the alluvial aquifers of the Rhône River and provide an overview of the regulation scheme. The second section discusses the structure of the river's interstitial system, including habitat diversity (Section II.A) and faunal assemblages (Section II.B). The main part of this last section concerns the understanding of distributional patterns on the three dimensions of the floodplain: longitudinal (Section II.B.3), lateral (Section II.B.4), and vertical (Section II.B.5). These results are explained in relation to a general hydrologic and geomorphologic hypothesis (Section II.B.2) and are synthesized in Section II.B.6. Finally the temporal modifications of these patterns are reviewed in relation to spates (Section III.A), gravel displacement (Section III.B), and long-term changes in surface habitats (Section III.C).

II. THE RHÔNE RIVER AND ITS AQUIFERS

The Rhône is a large river (ninth order; average annual flow at Lyon, 600 m³/sec⁻¹) and receives water from a complex montane environment. The valley is confined by alpine calcareous reliefs to the east, by great sedimentary basins to the north, and by ancient crystalline massifs and calcareous plateaus to the west. The hydrogeologic continuum involves surface and subsurface flow pathways. Karst (fissured), granite (fractured), and alluvial (porous) aquifers exist in association with the river course. The porous aquifers occur principally in the sedimentary basins of the Bresse and the Dombes (areas located north of the Rhône River) and in the Bas-Dauphiné (the area between the Rhône and the Isère Rivers). The Pliocene and Miocene formations are composed of clayey, sandy, and gravelly materials of low permeability.

As observed in many other large rivers (Stanford and Ward, 1993), the valley is composed of a series of plains alternating with canyons (Fig. 1). Consequently, the Quaternary alluvial filling is discontinuous, with each plain forming one aquifer (or more), which can be considered an entity. The natural hydrologic relationships between the river and its aquifers are mainly determined by the surface-water level: during floods the river recharges its aquifer by infiltration of surface water within the sediments; in contrast, during low water periods phreatic water flows toward the channel. These natural variations of the piezometric gradient have been strongly modified by artificial regulation of surface flow occurring on each plain. Generally a dam diverts most of the flow from the river to a hydroelectric power station; the original channel is bypassed and is used only as a spillway during periods of high water. Major effects of regulation on interstitial fauna are related to (1) frequent, strong, and rapid variations in flow rate (possible variations from 30 to 1500 m/sec⁻¹ in a few hours; Gaschignard-Fossati,

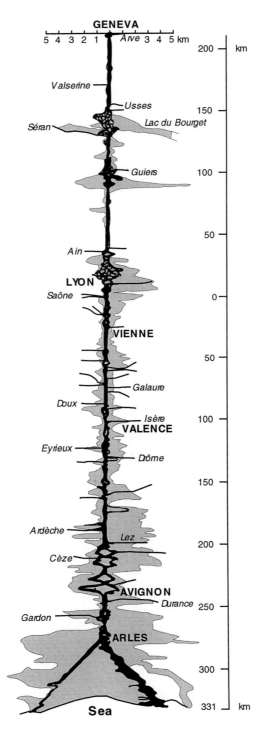

1986) and (2) permanent modifications of the groundwater table, which is raised upstream of the dam (*i.e.*, the alluvial aquifer is supplied by surface water) and lowered downstream (*i.e.*, permanent drainage of the aquifer).

However, the effect of the regulation scheme is highly influenced by the regional hydrogeologic configuration and behavior of groundwater flow within each alluvial plain. Indeed several types of alluvial aquifers can be distinguished. For example, in the upper section of the Rhône River, close to the high alpine reliefs, alluvial plains are deposited upon an ancient glaciolacustran substratum. The Holocene alluvium is less than 10 m thick, and the underflow of the river is limited. Farther down in the valley, near Lyon, the plains are characterized by a Quaternary accumulation of fluvio-glacial and fluvial materials over 30 m thick (Plate 1). The slopes of the hydrogeologic structures (*e.g.*, fluvioglacial hills or old terraces) range from 0.5 to 0.8‰, and the alluvium is highly permeable (10^{-2}–10^{-3} m/sec^{-1}). Hence, the Rhône River in the Lyon area is connected to extensive aquifers. In several localities bed rock outcrops limit alluvial deposition, and the aquifers are less than 10 m thick; the connectivity of the river is likewise reduced. In other areas the alluvial plain is bordered by karst, and groundwater enters the alluvial system as point sources from springs.

III. STRUCTURE OF THE RHÔNE RIVER INTERSTITIAL SYSTEM

A. Habitat Diversity

1. Local and Longitudinal Heterogeneity

The alluvial valley of the Rhône is a mosaic of many channels with many aquatic, semiaquatic, and terrestrial patches. Physical diversity is complemented by many plant patches. In contrast, the interstitial system connected to these large rivers seems to display a lower diversity of habitats, which are basically reduced to a single type, the sediment interstice (Gibert *et al.*, 1990). Consequently, the underground spatial units may appear less complex, as suggested by Turner *et al.* (1989). However, the interstitial volume is large, and habitat patchiness is important and linked to local conditions and to exchanges with surface habitats (Dole, 1984; Marmonier *et al.*, 1992). For example, in the surface main channel, the alluvium of the river bed is characterized by local irregularities and variations in slope, which produce different interstitial flow pathways, resulting in downwellings of surface water and upwellings of groundwater (White *et al.*, 1987; White, 1990). These hydrologic and geomorphic features generate different physical

FIGURE 1   The Rhône River from Geneva to the sea: defiles and alluvial plains illustrated by the centennial flood of 1856 (data collected by D. Poinsart in the "Archives Départementales du Rhône," Lyon, France).

Plate 1 The floodplain of the Rhône River (France) upstream of Lyon. This area shows different evolution and isolation of secondary channels. In the background the Jura mountains limit the extent of the floodplain.

patches in groundwater habitats. Moreover, differences in surface water quality (or dynamics) and in the distribution of basic resources (*e.g.,* available spaces and nutrients) promote groundwater heterogeneity.

2. Spatial and Temporal Floodplain Heterogeneity

In addition to longitudinal pathways complex habitats also exist in both horizontal and vertical dimensions: *i.e.,* from 0.5 m to 2 or 3 m under the substrate surface and from the channel axis to gravel bars near the banks or to the margins of the floodplain. Depending on the geological characteristics of the region, large rivers can develop important floodplains around their main channels. Hence by moving in the plain and leaving many side arms and former meanders (Fig. 2), they generate a large number of different aquatic ecosystems and wetlands (Schumm, 1977; Bravard *et al.,* 1986; Amoros *et al.,* 1987) and create distinct associated interstitial habitats, which correspond to the different evulsion stages of the surface channel (Gibert *et al.,* 1990). Generally side arms are progressively isolated from the main channel by sediment deposits at their upstream end. At this stage, the side arm is regularly flooded, and, during low-water periods, the current is strongly reduced. But good interstitial conditions may occur in the upstream part of the arm, where springs emerge (Gibert *et al.,* 1977, 1981), as in the downstream part the interstitial habitat is often clogged by fine sediment brought by the main channel. During low-water periods, the arm may be completely abandoned by the surface flow, and the downstream end may be closed by sediment deposits, forming a stagnant pond. Such a pond may be regularly flooded during high-water periods or gradually isolated from the main channel. However, groundwater flow may maintain well-developed interstitial habitats (Reygrobellet and Dole, 1982; Dole, 1983a), whereas the eutrophi-

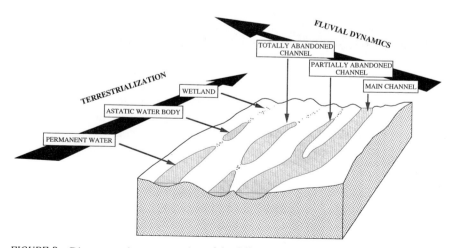

FIGURE 2   Diagrammatic representation of the different aquatic habitats of the river channel in the floodplain, and the dynamic relation between flooding and encroachment of terrestrial conditions.

cation of surface water and lowering of the groundwater table may allow succession of the surface biotopes from wetlands to riparia (Fig. 2). In this case, groundwater inputs are more irregular and the interface between groundwater and surface water is slowly modified in the transition zone between groundwater and the soil.

The diversity of interstitial habitats also increases where tributaries cross the floodplain. The confluence zones are generally characterized by rapid channel movements and strong groundwater fluxes, which generate highly diversified surface and interstitial habitats (Roux, 1986; Creuzé des Châtelliers, 1991a). In contrast, most of the human activities decrease the habitat diversity (*e.g.,* embankments), except in particular cases such as artificial gravel pits that are fed by groundwater, and generate new bodies of water and associated interstitial habitats.

B. Distributional Patterns of Interstitial Fauna

1. The Interstitial Fauna

Typical interstitial dwellers are stygobionts (ubiquitous stygobites and strict phreatobites; Chapter 1). Table I lists stygobite species collected since 1977 in the alluvial floodplains of the Rhône River from Geneva to the delta at Arles. The 38 identified species emphasize the biodiversity and richness of stygobites, which are distributed among 12 different taxonomic groups. Crustaceans dominate and include 23 species, of which 11 are amphipods (*Niphargus, Niphargopsis, Crangonyx,* and *Salentinella*). Most of the species are quite common, except for the copepod *Graeteriella unisetigera,* the cladoceran *Alona phreatica,* and the amphipod *Crangonyx subterraneus,* which were rarely found (Seyed-Reihani *et al.,* 1982a). *Troglochaetus beranecki* and *Siettitia avenionensis* are not rare but are isolated at a few particular sites in the hydrosystem (Dole, 1983a; Richoux and Reygrobellet, 1986).

We compared all the stygobite species from the Rhône with those reported from two other large European rivers, the Rhine (Table II) and the Danube (Table III). Many differences that could bias such a comparison exist among the three systems: (a) sampling effort (many studies of the Upper Rhône river were carried out, but very few of the Rhine), (b) sampling periods (the Rhine was studied very early in the century and many species listed could have been eliminated by pollution), and (c) differences in taxonomic resolution (Nematoda were well studied by Eder for the Danube, but not for the Rhône). Nonetheless, it appears that species richness is currently similar for the Rhône (38 stygobite species) and the Rhine (more than 35 stygobite species), but the Danube River has greater richness (more than 60 stygobite species). This difference could be related to the length of the Danube. Greater length suggests more complex biogeographical patterns, owing potentially to more variation in the human disturbance regime and spatial/temporal hydrologic patterns. On the other hand, the relatively short length of the Rhône river (812 km) compared with the lengths of the Rhine

TABLE I   Stygobite Species Collected in the
Alluvial Floodplain of the Rhône River
since 1977

Nematoda	*Onchulus nolli*
Mollusca	*Hauffenia minuta*
	Moitessieria lineolata
Archiannelida	*Troglochaetus beranecki*
Oligochaeta	*Potamodrilus fluviatilis*
	Aeolosoma gineti
	Rhyacodrilus balmensis
	Rhyacodrilus amphigenus
	Rhyacodrilus phreaticola
	Spiridion phreaticola
	Haber turquini
	Trichodrilus michaelseni
	Trichodrilus cernosvitovi
	Trichodrilus leruthi
Cladocera	*Alona phreatica*
Cyclopoida	*Graeteriella unisetigera*
Harpacticoida	*Elaphoidella elaphoides*
	Parastenocaris fontinalis
	Parastenocaris glareola
Ostracoda	*Fabaeformiscandona wegelini*
	Pseudocandona zschokkei
	Pseudocandona triquetra
	Cryptocandona kieferi
Isopoda	*Proasellus walteri*
	Microcharon reginae
Amphipoda	*Niphargus kochianus kochianus*
	Niphargus pachypus
	Niphargus rhenorhodanensis
	Niphargus renei
	Niphargus plateaui
	Niphargus laisi
	Salentinella juberthiae
	Salentinella delamarei
	Salentinella lescherae
	Niphargopsis casparyi
	Crangonyx subterraneus
Syncarida	*Bathynella* nov. sp.
Coleoptera	*Siettitia avenionensis*

(1298 km), and the Danube (2850 km) suggests greater biodiversity per kilometer of river. For the Rhône, species richness is higher within river segments that are more pristine or protected from pollution (*i.e.*, mainly upstream from Lyon).

2. Hydrologic Determinants of Distributional Patterns

In running waters, hydrology controls all secondary factors, such as geomorphology and the distribution of all types of resources. In alluvial

TABLE II   A Provisional List of Stygobite Species
Collected in the Alluvial Floodplain of the Rhine
River since 1933

Microturbellaria	*Protomonotresis centrophora*
Turbellaria	*Dendrocoelum agile*
tricladida	*Dendrocoelum remyi*
Nemertina	*Prostoma puteale*
Nematoda	*Onchulus fuscilabiatus*
Mollusca	*Potamopyrgus antipodarum*
Archiannelida	*Troglochaetus beranecki*
Oligochaeta	*Trichodrilus tenuis*
Cyclopoida	*Eucyclops graeteri*
	Acanthocyclops rhenanus (?)
	Paragraeteriella laisi
Harpacticoida	*Chappuisius inopinus*
	Nitrocrella omega
	Parastenocaris glareola
	Parastenocaris fontinalis
	Parastenocaris aedes
	Parastenocaris hippuris
Ostracoda	*Pseudocandona hetzogi*
	Pseudocandona zschokkei
	Pseudocandona schellenbergi
	Pseudocandona brisiaca
	Pseudocandona insueta
	Cryptocandona kieferi
	Mixtacandona laisi laisi
Isopoda	*Proasellus cavaticus*
Amphipoda	*Niphargus puteanus*
	Niphargus jovanovici bajuvaricus
	Niphargus jovanovici kieferi
	Niphargus laisi
	Niphargus fontanus
	Niphargus pachypus
	Niphargus kochianus kochianus
	Niphargus schellenbergi
	Bogidiella albertimagni
	Eucrangonyx vejdovskyi
	Niphargopsis casparyi
Syncarida	*Bathynella natans*

Note. Data from the "Stygofauna Mundi" (Botosaneanu,
1986), the works of Hertzog (1933, 1936, 1938), and J. H.
Stock (personal communication).

aquifers, pathways and fluxes between surface and interstitial habitats are
the major factors influencing composition, structure, and functioning of
the interstitial ecosystem. Water temperature patterns (White *et al.*, 1987),
nutrient dynamics (Valett *et al.*, 1990; Grimm, 1988), and microbiological
processes (Hendricks and White, 1991; Hendricks, 1993) are associated

TABLE III   A Provisional List of Stygobite Species Collected
in the Alluvial Floodplain of the Danube River

Microturbellaria	*Danubia antipa*
Turbellaria	*Dendrocoelum maculatum* var. *candidum*
tricladida	*Dendrocoelum romanodanubiale*
Nematoda	*Theristus franzbergeri*
	Theristus ruffoi
	Paramphidelus propinquus
	Stenonchulus troglodytes
	Pseudorhablolaimus limnophilus
Mollusca	*Bythiospeum geyeri*
	Lobaunia danubialis
Archiannelida	*Troglochaetus beranecki*
Oligochaeta	*Dorydrilus michaelseni*
Harpacticoida	*Parastenocaris fontinalis*
	Parastenocaris cf. *phyllura*
	Parastenocaris sp.
	Nitocrella hirta
	Eucyclops graeteri
Cyclopoida	*Austriocyclops vindobonae*
	Acanthocyclops rhenanus
	Acanthocyclops gmeineri
	Acanthocyclops sensitivus
	Graeteriella unisetigera
Ostracoda	*Pseudocandona bilobatoides*
	Pseudocandona pseudoparallela
	Pseudocandona aff. *szöcsi*
	Kovalevskiella sp.
	Fabaeformiscandona wegelini
	Cryptocandona kieferi
	Mixtacandona laisi vindobonensis
	Mixtacandona transleithanica
	Mixtacandona spandli
	Mixtacandona löffleri
Isopoda	*Proasellus slavus vindobonensis*
	Proasellus strouhali
Amphipoda	*Niphargopsis casparyi*
	Niphargus bajuvaricus
	Niphargus fontanus
	Niphargus inopinatus
	Niphargus cf. *kochianus*
	Niphargus rhenorhodanensis
	Niphargus valachicus
	Niphargus tauri?
	Niphargus jovanovici?
	Crangonyx subterraneus
	Bogidiella albertimagni
	Bogidiella skopljensis
Syncarida	*Bathynella vindobonensis*
	Parabathynella cf. *stygia*

(continues)

TABLE III (*Continued*)

Hydracarina	*Soldanellonyx chappuisi*
	Soldanellonyx monardi monardi
	Lobohalacarus weberi quadriporus
	Feltria cornuta paucipora
Hydrachnella	*Lethaxona pygmaea*
	Lethaxona cavifrons
	Aturus karamani
	Kongsbergia dentata
	Kongsbergia ruttneri
	Stygomomonia latipes
	Neoacarus hibernicus

Note. Data from the "Stygofauna Mundi" (Botosaneanu, 1986), the works of Schwoerbel (1964), Orghidan *et al.* (1979), Eder (1983), Danielopol (1976a,b), and Pospisil (Chapter 13).

with the rate and duration of flux. Interstitial assemblages are also structured in these hydrologic patterns (Godbout and Hynes, 1982; Marmonier and Dole, 1986).

The downwelling–upwelling sequence is the basic unit of water circulation pattern within the alluvial system (Fig. 3). Downwellings are located upstream of obstacles to surface water flow, and upwellings occur downstream (Thibodeau and Boyle, 1987). Obstacles may consist of rocks, hummocks of vegetation, and natural or artificial dams or more generally of river bed irregularities such as riffles and gravel bars (Vaux, 1968; Freeze and Cherry, 1979; Hendricks and White, 1988, 1991; Marmonier, 1988; Creuzé des Châtelliers, 1991a). As noted above water circulation may be modified by more general hydrogeologic features (*e.g.*, thickness of the aquifer and subsurface stratigraphy; Freeze and Cherry, 1979; Creuzé des

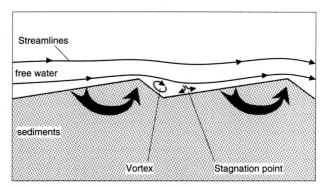

FIGURE 3 Over-bed and in-bed flow conditions occurring on stream bed irregularities [after Thibodeaux and Boyle (1987)].

Châtelliers, 1991a), geomorphologic characteristics (grain size and porosity of the sediments; Chapter 6), or particular conditions of surface flow (dried sections of streams; Grimm and Fisher, 1984; Valett *et al.*, 1990; Grimm *et al.*, 1991).

3.   The Main Channel

In the Rhône River, the longitudinal distribution of interstitial organisms was studied across several spatial scales, which correspond to different geomorphologic or geologic formations. The first scale concerned the riffle and the associated gravel bar. The riffle was considered the basic morphological unit of the channel, whereas the gravel bar is the lateral alluvial formation associated with the riffle–pool sequence. Second, a series of riffle–pool sequences occurring for several kilometers was considered a geomorphologic sector. As previously mentioned, each alluvial sector is impounded by a low-elevation dam, which creates a succession that is degraded and then aggraded within each sector (Creuzé des Châtelliers and Reygrobellet, 1990). The third scale is the fluvial hydrosystem, composed of several alluvial sectors, each with its own physical characteristics (slope, sinuosity, and gravel bed load) and different geomorphologic patterns (meandering, straightening, braiding, etc.; Bravard, 1987). All these scaled physical units generally correspond to hydrologic ones, which are upwellings or downwellings (first scale), large drainage or infiltration areas (second scale), and aquifers with different hydrogeologic characteristics (third scale). Marmonier (1988), working on several gravel bars, and Creuzé des Châtelliers (1991a), working on 46 riffles distributed along three distinct geomorphologic sectors, demonstrated relations between these hydrologic patterns and the longitudinal distribution of interstitial fauna.

A good example is illustrated in Fig. 4, which shows the faunal patterns at 0.5 m depth on the gravel bar scale (Marmonier, 1988). A clear faunal gradient was observed from the downwelling zone at the head of the bar to the upwelling zone at the tail of the bar. The distinctive taxa of the downwelling zone (0 and 20 m) were typical epigean fauna; taxa in the upwelling zone were mainly stygobite species (*Siettitia avenionensis, Salentinella* sp., *Parastenocaris* sp., *Proasellus walteri*, and *Pseudocandona zschokkei*). A third group was widely scattered along the gravel bar, including *Niphargopsis casparyi* and *Niphargus rhenorhodanensis*, which are ubiquitous stygobites. Similarly, larger-scale studies demonstrated that aggraded parts and limited aquifers are poorly colonized by stygobite fauna, but degraded parts and developed aquifers are colonized by rich and abundant stygobite assemblages.

Variants on this model existed on a gravel bar presenting an inverse hydrologic pattern (*i.e.*, upwelling head bar zone and downwelling tail bar zone; Dole-Olivier and Marmonier, 1992a). In this case, head bar and tail bar conditions (*e.g.*, sands and POM content) were quite independent of

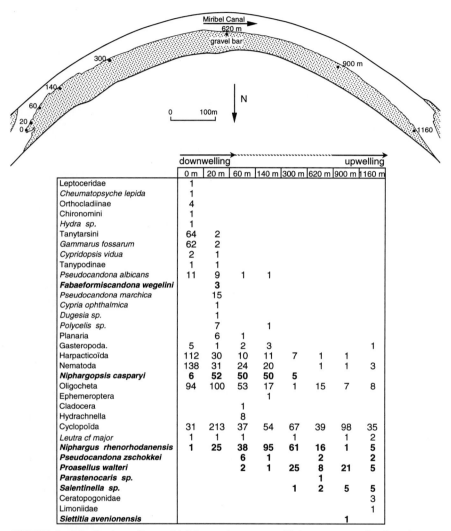

	downwelling						upwelling	
	0 m	20 m	60 m	140 m	300 m	620 m	900 m	1160 m
Leptoceridae	1							
Cheumatopsyche lepida	1							
Orthocladiinae	4							
Chironomini	1							
Hydra sp.	1							
Tanytarsini	64	2						
Gammarus fossarum	62	2						
Cypridopsis vidua	2	1						
Tanypodinae	1	1						
Pseudocandona albicans	11	9	1	1				
Fabaeformiscandona wegelini		3						
Pseudocandona marchica		15						
Cypria ophthalmica		1						
Dugesia sp.		1						
Polycelis sp.		7		1				
Planaria		6	1					
Gasteropoda.	5	1	2	3				1
Harpacticoïda	112	30	10	11	7	1	1	
Nematoda	138	31	24	20		1	1	3
Niphargopsis casparyi	6	52	50	50	5			
Oligocheta	94	100	53	17	1	15	7	8
Ephemeroptera				1				
Cladocera			1					
Hydrachnella			8					
Cyclopoïda	31	213	37	54	67	39	98	35
Leutra cf major	1	1	1		1		1	2
Niphargus rhenorhodanensis	1	25	38	95	61	16	1	5
Pseudocandona zschokkei			6	1		2		2
Proasellus walteri			2	1	25	8	21	5
Parastenocaris sp.						1		
Salentinella sp.					1	2	5	5
Ceratopogonidae								3
Limoniidae								1
Siettitia avenionensis							1	

FIGURE 4   Distribution of interstitial fauna along a gravel bar (-0.5 m under the substrate surface). Bold numbers and taxa correspond to hypogean fauna.

hydrologic patterns. Hence, some stygobite species are sensitive to hydrologic patterns, but others may be controlled by other variables, such as sediment condition (corresponding to head bar or tail bar positions). The longitudinal structure of the Rhône River, in relation to hydrology and geomorphology, is very generally described and must be further examined in the context of other parameters and discussed with respect to general unifying concepts, such as the hyporheic corridor concept (Stanford and Ward, 1993).

4. The Floodplain

A basic factor correlated with the distribution and abundance of interstitial fauna in the floodplain is distance from the main channel. Three stations were located on three channels corresponding to distinct evulsion stages (Fig. 5) (Dole,1983a): (1) the active channel (present stage), (2) a secondary channel cut off at the upstream end (intermediate stage), and (3) an abandoned old meander (old stage). The temporal variation of physical conditions (hydrology, temperature, chemical characteristics, etc.) decreased from the active channel to the old meander. The composition and structure of the interstitial assemblages were very different at the three stations. On the one hand, the density and species richness of interstitial assemblages were highest in the secondary arm, which was characterized by intermediate physical variability. In the main channel and the abandoned meander, the interstitial assemblages were less dense and less rich (Fig. 5). On the other hand, the diversity of stygofauna and the percentage of hypogean species increased from the present to the old stage (see "main taxa" in Fig. 5). These main structural characteristics were also demonstrated by the distribution of stygobite species at other places across the floodplain (Dole-Olivier *et al.*, 1993) (Fig. 5). Eleven species occurred in the margin of the floodplain, but only five were obtained in the main channel at 0.5 m depth. The 16 stygobite species collected in the floodplain were distributed on a transverse gradient from species restricted to the main channel and nearby areas (*Fabaeformis-candona wegelini, Niphargus rhenorhodanensis, Cryptocandona kieferi*, and *Pseudocandona zschokkei*) to species restricted to places situated far from the main channel (*e.g., Troglochaetus beranecki, Niphargus renei*, and *Microcharon reginae.*) Between these two extremes, some species such as *Niphargopsis casparyi* were widely distributed across the floodplain. In summary, the distance to the active channel corresponded to a gradient of physical stability (Dole and Chessel, 1986; Marmonier *et al.*, 1992), which structured the distribution of hypogean fauna of species resistant to the instability of the sediments to phreatobites that are very sensitive to instability (*e.g., Microcharon reginae*).

Another influence on spatial distribution of the interstitial fauna was the terrestrialization processes (Fig. 2) (Marmonier *et al.*, 1992). In the old meander (Fig. 6) at the upstream end, where terrestrial conditions prevailed (*e.g.*, due to decreased elevation of the water table), the interstitial assemblages were dominated by epigean fauna. Similar observations were made in a secondary braided arm (Marmonier *et al.*, 1992). Less interstitial flow of groundwater associated with terrestrialization strongly modified the physicochemical characteristics of the interstitial habitats (*e.g.*, decrease of water

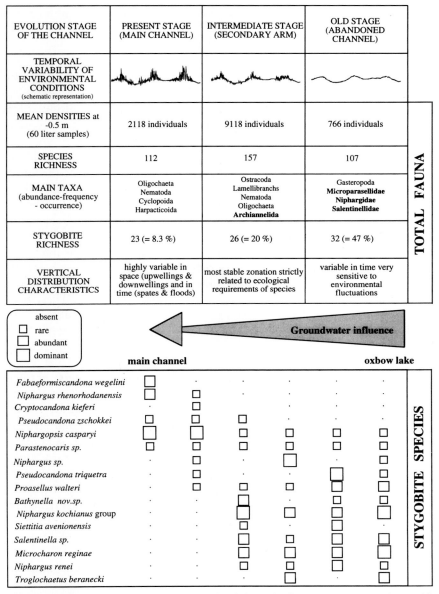

FIGURE 5 Faunal distribution across the floodplain (depth, 0.5 m) in comparison with general physical and biological characteristics of three channel types, and the transversal gradient of stygobite species across the Jons Floodplain [after Dole and Chessel (1986) and Dole-Olivier *et al.* (1993)]. Rare, <1 individual per sample; abundant, from 1 to 20 individuals per sample; dominant, >20 individuals per sample.

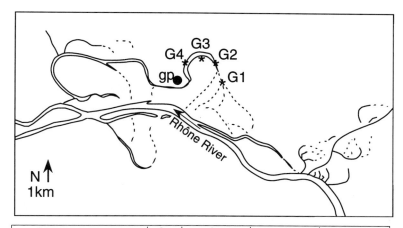

stations	G1		G2				G3				G4			
date	Apr	July	Apr		July		Apr		July		Apr		July	
depth (m)	0.5	1.0	0.5	1.0	0.5	1.0	0.5	1.0	0.5	1.0	0.5	1.0	0.5	1.0
Pseudocandona triquetra							1	2						
Niphargus renei							1							
Niphargopsis casparyi														
Bathynella sp.														
Crangonyx sp.											1			
Nlphargus gallicus											1		1	
Moitessiera lineolata							50	17	28	14			1	3
Troglochaetus beranecki							9	2						
Parastenocaris sp.											1			
B. diaphanum							23	17	19	4			2	2
Hauffenia minuta							48	10	18	3			28	6
Niphargus kochianus							17	3		6				
Salentinella sp.							19	5			23		54	37
Proasellus sp.							2			1				
Microcharon reginae							203	82	9	21	8	2		4
Niphargus pachypus	3								3		19	3	36	32
Cryptocandona kieferi					3	3								
Nematoda	48	39	2		1		58	2			45	1	15	2
Oligochaeta	6	16	2	5	9	31	15	9	4	4	19	4	94	23
Bivalvia													1	1
Gastropoda (epigean)	1	8											1	1
Hydracarina	5	2					1				2			
Cladocera			14		2	11					7		19	
Harpacticoida (epigean)	30	14	3	3	5		3	2			97	1	23	9
Cyclopoida (epigean)	7	12	4	3	22	276				6	13		34	3
Ostracoda (epogean)	15	10	13	1	31	3	6	4		1	2		9	1
Chironomidae	1	1	2	1	2	2				1	1	1	9	3
Diptera (others)														
Plecoptera									1					
Coleoptera														
Gammarus sp.											5		183	15
Planaria													8	
Asellidae (epigean)														
Ephemeroptera			3		8									
hypogean taxa (percent)														

FIGURE 6 Upstream/downstream gradient of the interstitial fauna (sampling depths, −0.5 and −1 m) in the Grand Gravier abandoned channel. The four stations (G1 to G4) were sampled in April and July 1987 (Station G1 clogged at −1 m deep). gp, location of the gravel pit studied for the long-term succession (see Section III.C).

exchanges, reduction of pore size, and modification of the thermal regime) and consequently altered composition and structure of interstitial fauna.

5. The Vertical Structure

a. Basic Concept   The "biological layer" (Dole, 1983a), containing higher densities of organisms and intense biological interactions (in opposition to the deep interstitial layer), generally extends 2 or 3 meters downward, depending on surface activity and dynamics, porosity of the sediment, and other characterisitics of the aquifer. In the Rhône, three vertical components can be distinguished (Fig. 7): (1) the benthic zone, corresponding to the

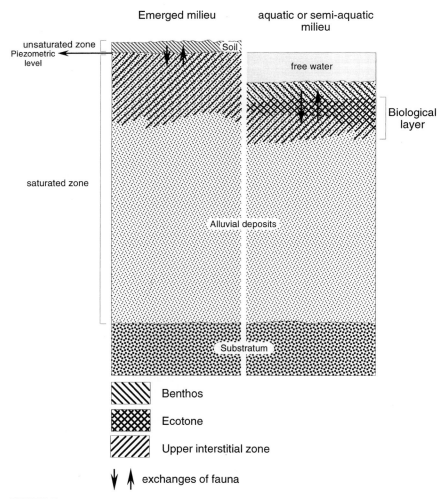

FIGURE 7   The "biological layer" in an alluvial aquifer of the Rhône River near Jons [after Dole (1984)].

bed sediments and exclusively colonized by benthic species, (2) the upper interstitial, or the top layer of alluvial deposits, in which communities are still dense and largely dominated by stygobite species, and (3) the transition zone between benthic and upper interstitial layers. Some stygobite species, some benthic species, and some additional ecotone specialists are found in the transition zone. As previously mentioned, the relative importance of these three components and their spatial extensions are highly variable. For this reason and because it is variously defined in other studies (Chapter 1), we do not include the term hyporheic in our conceptual model of the biological layer of the Rhône alluvial aquifer.

b. Vertical Zonation Changes in the Floodplain   The vertical zonation of interstitial assemblages was studied in three sites of the alluvial floodplain, corresponding to an increasing distance from the main channel and a decreasing importance of flood pulse action (see Section II.B.4). At the first site (main channel) the vertical distribution was highly variable in relation to spates (see Section III.A) and in relation to upwellings and downwellings (see Section II.B.5.C). At the second site (the secondary arm is still connected to the main channel at its downstream end) flood pulse and the spate effect were less important. Here the biological layer was very dense, with a strong vertical zonation characterized by a great density and diversity in subsurface zones (*e.g.,* 8840 individuals for 60-l samples and 20 taxa at 0.5 m) and a strong decrease with depth (*e.g.,* respectively 44 and 5 individuals corresponding to 12 and 3 taxa at 1.0 and 1.5 m depth respectively). The community could be divided into six groups of different vertical distributions from stygoxens and occasional hyporheos species, with high subsurface densities and null density at -1 and -1.5 m, to phreatobites, which were represented only in deep zones (Dole and Chessel, 1986). Such a vertical distribution was strictly linked to the ecological requirements of the species and was very stable in time (15 sampling dates). It corresponded to an expression of the resistance to flood pulse action. In contrast, in the third site (abandoned channel, without direct links to the fluvial axis), the vertical distribution of organisms is not so clear and is characterized by temporal variability. For example, stygobite and phreatobite species could be equally distributed from subsurface (-0.5 m) to deep (-2 or -3 m) zones or could be sometimes more abundant near the surface of the sediment. Organisms seemed to move without vertical constraints. The application of low artificial disturbance (modification of grain size by pumping; Dole and Chessel, 1986) demonstrated that the community was very sensitive to environmental variability. In conclusion (Fig. 5), in the main channel the vertical distribution of organisms is highly variable in space (in relation to upwellings and downwellings) and in time (in relation to spates and floods). In the floodplain margins, the vertical zonation is not strictly defined but is very sensitive to low-level disturbances, which rarely occur due to lack of connectivity with the main channel. The most stable and clearly defined vertical structure was observed in the secondary arm, where intermediate disturbance conditions persisted.

   c. Vertical Zonation Changes in Relation to Upwellings and Downwellings   The variation of environmental parameters along the vertical axis is very different in upwellings and downwellings, as illustrated in Fig. 8 (in which temperature and specific conductance are presented as examples). In this figure, each of the 21 vertical curves corresponded to one sampling date. The downwelling part was characterized by low variations with depth, indicating considerable surface water infiltrations to a depth of 2 m. The variability observed in surface water also occurred at −2 m. Specific conductance reached comparable high amplitude in surface water and in the −2-m layer (97 and 76 μS, respectively). Temperature amplitude was similar in surface water and in the −2-m layer (16.2 and 13.9°C, respectively). In contrast, the upwelling part was characterized by a clear vertical gradient and low variability in the −2-m zone (temperature presented only 3.6°C of annual amplitude variation) compared with great variability in surface waters (16.2°C). Faunal assemblages reacted to these different physical conditions (see examples in Fig. 8) by variable vertical distributions, except *Proasellus walteri,* which is not sensitive to water circulation patterns (Dole-Olivier and Marmonier, 1992c). *Niphargus rhenorhodanensis* was abundant in upwellings regardless of depth, but was rarest in downwellings. The densities of the phreatobite *Salentinella* were very low in the downwelling area, with a vertical distribution restricted to the deepest zone. In the upwelling zone the abundances of *Salentinella* increased with depth from 1.0 to 2 m. Both *Niphargopsis casparyi* and *Pseudocandona zschokkei* were very scarce, but these organisms were most abundant in subsurface zones, either in upwellings (*N. casparyi*) or in downwellings (*P. zschokkei*). These results demonstrated that the same species could display very different vertical distribution patterns in relation to water circulation. These differences are strongest for the total community in downwelling zones due to the infiltration of many epigean organisms to a depth of 2 m (M. J. Dole-Olivier, P. Marmonier, and J. L. Beffy, unpublished).

6. General Gradients in the Stygobite Community at the Floodplain Scale

   To explain the spatial structure of interstitial assemblages, we previously discussed the importance of water circulation patterns (upwelling/downwelling sequence) and the stability/instability of environmental conditions. The epigean component of these interstitial assemblages is also highly variable in relation to patch communities and the diversity existing in surface waters (Castella, 1987). In contrast, we observed that a small group of stygobite species occurred in the alluvial floodplain according to an ordinated distribution (Fig. 9). This ordination corresponded to a gradient habitat stability, which was repeated in the three spatial dimensions of the floodplain:

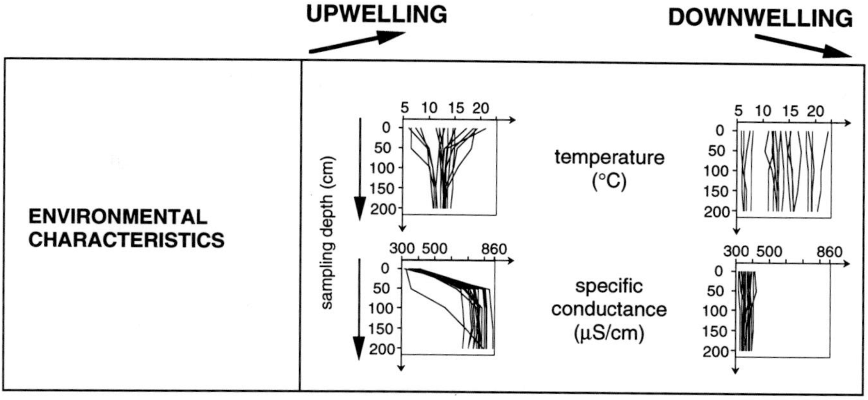

Niphargopsis casparyi	1	2	3	4	5	1	2	3	4	5
0,5 m	33		9	32	4	4	4	30		18
1 m	24	4	8	11		1			10	
1,5 m		2	1		2	1				1
2 m	1	3	1	2						

Niphargus rhenorhodanensis	1	2	3	4	5	1	2	3	4	5
0,5 m	244	6	655	274	42		7	13		19
1 m	91	177	621	129	86	10	6	7	27	18
1,5 m	23	119	252	40	641	4	1	3		
2 m	87	67	146	86	189	1	1			

Salentinella sp.	1	2	3	4	5	1	2	3	4	5
0,5 m										
1 m	6	1	20	16						
1,5 m	5	22	57	20	20					
2 m	28	74	142	72	5	4	8	3	5	3

Proasellus walteri	1	2	3	4	5	1	2	3	4	5
0,5 m										
1 m	1		1			4	5	1	2	3
1,5 m		1				1	4	2		
2 m										

Pseudocandona zschokkei	1	2	3	4	5	1	2	3	4	5
0,5 m	1						2			
1 m			1	1			4	2	19	2
1,5 m	1		1							1
2 m	1		1					1		1

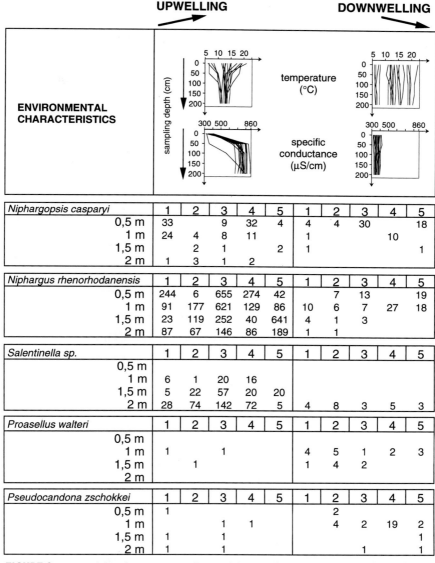

FIGURE 8 Vertical distribution in upwelling and downwelling zones. (top) Depth distribution of temperature and specific conductance (−0.5 to −2 m). (bottom) Density of the five major stygobite taxa occurring in the main channel; five replicate samples and four sampling depths are represented. Dates: 1, 13 August 1987; 2, 7 September 1987; 3, 5 November 1987; 4, 7 March 1988; 5, 9 November 1988 [after Dole-Olivier and Marmonier (1992c)].

(1) transversally, from the stable margins of the floodplain (e.g., oxbow lakes and abandoned channels) to the main channel, (2) vertically, from deep zones that are quite stable to subsurface zones that reflect surface variability, and (3) longitudinally, from upwellings that reflect stable ground-

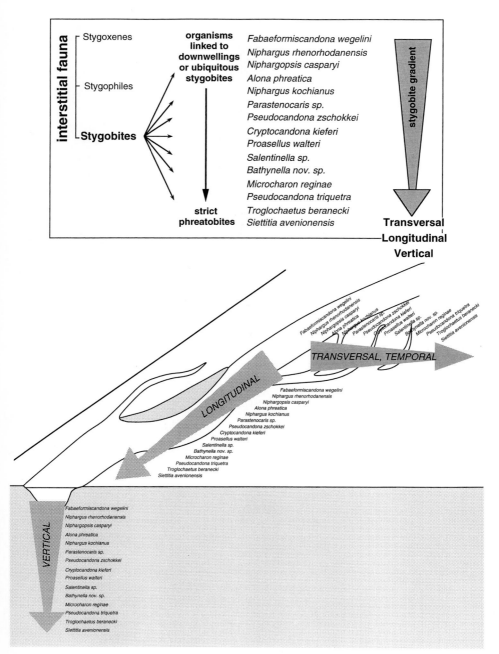

FIGURE 9   The four-dimensional nature of the stygobite gradient on the Rhône River system. Species are ordinated according to their sensitivity to environmental fluctuations. This qualitative gradient was repeated in the three spatial dimensions, longitudinal in the main channel (several scales), transversal across the floodplain, and vertical at a depth of a few meters [after Dole-Olivier *et al.* (1993)].

water conditions to downwellings that reflect variable surface water conditions. An additional dimension is time after disturbance (Fig. 9), and, as demonstrated in Marmonier and Dole (1986), the temporal succession of organisms after a spate produced the same stygobite gradient observed in the three spatial dimentions. Such a repetition of the same gradient response pattern in four dimensions is quite rare in nature and demonstrates the differential sensitivity of stygobite species to environmental fluctuations (Dole-Olivier *et al.*, 1993). This model (Fig. 9) on the floodplain scale is based only on hydrologic variability, which is a main structuring parameter in freshwater biology. The influence of other important variables, such as geomorphology, substrate characteristics, and nutrient dynamics, needs to be integrated.

IV. DYNAMICS OF THE INTERSTITIAL ASSEMBLAGES

A. Effect of Spates

The effect of hydrologic variation on the distribution and abundance of fauna was widely studied for stream benthos (e.g., Williams and Hynes, 1976; Delucchi, 1987, 1989; Scrimgeour *et al.*, 1988; Scrimgeour and Winterbourn, 1989). Models and suggestions for possible movements of epigean organisms included the interstitial milieu ("hyporheic route," Williams and Hynes, 1976; Williams, 1977; Cushing and Gaines, 1989; Delucchi, 1989). Although the inference that interstitial habitats may be refugia for zoobenthos is appealing, no studies have been forthcoming to demonstrate this function (Palmer *et al.*, 1992).

Initial studies of interstitial and groundwater assemblages of the Rhône River (Fig. 10) stressed the necessity to consider spatial heterogeneity for discussion of the effect of spates. Marmonier (1988) and Marmonier and Creuzé des Châtelliers (1991) demonstrated that this effect is different in downwelling and upwelling areas. In downwelling zones, surface water infiltration increases during spates and reduces the physical stability of subsurface sediments. A "washout effect" was observed: the abundance of animals and the number of species were strongly reduced. The postspate period was characterized by an enrichment of interstitial fauna with benthic invertebrates (at 0.5 m deep). In contrast upwellings, which are fed by groundwaters during period of low discharge, are also subjected to surface water infiltrations during spates (reversal of the piezometric gradient). In this case, a "trap effect" was observed in the sediments, which could be considered a stream "storage zone." The postspate period was characterized by an impoverishment of interstitial fauna, which is mainly colonized after a long period by stygobite species.

In order to observe the deformation of the vertical structure of the interstitial assemblages, a larger, thicker stratum was considered (deeper

STEPS →	1	2	3	4	5
added parameters	STATIONS FED BY GROUNDWATER = SINGLE HYDROLOGICAL TYPE	HYDROLOGICAL HETEROGENEITY	LARGER STICK STRATA (1.5 to 3 m)	SPATE CHARACTERISTICS (amplitude and duration)	SEASONAL EFFECT AND PRECISION OF SPATE CHARACTERISTICS
sampling characteristics	only few species one sampling level	total community one sampling level (0.5 m) upwelling and downwelling stations	total community several sampling levels (3 to 6) upwelling and downwelling stations	total community several sampling levels (4) upwelling and downwelling stations several spates (9) along the 4 seasons (16 months)	many spates several years running
bibliography	Gibert et al. 1981	Marmonier, 1988 Marmonier and Creuzé des Châtelliers, 1991	Marmonier & Dole, 1986	Dole-Olivier and Marmonier, 1992b	in perspective

FIGURE 10   Effect of hydrologic variations on interstitial assemblages: the successive steps in research on the Rhône River.

than the previous 0.5 m). The movements of both epigean and hypogean Amphipoda (*Gammarus, Niphargus, Niphargopsis,* and *Salentinella,* Marmonier and Dole, 1986) within the 1.5-m-thick substrate along a gravel bar (spate discharge, 745 m^3/sec^{-1}; minimum discharge, 30 m^3/sec^{-1}) were described. As previously described, the epigean species *Gammarus* invaded subsurface layers just after the spate and all along the gravel bar. The hypogean species *Niphargus,* which had the same distribution, was also represented in deep layers. In contrast, the phreatobite *Salentinella* was restricted to deep upwelling zones, in which it found very stable conditions. After a long period of low discharge, when the substrate was more stable, *Salentinella* could colonize subsurface layers of the upwelling zone or deep layers of the downwelling zone. In contrast *Gammarus* was gradually restricted to subsurface layers of the downwelling zone. These results demonstrated that the vertical movements of fauna are related to temporal changes in surface water flow, according to spatial hydrologic heterogeneity, and that they occurred with several meters of sediments and over several hundred meters along the gravel bar.

These vertical changes could be illustrated by a factorial correspondance analysis (Hill, 1974) (Fig. 11). In the prespate configuration, all the downwelling samples (sampled at dates 1, 2, 4, and 6) were characterized by the dominance of epigean species. They corresponded to positive values of the factorial scores, which decreased with depth (from 0.5 to 3.0 m), indicating a decrease of the epigean fauna. In contrast, upwelling zones (sampled at dates 3, 5, and 7) were characterized by the dominance of hypogean species. They corresponded to negative values, which were very small for the 0.5-m samples because these subsurface layers were partly colonized by epigean organisms. Hence, the prespate situation reflects the expected vertical structure, in relation to spatial heterogeneity. In the first postspate samples (date 8), strong modifications of vertical structure occurred and samples became similar in the upwelling and downwelling zones (i.e., high positive values of the factorial scores) for the first two levels (0.5 and 1.0 m). This situation reflected an infiltration of benthic fauna within the sediment to the 1.0 m depth. For this spate (745 m^3/sec^{-1}), the −1.5-m level remained undisturbed. Recolonization was accomplished 19 days after the spate (see date 9), when all the sampling levels presented negative values comparable to those of prespate upwelling samples.

Distortion of the vertical structure was also studied in relation to disturbance magnitude (amplitude and duration) over a 480-day period that included nine spates (Dole-Olivier and Marmonier, 1992b). Four sediment depths were sampled, from 0.5 to 2 m. All the samples from the first day after the spate were analyzed by correspondence analysis, and a qualitative gradient was defined from phreatobite and stygobite species to permanent hyporheos, occasional hyporheos, and stygoxens. The effect of each spate was measured by reference to this qualitative gradient (see Fig. 12, curved

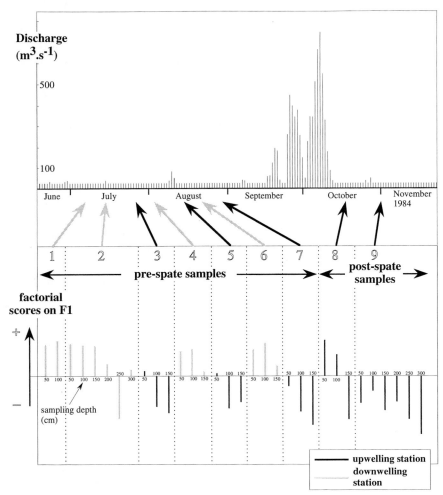

FIGURE 11   Effect of hydrologic variations on interstitial assemblages measured by correspondence analysis. (top) The hydrograph corresponding to the sampling period, with the position of the nine sampling dates. (bottom) Samples (10 l) are represented by their factorial scores on F1. Positive values correspond to samples dominated by stygoxens and species coming from occasional hyporheos; negative values correspond to samples characterized mainly by stygobites and phreatobites. The X axis corresponds to the nine sampling dates. For each sampling date, several levels are sampled (from 0.5 m to 1.5 or 3.0 m). Upwelling and downwelling stations are distinguished by two different frames.

arrows). Five classes of spates were recognized, corresponding to the five main positions of the four sampling depths along the gradient (Fig. 12). Except for one spate (spate 8), a good correspondence was observed between spate magnitude and the perturbation observed on interstitial assemblages. The impact of the spate increased with discharge and duration. The rate of

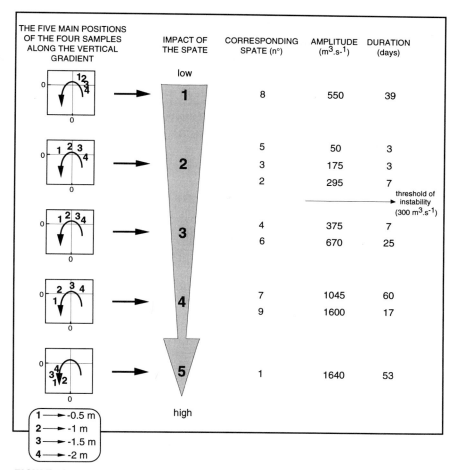

FIGURE 12 The five possible positions of the four samples along the vertical gradient (curved arrows) that measure the perturbation of interstitial assemblages on nine spates. Corresponding amplitude and duration of spates are mentioned. The discharge in "steady-state" conditions is 30 m^3/s^{-1}. The threshold of instability for the bed sediments corresponded to 300 m^3/s^{-1} (Darchis, 1979).

restructuring was also described by sampling 1, 7, and 17 days after each spate. Restructuration was measured by the displacement of the four sampling levels along the previous gradient (*i.e.*, toward the start of the curved arrow). No clear relation appeared between the dynamics of restructuration and the magnitude of spates. But it appeared that the disturbance regime (succession of spates) and the season are additional important parameters.

Studies carried out in upwelling zones were similarly conducted in downwelling zones. In this case, the infiltration of epigean organisms and hyporheos is greatest to a depth of 2 m (*Gammarus*, Chironomidae, copepods, insects larvae, etc.), with a rapid recovery observed seven days after the spate (P. Marmonier and M.-J. Dole-Olivier, unpublished results).

B. Effect of Gravel Displacement

A very common consequence of floods and spates is the displacement of bottom sediments (Lisle, 1982). In the Rhône River, the effects of these alluvial shifts have been studied in a bypassed channel of the river (Marmonier, 1991). From November 1983 to September 1984, the sediment of two gravel bars was displaced and the two sampling stations located on each gravel bar were progressively isolated from the main channel by alluvial deposits (Fig. 13): 32 m for the first station and 37 m for the second one.

At the start of the study, the interstitial and superficial waters were chemically very similar because the stations were located in a downwelling area. After isolation, the chemical characteristics of surface and interstitial

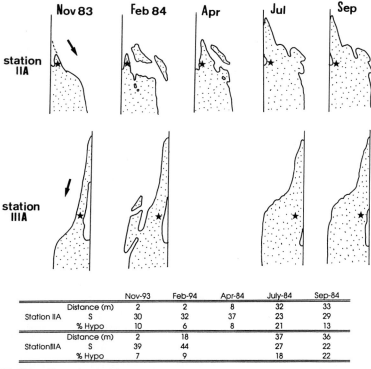

		Nov-93	Feb-94	Apr-84	July-84	Sep-84
	Distance (m)	2	2	8	32	33
Station IIA	S	30	32	37	23	29
	% Hypo	10	6	8	21	13
	Distance (m)	2	18		37	36
StationIIIA	S	39	44		27	22
	% Hypo	7	9		18	22

FIGURE 13 Effect of gravel bar displacement on interstitial assemblages: location of the sampling points (stars) in stations II-A and III-A and shape of the point bar during the study period [from November 1983 to September 1984; after Marmonier (1988)]. The table gives the distance from the sampling point to the channel, the taxonomic richness (S), and the percentage of hypogean species (% Hypo).

waters diverged (*e.g.*, the sulfate content decreased and the SiO_2 content increased in the interstitial water). These alluvial shifts produced modifications of interstitial water quality by geochemical processes (Marmonier, 1991).

The modification of gravel bar shape also induced modifications in the structure of interstitial assemblages (Fig. 13). When the stations were in contact with the river channel, there was a high rate of surface water infiltration, which brought with it numerous epigean invertebrates (such as Cladocera, Ostracoda, Ephemeroptera, and Trichoptera) within the sediments. Taxonomic richness exceeded 30 taxa and the percentage of stygobite species was low (from 6 to 10%). After isolation, the water infiltration from the river decreased and the interstitial habitats were influenced only by groundwater. The taxonomic richness decreased to less than 30 taxa and the percentage of stygobite species increased (between 13 and 22%); hence, the hypogean fauna developed more abundant and stable populations. This example demonstrated the dynamic nature of interstitial habitats and biocenoses.

C. Long-Term Changes

Studies of long-term changes in interstitial habitats are still rare. However, long-term studies are essential to understand groundwater system functioning because (1) the life cycle of hypogean animals is very long and (2) some interstitial habitats are very fragile (nonresilient) and recover slowly from disturbance (Dole and Chessel, 1986).

In the Rhône River floodplain, one example of the long-term succession of interstitial assemblages was studied in a gravel pit (J.-L. Reygrobellet, unpublished data). It was located near the abandoned Grand Gravier channel (Fig. 6), 1 km from the river, and was never flooded by the active channel. Samples were taken at 50 cm in depth, from 1979 (Seyed-Reihani, 1980) to 1985. During this period, the surface habitat was slowly modified by progressive eutrophication. Aquatic vegetation colonized the pit, and its bottom sediment was covered by a thin layer of fine sediment. The structure of interstitial fauna was also modified, as demonstrated by the ostracod assemblages (Marmonier, 1988; Fig. 14).

— From 1979 to 1983, two ostracod species were found in the interstitial layer: *Pseudocandona triquetra* (a hypogean organism) and *Pseudocandona albicans* (a crenobiont species frequent in the springs and in oligotrophic waters) (Bronshtein, 1947; Sywula, 1981).

— From 1983 to 1985, with the development of vegetation, a phytophilous species (*Cypridopsis vidua*) dominated the ostracod assemblage. During this period, the abundance of *P. albicans* decreased and it disappeared from the interstitial habitat. In contrast, the abundance of *P. triquetra* was not changed in samples until 1985.

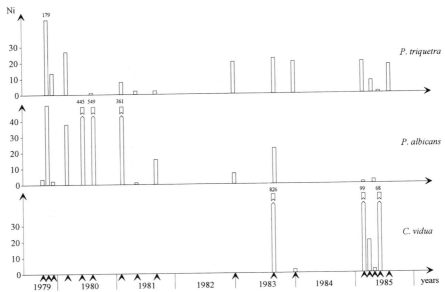

FIGURE 14 Abundance of three ostracod species in the interstitial habitat (-0.5 m deep) of a gravel pit (gp in Fig. 6) between 1979 and 1985 (Ni, number of individuals for 30-l samples).

This example illustrates the ability of interstitial assemblages to react to long-term modifications of surface water systems. These structural changes mostly concern the epigean component of the interstitial assemblages.

V. CONCLUSION: PERSPECTIVES FOR FUTURE INVESTIGATIONS

Review of the 15 years of research on the groundwater ecology of the Rhône suggests three main topics for future work. First, the impact of anthropogenic disturbances, such as pollution and river regulation, on groundwaters are still poorly documented from a biological point of view. Research on the use of stygofauna as a tool for river management is needed. Second, process-oriented studies are required to explain cause and effect in the distribution, abundance, and life histories of interstitial fauna. For example, the high diversity of habitat types and microhabitats in the floodplain is well known (Castella, 1987; Cellot *et al.*, 1994), but the corresponding groundwater habitats and distributional patterns of the interstitial fauna are yet to be described and understood. The next step in the study of spate effects could focus on seasonal variation and its role in the recolonization processes. Considering the role of the interstitial compartment in hydrosystem metabolism, research on nutrient cycling (exchanges of organic matter with the surface environment, degradation and transformation in the sediments by microbes, etc.) is needed. Finally, the interstitial assemblages of the Rhône River have to be widely studied in order to constitute a database with which to examine general ecological principles, such as the ecotone concept, the river habitat templet, and biodiversity gradients. Such research on groundwaters also should emphasize the participation of groundwater fauna in the global biodiversity and promote its protection and conservation.

ACKNOWLEDGMENTS

Financial support by CNRS (Program for Interdisciplinary Research on the Environment) is acknowledged. We thank Jack Stanford (Montana) for reviewing this chapter. The lists of stygobite species (Tables II and III) were drawn up with the helpful comments of P. Pospisil (Vienna), D. Danielopol (Mondsee, Austria), and S. Negrea (Bucharest, Romania) for the Danube River and of J. Stock (Amsterdam) for the Rhine River. Initial research on the groundwater fauna of the Rhône River was carried out by J. Gibert, R. Ginet, J. Mathieu, J.-L. Reygrobellet, and A. Seyed-Reihani. Results of these seminal studies are integrated in this chapter.

REFERENCES

Amoros, C., Roux, A. L., Reygrobellet, J.-L., Bravard, J.-P., and Pautou, G. (1987). A method for applied ecological studies of fluvial hydrosystems. *Regulated Rivers* 1, 17–36.
Botosaneanu, L., ed. (1986). "Stygofauna mundi." E. J. Brill, Dr. W. Backhuys, Leiden.
Bravard, J.-P. (1987). "Le Rhône- Du Léman à Lyon." La Manufacture, Colloq. l'Homme et la Nature, Lyon.
Bravard, J.-P., Amoros, C., and Pautou, G. (1986). Impact of civil engineering works on the successions of communities in a fluvial system. A methodological and predictive approach applied to a section of the Upper Rhône River, France. *Oikos* 47, 92–111.
Bronshtein, Z. S. (1947). "Freshwater Ostracoda," Fauna of the URSS (31), Vol. 2/1. Acad. Sci. URSS, Moscow.
Castella, E. (1987). Apport des macroinvertébrés aquatiques au diagnostic écologique des écosystèmes abandonnés par les fleuves. Recherche méthodologique sur le Haut-Rhône français. Thèse, Univ. Lyon.
Cellot, B., Dole-Olivier, M. J., Bornette, G., and Pautou, G. (1994). Temporal and spatial environmental variability in the upper Rhône River and its floodplain. *Freshwater Biol.* 31(3), 311–325.
Creuzé des Châtelliers, M. (1991a). Dynamique de répartition des biocénoses interstitielles du Rhône en relation avec des caractéristiques géomorphologiques (secteurs de Brégnier-Cordon, Miribel-Jonage et Donzère-Mondragon). Thèse, Univ. Lyon.
Creuzé des Châtelliers, M. (1991b). Geomorphologic processes and discontinuities in the macrodistribution of the interstitial fauna. A working hypothesis. *Verh. Int. Ver. Limnol.* 24(3), 1609–1612.
Creuzé des Châtelliers, M., and Reygrobellet, J.-L. (1990). Interactions between geomorphologic processes, benthic and hyporheic communities: First results on a bypassed canal of the french Upper-Rhône River. *Regulated Rivers* 5, 139–158.
Cushing, C. E., and Gaines, W. L. (1989). Thoughts on recolonization of endorheic cold desert spring-stream. *J. North Am. Benthol. Soc.* 8, 277–287.
Danielopol, D. L. (1976a). The distribution of the fauna in the interstitial habitats of riverine sediments of the Danube and the Piesting (Austria). *Int. J. Speleol.* 8, 23–51.

Danielopol, D. L. (1976b). Sur la distribution géographique de la faune interstitielle du Danube et de certains de ses affluents en basse Autriche. *Int. J. Speleol.* **8,** 323–329.

Danielopol, D. L. (1980). The role of the limnologist in ground water studies. *Int. Rev. Gesamten Hydrobiol.* **65,** 777–791.

Darchis, F. (1979). Relations entre la dynamique des fonds du Haut-Rhône français et la structure du peuplement benthique. Thèse, Ecole Nationale des Mines, Paris.

Delucchi, C. M. (1987). Comparison of community structure among streams with different temporal flow regimes. *Can. J. Zool.* **66,** 579–586.

Delucchi, C. M. (1989). Movement patterns of invertebrates in temporary and permanent streams. *Oecologia* **78,** 199–207.

Dole, M.-J. (1983a). Le domaine aquatique souterrain de la plaine alluviale du Rhône à l'Est de Lyon: Ecologie des niveaux supérieurs de la nappe. Thèse, Univ. Lyon.

Dole, M.-J. (1983b). Le domaine aquatique souterrain de la plaine alluviale du Rhône à l'Est de Lyon. 1. Diversité hydrologique et biocenotique de trois stations représentatives de la dynamique fluviale. *Vie Milieu* **33,** 219–229.

Dole, M.-J. (1984). Structure biocénotique des niveaux supérieurs de la nappe alluviale à l'Est de Lyon. *Mém. Biospéol.* **11,** 17–26.

Dole, M.-J., and Chessel, D. (1986). Stabilité physique et biologique des milieux interstitiels. Cas de deux stations du Haut Rhône. *Ann. Limnol.* **22,** 69–81.

Dole-Olivier, M.-J., and Marmonier, P. (1992a). Patch distribution of interstitial communities: prevailing factors. *Freshwater Biol.* **27,** 177–191.

Dole-Olivier, M.-J., and Marmonier, P. (1992b). Effects of spates on interstitial assemblages structure. Disturbance-perturbation relationship, rate of recovery. *Hydrobiologia* **230,** 49–61.

Dole-Olivier, M.-J., and Marmonier, P. (1992c). Ecological requirements of stygofauna in an active channel of the Rhône river. *Stygologia* **7**(2), 65–75.

Dole-Olivier, M.-J., Creuzé des Châtelliers, M., and Marmonier, P. (1993). Repeated gradients in subterranean landscape—Example of the stygofauna in the alluvial floodplain of the Rhône River (France). *Arch. Hydrobiol.* **127,** 451–471.

Eder, R. (1983). Nematoden aus dem Interstitial der Donau bei Fischamend (Niederösterreich). *Arch. Hydrobiol., Suppl.* **68,** 110–113.

Freeze, R. A., and Cherry, J. A. (1979). "Groundwater." Prentice-Hall, Englewood Cliffs, NJ.

Gaschignard-Fossati, O. (1986). Répartition spatiale des macroinvertébrés benthiques d'un bras vif du Rhône; rôle des crues et dynamique saisonnière. Thèse, Univ. Lyon.

Gibert, J., Ginet, R., Mathieu, J., Reygrobellet, J.-L., and Seyed-Reihani, A. (1977). Structure et fonctionnement des écosystèmes du Haut Rhône français. IV. Le peuplement des eaux phréatiques, premiers résultats. *Ann. Limnol.* **13**(1), 83–97.

Gibert, J., Ginet, R., Mathieu, J., and Reygrobellet, J.-L. (1981). Structure et fonctionnement des écosystèmes du Haut Rhône français. IX. Analyse des peuplements de deux stations phréatiques alimentant des bras morts. *Int. J. Speleol.* **11,** 141–158.

Gibert, J., Dole-Olivier, M.-J., Marmonier, P., and Vervier, P. (1990). Groundwater ecotones. *In* "Ecology and Management of Aquatic-Terresrial Ecotones" (R. J. Naiman and H. Décamps, eds.), Man & Biosphere Ser., Vol. 4, pp. 199–225. UNESCO, Paris.

Ginet, R. (1982). Structure et fonctionnement des écosystèmes du Haut Rhône français. XXIV. Les Amphipodes des eaux interstitielles en amont de Lyon. *Pol. Arch. Hydrobiol.* **29,** 231–237.

Godbout, L., and Hynes, H. B. N. (1982). The three dimensional distribution of the fauna in a single riffle in a stream in Ontario. *Hydrobiologia* **97,** 87–96.

Grimm, N. B. (1988). Role of macroinvertebrates in nitrogen dynamics of a desert stream. *Ecology* **69**(6), 1884–1893.

Grimm, N. B., and Fisher, S. G. (1984). Exchange between interstitial and surface water: Implications for stream metabolism and nutrient cycling. *Hydrobiologia* **111,** 219–228.

Grimm, N. B., Valett, H. M., Stanley, E. H., and Fisher, S. G. (1991). Contribution of the hyporheic zone to stability of an arid-land stream. *Verh. Int. Ver. Limnol.* **24,** 1595–1599.

Hendricks, S. P. (1993). Microbial ecology of the hyporheic zone: A perspective integrating hydrology and biology. *J. North Am. Benthol. Soc.* **12**, 70–78.

Hendricks, S. P., and White, D. S. (1988). Hummocking by lotic chara: Observations on alterations of hyporeic temperature patterns. *Aquat. Bot.* **31**, 13–22.

Hendricks, S. P., and White, D. S. (1991). Physicochemical patterns within a hyporheic zone of a Northern Michigan River, with comments on surface water patterns. *Can. J. Fish. Aquat. Sci.* **48**, 1645–1654.

Hertzog, L. (1933). *Bogidiella albertimagni sp. nov.*, ein neuer Grundwasseramphipode aus der Rheinebene bei Strassburg. I. Mitteilung. *Zool. Anz.* **102**(9/10), 225–227.

Hertzog, L. (1936). Crustaceen aus unterirdischen Biotopen des Rheintales bei Strassburg. I. Mitteilung. *Zool. Anz.* **114**(9/10), 271–279.

Hertzog, L. (1938). Crustaceen aus unterirdischen Biotopen des Rheintales bei Strassburg. III. Mitteilung. *Zool. Anz.* **123**(3), 45–56.

Hill, M. O. (1974). Correspondance Analysis: A neglected multivariate method. *J. R. Stat. Soc., Ser. C* **23**, 340–354.

Lisle, T. E. (1982). Effects of aggradation and degradation on riffle-pool morphology in natural gravel channels, Northwestern California. *Water Resour. Res.* **18**, 1643–1651.

Marmonier, P. (1988). Biocénoses interstitielles et circulation des eaux dans le sous-écoulement d'un chenal aménagé du Haut Rhône français. Thèse, Univ. Lyon.

Marmonier, P. (1991). Effect of alluvial shift on the spatial distribution of interstitial fauna. *Verh. Int. Ver. Limnol.* **24**(3), 1613–1616.

Marmonier, P., and Creuzé des Châtelliers, M. (1991). Effects of spates on interstitial assemblages of the Upper Rhône River. Importance of spatial heterogeneity. *Hydrobiologia* **210**, 243–251.

Marmonier, P., and Dole, M.-J. (1986). Les Amphipodes des sédiments d'un bras court-circuité du Rhône: logique de répartition et réaction aux crues. *Rev. Fr. Sci. Eau* **5**, 461–486.

Marmonier, P., Dole-Olivier, M.-J., and Creuzé des Châtelliers, M. (1992). Spatial distribution of interstitial assemblages in the floodplain of the Rhône river. *Regulated Rivers* **7**, 75–82.

Mathieu, J., and Amoros, C. (1982). Structure et fonctionnement des écosystèmes du Haut-Rhône français. XX. Evolution des populations de Copépodes Cyclopoïdes de deux stations phréatiques. *Pol. Arch. Hydrobiol.* **29**(2), 425–438.

Orghidan, T., Dancau, D., and Capuse, I. (1979). Cercetari asupra mediului acvatic hiporeic si a celui freatic din sectorul Defileului Dunarii inferioare. *In* "Speologia. Fauna acvatica subterana din zona defileului Dunarii" (T. Orghidan and S. Negra, eds.), pp. 87–104. Acad. R. S. Romania, Bucharest.

Palmer, M. A., Bely, A. E., and Berg, K. E. (1992). Response of invertebrates to lotic disturbance: A test of the hyporheic refuge hypothesis. *Oecologia* **89**, 182–194.

Reygrobellet, J.-L., and Dole, M.-J. (1982). Structure et fonctionnement des écosystèmes du Haut Rhône français, XVII. Le milieu interstitiel de la "lône du Grand Gravier"; premiers résultats hydrologiques et faunistiques. *Pol. Arch. Hydrobiol.* **29**(2), 485–500.

Reygrobellet, J.-L., Mathieu, J., Ginet, R., and Gibert, J. (1981). Structure et fonctionnement des écosystèmes du Haut Rhône français. VIII. Hydrologie de deux stations phréatiques dont l'eau alimente des bras morts. *Int. J. Speleol.* **11**, 129–139.

Richoux, P., and Reygrobellet, J.-L. (1986). First report on the ecology of the phreatic water beetle *Siettitia avenionensis* Guignot (Coleotera, Dytiscidae). *Entomol. Basiliensia* **11**, 371–384.

Roux, A. L. (1976). Structure et fonctionnement des écosystèmes du Haut Rhône français. I. Présentation de l'étude. *Bull. Ecol.* **7**(4), 475–478.

Roux, A. L. (ouvrage collectif publié sous la direction de) (1986). "Recherches interdisciplinaires sur les écosystèmes de la basse vallée de l'Ain (France): Potentialités évolutives et gestion." Univ. Sci. Méd. Grenoble, Grenoble.

Schumm, J. A. (1977). "The Fluvial System." Wiley (Interscience), New York.

Schwoerbel, J. (1964). Fauna freatica din vecinatatea Dunarii (fauna hiporeica). *Hidrobiologia* **5**, 157–164.

Scrimgeour, G. J., and Winterbourn, M. J. (1989). Effects of floods on epilithon and benthos macroinvertebrate populations in an unstable New Zealand River. *Hydrobiologia* **171**, 33–44.

Scrimgeour, G. J., Davidson, R. J., and Davidson, J. M. (1988). Recovery of benthic macroinvertebrate and epilithic communities following a large flood, in an unstable, braided, New Zealand River. *N. Z. J. Mar. Freshwater Res.* **22**, 337–344.

Seyed-Reihani, A. (1980). Etude écologique du milieu interstitiel lié au fleuve Rhône en amont de Lyon. Thèse, Univ. Lyon.

Seyed-Reihani, A., Gibert, J., and Ginet, R. (1982a). Structure et fonctionnement des ecosystèmes du Haut-Rhône français. XIII. Ecologie de deux stations interstitielles; influence de la pluviosité sur leur peuplement. *Pol. Arch. Hydrobiol.* **29**(2), 501–511.

Seyed-Reihani, A., Ginet, R., and Reygrobellet, J.-L. (1982b). Structure et fonctionnement des ecosystèmes du Haut-Rhône français. XXX. Le peuplement de trois stations interstitielles dans la plaine de Miribel Jonage (vallée du Rhône en amont de Lyon), en relation avec leur alimentation hydrogéologique. *Rev. Fr. Sci. Eau* **1**(2), 163–174.

Stanford, J. A., and Ward, J. V. (1993). An ecosystem perspective of alluvial rivers: connectivity and the hyporheic corridor. *J. North Am. Benthol. Soc.* **12**(1), 48–60.

Sywula, T. (1981). Ostracoda of the undergroundwater in Poland. *Rocz. Muz. Okr. Czestochova* **2**, 89–95.

Thibodeaux, L. J., and Boyle, J. D. (1987). Bedform-generated convective transport in bottom sediment. *Nature (London)* **325**, 341–343.

Turner, M. G., Dale, V. H., and Gardner, R. H. (1989). Predicting across scales: Theory development and testing. *Landscape Ecol.* **3**, 245–252.

Valett, H. M., Fischer, S. G., and Stanley, E. H. (1990). Physical and chemical characteristics of the hyporheic zone of a Sonoran Desert stream. *J. North Am. Benthol. Soc.* **9**(3), 201–215.

Vaux, W. G. (1968). Intragavel flow and interchange of water in a streambed. *Fish. Bull.* **66**(3), 479–489.

White, D. S. (1990). Biological relationships to convective flow patterns within stream beds. *Hydrobiologia* **196**, 149–158.

White, D. S., Elzinga, C. H., and Hendricks, S. P. (1987). Temperature patterns within the hyporheic zone of a northern Michigan River. *J. North Am. Benthol. Soc.* **6**(2), 85–91.

Williams, D. D. (1977). Movements of benthos during the recolonization of temporary streams. *Oikos* **29**, 306–312.

Williams, D. D., and Hynes, H. B. N. (1976). The recolonization mechanisms of stream benthos. *Oikos* **27**, 265–272.

13

The Groundwater Fauna of a Danube Aquifer in the "Lobau" Wetland in Vienna, Austria

P. Pospisil

Institute of Zoology, University of Vienna
Althanstrasse 14, 1090 Vienna, Austria

I. INTRODUCTION

In the Vienna area (Fig. 1), the Danube river is a mountain stream with high running velocity (up to 2.5 m/sec). After it breaks through the foothills of the Alps in the northern part of the city, it runs straight through the city area and, after a turn to the east, flows to the Slovakian border (Fig. 1). The straight river bed in Vienna is but 120 years old. Before the river was regulated in 1873, the Danube had shown strong furcation patterns, generating numerous side arms with lower running velocity. The area is covered with up to 20-m-thick layers of alluvial sediments accumulated by the river during and after the last ice age.

347

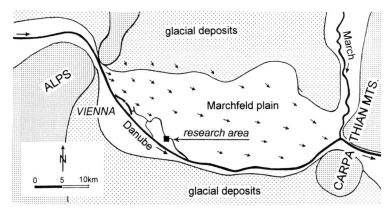

FIGURE 1   The Vienna Basin. The white area indicates postglacial river deposits. Arrows show the infiltration zones and direction of groundwater flow.

During the last 40 years the groundwater of these deposits was investigated occasionally by a few biologists, who often discovered new species. The first to take groundwater samples for faunal and ecological purposes was Spandl (1923). After World War II, Vornatscher (1972) investigated a few wells, partly those of riverine wetlands and partly those of the built-up area, both within the city borders. When examining his material, Löffler (1963) discovered a new ostracod subspecies, *Mixtacandona laisi vindobonensis,* and Kiefer (1964) established a new cyclopoid genus, *Austriocyclops vindobonae,* which has been rediscovered (Pospisil, 1994). During the mid-seventies, Danielopol (1976a,b, 1980, 1983, 1984, 1989, 1991) resumed the investigation of the groundwater of Vienna and discovered three new mite species of the genus *Schwiebea* beneath the very center of the city (Fain, 1982), the nematode *Theristus franzbergeri* (Schiemer, 1984), and a new syncarid crustacean, *Bathynella vindobonensis* Serban, 1989. Whereas Vornatscher (1972) initiated a faunal survey, Danielopol (1976b) investigated the abundance and depth profile of the groundwater fauna at the Danube riverbanks, as did Eder (1980, 1983) for the nematode fauna.

Until the late 1970s, most of the investigations focused on the gravel banks along the river. Danielopol (1983) started a project to study the groundwater near the "Eberschüttwasser," a Danube backwater 1 km away from the main channel. He chose as his location the Eberschüttwasser in the Lobau wetland. This area has since become the main research site of groundwater ecology in Austria.

Vienna holds a transitional position within Central Europe with regard to climatic, geomorphologic, botanical, and zoological factors (Ehrendorfer and Starmühlner, 1970). Even in the groundwater, species with their primary distribution in Western and Eastern Europe have been found to coexist (Danielopol, 1976a; Pospisil, 1989). Several species originally known from the Rhône–Rhine region also occur in the Vienna area, *e.g.*, *Mixtacandona laisi*, *Cryptocandona kieferi* (Ostracoda), and *Acanthocyclops sensitivus* (Cyclopoida). "Eastern" elements of the Viennese groundwater fauna are *Kovalevskiella* (Ostracoda) and *Proasellus slavus* (Isopoda).*Theristus franzbergeri* (Nematoda), on the other hand, belongs to an otherwise marine subgenus; it may be regarded as a relict of the Miocene when the Paratethys Sea covered this area.

II. GROUNDWATER HABITATS

Danielopol (1983) defined the groundwater ecosystem of the Danube aquifer using geomorphologic, geochemical, and hydrologic parameters. The area of this groundwater ecosystem generally coincides with the riverine forest ecosystem (within the Viennese city borders, the forest has been cleared except for the "Lobau"). Infiltration of groundwater into the aquifer takes place along the banks of the Danube throughout the city area. Smaller inputs to the groundwater body come from the aquifers of the Marchfeld plain and from precipitation. Output zones are the banks along the Danube downstream of Vienna, especially close to Devin and Hainburg at the Slovakian border; the evapotranspiration of the riverine forest; and withdrawal for agriculture and drinking water (Fig. 1) (Danielopol *et al.*, 1991; Pospisil and Danielopol, 1990).

The habitat of even a very small groundwater animal can theoretically have a wide extension, but during a research period it may live only within a limited area. For practical purposes, Danielopol (1983, 1989) chose different scales of habitats, in both space and time; the macrohabitat, the mesohabitat, and the microhabitat.

The macrohabitat is a large area (100 m to 1 km) in which most of the groundwater animals are expected to live all their life. This area is a representative part of the large groundwater ecosystem, containing several compartments such as the Danube riverbanks, the bed sediments of the Danube abandoned arms, and the different sediment layers within the alluvial aquifer. The macrohabitat concept is compatible with the scale of most hydrogeologic investigations (Hölting, 1984). In our case the Lobau wetland is regarded as a "macrohabitat."

The Lobau wetland is the Viennese part of the riverine wetland of the Danube. It has been severely affected by human activities: the area was cut off from the river 120 years ago, leaving the backwaters supplied only by groundwater most of the time, yet it is fed by river water during floods. Dropping of the groundwater table, caused by the withdrawal of water for municipal and agricultural use as well as the erosion of the Danube river bed in the course of the construction of a hydroelectric power plant upstream of Vienna, additionally affect this area.

There are several reasons for choosing this area for closer investigation of groundwater ecology; the most important are the vicinity of the city (providing the required infrastructure) and the fact that groundwater fauna is found regularly at this place (Pospisil and Danielopol, 1990). For the purpose of monitoring changes of the groundwater table due to a nearby drinking water plant, a monitoring well network has been implanted, making possible cheap investigations of the deeper, less easily accessible groundwater layers. Moreover, the rare occurrence of floods allows the investigation of undisturbed populations of groundwater fauna in the Lobau wetland over longer periods of time.

The mesohabitat concept refers to a smaller scale (meters). It makes it possible to trace the distribution of physicochemical factors within a more homogeneous area. Moreover, it is possible to observe the reaction of the meiomacrofauna to changing environmental parameters within shorter (seasonal) periods of time. In the Lobau wetland, the bed sediments of the Eberschüttwasser, an abandoned arm of the Danube, and the surrounding groundwater represent the mesohabitats.

The microhabitat has a small spatial and temporal extension (centimeters and hours). Within our framework, every single groundwater well can be regarded as a microhabitat and every single sample of meiomacrofauna, a representative of the fauna of a microhabitat.

III. RESEARCH SITES

So far three habitat types have been investigated in the area of the Eberschüttwasser: (A) the shallow groundwater (0 to 2 m deep) of sand–gravel bed sediments of the Eberschüttwasser (Figs. 2 and 3), (B) deeper groundwater layers (about 10 m below solid ground at a site located some 500 m away from the Eberschüttwasser and up to several 100 m distant from other surface waters) (Fig. 2), and (C) the groundwater layers (down to 12 m below ground) about 30 m from the banks of the Eberschüttwasser (Figs. 2 and 3). The two sites at the Eberschüttwasser area represent mesohabitats. The information gained from the wells at site B (Fig. 2), however, can be regarded only as isolated spot checks within a macrohabitat.

Sector A was studied by Danielopol (1983, 1984, 1991). He established a sampling grid of 24 groundwater pipes on a water-covered (0.5–1 m) area of 30 m^2 in the bed sediments of the Eberschüttwasser (Figs. 2 and 3). The standpipes had a diameter of 2.5 cm and were arranged in three rows with 4 pipes each reaching to a depth of 0.5 m below the sediment surface and in three rows with 2 pipes to a depth of 1.2 m. In addition, some pipes were fixed at random with depths of 2.65, 3, 4, 4.23, and 6.4 m. Samples were taken seasonally from the regular sampling grid and aperiodically from the other pipes. Samples (3 l) were removed with a piston pump fixed to

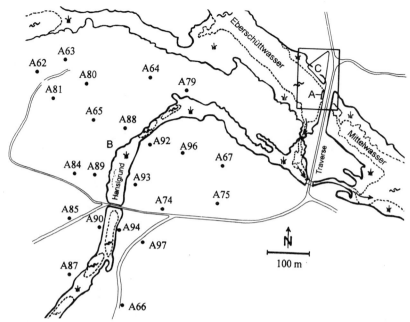

FIGURE 2 The sampling areas in the Lobau wetland. Groundwater piezometers of site B, reaching to about 10 m below the ground, are designated "A + No." The "Hanslgrund" water is almost completely covered by reeds and often dried up. For details of sites A and C, see Fig. 3.

the pipes (Bou–Rouch method). Oxygen and temperature were measured with an Orbisphere oxymeter. Particulate organic matter was determined with the loss-on-ignition method (2 hr at 520°C) and expressed as sediment weight loss; dissolved organic carbon was determined following the method of MacKereth *et al.* (1978). In order to map the spatial (*i.e.*, in a horizontal plane at one depth) distribution of the meiomacrofauna and chemical parameters, the Kriging geostatistical method was used (Delfiner and Delhomme, 1975; Delhomme, 1978).

Sector B was investigated by myself in 1988 and 1992 (Fig. 2) (Pospisil, 1989; Rogulj *et al.*, 1993). The multilevel iron wells were implanted several years ago in order to monitor the groundwater table around a nearby water withdrawal station of the Vienna Water Plants. The wells are situated at different locations with regard to the distance to surface waters (20 m to some 100 m) and the composition of surface vegetation (*Populus* forest, meadows, or *Crateagus* bush land). They are about 5 cm in diameter and are equipped with different types of filters according to videocamera investigations (P. Pospisil, unpublished). The double-packer sampler (Danielopol and Niederreiter, 1987) was used to take groundwater samples at a chosen depth in multilevel monitoring wells, as well as to measure the temperature

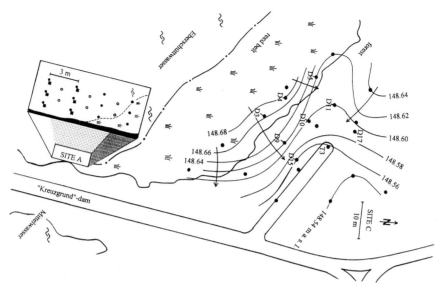

FIGURE 3 Regular sampling grids of sites A and C. (site A) Closed circles, 0.5-m-deep standpipes; closed squares, 1.2-m-deep standpipes; open circles, 2- to 3-m-deep standpipes. (site C) Wells designated "D + No." are transparent Plexiglas multilevel monitoring wells 5 to 13 m deep; the other black dots are hydrogeologic standpipes, 2 to 4 m deep with holes only near the tip. Diameters of monitoring wells and standpipes are 5 and 2.5 cm, respectively. Also shown are groundwater isohypses (meters above sea level) of January 15, 1992. Arrows indicate direction of flow. The water level of the Eberschüttwasser was 149 m above sea level; the shoreline was approximately located at the reed-forest border (a typical situation for most of the year).

and content of dissolved oxygen *in situ*. About 50 l of groundwater was usually taken and investigated for meiofauna. To gain greater amounts of animals for laboratory experiments, samples were taken without the double-packer head, thus yielding the fauna accumulating in the entire well together with large amounts of detritus.

Sector C, which was investigated in 1991–1993, is situated close to sector A in a mixed *Populus–Cornus* forest (Fig. 3). It consists of a regular sampling grid of three rows of multilevel monitoring wells 5 cm in diameter, reaching to a maximum depth of about 15 m below the surface. The pipes were implanted with a motor hammer, which is considered the least disturbing of available techniques. The grid covers an area of about 360 m² and extends from the reed belt of the Eberschüttwasser to a point some 18 m into the forest. The wells are made of transparent Plexiglas tubes and are equipped with numerous holes 5 mm in diameter over their entire surface. Scattered around the central sampling grid, about 15 smaller (2.5 cm in diameter) pipes, made of galvanized iron and reaching to depths of 2 to 4 m, were inserted mainly for hydrologic purposes, but were examined for animals as well. They are equipped with several 5-mm holes only near their tips. The total area covered by all wells is about 1600 m^2. In the small pipes, the groundwater table and the temperature (in 0.5-m-deep steps) were measured weekly; the oxygen content was determined every two weeks using a WTW electrode.

Within the main sampling grid the large wells were probed seasonally: the groundwater (5 l) and the meiofauna were extracted with the double-packer sampler; the temperature and the content of dissolved oxygen were measured *in situ* using a WTW electrode fixed to the packer head. A minivideocamera was used to survey the grain structure of the aquifer and to observe the groundwater animals in their natural environment (Niederreiter and Danielopol, 1991; Pospisil, 1992; Danielopol *et al.*, 1992, 1994).

Sampling site C is situated at a special location. The Eberschüttwasser is separated from the Mittelwasser by a dam. The Eberschüttwasser discharges into the Mittelwasser via a surface outlet to the right of the dam and via the groundwater to the left, exactly where the sampling grid is located (discharge flows indicated by arrows in Figs. 2 and 3). For that reason sector C can be regarded as a miniecosystem with hydrogeologically localized input (banks of the Eberschüttwasser) and output (Mittelwasser) areas (Danielopol *et al.*, 1992, 1994; Dreher *et al.*, 1994).

IV. CHARACTERISTICS OF GROUNDWATER HABITATS

A. Permeability

The heterogeneity of the sediments in the Lobau reflects the varying flow patterns of the river during its history. At sampling site A, gravel, sand, and silt predominate in the upper layers (above 0.5 m) of the sediments, partly causing low porosity. At a depth of 0.5 m, about 100 ml of sediment was found in 1 l of withdrawn groundwater (Danielopol, 1983). The "fine" fraction (sand and silt) considerably decreased at a depth of 3–4 m, as determined by Danielopol (1983).

Not much can be reported about the structure of the aquifer at site B. On the average, a considerably smaller amount of sediment was extracted here, only about 0.05 ml in 1 l of groundwater. The strong resistance felt during extraction of the groundwater with a piston pump could have been caused by clogging of the filter of the monitoring wells by iron bacteria, which was observed when examining one pipe with the groundwater videocamera.

A good impression of the composition (grain size) of the aquifer was gained using transparent wells and the videocamera at site C. Figure 4 shows well profiles derived from video observations. The largest part of the aquifer consists of mixed gravel and sand or silt with only very small interstices. However, coarse gravel with free interstices was found in the upper parts of many wells; the groundwater animals can probably spread rapidly in

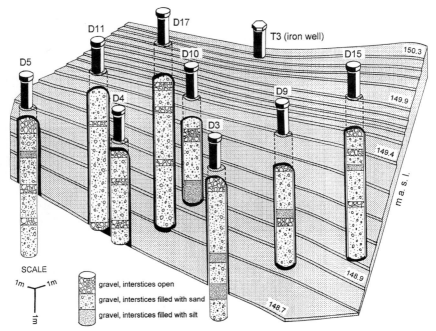

FIGURE 4 Depth profiles of Plexiglas wells of site C based on videocamera investigations. Mixed gravel–sand sediment prevails. However, some distinct layers with big interstices not filled with sand can be observed in comparable depths (most obvious in wells D3, D4, D5, D9, D10, and D15), which probably favor the spreading of the groundwater fauna. In wells D3 (4 and 6 m) and D10 (4 m) the interstices were completely filled with silt.

these layers. For some weeks after the implantation of the wells, the mixing of different groundwater layers often resulted in precipitation of iron oxides and considerable development of iron bacteria. Dark spots were observed, sometimes within densely packed sand, indicating anoxic microzones.

B. Temperature

Temperature curves for the three research sites exhibited different features. At site A, the temperature of the interstitial water of the bed sediments was closely dependent on the surface water, but showed slightly lower amplitudes. Most of the wells at site B, on the other hand, had comparatively stable temperatures (Fig. 5). At the wells at site C, which are comparable to the wells at site B as regards depth and (for T3) distance between the groundwater table and the soil surface, temperature fluctuated considerably in accordance with the surface water. During low-water periods, a division of the groundwater into two layers was sometimes recorded in the big multilevel monitoring wells. The temperature between the two groundwater layers then differed by up to 8°C in summer (the lower temperature being recorded in the deeper section).

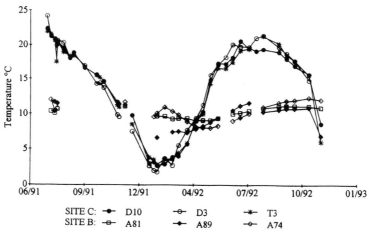

FIGURE 5   Comparison of temperature in wells D10, D3, and T3 (site C) and in wells A81, A89, and A74 (site B). For the location of these wells, see Figs. 2 and 3.

C. Specific Conductance

Measuring the specific conductance of the groundwater turned out to be a good method (along with temperature) for tracing the influence of the surface water on groundwater at site C. In most cases, the specific conductance of the groundwater was only slightly higher than that of the Eberschütt-wasser (about 430–490 μS/cm), but in well D5 (Fig. 3), below a persistent discontinuity layer at a depth of about 3.5 to 4 m, higher values (>500 μS/cm) were measured regularly. The specific conductance pattern at site C changed during a stagnant low water period (July–November 1992): sampling in October revealed very low specific conductance in the Eberschütt-wasser (about 320 μS/cm) on the one hand and high specific conductance (>500 μS/cm) not only in D5, but also this time in all wells, reflecting the rising influence of the inland groundwater.

D. Oxygen Concentration

The oxygen content of the interstitial water of the sediments at site A is always lower than that of the surface water. Figure 6 depicts seasonal values at different depths. At 0.5 m in depth, values greater than 1 mg/l dissolved oxygen were measured, whereas at greater depths the values were considerably lower and hypoxic conditions prevailed (between 0.1 and 0.5 mg/l). The oxygen values were strongly fluctuating at 0.5 m in depth, but were more stable at greater depths. Figure 7 depicts an example of the different horizontal oxygen concentrations at 0.5 and 2 m in sediment depth based on Kriging mapping. Interestingly, the oxygen content was comparatively high below the anoxic mud of the reed belt at 0.5 m.

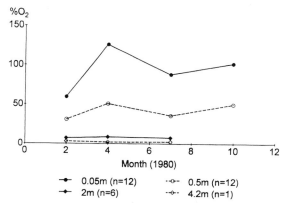

FIGURE 6   Seasonal fluctuation of oxygen concentration of the groundwater in different sediment depths at site A [data from Danielopol (1983)].

The deeper groundwater layers (site B) frequently exhibited very low oxygen concentrations during the summer sampling period in 1988. Values of about 0.2 mg/l dissolved oxygen and even less were found across the whole area, with the exception of well A81 (see Fig. 2), where higher values were recorded (about 1.5 mg/l O_2). During 1992 the same pattern was observed (including the peculiarity of well A81). A slight increase of oxygen

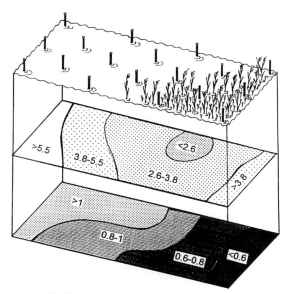

FIGURE 7   Horizontal distribution of oxygen concentration (mg/l) at site A in 0.5- and 2-m-deep sediment on February 12, 1980. The sampling grid and the approximate location of the reed belt are indicated. Based on maps generated with the Kriging method (Danielopol, 1983).

content was noticed in the cold season in some wells. The low oxygen content of the groundwater at site B seems to be a rather persistent condition.

At sampling site C, frequent measurements revealed values of about 0.2 mg/l dissolved oxygen from spring to autumn at all depths, in both multilevel monitoring wells and the small piezometers. As in some wells at site B, the oxygen content increased during the colder season (about 2 mg/l O_2 in January, but with short-timed peaks higher than 10 mg/l).

During periods of high oxygen content for the groundwater, only a slight vertical gradient was observed (a decrease with depth) in the large multilevel monitoring wells. Although these wells had high oxygen concentrations down to a depth of 10 m and more, the small hydrogeologic stand-pipes (maximum, 4 m deep) revealed mostly low values, except in two wells that have very close hydrologic connections with the surface water and in wells reaching only to the upmost groundwater stratum. The high values in the big wells recorded throughout their length, therefore, are in all likelihood a local phenomenon, caused by mixing of different groundwater layers (Pospisil, 1992).

E. Effects of Floods

Floods have an important influence on the physical parameters of the research sites. This influence depends on the position of the wells. In wells located close to a backwater, the temperature decreased rapidly because of an intrusion of surface water [some wells at site B, A89 and A90 (Fig. 2), and all wells at site C]. Wells that are located far (approximately 300 m) from backwaters [many wells at site B, A81 and A74 (Fig. 2)] showed almost no reaction with regard to the temperature, although the whole area was covered with water. Flooding also resulted in an increase of dissolved oxygen in the upper groundwater layers. The persistence of higher oxygen concentrations depended on the temperature of the groundwater. During the winter, higher oxygen concentrations prevailed for weeks at site C, whereas the groundwater temperature was low, at about 0.5°C. Summer floods led only to short-time peaks of oxygen, which was rapidly depleted owing to higher temperatures and increased microbial metabolic rates. This was demonstrated in May 1991; after an extraordinarily heavy rainfall and subsequent rising of the water table of the Eberschüttwasser, the oxygen content was comparatively high (about 2 mg/l) in wells D3 and D10. Three days later, the oxygen content had declined to about 0.7 mg/l (the groundwater temperature was about 15°C). In one groundwater well at site B, located far from surface waters [A81 (Fig. 2)], rather strange conditions were recorded after a November flood (1992). The oxygen concentrations rose drastically only in the deeper layer (6.5 m and deeper below the surface) two days after the flood, whereas the layers above 6.5 m remained hypoxic. One month later, this pattern was exactly reversed. All this time, the temperature remained almost constant (10–11°C) throughout the well (Pospisil, 1994).

V. DISTRIBUTION AND DYNAMICS OF GROUNDWATER ANIMAL ASSEMBLAGES

Table I lists the species discovered at sites A, B, and C. In the bed sediments of the Eberschüttwasser (site A), 86 species were discovered, of which only 7 were stygobites. In the deeper groundwater layers (site B), considerably fewer species were discovered, but most of the 21 identified were stygobites. At site C, an intermediate number of species was found (35 species).

The composition of the hypogean meiofauna differed remarkably among the three research sites: the ostracod *Mixtacandona spandli* Rogulj and Danielopol, the cyclopoid *Acanthocyclops gmeineri* Pospisil and *Austriocyclops vindobonae* Kiefer were discovered in wells at site B, but were never found in the bed sediments of the Eberschüttwasser (sector A) or at nearby site C. An analysis of the cyclopoids of the *Diacyclops languidus/languidoides* group at sites B and C (dominating the cyclopoid fauna at all sites) revealed five species altogether. Only one of them was present in both sampling areas (Pospisil, 1994). A preliminary examination of the cyclopoids collected by Danielopol (1983) at site A suggests the occurrence of the same cyclopoids as those at site C.

At site B, harpacticoids were represented only sporadically by the stygobite *Nitocrella hirta* Chappuis. It is noteworthy that Kiefer (1964), examining samples taken in the Vienna area by Vornatscher (1972), mentioned the "strange" fact that harpacticoids were completely absent. At site C, the harpacticoids were represented by small numbers of *Parastenocaris sp.*, *P. fontinalis* Schnitter and Chappuis, and *P. phyllura* Kiefer. They were found in sediment layers with low permeability in wells D10 and D15. Additionally, surface-dwelling harpacticoids were discovered three weeks after a strong flood in wells located in the reed belt (D3 and D4), together with planktonic Rotatoria. Surface-dwelling harpacticoids also occurred in well D3 after four months of very low water, which almost resulted in the drying up of the Eberschüttwasser.

On the whole, the ostracods and cyclopoids were the prevailing animal groups at all sites. *Cryptocandona kieferi* Klie was the most abundant ostracod throughout the research area. Among the cyclopoids, the *D. languidus/languidoides* group predominated. Sometimes other animals were highly abundant locally, *e.g.*, amphipods *(Niphargus sp.)* and hydrobiid groundwater snails (*Lobaunia danubialis* Haase and *Bythiospeum sp.*) (Haase, 1993) in wells at site B.

At sampling site A, the meiofauna abundance decreased with depth,

TABLE I List of Species at Sites A, B, and C

Site A	Site B	Site C
Turbellaria		
Bothrioplana semperi	ND	ND
Macrostomum obtusum		
Polycelis aff. *tenuis*		
Prorhynchus stagnalis		
Stenostomum aff. *leucops*		
Stenostomum aff. *unicolor*		
Tricladida sp.		
Typhloplanida sp.		
Oligochaeta		
Aelosoma sp.	ND	ND
Dorydrilus michaelseni		
Nais elinguis		
Pristina sp.		
Psammoryctides barbatus		
Tubifex tubifex		
Nematoda		
Achromadora terricola	ND	ND
Chromadorita bioculata		
Chromadorita leuckarti		
Dorylaimus oxycephaloides		
Dorylaimus sp.		
Hemicycliophora typica		
Ironus longicaudatus		
Monhystra dispar		
Monhystra sp.		
Monochus truncatus		
Plectus cirratus		
Plectus sp.		
Prismatolaimus dolichurs		
Teratocephalus terrestris		
Tobrilus gracilis		
Tobrilus pellucidus		
Tripyla filicaudata		
Tripyla glomerans		
Mollusca		
Bythiospeum sp.	*Bythiospeum* sp.	*Bythiospeum* sp.
Lobaunia danubialis		
Cyclopoida		
Acanthocyclops rhenanus[a]	*Acanthocyclops gmeineri*	*Acanthocyclops sensitivus*
Acanthocyclops robustus	*Acanthocyclops sensitivus*	*Acanthocyclops venustus*
Acanthocyclops sensitivus	*Acanthocyclops venustus*	*Diacyclops* aff. *languidoides* (one
Acanthocyclops sp.	*Austriocyclops vindobonae*	species)
Acanthocyclops venustus	*Diacyclops* aff. *languidoides*	*Diacyclops* aff. *languidus* (two
Acanthocyclops vernalis	(three species)	species)
Diacyclops bicuspidatus	*Eucyclops graeteri*	*Diacyclops bicuspidatus*
Diacyclops languidoides		*Graeteriella unisetigera*
Diacyclops languidus		*Macrocyclops albidus*
Eucyclops serrulatus		
Eucyclops sp.		
Mesocyclops leuckarti		
Paracyclops fimbriatus		
Paracyclops sp.		
Thermocyclops sp.		
Harpacticoida		
Attheyella crassa	*Nitocrella hirta*	*Bryocamptus minutus*
Bryocamptus minutus		*Canthocamptus staphylinus*

(*continues*)

TABLE I (continued)

Canthocamptus staphylinus		Elaphoidella gracilis
Elaphoidella gracilis		Parastenocaris fontinalis
Epactophanes richardi		Parastenocaris phyllura
Nitocra hibernica		Parastenocaris sp.
Parastenocaris sp.		Phyllognathopus vigiurei
Ostracoda		
Candona candida	Cryptocandona kieferi	Candona candida
Candonopsis kingsleii	Kovalevskiella sp.	Candona neglecta
Cryptocandona kieferi	Mixtacandona laisi vind.	Candonopsis kingsleii
Cyclocypris	Mixtacandona spandli	Cryptocandona kieferi
Cypria sp.	Pseudocandona eremita[b]	Cyprideis sp.[c]
Cypridopsis vidua	Pseudocandona lobipes	Cypridopsis vidua
Darwinula stevensoni		Fabaeformiscandona wegelini
Dolerocypris sp.		Kovalevskiella sp.
Eucypris sp.		Mixtacandona laisi vind.
Fabaeformiscandona fragilis		Mixtacandona spandli[b]
Isocypris sp.		Potamocypris sp.
Kovalevskiella sp.		Pseudocandona albicans
Limnocythere inopinata		Pseudocandona lobipes
Limnocythere stationis		
Mixtacandona laisi vindobonensis		
Potamocypris sp. (aff. variegata)		
Pseudocandona aff. albicans		
Pseudocandona compressa		
Pseudocandona lobipes		
Pseudocandona sp.		
Amphipoda		
Niphargus sp.	Niphargus bajuvaricus	Niphargus sp.
Niphargopsis sp.	Niphargus cf. kochianus	
	Niphargus inopinatus	
	Niphargus stygius/tatrensis	
Isopoda		
Asellus aquaticus	Proasellus slavus	Proasellus slavus
Proasellus slavus		Proasellus strouhali
Acari		
Arrenurus sp.	ND	Partly determined
Lobohalacarus weberi quadriporus		Schwiebea sp.
Lobohalacarus weberi weberi		
Neoacarus hibernicus		
Porolohmannella violacea		
Insect larvae		
Bidessus sp.	None	Cyphon palustris
Caenis maesta		
Ceratopogonidae gen. sp.		
Cyphon sp.		
Orthocladinae gen. sp.		
Tanipodinae gen. sp.		

Note. Because none of the sites has yet been completely analyzed, this table intends only to give an impression of the diversity of animal assemblages in the Lobau groundwater and of the differences at the three research sites [Site A from Danielopol (1983)]. ND, not determined.

[a] Uncertain determination.

[b] Valves only.

[c] Fossilized.

which is a common pattern in the bed sediments of surface waters [for data on the Danube river in Austria, see Bretschko (1987, 1992)]. The highest densities of animals occurred in samples taken from 0.5 m; at 1.2 m the density had decreased considerably; and at 2 to 6 m in depth the meiofauna was observed only occasionally. An analysis of the fauna demonstrated that both the epigean and the hypogean animals were most abundant in absolute numbers at a depth of 0.5 m, contrary to expectations that "true" hypogean animals are more abundant in deep groundwater layers. Hypogean ostracods sometimes outnumbered epigean ostracods even at 0.5 m in depth; the percentage of hypogean ostracods further increased with the depth of the sediment. On the other hand, the vertical distribution of epi- and hypogean cyclopoids was less distinct. The absolute numbers of ostracods and cyclopoids in 3-l samples from site A were often higher (100–800 specimens; Danielopol, 1983) than those in 50-l samples taken from the "deep" groundwater wells at site B (up to 50 specimens; Pospisil, 1993). This may be partly due to oligotrophic conditions in the deep groundwater. At site C, the abundance of meiofauna varied markedly between seasons. It was highest in the spring, when hundreds of animals were collected in several 5-l samples at some wells.

Insufficient oxygen supply may be another reason for the low abundance of the meiofauna. At site A, a significant correlation was observed between oxygen concentration and the abundance of meiofauna (Danielopol, 1983). On the contrary, no such correlation was evident for the deep groundwater wells: the abundance of the meiofauna was comparatively high in well A81 (situated at site B), which was better supplied with oxygen (about 1.5 mg/l), but similar densities of animals were recorded in some wells with a low oxygen content (about 0.1 mg/l; Pospisil, 1989). At site C, no correlation between oxygen and animal abundance was observed either, but the fluctuating oxygen dynamics caused difficulties in the evaluation of seasonal sampling data: the fauna may thrive during periods of better oxygen supply but, when the oxygen conditions deteriorate, the meiofauna can still survive for a considerable time.

At site C, the only indication for a marked depth distribution of animal abundance was observed in October 1992. The animals (cyclopoids and harpacticoids) had accumulated directly below the groundwater table, probably trying to avoid the anoxic conditions of the groundwater which had persisted for months.

The availability of oxygen may influence the horizontal distribution of some meiofauna species. At site A, the ostracod *C. kieferi* was most abundant in the wells with the highest oxygen supply throughout the research period (Danielopol, 1991). (Although this relation was significant, it has since been established that *C. kieferi* can also survive under low oxygen concentrations at site C.)

The small, slowly moving ostracod *Kovalevskiella sp.* was consistently found in a well-delimited area within the 30-m² sampling grid at site A, avoided by *C. kieferi*. An infiltration experiment revealed low porosity at the place preferred by *Kovalevskiella*. Apparently the interstices there were too small for a comparatively big ostracod like *Cryptocandona* (Danielopol, 1983). The size of the interstices may also limit the distribution of groundwater amphipods and isopods, which were discovered at all sampling sites but were rare at site A. The cyclopoids exhibited a fluctuating horizontal distribution that cannot easily be explained by the distribution of the parameters discussed above, such as the concentration of dissolved oxygen and organic matter, temperature, and porosity. It is likely that the response of this highly mobile animal group to varying environmental factors is too fast to be recorded by sampling on a seasonal basis.

The horizontal distribution of animals at site C is mainly determined by the influence of the surface water, which varies depending on the hydrologic situation. The wells in the reed belt are regularly inhabited by harpacticoids, cyclopoids, and ostracods typical of surface waters. Occasionally, larvae of the helodid beetle *Cyphon palustris* Thompson were discovered in many of the wells (most often in those in the reed belt), which they invaded from the banks of the Eberschüttwasser (Klausnitzer and Pospisil, 1991). Especially after floods, numerous surface water dwellers appear, even including plankton such as the cladoceran *Bosmina sp.* and the copepod *Diaptomus sp.*

Seasonal variations in the abundance of species at site A suggest that the animals move actively in search of better environmental conditions (Danielopol, 1991). Passive dispersal seemed to be indicated by the occurrence of groundwater amphipods in shallow surface water around well D10 during a flood at site C in August 1991.

VI. CONCLUSION

The data presented demonstrate that the Danube riverine aquifer in the Lobau wetland can tentatively be divided into two units: the upper bed sediments of the backwaters, with high fluctuations of physical and chemical parameters, and the deeper groundwater layers (exceeding about 2 m in depth) of the aquifer remote from and rarely influenced by surface waters, providing more stable environmental conditions.

The bed sediments are important distribution areas for some hypogean ostracods and cyclopoids, which generally exhibit high abundances. In the "deeper" groundwater, lower faunal densities are encountered.

The two units are also distinguished by the presence or absence of certain groundwater animals. The pattern of these distributions is not sufficiently understood. Ecophysiological characteristics (e.g., adaptation to low oxygen concentrations and low food supply) or competitive interactions are possible causes that have to be more closely investigated (Rogulj and Danielopol, 1993; Rogulj *et al.*, 1993; see also Chapter 8). Moreover, the distributional patterns of certain ostracods may reflect historic changes in the river bed (D. L. Danielopol, personal communication). The furcating river, the changing flow patterns, and the heterogeneity of the river aquifer in general may also lead to small-scale isolation and therefore may favor endemisms and speciation of genetically instable, radiating taxa, such as the *D. languidus/ languidoides* group. Far more extensive investigations are necessary to examine the general validity of this hypothesis.

Study of the transitional zone at site C is expected to give us some idea of how the faunal communities of the bed sediments and the deep inland groundwater are connected. However, to gain a comprehensive view, the observation area should be extended farther inland to a point at which the key parameters are similar to those of deep groundwater sector B. It will also be necessary to extend the investigations to other habitats of the same type in the Lobau.

The Lobau wetland is strongly affected by humans. Only small-scale floods occur there. The investigation of a more "natural" wetland, such as can be found only a few kilometers downstream of the present research area, might appear more promising. Mighty annual floods possibly have destructive effects on some groundwater habitats that have yet to be explored. Moreover, the intention to establish a national park on this valuable riverine forest wetland along the Danube river east of Vienna makes this point of view even more attractive (Pospisil and Danielopol, 1990).

On the other hand, the Lobau wetland is a very sensitive area: drinking water is withdrawn there, and it is threatened by oil pollution from a nearby Danube harbor and from the seepage of landfills some kilometers upstream. Currently a hydroelectric power plant is being constructed on the Danube in the vicinity of the research area. Its effects on the groundwater ecosystem are unpredictable. In the course of damming the Danube, river water will be available for groundwater enrichment measures. An artificial water enrichment project, currently undergoing a test phase, will again supply the backwaters with Danube water (Schiel, 1992; Imhof *et al.*, 1992; Schiemer *et al.*, 1992). The water level of the backwaters (*e.g.*, the Eberschüttwasser) is expected to rise, probably leading to conditions in the neighboring groundwater currently encountered only during floods. If research is continued on the Eberschüttwasser, this will be a unique opportunity to monitor the combined biological and hydrologic effects of water enrichment measures on groundwater. Groundwater ecological investigations in this area can thus be viewed in a more applied and practical context.

The discovery of at least four new or rare species of groundwater animals in the Lobau wetland highlights the importance of groundwater protection in the Vienna area. But we do not know enough about the effects of certain human interferences on the groundwater fauna; perhaps the "protected" bed sediments of the Lobau backwaters have a much higher abundance of groundwater organisms than the "natural" backwaters farther downstream, affected by floods. This view, however, may be true for the moment only. These floods, destructive as they may appear, prevent the clogging of the bed sediments by organic detritus, thus maintaining the water circulation, whereas the bed sediments of the Lobau backwaters will be clogged, sooner or later . The living conditions for the groundwater meiofauna will then deteriorate. Due to lack of comparative data from different locations one can only speculate about the significance of such effects on the groundwater fauna of the Danube riverine wetland.

ACKNOWLEDGMENTS

Many people contributed to the research presented in this chapter. H. Löffler provided the initial impetus to resume the examination of the groundwater of the Lobau. The investigations of site C were managed by D. Danielopol; J. Dreher was my coworker. T. Glatzel, M. Haase, G. Karaman, B. Rogulj, R. Rouch, and H. K. Schminke helped in determining the species of animals. I am especially obliged to the staff of the Lobau Forest Administration (Mr. Schreckeneder, Mrs. Placho, and Mr. Tomsic), who supported our work in every possible way. B. and G. Gollmann revised the English draft. Financial support was received from Project 7881-Bio of the Austrian "Fonds zur Förderung der wissenschaftlichen Forschung."

REFERENCES

Bretschko, G. (1987). Vorbereitende Untersuchungen zur Vertikalverteilung der Fauna in den Schottersedimenten der Stauwurzel Altenwörth. 26. *Arbeitstag. Internat. Arbeitsgem. Donauforsch., Passau,* pp. 103–108.

Bretschko, G. (1992). The sedimentfauna in the uppermost parts of the Impoundment "Altenwörth" (Danube, stream km 2005 and 2007). *Arch. Hydrobiol., Suppl.* **84,** (Veröeff. Arbeitsgem. Donauforsch. 8) (2–4), 131–168.

Danielopol, D. L. (1976a). Zoogeographische Probleme der Grundwasserfauna der Donau und ihrer Zuflüsse in Österreich. *Sitzungsber. Österr. Akad. Wiss., Math.-Naturwiss. Kl.* **13,** 203–208.

Danielopol, D. L. (1976b). The distribution of the fauna in habitats of riverine sediments of the Danube and Piesting (Austria). *Int. J. Speleol.* **8,** 23–51.

Danielopol, D. L. (1980). The role of the limnologist in groundwater studies. *Int. Rev. Gesamten Hydrobiol.* **65,** 777–791.

Danielopol, D. L. (1983). Der Einfluss organischer Verschmutzung auf das Grundwasserökosystem der Donau im Raum Wien und Niederösterreich. *Forschungsber., Bundesminist. Gesund. Umweltschutz* **5,** 155 pp.

Danielopol, D. L. (1984). Ecological investigations on the alluvial sediments on the Danube in the Vienna area—a phreatobiological project. *Verh. Int. Ver. Limnol.* **22,** 1755–1761.

Danielopol, D. L. (1989). Groundwater fauna associated with riverine aquifers. *J. North Am. Benthol. Soc.* **8,** 18–35.

Danielopol, D. L. (1991). Spatial distribution and dispersal of interstitial Crustacea in alluvial sediments of a backwater of the Danube at Vienna. *Stygologia* **6,** 97–110.

Danielopol, D. L., and Niederreiter, R. (1987). A sampling device for groundwater organisms and oxygen measurement in multi-level monitoring wells. *Stygologia* **3,** 252–263.

Danielopol, D. L., Pospisil, P., and Dreher, J. (1991). Ecological basic research with potential application for groundwater management. *IAHS Publ.* **202**, 215–228.

Danielopol, D. L., Dreher, J., Gunatilaka, A., Kaiser, M., Niederreiter, R., Pospisil, P., Creuzé des Châtelliers, M., and Richter, A. (1992). Ecology of organisms living in a hypoxic groundwater environment at Vienna (Austria); methodological questions and preliminary results. *In* "Ground Water Ecology" (J. A. Stanford and J. J. Simons, eds.), pp. 79–90. Am. Water Resour. Assoc. Bethesda, MD.

Danielopol, D. L., Rouch, R., Pospisil, P., Torreiter, P., and Moeszlacher, F. (1994). Ecotonal animal assemblages; their interest for groundwater studies. *In* "Groundwater/Surface Water Ecotones" (J. Gibert *et al.*, eds.). Cambridge Univ. Press, Cambridge, UK (in press).

Delfiner, P., and Delhomme, J. P. (1975). Optimum interpolation by Kriging. *In* "Display and Analysis of Spatial Data" (J. C. Davis, ed.), pp. 96–114. Wiley, London.

Delhomme, J. P. (1978). Kriging in the hydrosciences. *Adv. Water Res.* **1**, 251–266.

Dreher, J., Pospisil, P., and Danielopol, D. (1994). The role of hydrology in defining a groundwater ecosystem within the wetland of the Danube at Vienna. *In* "Groundwater/Surface Water Ecotones" (J. Gibert *et al.*, eds.). Cambridge Univ. Press, Cambridge, UK (in press).

Eder, R. (1980). Beiträge zur Kenntnis der interstitiellen Nematodenfauna am Beispiel eines Schotterkörpers der Donau bei Fischamend. Thesis, Univ. Wien.

Eder, R. (1983). Nematoden aus dem Interstitial der Donau bei Fischamend (Niederösterreich). *Arch. Hydrobiol., Suppl.* **68**, 100–113.

Ehrendorfer, F., and Starmühlner, F., eds. (1970). "Naturgeschichte Wiens," Vol. 1. Jugend und Volk, Wien.

Fain, A. (1982). Cinq espèces du genre *Schwiebea* Oudemans (Acari, Astigmata), dont trois nouvelles découvertes dans des sources du sous-sol de la Ville de Vienne (Autriche) au cours des travaux du métro. *Acarologia* **23**, 359–371.

Haase, M. (1993). *Hauffenia kerschneri* (St. Zimmerman 1930): 2 Arten zweier Gattungen. *Arch. Moll.* **121**, 91–109.

Hölting, B. (1984). "Hydrogeologie," 2nd ed. Enke, Stuttgart.

Imhof, G., Schiemer, F., and Janauer, G. A. (1992). Dotation Lobau-Begleitendes ökologisches Versuchsprogramm. *Österr. Wasserwirtsch.* **44**(11/12), 289–299.

Kiefer, F. (1964). Zur Kenntnis der subterranen Copepoden (Crustacea) Österreichs. *Ann. Naturhist. Mus. Wien* **67**, 477–485.

Klausnitzer, B., and Pospisil, P. (1991). Larvae of *Cyphon sp.* (Coleoptera, Helodidae) in Groundwater. *Aquat. Insects,* **13**, 161–165.

Löffler, H. (1963). Beiträge zur Fauna Austriaca: I. Die Ostracodenfauna Österreichs. *Sitzungsber. Österr. Akad. Wiss. Math.-Naturwiss. Kl., Abt. 1* **172**, 193–211.

MacKereth, F. J. H., Heron, J., and Talling, J. F. (1978). Water analysis: Some revised methods for limnologists. *Sci. Publ.—Freshwater Biol. Assoc.* **36**, 1–136.

Niederreiter, R., and Danielopol, D. L. (1991). The use of mini-videocameras for the description of groundwater habitats. *Mitt. Hydrogr. Dienst. Österr.* **65/66**, 85–89.

Pospisil, P. (1989). *Acanthocyclops gmeineri* n. sp. (Crustacea, Copepoda) aus dem Grundwasser von Wien (Österreich): Bemerkungen zur Zoogeographie und zur Sauerstoffsituation des Grundwassers am Fundort. *Zool. Anz.* **223**, 220–230.

Pospisil, P. (1992). Sampling methods for groundwater animals of unconsolidated sediments. *In* "The Natural History of Biospeleology" (A. I. Camacho, ed.), Monogr. 7, pp. 109–134. Mus. Nac. Cienc. Nat. C.S.I.C., Madrid.

Pospisil, P. (1994). Die Grundwassercyclopiden der Lobau in Wien (Österreich); faunistische, taxonomische und ökologische Untersuchungen. Thesis, Univ. Vienna.

Pospisil, P., and Danielopol, D. L. (1990). Vorschläge für den Schutz der Grundwasserfauna im geplanten Nationalpark "Donauauen" östlich von Wien, Österreich. *Stygologia* **5**, 75–85.

Rogulj, B., and Danielopol, D. L. (1993). Three new *Mixtacandona* (Ostracoda) species from Croatia, Austria and France. *Vie Milieu* **43**(2–3), 145–154.

Rogulj, B., Danielopol, D. L., Marmonier, P., and Pospisil, P. (1993). Adaptive morphology, ecology and biogeographical distribution of the species group *Mixtacandona hvarensis* (Ostracoda, Candoninae). *Mém. Biospéol.* **20**, 195–207.

Schiel, W. (1992). Dotation Lobau-Perspektiven einer interdisziplinären Langzeitplanung. *Österr. Wasserwirtsch.* **44**(11/12), 287–288.

Schiemer, F. (1984). *Theristus franzbergeri* n. *sp.*, a groundwater nematode of marine origin from the Danube. *Arch. Hydrobiol.* **101**, 259–263.

Schiemer, F., Pokorny, J., Gätz, N., Pospisil, P., and Christoff-Dirry, P. (1992). Limnologische Gesichtspunkte bei der Beurteilung von Augewässerdotationen. *Österr. Wasserwirtsch.* **44**(11/12), 300–306.

Serban, E. (1989). Taxa nouveaux des Bathyllenides d'Europe (Bathynellacea, Podophallocarida, Malacostraca). *Trav. Inst. Spéol. "Emile Racovitza"* **28**, 3–17.

Spandl, H. (1923). Über die Fauna der unterirdischen Gewässer. *Verh. d. Zool. Bot. Ges.* **73**, 43–45.

Vornatscher, J. (1972). Die Tierwelt des Grundwassers—Leben im Dunkeln. *In* "Naturgeschichte Wiens" (F. Ehrendorfer and F. Starmühlner, eds.), Vol. 2, pp. 659–669. Jugend und Volk, Wien.

14

Ecology of the Alluvial Aquifers of the Flathead River, Montana

J. A. Stanford,* J. V. Ward,† and B. K. Ellis*

* Flathead Lake Biological Station
The University of Montana
Polson, Montana 59860

† Department of Biology
Colorado State University
Fort Collins, Colorado 80523

I. INTRODUCTION

Stanford and Gaufin (1974) reported the presence of abundant insect larvae deep in the alluvia of the Tobacco River, a small gravel bed river in northwestern Montana. The insects were entrained in an infiltration gallery that provided potable water from the alluvial aquifer to the town of Eureka, Montana. The collection system was located ca. 150 m from the river and 3 m below the floodplain surface. The most abundant forms were stone flies (Plecoptera) of the genera *Isocapnia* and *Paraperla*. However, many other taxa, traditionally thought of as river macrobenthos, were also present, either actively or passively entering the collection system from the interstitial aquifer of the river. The discovery, coupled with the wide range in western North America of adult *Isocapnia* and *Paraperla,* led the authors to propose that the deep penetration of benthos into the saturated alluvia of gravel bed rivers was likely a widespread phenomenon that had significant ramifications for river ecology.

A decade later Stanford and Ward (1988) confirmed the occurrence of the same Plecoptera, along with over 70 other invertebrate taxa, in the shallow alluvial aquifer of the much larger Flathead River in the Kalispell Valley (Fig. 1); Plecoptera larvae were pumped from drilled wells on the valley bottom up to 3 km from the river channel. Others working on interstitial fauna of gravel bed streams and rivers in Europe (Bretschko and Klemens, 1986; Danielopol, 1976; see also Chapters 12 and 13), Canada (Coleman and Hynes, 1970; Pugsley and Hynes, 1986; Williams and Hynes, 1974), and elsewhere in the United States (Boulton *et al.,* 1991, 1992; Palmer, 1990; Pennak and Ward, 1986; Ward and Voelz, 1990; see also Chapter 15) also demonstrated speciose interstitial assemblages, but at the scale of meters or less. Based on our work summarized below and research on the Rhône River in France (Marmonier *et al.,* 1992; Dole-Olivier *et al.,* this volume), it is clear that a diverse and abundant interstitial fauna may inhabit the alluvium of entire floodplains kilometers from the river channel. Our long-term studies in the Flathead catchment now emphasize the four-dimensional nature of the riverine and floodplain landscapes (Ward, 1989) and the processes and implications related to water and materials exchange between floodplain alluvial aquifers and the river (Stanford and Ward, 1993).

The purpose of this chapter is to summarize the data and interpretations linking the groundwaters and surface waters in the Flathead catchment. Much of the work remains unpublished, and some original data are included herein.

II. BIOPHYSICAL SETTING OF THE CATCHMENT ECOSYSTEM

The Flathead River–Lake Catchment encompasses 22,241 km^2 in western Montana and British Columbia (Canada) (Fig. 1). The altitudinal gradient ranges from 750 m above sea level (a.s.l.) at confluence with the Clark Fork of the Columbia River to 882 m a.s.l. at Flathead Lake and over 4000 m a.s.l. on the Continental Divide. The geology is complex. Rock ages and composition vary from the very old and nutrient-poor Belt Series of the Precambrian age in Glacier National Park (drainage area between the Continental Divide and the North and Middle Forks; Fig. 1) to much younger limestones and other Mesozoic strata in the headwater reaches of the North, Middle, and South Forks. All of these formations were modified and molded into the present landscape by the cordilleran orogeny and subsequent erosion

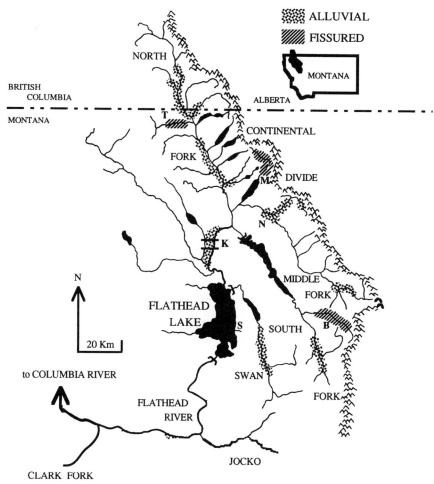

FIGURE 1 Flathead River catchment in Montana and British Columbia, Canada, showing alluvial floodplains (stippled), including the primary study sites at Nyack (N, a montane floodplain) and Kalispell Valley (K, a piedmount valley floodplain). Well transects at the Kalispell Valley site are shown as bold lines. Lakeshore alluvia containing groundwater biota occur at site S (Flathead Lake Biological Station). Cave streams in fissured bedrock substrata were studied at site T (Trail Creek) and were reported at sites M (McDonald Creek) and B (Spotted Bear River) [after Stanford and Ward (1993)].

by Pleistocene glaciation and fluvial outwash processes (Locke, 1990; Ross, 1963).

The land mass is drained by a network of mostly pristine streams and lakes. Streams generally begin as spring brooks emerging from fissured aquifers in the bedrock or from alluvia usually associated with the glacial history (see Fig. 6 in Chapter 1). The annual hydrologic pattern of surface flow is distinctively maximized in May–July by spring snowmelt in the mountain headwaters of each subcatchment. Streams, lakes, and near-surface aquifers fill to capacity or flood, depending on the intensity and duration of the snowmelt spate. Lower volume floods may occur at other times of the year, resulting from short-term, heavy precipitation created by the collision of Pacific maritime and continental air masses within the catchment. Average annual flow of the sixth-order (Strahler, 1965) Flathead River at confluence with 480-km^2 Flathead Lake is 340 m^3/sec. Three dams regulate the flow of the mainstem river above and below Flathead Lake and the top 3 m of the lake (Fig. 1).

Glaciation undoubtedly was responsible for the initial formation of the larger alluvial reaches (Fig. 1), although local Quartenary alluvial structures in the catchment have not been studied in detail. Our primary study sites are the Kalispell Valley alluvial aquifer (K; Fig. 1), between the Whitefish (mean Q = 20m^3/sec) and Flathead River channels upstream from Flathead Lake and the Nyack Valley aquifer on the Middle Fork (N; Fig. 1). The Nyack Valley is an aggraded segment that constricts into steep canyons at the upstream and downstream ends. In both cases the valleys apparently were shaped by glaciation, and postglacial fluvial outwash left an expansive deposit of cobble and gravel alluvia about 10 m thick, overlying an impermeable layer of Tertiary clay. Subsequent flooding and migration of the active river channel across the valley in response to centuries of cut and fill alluviation and, perhaps, block fault tectonics deposited 1–3 m of fine sediments on the ancient floodplain surface. Hence, the extant landscape of the Kalispell Valley is composed of the modern floodplain and channel, low terraces of the ancient floodplain, higher, lateral terraces created by glaciation (see Fig. 3), and a shallow, unconfined alluvial aquifer. Ancient river channels (paleochannels) are visible as depressions in the ancient floodplain surface, and the buried channels are remarkably evident in traces done with ground penetrating radar (G. Poole, Flathead Lake Biological Station, unpublished). The montane valley (N) differs from the piedmont valley (K) only in being smaller, and it lacks the high lateral terraces.

Rubble bottom streams also occur in karstic caves in Mesozoic limestones (B and T in Fig. 1) (J. A. Stanford, *et al.*, unpublished) and in the Belt Series substrata (M; Fig. 1) (G. Gregory, Glacier National Park, West Glacier, Montana, personal communication). The drainage patterns of the cave streams have not been documented.

The waters of the Flathead are cold, and concentrations of dissolved nutrients, such as nitrogen, phosphorus, and organic matter (DOM), in surface waters are naturally near or below detection limits; so the lakes and streams naturally do not produce much plant and animal material compared with most other waters in the United States (Ellis and Stanford, 1988; Stanford et al., 1992).

An important feature of the Flathead is that many aquatic species are present, owing in part to the pristine, oligotrophic nature of the waters and the wide variety of aquatic habitats along the altitudinal gradient (Hall *et al.*, 1992). Moreover, the Flathead exists midway in the north–south continuum of the Rocky Mountains, and three major river systems of the North America continent share adjacent headwaters in Glacier National Park (*i.e.*, the area between the North and Middle Forks and the Continental Divide; Fig. 1). these include the Saskatchewan, Missouri, and Columbia Rivers, which drain into the Arctic, Atlantic, and Pacific Oceans, respectively. Hence, many species on the fringe of their continental distribution are present, and the area is a "melting pot" of biodiversity. For example, 67% (101 species) of Plecoptera reported for the Rocky Mountains from Alaska to New Mexico have been collected in the Flathead catchment [updated from Stanford and Ward (1983)]. Native biota include species [e.g., bull charr (Fraley and Shepard, 1989) and west-slope cutthroat trout (Marnell, 1988)] that are relatively rare in North America, having been eliminated from less pristine areas by humans or existing as relict populations since Pleistocene glaciation (Stanford and Prescott, 1988).

Biophysical features of the catchment are naturally interrelated; processes in one place or time may be influenced or controlled by adjacent processes. For example, the riparian plant assemblages of the alluvial floodplains appear to be closely linked to piezometric gradients associated with groundwater down- and upwelling (Stanford and Ward, 1993). Although the river–lake ecosystem is reported to be relatively pristine for similarly sized catchments in the human-dominated latitudes of the world, stream- and lake-level regulation by dams (Stanford and Hauer, 1992) and other human sources of disturbance have exerted measured influences on natural ecosystem integrity (Spencer *et al.*, 1991; Stanford and Ward, 1992). It is within this complex biophysical setting that we wish to summarize what currently is known about the groundwater ecology of the alluvial aquifers of the Flathead River.

III. METHODS

Research wells, installed with a hollow auger drilling rig (Fig. 2), consisted of slotted PVC pipe. We used an electric saw to cut 5 mm × 5 cm alternating slots in the pipe before installation. The wells were cappped on the top and bottom, although bottom caps were also slotted to allow fauna that collected in the wells between sampling periods to escape. Wells were arrayed across the modern and ancient floodplains of the Kalispell Valley, including two transects (Fig. 3) that were intensively studied. Most wells penetrated at least 5 m into the saturated alluvia.

Over 200 wells were installed to determine aquifer characteristics, such as flow direction and basic ion chemistry, in comparison with those of the river. Several large wells were installed by the Montana Bureau of Mines and

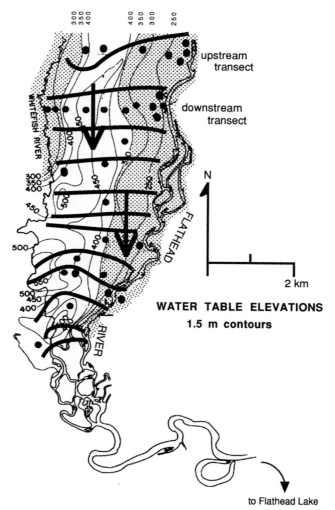

300 350 400   400 350 300   250

upstream
transect

downstream
transect

N

2 km

WATER TABLE ELEVATIONS
1.5 m contours

to Flathead Lake

FIGURE 3 Water table elevations (1.5-m isopleths, bold contours) obtained from time series measures of depth to water in 271 monitoring wells installed in the Kalispell Valley alluvial aquifer. Thin contours are isopleths of specific conductance (μS/cm) of the groundwater. Arrows indicate the direction of groundwater flow (declining elevation of water table). Solid circles are wells (including two transects; see Fig. 5) used to sample groundwater biota; the stippled area indicates the portion of the aquifer in which amphibites were routinely collected [ca. corresponding to the 350 μS/cm contour; after Stanford and Ward (1988)].

FIGURE 2 (top) Hollow auger drilling rig used to install sampling wells (slotted PVC pipe) in the alluvial floodplains of the Flathead River. A finished well is shown in the foreground. (bottom) Janine Gibert (left) and Bonnie Ellis (right) remove small rocks that have been left in the well to allow colonization of epilithic biofilm.

Geology to allow pump tests for the estimation of hydraulic conductivity, transmissivity, and flow rates; others we fitted with temperature and pressure sensors and data loggers to document thermal patterns and water table dynamics. Wells plotted in Fig. 3, in particular the upstream and downstream transects, were sampled to determine the distribution of groundwater fauna. Data loggers for temperature and flow also were installed at river sites, and additional data were available from the U.S. Geological Survey.

On each sampling date, the depth to the water table was recorded, and then wells were purged using a gas-powered diaphragm pump connected to a 5-cm-diameter tube inserted into the bottom of the wells. The pump rate for most wells was ca. 40 l per minute. All water, fauna, and detritus pumped out of a specific well during the first 10 min were filtered with a 100-μm-mesh plankton net. Pumping was continued until a clear water flow was attained, and then water samples for chemical constituents (nitrogen and phosphorus forms and DOM) and microbial determinations [acetone-extractable chlorophyll and epiflouresence (DAPI) cell counts] were taken from the pump stream. Temperatures, specific conductance values, and dissolved oxygen concentrations were recorded in each well before and after pumping. Samples for dissolved oxygen, specific conductance determinations, chemical analyses, and benthos also were collected in the river on specific sampling dates for comparison to well data. Biophysical data summarized herein were obtained from quarterly (spring, summer, fall, and winter) samplings of the wells in the two transects (Fig. 3) during the period January 1988–April 1989. Two of the wells, C and E, were sampled less frequently due to access problems. The transects were installed in an attempt to document the flux of water and materials through the aquifer and to avoid the urbanized area of the aquifer farther downstream, where groundwater effects were evident (Noble and Stanford, 1986). Wells were drilled in a grid on the Nyack Valley and sampled in a similar fashion.

IV. HYDRAULIC CONNECTIVITY OF THE ALLUVIAL AQUIFERS WITH THE RIVER

The boundaries of the Kalispell Valley alluvial aquifer are not clearly delineated, but the system is ca. 5–6 km wide at the upstream end and ca. 13 km long. The aquifer apparently is fed by the Flathead River and perhaps to a lesser extent by the much smaller Whitefish River and other small tributaries that flow into the alluvial domain of the valley from the Whitefish Range at the upstream end (Konizeski *et al.*, 1968). Additional geohydrological investigations are needed to precisely determine the extent of the aquifer and the relative importance of different sources of water. However, we demonstrated that the aquifer is hydraulically connected to the Flathead River; seasonal and daily dynamics of the water table are clearly correlated with river flow (Fig. 4). Water flows through the aquifer from north to south and is highly interactive with the river (Fig. 3). Upwelling groundwaters are evident as spring brooks at many locations on the active and ancient floodplain; most erupt in paleochannels (Fig. 5). The aquifer has an average slope of 2°. Hydraulic conductivity and, by inference, interstitial porosity are variable because some wells and some layers within wells yielded water at higher rates than others. However, in general, hydraulic conductivity estimates (0.1–10 cm/sec) from pump tests were among the highest recorded in unconsolidated alluvia (Freeze and Cherry, 1979). Wells in or near paleo-channels always produced the highest pump rates, which inferred that the paleochannels were zones of the highest porosity. The aquifer discharges into the Flathead River at or before the anastomosed zone downstream of the confluence of the Whitefish River (Fig. 3), where the aquifer terminates in contact with fine-grained paleodeltaic sediments of postglacial Flathead Lake.

More detailed geohydraulic investigations were performed at the Nyack Valley site (J. A. Stanford *et al.*, unpublished data). The valley is 3 km across at the widest point by 8 km long, and the alluvial aquifer occupies

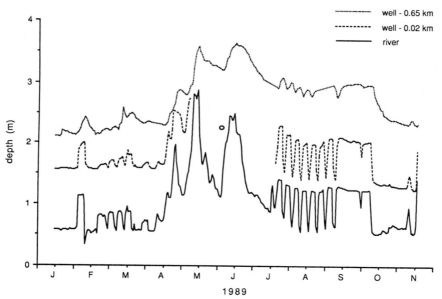

FIGURE 4 Hydrographs obtained in the Flathead River channel and in two wells located 0.02 and 0.65 km laterally (west) from a channel in the Kalispell Valley (sites Ru, A, and D in Fig. 5). Nonseasonal fluctuations were caused by discharges from a large hydroelectric dam on an upstream tributary on the South Fork (Fig. 1). Data are depths to water surface in the wells and the river channel relative to surface benchmarks as measured by hydrostatic sensors reporting on the hour to data loggers during 1989 [from Stanford and Ward (1993)].

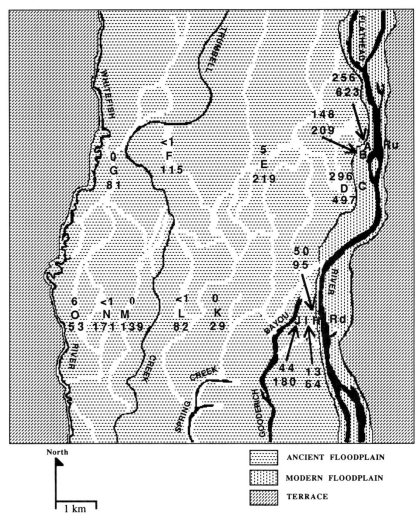

FIGURE 5 Mean abundance of amphibites (upper numbers) and stygobites (lower numbers) in wells (letters) in two transects (see Fig. 3) across the alluvial aquifer in the Kalispell Valley (upstream transect, wells A–G from river sampling site Ru; downstream transect, wells H–O from river sampling site Rd). Paleochannels on the ancient floodplain (see text) are indicated by pathways in the stippling. Spring Creek and Gooderich Bayou are perennial spring brooks that emerge from paleochannels. No biological data were collected in well C.

the entire valley bottom. Hydrographs in wells near the valley walls were directly correlated with the river flow pattern. Maps of piezometric gradients and accretion studies showed that 30% of the river flow enters the alluvium at the upstream end of the valley. Interstitial water upwells from the middle to the end of the aquifer in relation to ponding caused by the constriction of the valley at the downstream end. All of the interstitial flow enters the river channel directly or by spring brooks. Upwelling groundwaters create a network of spring brooks on the modern and ancient floodplain and are a primary determinant of riparian plant assemblages (Stanford and Ward, 1993).

V. TEMPERATURE AND SOLUTE DYNAMICS

Interaction between the Kalispell Valley aquifer and the Flathead River was inferred by mapping isolines of specific conductance (Fig. 3) relative to the river (190 μS/cm). Values in the aquifer were similar near the river, indicating a large hyporheic or transition zone between the river and the aquifer; values were over two times higher in wells most distant from the river, indicating a longer residence time of phreatic water.

The annual temperature amplitude in the river was 2–20°C; the river did not freeze in spite of very cold winter air temperatures (-30°C), owing to regulated flows from Hungry Horse Dam and the discharge of warm water from the aquifer into the river. The annual temperature amplitude in the aquifer was 6–11°C, and the amplitude was most pronounced in wells nearest the river. Dissolved oxygen concentrations in the wells were consistently >50% saturation (4–10 mg/l). In the river dissolved oxygen was always near saturation or supersaturated (J. A. Stanford *et al.*, unpublished data).

We observed a strong tendency for dissolved constituents, especially nitrate to concentrate in the center of the aquifer (Fig. 6). Mass balance calculations suggested that 12–17% of the base flow nitrate load of the river downstream from the aquifer was derived from nitrate generation, probably due to nitrification, within the aquifer. Soluble reactive phosphorus was an order of magnitude higher in well samples than in the river, and we estimated that 4–25% of the river load during base flow was due to aquifer discharge (Stanford and Ward, 1988). High solute loads in groundwaters discharging into the oligotrophic surface waters produced mats of green algae, *Ulothrix zonata*, and diatoms that otherwise would not be present, and thereby they influenced the distribution of benthic consumers (Hauer and Stanford, 1981, 1982, 1986) and may strongly influence bioproduction in surface waters (Stanford and Ward, 1993). On the other hand, DOM values were consistently low (<2 mg/l) in the aquifer, suggesting the possible carbon limitation of microbial metabolism relative to other dissolved constituents within the aquifer (Stanford and Ward, 1988).

We initially established the alternate study site at Nyack because we were concerned about contamination of the aquifer in the Kalispell Valley by human activities, especially fertilizers from farms, and the possible interactive effects of the regulation of river flow by Hungry Horse Dam. In

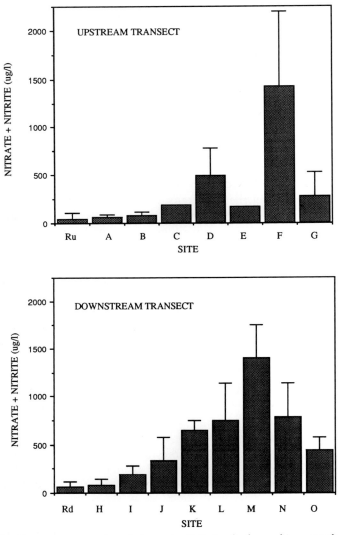

FIGURE 6   Average concentration of nitrate + nitrite (μg/l) observed in quarterly (see text) samples obtained during the period January 1988–April 1989 in the Flathead River and wells located in the transects shown in Fig. 5 (bars indicate ranges).

comparison, the Nyack Valley is virtually pristine. However, we observed no significant differences in solute or dissolved oxygen patterns. Thermal patterns were also similar, except that the river was near 0°C in midwinter and froze over at the upstream (downwelling) end for long periods (J. A. Stanford *et al.*, unpublished data). We concluded that contamination problems in the Kalispell Valley were minimal, at least on the two well transects (Figs. 3 and 5). But, contamination of wells was highly probable (*i.e.*, highest specific conductance, lowest dissolved oxygen readings, highest DOM and nitrate concentrations, high coliform bacteria counts, and fewer groundwater biota) in a limited area of intensive housing development and individual sewage drain fields on the ancient floodplain near the town of Kalispell at the lower end of the aquifer (Noble and Stanford, 1986).

VI. DISTRIBUTION AND ABUNDANCE OF GROUNDWATER BIOTA

A. Microbial Ecology

The main objective was to determine the relative importance of free living and attached bacteria, fungi, and Protozoa in the groundwater food web of the alluvial aquifers. We also examined the entrainment of riverine algae and bacteria in the aquifers. This section is based primarily on the work of Ellis *et al.* (1995).

In the Kalispell Valley, diatoms common in the Flathead River, such as *Synedra, Achnanthes, Navicula, Gomphonema,* and *Cymbella* were present in well samples up to 0.35 km from the river, with autofluorescing chromatophores very visible. Green algae were found primarily in wells far from the river channel (0.65–2.55 km), whereas blue–green algae were rather ubiquitous. This strongly suggested that riverine particulate organic carbon was entrained in the aquifer. Indeed, measurable concentrations of chlorophyll *a* were detected in samples from many wells; values were consistently highest in the upstream transect (where the river is influent to the aquifer) in wells nearest the river and reached a maximum of 14 ng/l.

However, in spite of the strong hydraulic connectivity between the aquifer and the river in the Kalispell Valley (Fig. 4), the total microbial density in wells near the river was about 80% lower than that in the river, and only 2–3% of the riverine microbial density occurred in the most distant wells. In the remote wells densities averaged ca. 9.0×10^3 cells/ml, which is at the low end of reported values for groundwater. The immediate decline of bacteria relative to the river may be due in part to a similar decline in low-molecular-weight utilizable carbon. Multiple regression analysis indicated that SRP, ammonium, nitrate + nitrate, and DOM concentrations (in that order) explained about half the variance in the free living bacteria assemblages in the aquifer.

Protozoans were detected at all river sites and in all the wells. Flagellated and nonflagellated protozoans, including amoebae and a single ciliate, were observed. Densities of all forms ranged from 0 to 213/ml. Replicate samples indicated high variances in the density estimates (*i.e.*, standard deviations averaged 26% of the mean).

In both the river and the aquifer, solitary cocci dominated densities of free living bacteria. Bacilli dominated the biomass in the river and many of the wells. However, actinomycetes, fungi, and cocci dominated the biomass in some wells. Fungi have been found in low numbers in a few aquifers (Madsen and Ghiorse, 1993). The presence of substantial populations of fungi suggests a strong connection to the surface. Infiltration of precipitation carrying soil detritus and microbiota (perhaps including algae) may be an important process in spite of the river being by far the dominant water source for the aquifer. Moreover, most of the attached bacteria observed in water samples from the aquifer were assoicated with organic detrital particles. The fine mineral sediment particles flushed from the aquifer during pumping appeared to be free of any bacterial colonization. This suggested that epilithic biofilms may be important in the aquifer.

In environments characterized by low nutrient concentrations, organisms attached to solid surfaces have an advantage in nutrient uptake in relation to free living forms, owing to continuous delivery of nutrients by water movement and because of nutrient release and feedback (looping) between bacteria and primary consumers (Protozoa and very small metazoans) that compose biofilms. In an effort to understand the epilithic microbial community of the Kalispell Valley aquifer, artifical ceramic tiles and rocks obtained from the drilling of the wells were suspended in wells with high flow rates and high oxygen concentrations for 38 days (Fig. 2). They were then removed, and [^{3}H]thymidine incorporation into DNA was determined. Results from all incubations ($n = 11$) indicated that production was significantly greater on the natural substrata than on artificial tiles ($p < 0.01$, ANOVA). Protozoan densities were also higher on rocks than on tiles, with values ranging from 99 to 6399/cm^2 on rock substrates versus 2.6 to 79/cm^2 on tiles. Although it is difficult to compare bacterial numbers per unit surface area with those per unit volume of water, it is evident that extensive growths occurred on rock substratum in comparison with densities in the interstitial waters (Table I). The same pattern also appeared to be true for protozoan densities: 99 to 6399/;cm^2 on the rock substratum versus 0 to 213/ml in interstitial waters.

The incubation time selected was arbitrary, and we surmise that the populations were early stages in the colonization of the experimental substratum. Very small size classes of bacteria dominated the rock substratum, as well as goundwater and the river. But large bacteria were more common on rocks than in the interstitial waters, suggesting that biofilms provide more optimal environmental conditions. High numbers of fungi and Protozoa on the rocks also have important implications. Fungi and actinomycetes may be very active in the aerobic heterotrophic metabolism of the groundwater microbial assemblage. Even when they are present in low numbers, their contribution to total subsurface metabolic activity may be significant (Madsen and Ghiorse, 1993). The predominant form of nutrition for free living Protozoa is the ingestion of bacterial prey (Fenchel, 1987), and predation by Protozoa has been shown to accelerate the cycling of carbon and other nutrients (Stout, 1980; Fenchel, 1987).

TABLE I Bacterial Abundance and Biomass in the Interstitial Water and on Natural Substratum Incubated in Wells (see Fig. 5) in the Kalispell and Nyack Alluvial Aquifers (see Fig. 1)

Well	Distance from river (km)	Density		Volume	
		Water (number/ml)	Rocks (number/cm^2)	Water ($\mu m^3/ml$)	Rocks ($\mu m^3/cm^2$)
Kalispell Valley					
D	0.65	$6.19 \pm 0.34 \times 10^4$	$7.49 \pm 1.11 \times 10^6$	$3.18 \pm 0.23 \times 10^3$	$7.59 \pm 0.62 \times 10^5$
E	1.96	$7.51 \pm 0.34 \times 10^4$	$3.24 \pm 0.50 \times 10^6$	$5.12 \pm 0.22 \times 10^2$	$2.86 \pm 0.56 \times 10^5$
H	0.03	$1.11 \pm 0.04 \times 10^5$	$3.94 \pm 0.55 \times 10^6$	$5.56 \pm 0.28 \times 10^3$	$2.66 \pm 0.30 \times 10^5$
J	0.35	$2.50 \pm 0.18 \times 10^4$	$1.22 \pm 0.22 \times 10^6$	$1.38 \pm 0.14 \times 10^3$	$1.08 \pm 0.27 \times 10^5$
Nyack Valley					
Nyack Down	0.28	$3.80 \pm 0.15 \times 10^4$	$2.01 \pm 0.29 \times 10^6$	$1.91 \pm 0.08 \times 10^3$	$2.04 \pm 0.18 \times 10^5$

Although the database is not as extensive, microbial dynamics appeared to be similar at the Nyack site, in that the densities and biomass of bacteria were higher and more variable in the river than in the wells and that the numbers and biomass declined with distance from the river channel. Protozoa were detected in wells 0.28 and 0.3 km from the river. Abundance ranged from 0 to 2.0/ml in the river and from 0 to 2.4/ml in the aquifer. Although flagellates were observed, the majority had no flagella and were probably cysts. No ciliates or amoebae were detected. Chlorophyll-containing algal cells were present in a well 0.28 km from the river.

We concluded that a rich, but perhaps carbon-limited, microbial biofilm is the primary base of the food web in both aquifers. The existence of a microbial loop, which tightens the efficiency of nutrient utilization, seems highly likely and probably explains how a diverse metazoan biocenosis can exist in these aquifers.

B. Invertebrate Ecology

To date we have found over 80 taxa of groundwater animals (*i.e.*, stygophiles and stygobites; Table II) in the two alluvial aquifers of the Flathead River. Most of these were members of the very speciose benthos (200 + species) that occasionally showed up in samples from wells near the river or other groundwater collections within a few meters of water flowing in the river channel (*i.e.*, occasional hyporheos). However, 10 crustacean (Ward *et al.*, 1994), 14 insect, and several other species (J. A. Stanford *et al.*, unpublished data) were routinely collected in wells far from the river channel (stygophiles and stygobites; Table II). Of these, a copepod species new to science has been described (Reid *et al.*, 1991), and perhaps as many as 10 other crustacean and 2 insect species remain to be described. The Acarina and Oligochaeta also require further taxonomic resolution. Nematodes, segmented worms, and other unknown organisms with DAPI-stained DNA were observed in epilithic biofilms in wells and in the river and were classified as permanent hyporheos; nematode densities ranged from 0 to 7.9/ml in wells and from 0 to 2.0/ml in river samples (Ellis *et al.*, 1995).

A unique aspect of the groundwater fauna of the Flathead River is the presence of amphibitic Plecoptera (Table II). At least six species, some with multiple morphs, are involved. These insects are large bodied (2–3.5 cm in length as mature nymphs) and reside deep in the alluvia as larvae, moving to the river or spring brooks during the spring to emerge as flying adults. They mate and produce eggs, which are deposited into the river, where presumably they are entrained in the aquifer before hatching or from which the first instar larvae immediately migrate into the interstitial environment (Stanford and Gaufin, 1974; Stanford and Ward, 1988). We now think that the larvae follow thermal gradients to sites of epigean ecdysis during the emergence process, because the wells across the valley floor

TABLE II   Types, Common Species, and Total Numbers of Species of Epigean and Hypogean Invertebrates Found to Date in the Flathead River, Based on the Typology of Gibert *et al.* (Chapter 1)

Life history type	Numeric dominants	Total number of species

Stygoxene	Trichoptera
	  Archtopsyche grandis
	Plecoptera
	  Taenionema pacificum
	  Suwallia pallidula
	  Isoperla fulva
	Ephemeroptera
	  Ephemerella inermis
	  Rhithrogena spp.
	  Baetis tricaudatus
	Diptera
	  Simulium spp.
	  Procladius sp.
		137
		
		137

Stygophile
  Occasional hyporheos
	Ephemeroptera
	  Ameletus 2 spp.
	  Paraleptophlebia sp.
	Plecoptera
	  Capnia confusa
	  Suwallia pallidula
	  Sweltsa coloradensis
	  Hesperoperla pacifica
	Diptera
	  Cricotopus sp.
		46

Stygophile
  Amphibite
	Plecoptera
	  Isocapnia crinita
	    I. grandis
	    I. vedderensis
	    I. missouri
	  Paraperla frontalis
	  Kathroperla perdita
		6

Stygophile
  Permanent hyporheos
	Copepoda
	  Acanthocyclops montana
	  Diacyclops crassicaudus
	    brachycercus
	  D. languidoides
	Ostracoda
	  Cavernocypris wardi
	Oligochaeta
	  Lumbricillus sp.
	  Lumbriculidae indet sp.
		20

(continues)

TABLE II (continued)

	Acarina	
	Oribatei indet sp.	
	Halacaridae spp.	
Stygobite	Amphipoda	8
Ubiquitous[a]	*Stygobromus* 2 spp.[a]	
	Isopoda	
	Salmasellus steganothrix	
	Microcharon sp.	
	Bathynellacea	
	Bathynella 2 spp.	
	Copepoda	
	Parastenocaris sp.	
	Archiannelida	
	Troglochaetus beranecki	
Stygobite	?	?
Phreatobite		

[a] Only *Stygobromus* sp. has been found in cave streams and alluvial aquifers.

sequentially become depauperate of larvae when river water penetrates the aquifer in association with the spring spate, when river temperatures are up to 4°C warmer than groundwater. Other explanations exist, including the possibility that the larvae can find surface water by following ion gradients. However, large numbers climb up the well pipes to undergo ecdysis above the water table, and water temperatures are significantly higher at the top of the wells during warm weather in the spring.

The distribution of stygobites and amphibites within the aquifer is clear. Stygobites are ubiquitous; amphibites are most common near the river (Figs. 3 and 5), although they were collected in all except three wells in the two transects (Fig. 5). By examining the distribution of the crustacean fauna using gradient and correspondence analyses, Ward *et al.* (1994) showed that crustacean stygobites attained maximum relative abundance near the center of the ancient floodplain, whereas permanent hyporheos (*e.g.*, copepods and ostracods; Table II) dominated closer to the river or in wells on the modern floodplain. Areas in the center of the aquifer (*i.e.*, about midway in the two transects across the ancient and modern floodplains) are associated with higher ion content (Fig. 3). We think that higher ion content is indicative of the longer residence time of water in the aquifer and the presence of zones of lower hydraulic conductivity and, by inference, lower porosity. Hence, faunal patterns to some extent may be attributed to variations in porosity. However, well G (Fig. 5) penetrated very porous substrata based on high pump rates (>40 l/min), yet it was characterized by high ion content, fairly abundant styobites, and no amphibites. It is likely the high-ion waters are phreatic and not very interactive with the river for reasons that relate to the complexity of the interstitial milieu of this aquifer. As did Marmonier *et al.* (1992) for an alluvial floodplain aquifer on the Rhône River, Ward *et al.* (1994) concluded that the distribution and abundance of the groundwater fauna on the floodplain scale are structured by hydrogeologic and geomorphic processes that are not necessarily related in any meaningful way to distance from the river owing to spatial discontinuities in the lattice-like alluvium. An analysis of the relation of paleochannels, which are known to be areas of higher hydraulic conductivity, to faunal patterns (*e.g.*, installation of wells in high-porosity sites determined by ground penetrating radar or seismic techniques) may be more revealing.

The groundwater fauna of the Nyack Valley aquifer is very similar to that of the Kalispell Valley, except that some of the occasional hyporheos are different and are found in wells near the valley wall. Phreatic waters may be more limited. Spring-brook fauna also were studied to determine affinities with the groundwater biocenosis. Although permanent hyporheos (mainly copepods) were abundant in the spring brooks, stygobites were never found and amphibites were rare. The spring-brook fauna was primarily composed of stygoxenes with temperature and substratum as primary determinants (Case, 1994).

C. Food Web Relationships

The energy base of the food web of the Flathead River alluvial aquifers appears to be the speciose microbial biofilm (bacteria and fungi) and associated microconsumers (Protozoa, nematodes, and other microscopic fauna). The productivity of this microbial loop is likely limited by the availability of short-chain DOM, because other nutrients appear to increase in concentration relative to those in the river. Although POM is entrained in the aquifer, based on the presence of algal cells with viable chromatophores and measurable concentrations of chlorophyll, the importance of detritus relative to DOM is not clear. Although others have shown that microbial metabolism in groundwaters is controlled by entrainment of POM (Chapter 7), most of the work has been done in situations in which some or most of the organic matter is derived from pollution. The pristine Flathead systems are extremely oligotrophic, and carbon limitation relative to the supply of other nutrients (N, P, and trace metals) remains at issue. Because nitrate accumulates in aquifers and oxic conditions persist in the wells, nitrification is an important process within the groundwater microbial loop. Our work is now emphasizing the assessment of N flux and the importance of the groundwater N subsidy to biofilms in surface waters.

Higher level consumers in the food web are composed of a wide variety of sizes and taxa of stygobite and stygophile consumers. Amphibitic stone flies seem to be the top consumers in the groundwater food web, at least in the hyporheic waters. However, in phreatic waters, where large stone flies are rare, *Stygobromus* spp. likely is the top consumer.

With R. Wissmar (University of Washington, Seattle, unpublished), we analyzed the stable isotopic enrichment of ^{15}N to trace trophic relations of the groundwater food web in comparison with that of the river (Peterson and Fry, 1987). The isotopic enrichment values for river and aquifer biofilms were significantly higher in the aquifer, supporting the notion of different conversion pathways for organic compounds. Nitrogen compounds incorporated into the food web by biofilm uptake were different in the river and in the aquifer, which supports the inference of nitrification as an important biotic pathway in groundwater. As a consequence, high-level consumer groups in the aquifer showed high enrichment values; *Paraperla* and *Stygobromus* had the highest enrichment values ($+8-11$ ‰), supporting the notion of these genera as high or top-end consumers in the food web. However, they may not be the top consumers in the interstitial food web, because a small (5 cm) salamander, tentatively identified in the genus *Ambystoma*, was collected from well D (Fig. 5).

In addition to the importance of nutrient conversions in groundwater food webs, transport of groundwater macrofauna to surface waters may also provide an important food source for riverine consumers. The amphibitic stone flies are very abundant during emergence in the river and on the shoreline, relative to most of the benthic species (Stanford, 1975). The large volume of the groundwater system apparently allows greater production of the amphibitic stone flies, and, when they move from the groundwater into the river during the emergence process, they dominate the drift and are a major food source for fish and other top consumers in the river (Perry and Perry, 1986).

Amphibitic assemblages similar to those in the Flathead likely exist in many well-oxygenated interstitial aquifers fed by river interflow in North America and elsewhere, depending on the origin and attributes of the porous milieu (Stanford and Ward, 1993). We have found the same interstitial fauna in the Methow River in Washington, and others in western North America will likely be reported in the future.

D. Cave Streams

We know only that the stygobite *Stygobormus* spp. is present in cave streams in the Flathead catchment. Several specimens were collected with a kick net at the Trail Creek site. The stream intersects the cave about 300 m from the dry portal. Presumably the stream flows into the alluvium of Trail Creek, but additional work is needed. Hence, we conclude that *Stygobromus* is a ubiquitous genus (Table II), but we do not know if phreatobites are present in the alluvial aquifers, owing to a lack of species-specific associations with habitat types. However, this is the first report of groundwater fauna in cave systems in the Belt Series substrata of the Rocky Mountains.

VII. CONCLUSION

Shallow alluvial aquifers biophysically connected to the main channel are fundamental units of the geohydraulic continuum of the Flathead River. They are not delineated by the active river floodplain; they also include adjacent low terraces composed of ancient glacial outwash floodplains and modified by recent cut and fill alluviation mediated by flooding. Paleochannels are evident on the ancient floodplain surface and provide clues to the complicated lattice-like structure of the alluvial aquifer. Interactions between river inflow and aquifer discharge produce biotic patterns that manifest as complex groundwater food webs and riparian plant assemblages, although detailed study of the latter remains to be performed. Over 80 taxa of stygophiles and stygobites are present in the groundwater food webs. Considerable taxonomic evaluation of the groundwater fauna also remains to be performed, and the list of stygobites will likely increase. Clearly, riverine biodiversity is in large measure influenced by the groundwater biotope. Bioproduction may be controlled by the availability of low-molecular-weight DOM, although nitrification and other microbially mediated conversions increase the solute load of the water as it fluxes from the river, through the aquifer, and back into the river. These and other process–response mechanisms also require greater elucidation. However, it is clear that the interaction zone between the river and the aquifer is expansive (*i.e.*, involving amphibitic and stygobitic forms; Figs. 3 and 5), and the flux of materials in this zone probably controls the biophysiology of the river and also may influence advective processes in Flathead Lake (*e.g.*, pelagic autotrophy in response to nutrient loads from the river).

These findings have profound ramifications for the management of river corridors in the Flathead and other gravel bed rivers, especially those that remain relatively pristine. The expansive interaction zone and the flux of water and materials between surface water and groundwater require a broader conception of the floodplain (riparian) corridor than usually is applied in river management plans. In the Kalispell Valley human activities on the ancient floodplain are beginning to subsidize the nutrient load of the aquifer and alter faunal distribution patterns. The sensitive nature of these groundwater environments requires greater appreciation in the context of pollution control. In the Flathead and elsewhere these shallow alluvial aquifers are strategic sources of potable supplies. Moreover, we are beginning to recognize important interactions between wetlands created by groundwater upwelling and wildlife. For example, for the Nyack Valley, we have observed that spring brooks are used as spawning and rearing habitats for fishes and are modified by dam building activities of the beaver (*Castor canadensis*). A variety of ungulates and their predators rely on the riparian plant communities produced by the interactions between interstitial and overland flow processes. We conclude that conservation of groundwater aquifers is important not only from the standpoint of hypogean flora and fauna that are not well known, but also from the standpoint of influences on terrestrial flora and fauna on scales and in time frames that are not understood at all.

ACKNOWLEDGMENTS

This study was supported by a National Science Foundation grant (BSR-8705269) and by funding from the National Park Service. We give special thanks to John Dalimata and Neil and Frances Graham for access and logistical support on the Nyack and Kalispell Valley floodplains. Thanks also to Jim Craft and Roger Noble for technical help.

REFERENCES

Boulton, A. J., Stibbe, S. E., Grimm, N. B., and Fisher, S. G. (1991). Invertebrate recolonization of small patches of defaunated hyporheic sediments in a Sonoran Desert stream. *Freshwater Biol.* **26**, 267–277.

Boulton, A. J., Valett, H. M., and Fisher, S. G. (1992). Spatial distribution and taxonomic composition of the hyporheos of several Sonoran Desert streams. *Arch. Hydrobiol.* **125**, 37.

Bretschko, G., and Klemens, W. E. (1986). Quantitative methods and aspects in the study of the interstitial fauna of running waters. *Stygologia* **2**, 297–316.

Case, G. L. (1994). Benthic and interstitial faunal patterns on a floodplain of an alluvial river. Thesis, University of Montana, Missoula.

Coleman, M. J., and Hynes, H. B. N. (1970). The vertical distribution of the invertebrate fauna in the bed of a stream. *Limnol. Oceanogr.* **15**, 31–70.

Danielopol, D. L. (1976). The distribution of the fauna in the interstitial habitats of riverine sediments of the Danube and the Piesting (Austria). *Int. J. Speleol.* **8**, 23–51.

Ellis, B. K., and Stanford, J. A. (1988). Nutrient subsidy in montane lakes: Fluvial sediments versus volcanic ash. *Verh. Int. Ver. Limnol.* **23**, 327–340.

Ellis, B. K., Stanford, J. A., and Ward, J. V. (1995). Microbial ecology of the alluvial aquifers of the Flathead River, Montana (USA). *J. North Am. Benthol. Soc.* (in review).

Fenchel, T. (1987). "Ecology of Protozoa." Science Tech. Publisher, Madison, WI.

Fraley, J. J., and Shepard, B. B. (1989). Life history, ecology and population status of migratory bull trout (*Salvelinus confluentus*) in the Flathead Lake and River system, Montana. *Northwest Sci.* **63**, 133–143.

Freeze, R. A., and Cherry, J. A. (1979). "Groundwater." Prentice-Hall, Englewood Cliffs, NJ.

Hall, C. A. S., Stanford, J. A., and Hauer, F. R. (1992). The distribution and abundance of organisms as a consequence of energy balances along multiple environmental gradients. *Oikos* **65**, 377–390.

Hauer, F. R., and Stanford, J. A. (1981). Larval specialization and phenotypic variation in *Arctopsyche grandis* (Trichoptera: Hydropsychidae). *Ecology* **62**, 645–653.

Hauer, F. R., and Stanford, J. A. (1982). Bionomics of *Dicosmoecus gilvipes* (Trichoptera: Limnephilidae) in a large western montane river. *Am. Midl. Nat.* **108**, 81–87.

Hauer, F. R., and Stanford, J. A. (1986). Ecology and coexistence of two species of *Brachycentrus* (Trichoptera) in a Rocky Mountain river. *Can. J. Zool.* **64**, 1469–1474.

Konizeski, R. L., Brietkrietz, A., and McMurtrey, R. G. (1968). Geology and groundwater resources of the Kalispell Valley, northwestern Montana. *Mont., Bur. Mines Geol., Bull.* **68**, 1–42.

Locke, W. W. (1990). Late Pleistocene glaciers and climate of western Montana, U.S.A. *Arct. Alp. Res.* **22**, 1–13.

Madsen, E. L., and Ghiorse, W. C. (1993). Groundwater microbiology: Subsurface ecosystem processes. *In* "Aquatic Microbiology: An Ecological Approach" (T. E. Ford, ed.), pp. 167–213. Blackwell, Boston.

Marmonier, P., Dole-Olivier, M. J., and Creuzé des Châtelliers, M. (1992). Spatial distribution of interstitial assemblages in the floodplain of the Rhône River. *Regulated Rivers* **7**, 75–82.

Marnell, L. F. (1988). Status of the westslope cutthroat trout in Glacier National Park, Montana. *Am. Fish. Soc. Symp.* **4**, 61–70.

Noble, R. A., and Stanford, J. A. (1986). Groundwater resources and water quality of the unconfined aquifers in the Kalispell Valley, Montana. Open File Report 093-86. Flathead Lake Biological Station, The University of Montana, Polson, Montana.

Palmer, M. A. (1990). Temporal and spatial dynamics of meiofauna within the hyporheic zone of Goose Creek, Virginia. *J. North Am. Benthol. Soc.* **9**, 17–25.

Pennak, R. W., and Ward, J. V. (1986). Interstitial faunal communities of the hyporheic and adjacent groundwater biotopes of a Colorado mountain stream. *Arch. Hydrobiol., Suppl.* **74**, 356–396.

Perry, S. A., and Perry, W. B. (1986). Effects of experimental flow regulation on invertebrate drift and stranding in the Flathead and Kootenai Rivers, Montana, USA. *Hydrobiologia* **134**, 171–182.

Peterson, B. J., and Fry, B. (1987). Stable isotopes in ecosystems analysis. *Annu. Rev. Ecol. Syst.* **18**, 293–320.

Pugsley, C. W., and Hynes, H. B. N. (1986). Three-dimensional distribution of winter stonefly nymphs, *Allocapnia pygmaea*, within the substrate of a southern Ontario river. *Can. J. Fish. Aquat. Sci.* **43**, 1812–1817.

Reid, J. W., Reed, E. B., Ward, J. V., Voelz, N. J., and Stanford, J. A. (1991). *Diacyclops languidoides* (Lilljeborg, 1901) s.l. and *Acanthocyclops montana*, new species (Copepoda, Cyclopoida), from groundwater in Montana, USA. Hydrobiologia **218**, 133–149.

Ross, C. P. (1963). The belt series in Montana. *Geol. Surv. Prof. Pap. (U.S.)* **346**.

Spencer, C. N., McClelland, B. R., and Stanford, J. A. (1991). Shrimp stocking, salmon collapse, and eagle displacement: Cascading interactions in the food web of a large aquatic ecosystem. *BioScience* **41**, 14–21.

Stanford, J. A. (1975). Ecological studies of Plecoptera in the Upper Flathead and Tobacco Rivers, Montana. Ph.D. Dissertation, University of Utah, Salt Lake City.

Stanford, J. A., and Gaufin, A. R. (1974). Hyporheic communities of two Montana Rivers. *Science* **185**, 700–702.

Stanford, J. A., and Hauer, F. R. (1992). Mitigating the impacts of stream and lake regulation in the Flathead River Catchment, Montana, USA: An ecosystem perspective. *Aquat. Conserv.* **2**, 35–63.

Stanford, J. A., and Prescott, G. W. (1988). Limnological features of a remote alpine lake in Montana, including a new species of Cladophora (Chlorophyta). *J. North Am. Benthol. Soc.* 7(2), 140–151.

Stanford, J. A., and Ward, J. V. (1983). Insect species diversity as a function of environmental variability and disturbance in stream systems. *In* "Stream Ecology: Application and Testing of General Ecological Theory" (J. R. Barnes and G. W. Minshall, eds.), pp. 265–278. Plenum, New York.

Stanford, J. A., and Ward, J. V. (1988). The hyporheic habitat of river ecosystems. *Nature (London)* **335**, 64–66.

Stanford, J. A., and Ward, J. V. (1992). Management of aquatic resources in large catchments: Recognizing interactions between ecosystem connectivity and environmental disturbance. *In* "Watershed Management" (R. J. Naiman, ed.), pp. 91–124. Springer-Verlag, New York.

Stanford, J. A., and Ward, J. V. (1993). An ecosystem perspective of alluvial rivers: Connectivity and the hyporheic corridor. *J. North Am. Benthol. Soc.* **12**, 48–60.

Stanford, J. A., Ellis, B. K., Chess, D. W., Craft, J. A., and Poole, G. C. (1992). "Monitoring Water Quality in Flathead Lake, Montana. 1992 Progress Report." Open File Rep. 128-92. Flathead Lake Biological Station, University of Montana, Polson.

Stout, J. D. (1980). The role of protozoa in nutrient cycling and energy flow. *Adv. Microb. Ecol.* **4**, 1–49.

Strahler, A. N. (1965). "Introduction to Physical Geography." Wiley, New York.

Ward, J. V. (1989). The four-dimensional nature of lotic ecosystems. *J. North Am. Benthol. Soc.* **8**, 2–8.

Ward, J. V., and Voelz, N. J. (1990). Gradient analysis of interstitial meiofauna along a longitudinal stream profile. *Stygologia* **5**, 93–99.

Ward, J. V., Stanford, J. A., and Voelz, N. J. (1994). Spatial distribution patterns of Crustacea in the floodplain aquifer of an alluvial river. *Hydrobiologia* (in press).

Williams, D. D., and Hynes, H. B. N. (1974). The occurrence of benthos deep in the substratum of a stream. *Freshwater Biol.* **4**, 233–256.

15

Groundwater Fauna of the South Platte River System, Colorado

J. V. Ward and N. J. Voelz[1]

Department of Biology
Colorado State University
Fort Collins, Colorado 80523

I. INTRODUCTION

The groundwater fauna associated with alluvial aquifers of North American rivers has remained virtually uninvestigated. Aside from studies of the distribution of riverine animals in surficial bed sediments, most research on the hypogean fauna has been conducted during the past decade (Pennak and Ward, 1986; Stanford and Ward, 1988; Strayer, 1988; Williams, 1989; Palmer, 1990; Ward and Voelz, 1990; Boulton *et al.*, 1992). This is in marked contrast to the intensive and extensive work on the groundwater

---

[1] Present address: Department of Biological Sciences, St. Cloud State University, St. Cloud, Minnesota, 56301.

fauna of European alluvial rivers (Orghidan, 1959; Gibert *et al.*, 1977; Dole, 1985; Danielopol, 1989; Marmonier *et al.*, 1992).

This chapter summarizes results of research that examined the fauna associated with the alluvial aquifer system of the South Platte River in Colorado. The South Platte River traverses an extensive elevation gradient (>2000 m), first as a Rocky Mountain stream and then as a Great Plains river. Patterns of faunal abundance, biodiversity, and community composition are examined in phreatic, hyporheic, and surface gravel biotopes along the elevation gradient.

II. THE GEOMORPHIC SETTING

The South Platte River has its headwaters high in the Colorado Cordillera, descends 2000 m in about 200 km through the Rocky Mountains, and then descends another 900 m in 360 km as it crosses the Great Plains before leaving Colorado to join the North Platte River. The river catchment encompasses 62,238 km², most of which (49,262 km²) lies within the state of Colorado.

The South Platte River traverses two major physiographic provinces. The mountain stream segment is in the Southern Rocky Mountain Province; the plains river flows through the Great Plains Province.

A. The Mountain Stream

The mountain stream has its headwaters at 3700 m above sea level (a.s.l.) near the Continental Divide in the Mosquito Range, crosses South Park, a high intermontane basin (ca. 2700 m), and then descends precipitously through the Front Range, the easternmost unit of the Colorado Rockies. This portion of the Cordilleran chain was formed by vertical uplift during the Laramide orogeny, which began in the late Mesozoic, some 70–65 million years BP, and its formation continued into the Eocene (Tweto, 1980). Uplifted Paleozoic and Mesozoic sedimentary rocks initially capped the Mosquito and Front Ranges, but were eroded before orogeny ended, exposing the Precambrian cores of igneous and metamorphic rocks (Bradley, 1987; Madole, 1987).

The South Park intermontane basin was formed by synclinal downwarping between the rising Mosquito and Front Ranges (Tweto, 1980). Older marine and freshwater deposits (sandstones and shales) are overlain by Precambrian granitic materials derived from the surrounding mountains.

Mountain valleys were formed by stream incision in the early Tertiary and sculptured by glacial action in the reach above South Park. This is reflected in the mountain landscape by the broader valleys of the headwaters and the narrow canyon terrain in the lower mountains and foothills.

Four major glaciations occurred in the Colorado Rockies (Meierding and Birkeland, 1980). Pre-Bull Lake and Bull Lake glacial tills are sparse and of uncertain age. Pinedale glaciation began >30,000 years BP and glaciers began receding about 14,000 years ago. Glaciers were eliminated from this region during the altithermal maximum (ca. 7500–6000 years BP). At least four minor advances of cirque glaciers occurred during the Neoglaciation, over the last 6000 years. Glacier remnants exist in cirques at the sources of several high elevation tributaries.

The deglaciation phase of each glacial episode yielded large amounts of sediment to stream valleys, resulting in glaciofluvial deposition and transport (Ritter, 1987). The coarse-grained sediment that eroded from the mountains and was deposited in adjacent basins (South Park and Denver Basins) came to dominate the surficial alluvium because the gravels are more resistant to fluvial transport than the older fine-grained basin fill sediment.

B. The Plains River

As the Front Range uplifted, marine and freshwater sedimentary strata, ranging in age from the late Paleozoic to the early Cenozoic, were sharply upturned to form the foothills between the now exposed Precambrian mountain core to the west and the Great Plains to the east. The South Platte River emerges from the foothills at an elevation of ca. 1650 m and becomes a high plains river.

Fluvial processes have dominated Quaternary landform development in this portion of the Great Plains (Hadley and Toy, 1987). The South Platte is a snowmelt river (Poff and Ward, 1989), with about 70% of the annual discharge occurring during spring runoff. The Great Plains are semiarid, with evapotranspiration typically exceeding annual precipitation. Whereas the North Platte River exhibited historical reductions of annual mean and peak discharges, those changes have not been observed for the plains segment of the South Platte River (Hadley and Toy, 1987). This relates to the much more extensive reservoir storage and flow diversion of the North Platte and transbasin diversions of water into the South Platte catchment. Nonetheless, there have been significant hydrologic and morphologic changes (Nadler and Schumm, 1981). The plains river, historically intermittent, is now perennial. Bottomland gallery forests that became established during the drought of the 1930s, when discharge was greatly reduced, stabilized the channel, which resulted in declining channel widths. By 1952 channel widths averaged only 15% of 1867 values.

C. Alluvial Aquifers

The alluvial aquifers of the mountain stream occur in shallow deposits of coarse-grained sediment lying directly on crystalline bedrock. Alluvium is absent in the steepest locales within canyon sections in which bedrock is exposed on the stream bed. Virtually no published data are available on the alluvial deposits of the mountain stream. An exception is a study of the foothills segment of the South Platte (Scott, 1963), for which alluvial deposits are up to 6 m thick. Most stream incision was completed by early Wisconsin (Würm) time. Since then fluvial action has reworked the outwash deposited by mountain glaciers. Erosion–deposition cycles have alternately transported and deposited floodplain alluvium. Most surficial stream bed alluvium, of both the mountain stream and the plains river, consists of coarse-grained late Holocene (<3000 years BP) deposits eroded from the Precambrian core of the Rocky Mountains (Madole, 1994, personal communication).

More information is available for the groundwater aquifers of the Denver Basin, the Great Plains portion of the South Platte catchment (Pearl, 1974; Hurr *et al.*, 1975; Tweto, 1980; Norris *et al.*, 1985; Hurr and Hearne, 1985; Robson, 1987). The unconfined alluvial aquifer of the plains river consists of sediment deposited in broad valleys cut into the underlying sedimentary bedrock. The alluvium is mainly gravel and sand with smaller amounts of silt and clay and some pebbles, cobbles, and boulders. The width of the alluvial aquifer ranges from 1.6 to 16 km and the thickness of the saturated zone ranges from <10 to >75 m. Estimated water storage is 10,271 hm^3. The depth of the water table varies from 0 to 24 m below the surface, and the water table slopes downstream at an average rate of 1.4 m per kilometer. Transmissivity ranges from 186 to 18,600 m^2/day.

The alluvial aquifer overlies an extensive bedrock aquifer system, contained within sedimentary strata of late Cretaceous and early Tertiary age (Robson, 1987). Transmissivity typically is at least an order of magnitude lower than that in the alluvial aquifer. The alluvial aquifer is hydraulically interconnected with the river, but there is little exchange between the alluvial and the bedrock aquifers.

Groundwater of the alluvial aquifer is of the calcium bicarbonate or calcium sulfate type (Norris *et al.*, 1985). Total dissolved solids (TDS) in the aquifer increase in the downstream direction, with concentrations <100 mg/l in the mountain stream and values ranging from 100 to 2,500 mg/l or more in the plains aquifer.

III. SITE LOCATIONS

Sampling sites were established at nine locations along a 475-km segment of the longitudinal profile with an elevation drop of 1995 m (Fig. 1). Sites consisted of riffles and adjacent point or lateral bars at locations not directly influenced by known sources of pollution. Domestic and industrial pollutants from Denver and other municipalities enter the river down-

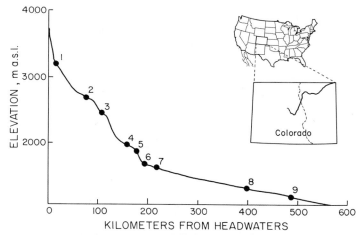

FIGURE 1   Longitudinal profile of the South Platte River in Colorado (impoundments omitted), showing sampling locations. The river's course is shown as a solid line on the Colorado map. The dashed line delimits the Rocky Mountains (to the west) and the Great Plains (to the east).

stream from Site 7, thus accounting for the large distance between Site 7 and Site 8.

Site 1 (3194 m a.s.l.) was on the Middle Fork of the South Platte River near the top of Hoosier Pass. The headwater stream meanders across a narrow floodplain of Holocene alluvium bounded by a high terrace of late Pleistocene glaciofluvial sediment (Madole, 1994).

Site 2 (2670 m a.s.l.) was located about 1 km below the confluence of the Middle Fork and the South Fork in South Park, a broad treeless intermontane basin. Site 3 (2451 m a.s.l.) was located in Eleven Mile Canyon ca. 11 km downstream from Eleven Mile Canyon Reservoir. The mean annual discharge (56 years of record) was 2.2 m³/sec. Site 4 (1951 m a.s.l.) was located near the mouth of Cheesman Canyon, 8.5 km below Cheesman Reservoir. The discharge (61 years of record) averaged 4.7 m³/sec. Site 5 (1863 m a.s.l.) was located in the foothills at the confluence of the North Fork and the South Platte River and was the only location in Colorado at which studies of groundwater fauna had previously been conducted (Pennak and Ward, 1986).

Site 6 (1670 m a.s.l.) in the lower foothills was located in Waterton Canyon, ca. 8 km below Strontia Springs Reservoir. The site location was just upstream from the transition from Precambrian to sedimentary bedrock. The Precambrian bedrock of biotite and muscovite granite is overlain with up to 6 m of coarse alluvium (Scott, 1963).

Site 7 (1637 m a.s.l.), in a broad valley downstream from the mouth of Waterton Canyon, was located about 5 km upstream from the greater

Denver metropolitan area. The alluvium beneath the river is ca. 10 m thick and overlies sedimentary bedrock (Scott, 1963). The discharge (33 years of record) averaged 6.6 m^3/sec.

Sites 8 and 9 (1332 and 1199 m a.s.l., respectively), were located in the Great Plains Physiographic Province. The plains segment of the South Platte River "exhibits a uniform pattern throughout eastern Colorado" (Nadler and Schumm, 1981). In the early 1800s the river was wide and straight with numerous transient bars; flow was intermittent and floodplain vegetation was sparse. Summer irrigation and flow regulation resulted in perennial flow and dense vegetation now occurs on the floodplain. Bars have become islands, sinuosity has increased, and the river, although braided, is dominated by a single thalweg. At Site 8 the channel width decreased from 535 m in 1867 to 109 m in 1977. Sediment characteristics have not changed appreciably over the last 150 years, nor has the mean gradient (ca. 0.0013). Site 8 was about 2 km south of the small village of Goodrich. Site 9 was about 2 km southeast of the town of Sterling. Discharge (22 years of record) at a gauging station between these two sites averaged 16.2 m^3/sec.

The woody riparian vegetation of Sites 1–7 was dominated by willow shrubs (*Salix* spp.). Scattered cottonwood trees (*Populus sargentii*) occurred at Site 7. The plains river (Sites 8 and 9) floodplain contained bottomland gallery forests of cottonwoods, willows, and box elder (*Acer negundo*).

IV. PHYSICOCHEMICAL CONDITIONS

Mountain stream riffles were dominated by rubble. The coarsest alluvium occurred in canyon-constrained reaches (Sites 3, 4, and 6) that were characterized by high gradients (*e.g.*, 13.3 m/km at Site 6). Rubble was also the predominant size class in alluvial bars at most mountain stream sites. Rubble was largely absent from the gravel bar at Site 5, however, and the bar at Site 7 contained less rubble and more sand than did the adjacent riffle. Gravel and sand constituted the majority of the surficial alluvium of the plains river.

Mineral substratum up to 64 mm in diameter was collected from the bed (hyporheic) and bars (phreatic) at each site during autumn base flow conditions (Table I). The hyporheic substratum was similar at Sites 1–7, with pebbles and gravel constituting the majority of sediment <64 mm in diameter. In the plains river (Sites 8 and 9) pebbles are less abundant and sand is more abundant than those in the mountain stream. At mountain stream sites, sand was generally more abundant in phreatic than in hyporheic habitats. Silts and clays constituted <2% of the hyporheic and phreatic substratum at all sites. The organic content was uniformly <1%.

A limited array of other physicochemical variables was measured in the field. Spot water temperatures were taken from the three habitat types at

TABLE I Percentage Composition of the Mineral Size Classes (Based on Particles <64 mm in Diameter) and the Organic Content of the Substratum at Phreatic (P) and Hyporheic (H) Habitats at Each Site

													Sampling site									
	1		2		3		4		5		6		7		8		9	
Category	P	H	P	H	P	H	P	H	P	H	P	H	P	H	P	H	P	H
Pebble	47.2	54.2	44.0	56.5	31.9	45.5	36.2	60.6	20.4	28.3	48.1	49.7	19.9	52.2	20.8	2.1	3.6	3.6
Gravel	25.4	35.5	22.1	30.9	37.3	29.5	38.5	31.5	52.4	44.1	33.9	38.0	38.8	43.0	40.9	43.7	48.2	47.4
Sand	24.9	9.8	32.5	11.9	29.3	23.9	22.0	7.5	26.1	26.2	16.6	11.9	38.5	4.6	37.2	52.6	46.9	47.4
Silt	0.8	0.1	0.3	0.1	0.3	0.2	1.5	<0.1	1.1	1.4	0.5	0.1	1.2	<0.1	0.4	1.1	1.0	0.5
Clay	1.6	0.4	1.0	0.5	1.2	0.7	1.8	0.3	<1.0	<1.0	0.7	0.2	1.7	0.2	0.8	0.5	0.5	1.0
Organic[a]	0.8	0.2	0.2	0.3	0.3	0.5	0.9	0.3	0.1	0.1	0.4	0.2	0.2	0.2	<0.1	<0.1	<0.1	<0.1

[a] As a percentage of the sand and finer fraction.

each site on each date. Chemical analyses of surface, hyporheic, and phreatic waters were conducted during the autumn sampling period. The physico-chemical data in Table II indicate only general conditions, and even broad patterns are not always consistent. For example, spot temperatures reflect the time of day and weather conditions on the day a given site was sampled (field sampling required several days). Surface waters exhibited oxygen levels near saturation, but hyporheic or phreatic habitats at four sites had concentrations <4.0 mg/l. Total hardness values were considerably higher in the plains river than in the mountain stream. The pH varied from circumneutral to basic. The lowest pH value (6.8 at Site 4, phreatic) occurred in a habitat with low oxygen and very high free CO_2 levels. Excessively high free CO_2 may indicate microsites of high microbial activity (Pennak and Ward, 1986).

V. FAUNAL COMPOSITION AND BIODIVERSITY

Samples were collected with a Bou–Rouch pump (Bou, 1974) and pro-cessed as described in Pennak and Ward (1986). A plankton bucket equipped with 48-μm mesh was used to concentrate samples in the field and the laboratory. Six 5-l samples were collected during spring, summer, and au-tumn at each site. Three "replicates" were collected 30 cm below the stream bed (the hyporheic habitat or the underflow) and three were collected 30 cm below the water table in adjacent (2–20 m from the water's edge) bars (phreatic habitat). Because of the inordinate amount of time required to process samples in the laboratory, all three replicates were analyzed only from selected locations (Sites 1, 7, and 9). For comparative purposes, a third habitat, "surface gravel," was examined by taking cores of the top 10 cm of surficial bed sediments (largely sand and gravel) until a sufficient composite sample (1.5–3.0 l) was obtained. Site 5 was sampled monthly for one year in a previous study (Pennak and Ward, 1986) that did not include collections of surface gravel; data from spring, summer, and autumn hyporheic and phreatic samples (30-cm depths) are included herein.

Community composition varied as a function of site location and habitat type (Fig. 2; Table III). Some groups, such as bathynellaceans and archianne-lids, exhibited disjunct distribution patterns along the longitudinal profile. Others, such as cyclopoid copepods and chironomids, were represented at all sampling locations. Yet others were confined to a particular riverine segment (*e.g.,* amphipods and plecopterans).

Based on habitat occurrences, individual taxa were categorized as hypo-gean, epigean, transitional, or eurytopic (as defined in Table IV). The rela-tionship between biotope occurrence and site distribution pattern (Table IV) demonstrates (1) the absence of a distinctive plains faunal assemblage, (2) the restriction of many hypogean and epigean faunal elements to the mountain stream, and (3) the euryzonal character of the eurytopic fauna.

TABLE II Spot Measurements of Temperature and Chemical Variables for Surface (S), Hyporheic (H), and Phreatic (P) Waters

						Sampling site				
Variable		1	2	3	4	5	6	7	8	9
Temperature (°C)	Autumn									
	S	8.0	14.0	12.5	14.0	—	7.5	10.0	9.0	11.5
	H	7.5	13.0	12.0	12.5	9.0	8.5	11.0	10.5	9.0
	P	5.0	12.0	12.0	11.5	12.5	8.5	11.5	8.0	12.0
	Spring									
	S	4.0	6.0	7.0	5.0	—	6.0	9.0	18.0	17.0
	H	5.0	7.0	6.5	4.5	7.0	8.0	10.0	17.0	14.0
	P	7.0	6.5	7.0	4.0	3.0	11.0	10.0	12.0	11.5
	Summer									
	S	10.5	15.0	18.5	18.0	—	16.0	24.0	25.5	26.0
	H	10.5	15.5	17.5	18.0	17.0	15.5	18.5	20.5	23.0
	P	9.0	17.5	18.0	15.0	20.0	20.0	19.5	22.0	22.0
Dissolved oxygen (mg/l)	S	9.4	9.4	8.6	9.1	—	10.5	10.0	9.4	10.2
	H	7.1	6.1	<1.0	5.6	10.7	9.8	2.9	4.6	7.7
	P	9.4	4.0	<1.0	6.1	3.1	7.2	3.5	4.3	0.6
Total hardness (mg/l $CaCO_3$)	S	112	260	222	166	—	160	200	544	746
	H	120	276	252	174	—	140	242	602	756
	P	120	264	240	136	—	156	282	652	580
pH units	S	7.7	8.2	8.4	8.3	—	8.2	8.2	8.1	8.2
	H	7.5	8.3	7.0	7.8	7.0	7.8	7.0	7.3	8.2
	P	7.3	7.6	6.8	7.2	7.3	7.6	7.1	7.3	7.6
Free CO_2 (mg/l)	S	2.5	3.0	0.0	0.0	—	1.0	2.0	3.0	2.0
	H	3.0	7.5	—	4.0	2.5	11.0	15.0	16.0	4.5
	P	4.5	11.0	35.0	6.0	3.5	6.0	—	15.0	15.0

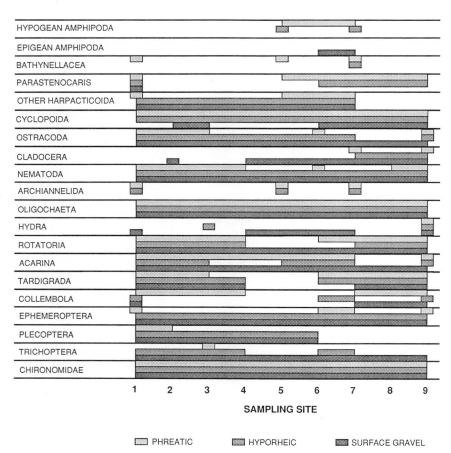

FIGURE 2   Distribution of major faunal groups, by habitat, along the elevation gradient. Surface gravel was not sampled at Site 5. If a taxon occurred in surface gravel at Sites 4 and 6, it was assumed to also be present at Site 5 for the purposes of this figure. Single-site occurrences (e.g., Bathynellacea at Sites 1, 5, and 7) are plotted as a narrow bar extending on either side of the site location.

Genera with the greatest hypogean affinities include *Troglochaetus* (Archiannelida), *Stygobromus* (Amphipoda), *Stygothrombium* and *Speleorchestres* (Acarina), an unknown Lumbriculidae (Oligochaeta), *Bathynella* (Bathynellacea), and *Parastenocaris* (Copepoda).

Many insects, although considered typical surface dwellers, were also collected from hyporheic habitats, but only certain collembolans, chironomids, and the ephemeropteran *Ameletus* penetrated the phreatic habitats in appreciable numbers. We did not encounter plecopterans that were confined to groundwaters for their entire nymphal existence as reported by Stanford and Ward (1988).

TABLE III Faunal List and Abundance Based on Means of All Sampling Dates as Follows: R (Rare), <10; C (Common), 10–100; and A (Abundant), >100 Organisms per 5-l Sample (Phreatic and Hyporheic) or per Liter Gravel (Surface Gravel)

Site and habitat

Taxon	1 P	1 H	1 G	2 P	2 H	2 G	3 P	3 H	3 G	4 P	4 H	4 G	5 P	5 H	5 G	6 P	6 H	6 G	7 P	7 H	7 G	8 P	8 H	8 G	9 P	9 H	9 G
Coelenterata																											
Hydra spp.	—	R	—	—	—	—	—	R	—	—	—	A	—	—	—	—	—	R	—	—	C	—	—	—	—	R	R
Turbellaria																											
Dugesia dorotocephala	—	R	—	—	—	—	—	—	—	—	R	—	—	—	—	—	—	—	—	—	—	—	—	—	—	—	—
Polycelis coronata	—	R	—	—	—	—	—	—	—	—	—	C	—	—	—	—	—	—	—	—	—	—	—	—	—	—	—
Microturbellaria	—	—	—	—	—	—	—	—	—	—	—	—	—	—	—	—	—	—	—	—	—	—	—	—	—	R	R
Nematoda	C	A	A	C	A	A	C	C	C	C	A	A	—	R	R	—	A	C	—	A	A	C	C	C	C	R	R
Archiannelida																											
Troglochaetus beranecki	A	C	—	—	—	—	—	—	—	—	—	—	R	C	—	—	—	—	C	R	—	—	—	—	—	—	—
Oligochaeta																											
Chaetogaster sp.	—	—	C	—	C	—	—	—	—	—	—	C	—	—	—	—	—	—	—	—	—	—	—	—	—	—	—
Nais barbata	—	—	—	—	—	—	—	—	—	—	—	—	—	—	—	—	—	C	—	R	R	—	—	—	—	R	—
Nais communis	—	—	—	—	—	—	—	—	—	—	—	—	—	R	C	—	—	R	—	R	C	—	R	R	—	R	R
Nais elinguis	—	—	—	—	—	—	—	—	—	—	—	R	—	—	—	—	—	—	—	—	—	—	—	—	—	—	—
Nais variabilis	—	—	—	—	R	—	—	—	—	—	R	—	—	—	—	—	—	—	—	R	R	—	—	—	—	R	R
Pristina aequiseta	—	R	—	—	R	—	—	—	—	—	C	C	—	—	—	—	—	—	—	R	R	—	—	—	—	—	—
Pristina osborni	—	R	—	—	R	R	—	—	—	—	R	R	—	—	—	—	—	—	—	—	R	—	—	—	—	—	—
Pristina sima	—	—	—	—	—	—	—	—	—	—	R	R	—	—	—	—	—	—	—	—	—	—	—	—	—	—	—
Pristinella jenkinae	—	—	C	—	—	C	—	—	—	—	R	R	—	—	—	—	—	—	—	—	—	—	—	—	—	—	—
Ophidonais serpentina	—	—	—	—	—	—	—	—	—	—	R	—	—	—	—	—	C	R	—	R	R	—	R	R	—	—	—
Vejdovskyella intermedia	—	—	—	—	—	—	—	—	—	—	R	R	—	—	—	—	R	R	—	R	C	—	—	—	—	R	R
Aeolosoma sp.	—	—	—	—	—	—	—	R	—	R	R	A	—	—	—	—	C	C	—	C	C	—	R	C	—	R	—
Limnodrilus hoffmeisteri	—	—	—	—	R	—	—	—	—	—	—	—	—	—	—	—	C	C	—	R	C	—	C	R	—	R	R
Psammoryctides californianus (?)	R	R	—	—	—	—	—	—	R	—	R	R	—	—	—	—	—	—	—	—	—	—	R	—	—	—	—

(continues)

TABLE III (Continued)

Site and habitat

Taxon	1 P	1 H	1 G	2 P	2 H	2 G	3 P	3 H	3 G	4 P	4 H	4 G	5 P	5 H	5 G	6 P	6 H	6 G	7 P	7 H	7 G	8 P	8 H	8 G	9 P	9 H	9 G
Aulodrilus limnobius	—	C	—	—	R	—	—	R	—	—	C	C	—	—	—	—	—	—	—	—	R	—	—	—	—	—	—
Lumbricillus sp.	—	C	—	—	R	C	—	R	R	—	C	C	—	—	—	—	R	R	—	R	R	R	—	R	—	—	—
Haplotaxis gordioides	—	—	—	—	—	—	—	—	—	—	—	R	—	—	—	—	—	—	R	R	—	—	—	—	—	—	—
Lumbriculidae indet. sp.	—	—	—	—	—	—	R	—	R	R	R	R	—	—	—	—	—	—	—	—	—	R	—	—	—	—	—
Eiseniella tetraedra	—	—	—	—	—	R	—	—	R	R	R	R	R	A	R	R	—	R	—	—	—	—	R	—	—	—	—
Hirudinea																											
Erpobdella punctata	—	—	—	—	—	—	—	—	—	—	R	—	—	—	—	—	—	—	—	—	—	—	—	—	—	—	—
Tardigrada																											
Macrobiotus spp.	R	R	R	C	R	C	R	C	R	C	C	C	R	—	R	R	R	R	C	C	R	R	R	C	R	R	R
Rotatoria																											
Cephalodella sp.	—	C	C	C	R	R	R	R	—	R	—	R	—	—	—	R	—	R	R	C	—	C	C	R	C	R	C
Notholca sp.	R	R	—	—	—	—	—	—	R	—	—	—	—	—	—	—	—	—	C	R	C	A	A	R	A	A	A
Brachionus sp.	R	R	R	R	R	—	—	—	—	—	R	—	—	—	—	C	C	—	C	R	C	C	C	C	—	—	—
Kellicottia longispina	—	—	—	C	C	—	C	—	—	—	—	—	R	R	R	C	—	C	C	C	R	C	C	—	—	—	—
Polyarthra sp.	R	—	—	—	—	—	—	—	R	—	—	—	—	—	—	—	—	—	—	—	—	—	—	—	—	—	—
Philodina sp.	R	C	—	—	—	—	—	—	—	—	—	—	—	—	—	—	—	—	—	—	—	—	—	—	—	—	—
Dissotrocha sp.	R	R	—	—	—	—	—	—	—	—	—	—	—	—	—	—	—	—	—	—	—	—	—	—	—	—	—
Amphipoda																											
Stygobromus spp.	—	—	—	—	—	—	—	—	—	—	—	—	R	R	R	R	—	R	R	R	—	R	R	R	—	—	—
Crangonyx gracilis complex	—	—	—	—	—	—	—	—	—	—	—	—	—	R	—	—	R	—	—	R	—	—	R	—	—	—	—
Bathynellacea																											
Bathynella riparia	R	—	—	—	—	—	—	—	—	—	—	—	C	—	C	C	—	—	C	R	C	—	—	—	—	—	—
Harpacticoida																											
Parastenocaris spp.	A	A	—	—	—	—	—	—	—	—	—	—	R	R	R	C	A	C	C	C	C	C	R	C	—	C	C
Attheyella idahoensis	C	C	—	—	—	C	—	—	—	C	C	A	A	A	A	A	C	C	C	C	—	R	R	R	—	—	—
Bryocamptus spp.	A	A	C	—	—	—	—	—	—	A	A	A	—	—	—	C	C	—	—	—	—	—	—	—	C	—	—
Canthocamptus vagus	C	—	—	—	—	—	—	—	—	—	—	—	—	—	C	—	—	—	R	R	R	—	—	—	—	—	—

402

Canthocamptidae indet. sp.	—	—	—	—	C	—	—	—	—	—	—	—	—	—	—	—	—	—	—	—	—	—	—	—	—
Maraenobiotus insignipes	—	—	—	—	—	—	R	C	—	—	—	—	—	—	—	—	—	—	—	—	—	—	—	—	—
Moraria mrazeki	—	—	—	—	—	—	—	—	—	C	—	—	—	—	—	—	—	—	—	—	—	—	—	—	—
Nitocrella incerta (?)	—	—	—	—	C	—	—	—	—	—	—	—	—	—	—	—	—	—	—	—	—	—	—	—	—
Onychocamptus mohammed	—	—	—	—	R	—	—	—	—	—	—	—	—	—	—	—	—	—	—	—	—	—	—	—	—
Phyllognathopus viguieri	—	—	—	—	—	—	—	—	—	—	—	A	A	—	—	—	—	—	—	—	—	—	—	—	—
Cyclopoida																									
Acanthocyclops pennaki	—	—	—	—	—	—	—	—	—	—	R	R	—	—	—	—	—	—	—	—	—	—	—	—	—
Acanthocyclops vernalis	R	C	—	—	—	C	C	—	—	—	—	C	A	—	—	—	—	—	—	—	—	R	—	—	R
Diacyclops crassicaudis brachycercus	—	—	—	—	—	R	C	—	—	—	—	—	—	—	—	—	—	—	—	C	—	C	C	C	R
Diacyclops nearcticus	—	—	—	—	—	—	—	—	C	C	—	—	—	—	—	—	—	—	—	—	—	—	—	—	—
Diacyclops thomasi	—	—	—	—	—	—	C	—	—	—	—	—	—	—	C	R	C	—	—	—	—	—	—	—	—
Macrocyclops albidus	—	—	—	—	—	—	—	—	—	—	—	C	—	—	—	—	—	—	—	—	—	—	—	—	—
Microcyclops pumilis	—	—	—	—	—	—	—	—	—	—	C	A	—	—	—	—	—	—	—	—	—	—	—	—	—
Microcyclops varicans rubellus	—	R	—	R	C	—	—	—	—	—	—	C	—	R	—	—	—	—	R	—	—	—	—	—	—
Microcyclops sp.	—	—	—	—	—	—	—	—	—	—	—	R	—	—	—	—	—	—	—	—	—	—	—	—	—
Paracyclops fimbriatus chiltoni	R	—	—	—	—	—	—	—	C	—	—	—	A	R	—	C	R	—	—	—	—	—	—	—	—
Eucyclops sp.	—	—	—	—	—	—	—	R	—	—	—	—	—	—	—	—	—	—	—	—	—	—	—	—	—
Calanoida	—	—	—	—	—	R	—	—	—	—	—	—	—	—	—	—	—	—	R	R	—	—	—	—	—
Ostracoda																									
Eucypris sp.	—	—	—	—	—	—	—	—	—	—	—	R	—	—	—	—	—	—	—	—	—	—	—	—	—
Cavernocypris wardi	C	C	C	—	C	—	—	—	—	—	C	—	R	C	—	R	—	—	—	—	—	—	—	—	—
Potamocypris spp.	—	—	—	—	—	—	—	—	—	—	—	—	—	—	—	—	—	—	—	—	R	—	R	R	—
Strandesia canadensis	—	R	C	—	—	—	—	—	—	—	—	—	—	—	—	—	—	—	—	—	—	—	—	—	—
Cypria sp.	—	R	R	—	—	—	—	—	—	—	—	—	—	—	—	—	—	—	—	—	—	—	—	—	—
Ilyocypris sp.	—	R	—	—	—	—	—	—	—	—	—	—	—	—	—	R	—	—	—	—	—	R	—	—	—
Candona candida	—	—	—	—	R	—	—	—	—	—	—	—	—	—	—	—	—	—	—	—	—	—	—	—	—
Candona sp. 1	—	R	—	—	R	—	—	C	R	—	—	—	—	—	—	—	—	—	—	—	—	—	—	—	—
Fabaeformiscandona wegelini	—	—	—	—	—	R	—	—	—	—	—	—	—	—	—	R	R	—	—	—	—	—	—	—	—
Nannocandona faba	R	R	—	—	—	—	—	—	—	—	—	—	—	—	—	—	—	—	—	—	—	—	—	—	—
Pseudocandona sp. 1	—	—	—	—	—	—	—	—	—	—	—	—	R	R	R	—	—	—	—	—	—	—	—	—	—
Fabaeformiscandona pennaki	—	—	—	—	R	—	—	—	—	—	C	—	R	—	—	—	—	—	—	—	—	—	—	—	—
Pseudocandona sp. 2	—	—	—	—	R	C	C	—	—	—	—	—	—	—	—	—	—	—	—	—	—	—	—	—	—
Pseudocandona sp. 3	—	R	—	—	—	—	—	—	—	—	—	—	—	—	—	R	—	—	—	—	—	R	—	—	—
Pseudocandona cf. *albicans*	—	—	—	—	—	—	—	—	R	C	—	—	—	—	—	—	—	—	—	—	—	—	—	—	—

(continues)

TABLE III (Continued)

Taxon	1 P	1 H	1 G	2 P	2 H	2 G	3 P	3 H	3 G	4 P	4 H	4 G	5 P	5 H	5 G	6 P	6 H	6 G	7 P	7 H	7 G	8 P	8 H	8 G	9 P	9 H	9 G
Cladocera																											
Acantholeberis sp.	—	—	—	—	—	—	—	—	—	—	—	—	—	—	—	—	—	—	—	—	—	—	R	R	R	—	R
Alona spp.	—	—	—	—	—	—	—	—	—	—	C	—	—	—	—	—	—	C	—	—	—	R	R	—	—	R	R
Ceriodaphnia sp.	—	—	—	—	—	—	—	—	—	—	—	—	—	—	—	—	—	R	—	—	—	—	—	—	—	—	—
Chydorus sp.	—	—	—	—	R	—	—	—	—	—	—	—	—	—	—	—	—	R	—	R	C	—	—	—	—	—	—
Daphnia spp.	—	—	—	—	—	—	—	—	—	—	—	—	—	—	—	—	—	—	—	R	R	—	—	—	—	—	—
Myriopoda																											
Acarina																											
Arctoseius cetratus	—	R	R	—	—	—	—	—	—	—	—	—	—	—	—	—	—	—	—	—	—	—	—	—	—	—	—
Dendrolaelaps sp.	—	C	—	—	—	—	—	—	—	—	—	—	—	—	—	—	—	—	—	—	—	—	—	—	—	—	—
Stygothrombium sp.	—	—	—	—	—	C	—	—	—	—	—	—	—	—	—	—	—	—	—	—	—	—	—	—	—	—	—
Limnohalacaridae[a]	—	—	—	—	R	R	—	—	—	—	—	—	R	R	—	A	A	A	R	R	—	—	—	—	—	—	—
Speleorchestres sp.	—	—	—	—	—	—	R	—	—	—	—	—	—	—	—	R	R	R	—	—	—	—	—	—	—	—	—
Bryobia sp.	—	R	C	—	—	—	—	—	—	—	—	—	—	—	—	—	—	—	—	—	—	—	—	—	—	—	—
Alicorhagia sp.	—	C	—	—	—	—	—	—	—	—	—	—	—	—	—	—	—	—	—	C	R	—	—	—	R	R	—
Teutonia sp.	—	—	R	—	—	—	—	R	—	—	—	—	—	—	—	—	—	—	—	—	—	—	—	—	—	—	—
Feltria spp.	—	—	R	—	R	—	—	—	—	—	—	—	—	—	—	—	R	R	—	—	R	—	—	—	—	—	—
Atractides sp.	R	—	—	—	—	—	—	—	—	—	R	—	—	R	R	—	—	R	—	—	—	—	—	—	—	—	—
Lebertia sp.	—	—	—	—	—	—	—	—	—	—	R	—	—	—	—	—	—	—	—	—	—	—	—	—	—	—	—
Sperchon spp.	—	—	—	—	—	—	R	—	—	—	—	—	—	—	—	—	R	R	—	—	—	—	—	—	—	—	—
Neoacarus sp.	—	—	R	—	—	—	R	—	—	—	—	—	—	—	—	—	R	R	—	—	—	—	—	—	—	—	—
Aturus sp.	—	R	—	—	—	—	—	—	—	—	—	—	—	—	—	—	—	R	—	—	—	—	—	—	—	—	—
Trimalaconothrus sp.	—	—	—	—	R	—	—	—	—	R	—	—	—	—	—	—	R	R	—	—	—	—	—	—	—	—	—
Malaconothrus sp.	—	—	—	—	R	—	—	—	—	—	—	—	—	—	—	—	—	—	—	—	—	—	—	—	—	—	—
Gastropoda																											
Physella sp.	—	—	—	—	—	—	—	—	—	—	R	—	—	—	—	—	—	—	—	—	—	—	—	—	—	—	—

Taxon																										
Pelecypoda																										
Pisidium spp.	—	R	—	—	—	—	—	—	—	—	—	—	—	—	—	—	—	—	—	—	—	—	—	—	—	—
Collembola																										
Agrenia bidenticulata	R	R	—	—	—	—	—	—	—	—	—	—	—	—	—	—	—	—	—	—	—	—	R	—	R	—
Isotomurus binus	—	—	—	—	—	—	—	—	—	—	—	—	—	—	—	—	—	—	—	R	R	R	—	R	—	—
Isotomiella minor	—	—	—	—	—	—	—	—	—	—	—	—	—	—	—	—	—	—	R	R	R	R	—	—	—	—
Entomobrya sp.	—	—	R	—	—	—	—	—	—	—	—	—	—	—	—	—	—	—	—	—	—	R	—	—	—	—
Hypogastrura denticulata (?)	—	—	—	—	—	—	—	—	—	—	—	—	—	R	—	R	—	—	—	R	—	R	—	—	—	—
Onychiurus armatus complex	—	—	R	—	—	—	—	—	—	—	—	—	—	R	—	R	C	R	C	R	R	R	R	R	R	—
Proisotoma sp.	—	C	—	—	—	—	—	—	—	—	—	—	—	—	—	—	—	—	—	—	—	—	—	—	—	—
Sinella sp.	C	—	—	—	—	—	—	—	—	—	—	—	—	—	—	—	—	—	—	R	—	R	R	R	R	—
Tullbergia macrochaeta	C	—	—	—	—	—	—	—	—	—	—	—	—	—	—	—	—	—	—	—	—	R	R	R	R	—
Folsomia sp.	C	—	—	—	—	—	—	—	—	—	—	—	—	—	—	—	—	—	—	R	R	—	—	—	—	—
Paranura cf *anops*	—	—	—	—	—	—	—	—	—	—	—	—	—	—	—	—	—	—	—	R	—	—	—	—	—	—
Anurida granularia	—	—	—	R	—	—	—	—	—	—	—	—	—	—	—	—	—	—	—	—	—	—	—	—	—	—
Ephemeroptera																										
Baetis spp.	C	R	—	R	R	—	—	—	—	C	R	—	C	R	C	R	—	—	R	R	R	R	—	R	R	R
Ameletus spp.	C	C	—	R	R	—	—	—	—	R	R	—	R	R	R	R	—	—	R	R	R	—	R	R	R	R
Choroterpes sp.	—	—	—	—	—	—	—	—	—	—	—	—	—	—	—	—	—	—	—	—	—	—	—	R	—	—
Paraleptophlebia sp.	—	R	—	R	R	—	R	—	—	R	R	—	—	—	R	—	R	—	R	—	R	C	R	—	—	—
Drunella doddsi	—	R	—	R	R	—	C	—	R	R	R	—	R	—	R	—	R	—	R	—	—	R	—	C	R	—
Ephemerella infrequens/inermis	—	—	—	—	—	—	—	—	R	C	—	—	R	R	—	—	—	—	C	—	—	—	—	—	—	—
Cinygmula sp.	—	R	—	C	R	—	—	—	—	R	—	—	C	—	—	—	C	—	—	—	—	C	—	R	—	—
Heptagenia elegantula	—	—	—	R	—	—	R	—	—	R	—	—	R	R	—	—	—	—	—	—	—	R	—	—	—	—
Rhithrogena hageni	—	R	—	R	R	—	C	—	R	C	—	—	C	—	—	—	C	—	—	—	—	C	—	R	—	—
Tricorythodes minutus	—	—	—	—	—	—	—	—	—	—	—	—	R	—	—	—	R	—	—	—	—	R	—	—	—	—
Odonata																										
Ophiogomphus severus	—	—	—	R	—	—	—	—	—	—	—	—	R	—	—	—	—	—	—	—	—	—	—	—	—	—
Plecoptera																										
Capnia/Utacapnia spp.	—	R	—	R	R	—	R	—	—	R	R	—	R	R	R	R	—	—	R	—	R	R	—	—	—	—
Paraleuctra sp.	—	R	—	R	—	—	—	—	—	R	—	—	R	—	—	—	—	—	—	—	—	—	—	—	—	—
Taenionema sp.	—	R	—	—	—	—	—	—	—	—	—	—	—	—	—	—	R	—	—	—	—	R	—	—	—	—
Zapada oregonensis	—	—	—	R	—	R	—	—	—	—	—	—	—	—	—	—	R	—	—	—	—	—	—	—	—	—
Claassenia sabulosa	—	—	—	—	—	R	—	—	—	—	—	—	—	—	—	—	R	—	—	—	—	—	—	—	—	—
Cultus sp.	—	—	—	—	—	—	—	—	—	—	—	—	—	—	—	—	—	—	—	—	—	—	—	—	—	—

(continues)

405

TABLE III (Continued)

	Site and habitat																										
	1			2			3			4			5			6			7			8			9		
Taxon	P	H	G	P	H	G	P	H	G	P	H	G	P	H	G	P	H	G	P	H	G	P	H	G	P	H	G
Isoperla fulva	—	—	—	—	—	—	R	R	—	—	—	—	—	—	—	—	—	—	—	—	—	—	—	—	—	—	—
Alloperla sp.	—	—	R	—	—	—	—	—	—	—	—	—	—	—	—	—	—	—	—	—	—	—	—	—	—	—	—
Suwallia spp.	—	—	R	—	—	—	—	—	—	—	—	—	—	—	—	—	—	—	—	—	—	—	—	—	—	—	—
Sweltsa spp.	R	R	R	—	—	R	—	—	—	—	—	R	—	—	R	—	R	R	—	—	—	—	—	—	—	—	—
Trichoptera																											
Rhyacophila spp.	R	R	—	R	R	—	R	R	—	—	—	—	—	—	—	—	—	—	—	—	—	—	—	—	—	—	—
Glossosoma spp.	R	R	—	R	—	—	R	R	—	—	—	R	—	—	—	—	R	—	—	—	—	—	—	—	—	—	—
Brachycentrus sp.	—	—	—	—	—	—	—	—	—	—	—	R	—	—	—	—	—	R	—	—	—	—	—	—	—	—	—
Lepidostoma sp.	—	—	—	—	—	—	—	—	—	—	R	C	—	—	—	—	—	—	—	—	—	—	—	—	—	—	—
Oecetis sp.	—	—	—	—	—	—	—	R	—	—	R	C	—	—	—	—	—	—	—	—	—	—	—	—	—	—	—
Cheumatopsyche spp.	—	—	—	—	—	—	—	R	—	—	R	C	—	—	—	—	R	R	—	R	R	—	—	—	—	—	—
Hydropsyche spp.	—	—	—	R	R	—	R	R	—	—	R	—	—	—	—	R	R	—	—	R	R	—	R	—	—	—	R
Hydroptila sp.	—	—	—	—	—	—	R	—	—	—	—	—	—	—	—	—	—	—	—	—	—	—	—	—	—	—	—
Psychomyia flavida	—	—	—	—	—	—	—	—	—	—	—	—	—	—	—	—	—	—	—	R	—	—	—	—	—	—	—
Coleoptera																											
Optioservus spp.	—	—	—	—	—	—	—	R	—	—	R	C	—	C	—	—	—	R	—	R	—	—	R	—	—	—	—
Heterlimnius corpulentus	—	—	—	—	R	—	—	—	—	—	R	—	—	R	—	—	—	—	—	—	—	—	R	—	—	—	—
Ceratopogonidae																											
Alluaudomyia sp.	—	—	—	—	—	—	—	—	—	—	—	—	—	—	—	—	—	—	—	R	—	—	R	R	—	R	—
Bezzia/Palpomyia spp.	R	—	—	—	—	—	—	R	—	—	—	—	—	C	—	—	—	—	—	—	—	—	R	R	—	—	—
Probezzia sp.	C	—	—	R	—	—	—	—	—	—	—	—	—	—	—	—	—	—	—	—	—	—	—	—	—	—	—
Culicoides sp.	—	—	—	—	—	—	—	—	—	—	—	—	—	—	—	—	—	—	—	—	—	—	—	—	—	—	—
Dasyhelea sp.	—	—	—	—	—	—	—	R	—	—	—	—	—	—	—	—	—	—	—	—	—	—	—	—	—	—	—
Monohelea sp.	—	—	—	—	—	—	—	R	—	—	—	—	—	—	—	—	—	—	—	—	—	—	—	—	—	—	—
Athericidae																											
Atherix pachypus	R	R	—	—	—	—	—	—	—	—	—	—	—	—	—	—	R	—	—	—	—	—	—	—	—	—	—

Taxon																							
Tipulidae																							
Dicranota sp.	—	—	—	—	—	—	—	—	—	—	—	—	—	—	—	—	—	—	—	—	—	R	R
Erioptera sp.	—	R	—	—	—	—	—	—	—	—	—	—	—	—	—	—	—	—	—	—	—	C	R
Hesperoconopa sp.	—	R	R	—	—	—	—	—	—	—	—	—	—	—	—	—	—	—	—	—	—	R	R
Hexatoma sp.	—	R	R	—	—	—	—	—	—	—	—	—	—	—	—	—	—	—	—	—	—	—	—
Limnophila sp.	—	—	—	—	—	—	—	—	—	—	—	—	R	—	—	—	—	—	—	R	—	R	—
Molophilus sp.	R	—	—	—	—	—	—	—	—	—	—	—	R	—	—	—	—	—	—	R	—	R	—
Ormosia sp.	—	—	—	R	—	—	—	—	—	—	—	—	R	—	—	—	—	—	—	R	—	C	—
Pilaria sp.	—	—	—	—	—	—	—	—	—	—	—	—	—	—	—	—	—	—	—	—	—	C	—
Polymera sp.	—	—	R	—	—	—	—	—	—	—	—	—	—	—	—	—	—	—	—	—	—	R	—
Tanyderidae																							
Protanyderus margarita	—	—	—	—	—	—	—	—	—	—	—	—	—	—	R	—	—	—	—	—	—	—	—
Empididae																							
Chelifera sp.	—	—	—	—	—	R	R	—	—	—	—	—	—	R	—	—	—	—	—	—	—	—	—
Hemerodromia sp.	—	—	—	—	R	R	—	—	—	—	—	—	—	—	—	R	—	—	—	—	—	—	—
Muscidae																							
Limnophora sp.	—	—	—	—	—	—	—	—	—	—	—	—	—	—	—	—	R	—	—	R	—	R	R
Simuliidae																							
Simulium spp.	—	—	—	—	—	R	R	—	—	—	—	—	—	R	R	—	R	—	—	C	—	C	R
Prosimulium sp.	—	—	—	—	—	—	—	—	—	—	—	—	—	R	—	—	—	—	—	—	R	R	—
Psychodidae																							
Pericoma sp.	—	—	—	—	—	—	—	—	—	—	—	—	—	—	—	P	—	—	—	—	—	—	—
Chironomidae																							
Natarsia sp.	R	—	—	—	—	—	—	—	—	—	—	—	—	—	R	R	—	R	—	—	—	—	—
Thienemannimyia gr.	—	R	—	C	C	R	—	—	—	R	R	R	R	R	R	R	—	C	—	R	R	—	—
Diamesa spinacies	—	—	—	R	R	R	R	—	—	—	R	R	—	R	R	R	—	R	—	—	—	—	—
Pagastia sp.	—	—	—	R	R	R	R	—	—	R	—	—	—	—	—	—	—	R	—	—	—	—	—
Potthastia longimanus	R	—	R	—	C	R	R	—	—	C	C	—	—	—	—	R	—	R	—	—	C	C	R
Corynoneura celeripes	C	R	R	C	R	R	R	—	—	C	C	—	—	—	R	C	—	R	—	R	R	C	R
Corynoneura sp.	R	R	C	R	R	R	A	R	R	A	C	R	—	R	R	A	—	C	R	R	A	R	R
Cricotopus bicinctus gr.	R	—	R	R	R	—	R	—	—	R	R	—	—	—	—	R	—	R	C	R	C	R	R
Cricotopus spp.	—	R	R	C	C	R	C	—	—	C	C	—	—	—	—	R	—	—	R	C	A	R	R
Cricotopus/Orthocladius spp.	R	—	—	—	—	—	—	—	—	R	—	—	—	—	—	—	—	—	—	R	—	—	R
Eukiefferiella pseudomontana gr.	—	—	R	—	—	—	—	—	—	—	—	—	—	—	—	—	—	R	—	R	R	R	—
Eukiefferiella sp.	—	—	—	—	—	R	R	—	—	R	R	R	—	—	—	—	—	R	—	R	R	R	R

(continues)

TABLE III (Continued)

Taxon	1 P	1 H	1 G	2 P	2 H	2 G	3 P	3 H	3 G	4 P	4 H	4 G	5 P	5 H	5 G	6 P	6 H	6 G	7 P	7 H	7 G	8 P	8 H	8 G	9 P	9 H	9 G
Heleniella sp.	—	R	C	—	R	—	—	—	—	—	—	—	—	—	—	—	—	—	—	—	—	—	—	—	—	—	—
Heterotrissocladius sp.	R	R	R	—	R	—	—	—	—	—	—	—	—	—	—	—	R	—	—	—	—	—	—	—	—	—	—
Krenosmittia sp.	R	R	R	—	—	—	—	—	—	—	—	—	—	—	—	—	—	—	—	—	—	—	—	—	—	—	—
Lopescladius sp.	—	—	—	—	—	—	—	R	C	—	R	R	—	—	—	—	R	R	—	—	C	—	—	—	—	—	—
Nanocladius distinctus	—	—	—	—	—	—	—	R	—	—	R	A	—	R	—	—	R	R	—	R	R	—	R	—	—	—	—
Orthocladius spp.	—	—	—	—	R	—	—	R	—	—	—	R	—	R	—	—	—	—	—	R	—	—	—	—	—	—	—
Parakiefferiella sp.	—	R	R	—	R	C	—	R	R	—	R	R	—	—	—	—	—	—	—	—	—	—	—	—	—	—	—
Parametriocnemus sp.	R	R	R	—	—	—	—	R	R	—	R	R	—	—	—	—	—	—	—	—	—	—	—	—	—	R	R
Rheocricotopus sp.	R	R	—	—	R	—	—	—	—	—	—	—	—	R	—	—	—	—	—	—	—	—	—	—	—	—	R
Rheosmittia sp.	—	—	—	—	—	—	—	R	A	—	—	—	—	—	—	—	—	—	—	—	—	—	—	—	—	—	—
Rheotanytarsus sp.	—	—	—	—	—	—	—	R	A	—	—	—	—	R	—	—	—	—	—	—	—	—	R	—	—	—	R
Synorthocladius sp.	R	R	—	—	—	—	—	R	R	—	—	—	—	—	—	—	—	R	—	—	—	—	—	—	—	—	R
Thienemanniella nr *fusca*	R	R	C	—	C	R	—	R	R	—	—	C	—	—	R	—	—	—	—	—	—	—	—	—	—	—	—
Thienemanniella (prob. *xena*)	—	—	—	—	R	—	—	—	—	—	R	—	—	—	R	—	—	R	—	—	—	—	—	—	—	—	—
Tvetenia discoloripes gr.	—	—	—	—	—	—	—	R	R	—	R	—	—	—	—	—	R	R	—	—	—	—	—	—	—	—	—
Chironomus sp.	—	—	—	—	—	—	—	—	—	—	—	—	—	—	—	—	—	—	—	—	—	—	—	—	R	R	R
Cladotanytarsus sp.	R	R	C	—	C	R	—	C	C	—	C	A	—	—	—	—	C	—	—	R	R	—	—	—	—	—	R
Cladotanytarsus (*Lenziella*) sp.	—	—	—	—	C	—	—	C	—	—	R	R	—	—	—	—	C	C	—	R	R	—	R	—	—	—	R
Cryptochironomus sp.	—	—	—	—	—	—	—	C	—	—	R	R	—	—	—	—	R	R	—	R	R	—	R	—	—	—	—
Dicrotendipes neomodestus	—	—	—	—	—	—	—	—	—	—	R	R	—	—	—	—	—	—	—	R	R	—	—	—	—	—	—
Micropsectra sp.	—	C	—	—	C	—	—	—	—	—	C	C	—	—	—	—	—	—	—	—	—	—	—	—	—	—	—
Microtendipes pedellus gr.	—	—	—	—	—	—	—	R	R	—	R	R	—	—	—	—	C	—	—	R	R	—	R	—	—	—	R
Microtendipes sp.	—	—	—	—	—	—	—	R	R	—	R	C	—	R	R	—	R	R	—	R	R	—	—	—	—	—	R
Paratanytarsus sp.	C	—	—	—	—	—	—	—	—	—	—	C	—	R	—	—	—	—	—	—	R	—	—	—	—	—	—
Paratendipes sp.	—	—	—	—	—	—	—	C	—	—	—	—	—	—	—	—	—	—	—	—	—	—	—	—	—	—	—
Phaenopsectra (probably *dyari*)	—	—	—	—	—	—	—	R	R	—	R	R	—	—	—	—	R	—	—	R	R	—	—	—	—	—	R
Polypedilum convictum	R	R	—	R	C	—	R	C	C	—	R	C	—	—	—	R	C	—	R	R	R	—	—	—	—	R	R

408

Taxon																				
Polypedilum fallax gr.	—	—	—	—	—	—	—	—	—	—	—	—	R	—	—	—	—	—	—	—
Polypedilum sp. C	—	—	R	C	—	R	C	—	R	C	—	—	R	—	—	R	—	—	—	—
Polypedilum sp.	—	—	—	—	—	—	—	—	—	—	—	R	—	—	R	R	—	R	R	—
Robackia sp.	—	—	—	—	—	—	—	—	R	—	—	—	R	—	R	R	—	R	—	—
Saetheria sp.	—	—	—	—	—	—	—	—	—	—	—	—	—	—	R	C	—	—	R	R
Stempellina sp.	—	—	—	—	—	—	—	—	R	—	R	—	—	—	—	—	—	—	—	—
Stictochironomus sp.	R	—	—	—	—	—	—	—	R	—	R	—	—	R	R	—	R	—	—	R
Tanytarsus sp.	—	—	—	A	—	—	C	—	C	—	—	—	R	—	R	—	—	—	—	—

Note. Habitats designated by P (phreatic), H (hyporheic), and G (surface gravel).

[a] *Lobohalacarus* sp. and *Soldanellonyx* sp.

TABLE IV   Relationship between Biotope Occurrence and Altitudinal Distribution Pattern Based on Percentage Composition by Number of Common Taxa[a]

	Mountain fauna	Plains fauna	Ecotonal fauna	Euryzonal fauna
Hypogean fauna	67	—	6	28
Epigean fauna	57	—	14	29
Transitional fauna	59	4	11	26
Eurytopic fauna	33	—	3	63

[a] Excluding rare taxa taken at only one site. Mountain fauna, taxa restricted to Sites 1–6; plains fauna, taxa restricted to Site 8, Site 9, or both; ecotonal fauna, taxa restricted to Sites 5–7 in the foothills/plains ecotone; euryzonal fauna, taxa occurring at one or more mountain sites and one or more plains sites; hypogean fauna, taxa restricted to phreatic and/or hyporheic habitats, rarely if ever collected from surficial stream sediment; epigean fauna, taxa restricted to surface gravel, rarely if ever found in the hyporheic habitat and never collected from the phreatic habitat; transitional fauna, commonly occurring in both hyporheic and surface gravel habitats, but rarely if every taken from the phreatic habitat; eurytopic fauna, commonly occurring in all three habitats.

Mean faunal densities at different sites, all dates combined, ranged from 163 to 1871 animals per 5-l sample for hyporheic habitats and from 83 to 441 animals per 5-l sample for phreatic habitats. Abundance levels were higher in hyporheic than in phreatic habitats, except at plains sites, at which values were similar (Fig. 3). Mean densities in surface gravels ranged from 306 to 4886 animals per liter gravel. The densities in all three habitats exhibited similar patterns along the course of the river. The groundwater fauna was most abundant at the high elevation location (Site 1) and at locations in the foothills/plains ecotone (Sites 5–7).

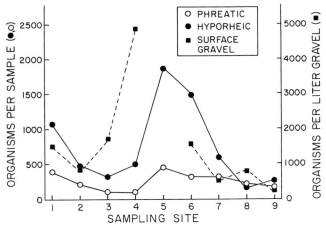

FIGURE 3   Mean abundance levels, all dates combined, of the total fauna collected from each habitat.

The eight major taxa in Fig. 4 account for 97.6 to 100% of the total fauna collected from habitats at each site. Insects and crustaceans collectively contributed the majority of organisms at most sites in most habitats. The relative abundance of crustaceans typically declined, concomitant with an increase in the relative abundance of insects, in the series phreatic–hyporheic–surface gravel habitats (Fig. 4). Rotifers were abundant in the plains river, especially Site 9, at which densities were greater than those of crustaceans or insects in all three habitats.

A total of 213 taxa have been identified, 89 of which were collected from phreatic habitats; 142, from hyporheic habitats; and 173, from surface gravel; 56 taxa were found only in surface gravel. Of the 213 total taxa, 114 are insects and 46 are crustaceans. Insect diversity declined from surface gravel (102 taxa), through the hyporheic (69 taxa), to the phreatic habitats (34 taxa), whereas crustacean diversity was not markedly different in the three habitats (32, 35, and 24 taxa, respectively). Faunal diversity exhibited no distinct pattern along the longitudinal profile, but there was a clear trend of progressively higher diversity from phreatic to hyporheic to surface gravel habitats (Fig. 5).

Alluvial aquifers contain a reservoir of biodiversity that remains largely unexplored in North America. Seven species from the South Platte River new to science have been described. All new species are crustaceans and six are endemic to the river system.

During an earlier study in the vicinity of Site 5, four blind amphipods were discovered in benthic stream samples. Additional sampling deeper in the stream bed recovered additional specimens, consisting of two new species

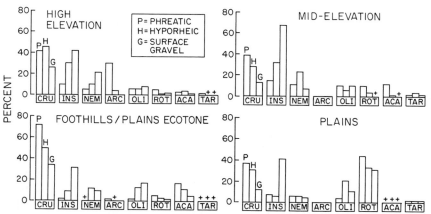

FIGURE 4 Relative abundances of the eight major faunal groups by habitat and elevation zones. High elevation, Site 1; midelevation, Sites 2, 3, and 4; foothills/plains ecotone, Sites 5, 6, and 7; and plains, Sites 8 and 9. CRU, crustacean; INS, insects; NEM, nematodes, ARC, archiannelids; OLI, oligochaetes; ROT, rotifers; ACA, acarines; TAR, tardigrades; +, present at <1.0%.

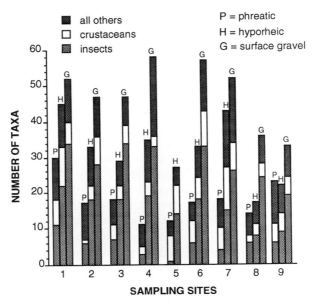

FIGURE 5   Number of animal taxa identified from each habitat at each site.

of *Stygobromus,* the first record of subterranean amphipods from Colorado (Ward, 1977). This discovery suggested that a groundwater community might inhabit the alluvial aquifer of the river and led to a detailed study of the remarkable subterranean fauna at this location (Pennak and Ward, 1986) and to the extensive longitudinal study reported herein.

Among the ostracods, two new species have been described (*Caverno-cypris wardi* and *Fabaeformiscandona pennaki*), and two species previously known only from Europe (*F. wegelini* and *Nannocandona faba*) have been recorded (Marmonier *et al.,* 1989; Marmonier and Ward, 1990). *C. wardi* is widely distributed in the South Platte River (Table III) and has subsequently been collected from the alluvial aquifer of the Flathead River, Montana, where it is the most abundant ostracod in well samples (Ward *et al.,* 1994).

The phreatic habitat of Site 5 also contained *Bathynella riparia,* only the second bathynellacean from North America described (Pennak and Ward, 1985a). *B. riparia* is an obligate stygobite largely confined to phreatic habitats. Two additional species of bathynellaceans from phreatic habitats in California were described (Schminke and Noodt, 1988). The species from southern California (*Iberobathynella californica*) is related to European bathynellaceans, whereas the species from Northern California (*Pacificaba-thynella sequoiae*) is related to East Asian forms. The zoogeographic affinity of *B. riparia* is unknown. Bathynellaceans collected from other North American localities await description. This interesting group of subterranean crustaceans may prove to be widespread in phreatic habitats in unglaciated regions of North America.

Two endemic cyclopoid copepods occur in phreatic and hyporheic habitats at Site 5. *Acanthocyclops pennaki* is large compared with hypogean congeners and, although exhibiting some morphological features typifying subterranean cyclopoids, may not be strictly hypogean (Reid, 1994). In contrast, *Microcyclops pumilis* is one of the smallest known freshwater cyclopoids (Pennak and Ward, 1985b), with total lengths of 0.51–0.61 mm (females) and 0.40–0.47 mm (males).

VI. THE MICROHABITAT GRADIENT

On the microhabitat scale faunal distribution patterns were examined along habitat transects at each site (epigeic–hyporheic–phreatic) and at depths of 15, 30, and 50 cm (Site 5 only; Pennak and Ward, 1986).

A detailed analysis of habitat transect data is presented elsewhere (Ward and Voelz, 1995). Here we focus on the common members of the stygofauna. The four most abundant stygobiontic taxa collectively contributed a large proportion of the total fauna in phreatic and hyporheic habitats at sites where hypogean forms were well represented (Fig. 6). None of these four taxa occurred in epigean samples (top 10 cm of bed), except a very few specimens of *Parastenocaris* taken at Site 1.

The amphipod *Stygobromus*, the largest hypogean animals encountered, did not exhibit high numerical abundances in any habitat. Individuals were patchily distributed and many samples lacked specimens. *Bathynella* was the most strictly hypogean form, being largely confined to the phreatic habitat. The vermiform harpacticoid copepod *Parastenocaris* was the most abundant and widely distributed member of the stygofauna. *Parastenocaris* was abundant in both phreatic and hyporheic habitats, but was rarely collected from Site 5, at which other hypogean forms were common. *Parastenocaris* is also rare in the alluvial aquifer of the Flathead River (Ward *et al.*, 1994). The gravel bar at Site 5 is highly porous (Pennak and Ward, 1986), and the Flathead River aquifer exhibits high hydraulic conductivity (Stanford and Ward, 1988). It may be that high densities of *Parastenocaris* are associated with fine-grained sediments. This copepod is the only one of the four stygobionts in Fig. 6 to occur in the plains river (Sites 8 and 9), for which the substratum is dominated by sand and gravel.

Microhabitat distribution patterns were most intensively examined at Site 5, at which sampling was conducted monthly along a habitat transect and seasonally along a depth transect for each habitat type (Pennak and Ward, 1986). Three "habitats" were sampled: the hyporheic habitat, beneath the stream; the shore habitat, 1–2 m from the water's edge, and the phreatic habitat, about 20 m from the stream. Monthly samples were taken

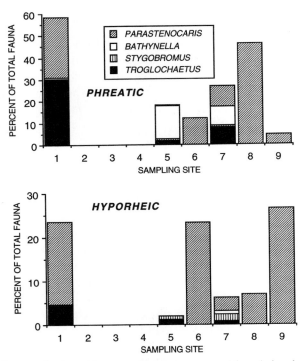

FIGURE 6   The contribution of four stygobiontic taxa to total faunal abundance in phreatic and hyporheic habitats.

50 cm below the stream bed (hyporheic) and 50 cm below the water table (shore and phreatic). Three depths (15, 30, and 50 cm) were sampled seasonally.

The archiannelid *Troglochaetus beranecki* is used as an example of the distribution of a hypogean species along the habitat and depth gradients (Fig. 7). The richest population occurred in the shore habitat, in which there is a mixing of hyporheic and phreatic waters. In the hyporheic habitat, population density increased with depth, with very few specimens collected from 15 cm. In the shore habitat, the densest population occurred at the intermediate depth. Too few specimens occurred in the phreatic habitat to discern an abundance pattern with depth.

The relative abundance of *Troglochaetus* across the three habitats differed for different months (Fig. 7). Density maxima were attained during May (hyporheic), November (shore), and March (phreatic). On some sampling dates no specimens were collected from one or more habitats. Some 5-l samples contained hundreds of individuals; others contained none. Therefore, it appears that many replicate samples would be required to accurately determine the temporal and spatial population patterns of *Troglochaetus,*

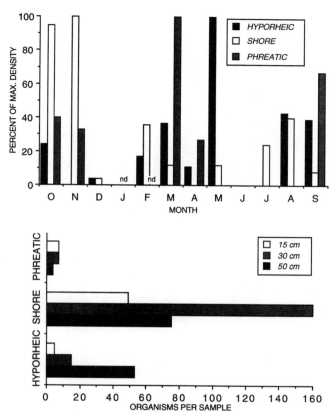

FIGURE 7   Relative abundance of *Troglochaetus beranecki* along habitat transects at Site 5 for each month and density levels for three depths. Frozen ground precluded sampling of the shore habitat in January and the phreatic habitat in February (indicated by nd). Data from Pennak and Ward (1986).

despite being a common member of the stygofauna at Site 5, and this undoubtedly applies to other groundwater animals.

VII. THE LONGITUDINAL GRADIENT

Four gradient analysis techniques, faunal congruity (Terborgh, 1971), percentage similarity (Whittaker, 1975), TWINSPAN classification (Hill, 1979a), and detrended correspondence analysis (Hill, 1979b), were employed to further examine distribution patterns along the longitudinal profile of the South Platte River. These analyses included only taxa with seasonal means ≥5 individuals per 5-l Bou–Rouch sample (phreatic and hyporheic habitats) or ≥5 individuals per liter of gravel (surface gravel habitat) at one or more sampling sites. All stygobionts, irrespective of abundance, were

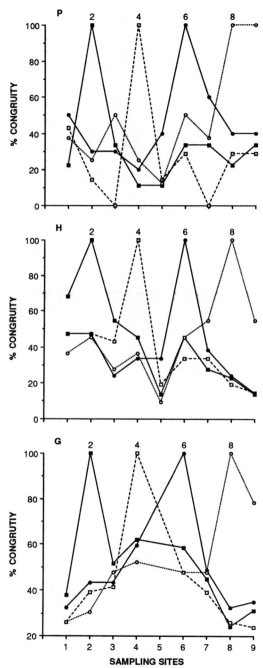

FIGURE 8 Faunal congruity curves for the even-numbered sites of each habitat. P, phreatic; H, hyporheic; G, surface gravel.

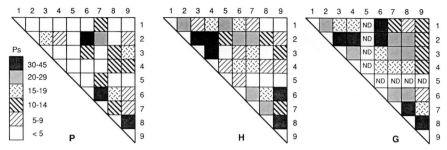

FIGURE 9 Matrices of percent similarity (PS), a measure of faunal overlap, between sites based on numerical abundances of animal taxa in each habitat. P, phreatic; H, hyporheic; G, surface gravel; ND, no data.

included in the analyses. The fauna meeting these criteria included 36, 70, and 95 taxa, respectively, for phreatic, hyporheic, and surface gravel habitats. Nematoda, not identified further, were excluded from the analyses. Most of the taxa included were species.

Faunal congruity curves (Fig. 8) compare the fauna present at a given site with the faunal composition at every other site. If species distribution patterns are correlated with the elevation gradient, each curve should progressively decline toward the abscissa (*i.e.*, progressively lower values at sites that are increasingly distant from the reference site). Such a progressive decline is exhibited only by certain curves of the surface gravel habitat (*e.g.*, Site 4). Moreover, very low faunal congruity is apparent, even between some adjacent sites, for hyporheic and especially phreatic habitats (note the different vertical scale for surface gravel). There was, however, perfect faunal congruity between the phreatic habitats of the two plains river locations (Sites 8 and 9). This means that all designated taxa present in the phreatic habitat at Site 8 also occurred in that habitat at Site 9 (as did a few additional taxa not found at Site 8).

Percentage similarity (PS) matrices were used to determine faunal overlap between sites (Fig. 9). There is little faunal overlap between sites in hypogean biotopes, and this is especially pronounced for the phreatic habitat; adjacent sites (along diagonals in Fig. 9) do not necessarily exhibit greater faunal overlap than more distant sites. Clearly, there is little indication of altitudinal trends for either hypogean biotope. In all three habitats the greatest faunal overlap (38–43%) occurred between the two plains river sites. Although the absence of data for surface gravel at Site 5 makes it difficult to discern the overall pattern of faunal overlap for the epigean habitat, there is some evidence of a weak altitudinal gradient. A much stronger altitudinal gradient was, however, demonstrated for macrobenthic riffle fauna (Ward, 1986) than for the interstitial surface gravel fauna of the present study.

A hierarchical classification (TWINSPAN) was performed to cluster sampling sites based on community composition and relative abundance of taxa (Fig. 10). In the phreatic habitat the first level separation, for which *T. beranecki* proved to be the key indicator, isolated Sites 1, 5, and 7 from the remaining sites. These three sites appear to possess especially favorable features for stygobionts. *Troglochaetus* and *Bathynella* occurred only at Sites 1, 5, and 7. However, whereas *Troglochaetus* occurred in both phreatic and hyporheic habitats, *Bathynella* was essentially confined to the phreatic.

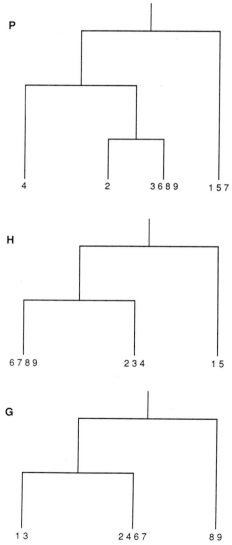

FIGURE 10   TWINSPAN dendrograms of site clusters based on numerical abundances of animal taxa in each habitat. P, phreatic; H, hyporheic; G, surface gravel.

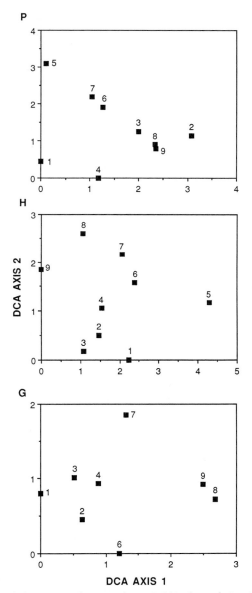

FIGURE 11 Detrended correspondence analysis (DCA) plots of sites in two-dimensional space based on presence–absence of animal taxa in each habitat. P, phreatic; H, hyporheic; G, surface gravel. Eigenvalues for the first three DCA axes, respectively, are 0.546, 0.439, and 0.039 (phreatic); 0.559, 0.338, and 0.070 (hyporheic); and 0.406, 0.149, and 0.031 (surface gravel).

In the hyporheic habitat the first level separation, for which the cyclopoid *Acanthocyclops vernalis* was the key indicator, isolated Sites 1 and 5 from the remaining sites. *A. vernalis* occurred in both phreatic and hyporheic habitats at Sites 1 and 5, but was never collected at Site 7. Cluster analysis gave no indication of an altitudinal gradient based on the phreatic habitat, but hyporheic clusters suggest a weak gradient structure. A stronger elevation gradient is apparent for the epigean habitat clusters. The first level separation of the surface gravel habitat isolates the plains river sites from the mountain stream sites.

Detrended correspondence analysis (DCA) is an ordination technique that plots the relationship between samples (sites) in low-dimensional space (Fig. 11). Eigenvalues (see the legend to Fig. 11) indicate a two-dimensional spatial representation of the site relationships for all three habitats. DCA plots of the phreatic and hyporheic habitats give no indication that elevation-related variables have structured faunal assemblages. Sites are not sequentially arrayed along either DCA axis, although there is a significant negative correlation between altitude and site scores on the secondary axis for the hyporheic ($r_s = -0.93$, $p < 0.01$). Sites are widely dispersed in two-dimensional space (note the different scales of axes for the different habitats). Adjacent field sites do not necessarily plot in close proximity. Only for the epigean habitat is there a significant correlation between altitude and site scores along the primary axis ($r_s = -0.95$, $p < 0.01$).

The position of phreatic sites on the left side of Fig. 11 (especially Sites 1, 5, and 7) is influenced by the presence of key hypogean taxa (*B. riparia*, *Stygobromus* spp., *T. beranecki*, *Parastenocaris* spp., and *Canthocamptus vagus*) and the eurytopic ostracod *Cavernocypris wardi*. The anomalous position of Site 2 is a result of two collembolans (*Tullbergia macrochaeta* and *Paranura anops*) that occur only at that site. The positions of Sites 3, 8, and 9 are partially influenced by *Diacyclops crassicaudis brachycercus*, which occurred only at those sites, and by *Cephalodella sp.*, which also occurred in the phreatic at Site 2.

VIII. CONCLUSION

Because we are not aware of previous research on groundwater fauna collected along an extensive elevation gradient, there is nothing with which to directly compare the patterns described herein. Altitudinal distribution patterns of the macrobenthic riffle fauna over a comparable elevation gradient (1870 m) in another Rocky Mountain River were largely structured by direct correlates of elevation, such as temperature (Ward, 1986). In the present study, a weak altitudinal pattern is indicated by the interstitial fauna in surface gravel, but the distribution of the hypogean fauna is apparently decoupled from altitude.

In an earlier paper (Ward and Voelz, 1990) we suggested that site-specific geomorphic features may be more important than elevation-related variables in structuring the groundwater communities of alluvial rivers.

Creuzé des Châtelliers (1991) also hypothesized "a link between geomorphological and hydrogeological processes and the distribution of interstitial fauna." Application of a variety of analytical techniques to the faunal data set of the South Platte River provides yet further support for this viewpoint. The implications are clear: geomorphic and hydrogeologic data should be used in concert with faunal data if we are to elucidate the factors and interactions that are responsible for structuring the groundwater communities of alluvial rivers.

ACKNOWLEDGMENTS

The following specialists graciously provided taxonomic assistance: Drs. J. W. Reid and R. L. Whitman (Copepoda); Dr. M. J. Wetzel (Oligochaeta); Dr. P. Marmonier (Ostracoda); Dr. L. C. Ferrington (Chironomidae); Drs. G. W. Krantz, R. A. Norton, B. P. Smith, and D. E. Walter (Acarina); Dr. B. C. Kondratieff (Insecta); and Dr. R. D. Waltz (Collembola). The Colorado State University Soil Testing Laboratory conducted particle size analyses and determined the organic content of substrate samples. We thank Professor Janine Gibert and two anonymous reviewers for comments and thank Mrs. Nadine Kuehl for typing the manuscript. Support was provided in part by the Colorado Water Resources Research Institute.

REFERENCES

Bou, C. (1974). Les méthodes de récolte dans les eaux souterraines interstitielles. *Ann. Spéléol.* **29**, 611–619.

Boulton, A. J., Valett, H. M., and Fisher, S. G. (1992). Spatial distribution and taxonomic composition of the hyporheos of several Sonoran Desert streams. *Arch. Hydrobiol.* **125**, 37–61.

Bradley, W. C. (1987). Erosion surfaces of the Colorado Front Range: A review. *In* "Geomorphic Systems of North America" (W. L. Graf, ed.), pp. 215–110. Geol. Soc. Am., Boulder, CO.

Creuzé des Châtelliers, M. (1991). Geomorphological processes and discontinuities in the macrodistribution of the interstitial fauna. A working hypothesis. *Verh. Int. Ver. Limnol.* **24**, 1609–1612.

Danielopol, D. L. (1989). Groundwater fauna associated with riverine aquifers. *J. North Am. Benthol. Soc.* **8**, 18–35.

Dole, M. J. (1985). Le domaine aquatique souterrain de la plaine alluviale du Rhône à l'est de Lyon. 2. Structure verticale des peuplements des niveaux supérieurs de la nappe. *Stygologia* **1**, 270–291.

Gibert, J., Ginet, R., Mathieu, J., Reygrobellet, J. L., and Seyed-Reihani, A. (1977). Structure et fonctionnement des écosystèmes du Haut-Rhône français. IV. Le peuplement des eaux phréatiques: Premiers résultats. *Ann. Limnol.* **13**, 83–97.

Hadley, R. F., and Toy, T. J. (1987). Fluvial processes and river adjustments on the Great Plains. *In* "Geomorphic Systems of North America" (W. L. Graf, ed.), pp. 182–188. Geol. Soc. Am., Boulder, CO.

Hill, M. O. (1979a). "TWINSPAN—A FORTRAN Program for Arranging Multivariate Data in an Ordered Two-Way Table by Classification of the Individuals and Attributes." Cornell University, Ithaca, NY.

Hill, M. O. (1979b). "DECORANA—A FORTRAN Program for Detrended Correspondence Analysis and Reciprocal Averaging." Cornell University, Ithaca, NY.

Hurr, R. T., and Hearne, G. A. (1985). Ground-water resources. National water summary—Colorado. *Geol. Surv., Water-Supply Pap.* (*U.S.*) **2275**, 153–160.

Hurr, R. T., Schneider, P. A., Jr., and Minges, D. R. (1975). Hydrology of the South Platte River Valley, northeastern Colorado. *Colo. Water Resour. Circ.* **28**.

Madole, R. F. (1987). Rocky Mountains—summary and conclusions. *In* "Geomorphic Systems of North America" (W. L. Graf, ed.), pp. 247–249. Geol. Soc. Am., Boulder, CO.

Madole, R. F. (1994). Distribution of Pleistocene glaciers and nature of glacier deposits on the East Slope of the Medicine Bow Mountains, Front Range, and Mosquito Range, Colorado. *U.S. Geol. Surv., Invest. Map Ser.* (in press).

Marmonier, P., and Ward, J. V. (1990). Surficial and interstitial Ostracoda of the South Platte River (Colorado, U.S.A.)—Systematics and biogeography. *Stygologia* **5**, 225–239.

Marmonier, P., Meisch, C., and Danielopol, D. L. (1989). A review of the genus *Cavernocypris* Hartmann (Ostracoda, Cypridopsinae): Systematics, ecology and biogeography. *Bull. Soc. Nat. Luxemb.* **89**, 221–278.

Marmonier, P., Dole-Olivier, M.-J., and Creuzé des Châtelliers, M. (1992). Spatial distribution of interstitial assemblages in the floodplain of the Rhône River. *Regulated Rivers* **7**, 75–82.

Meierding, T. C., and Birkeland, P. W. (1980). Quaternary glaciation of Colorado. *In* "Colorado Geology" (H. C. Kent and K. W. Porter, eds.), pp. 165–173. Rocky Mt. Assoc. Geol., Denver, CO.

Nadler, C. T., and Schumm, S. A. (1981). Metamorphosis of South Platte and Arkansas rivers, eastern Colorado. *Phys. Geogr.* **2**, 95–115.

Norris, J. M., Robson, S. G., and Parker, R. S. (1985). Summary of hydrologic information for the Denver coal region, Colorado. *Water-Resour. Invest.* (*U.S. Geol. Surv.*) **84-4337**.

Orghidan, T. (1959). Ein neuer Lebensraum des unterirdischen Wassers der hyporheischen Biotope. *Arch. Hydrobiol.* **55**, 392–414.

Palmer, M. A. (1990). Temporal and spatial dynamics of meiofauna within the hyporheic zone of Goose Creek, Virginia. *J. North Am. Benthol. Soc.* **9**, 17–25.

Pearl, R. H. (1974). "Geology of Ground Water Resources in Colorado," Spec. Publ. No. 4. Colorado Geological Survey, Denver.

Pennak, R. W., and Ward, J. V. (1985a). Bathynellacea (Crustacea: Syncarida) in the United States, and a new species from the phreatic zone of a Colorado mountain stream. *Trans. Am. Microsc. Soc.* **104**, 209–215.

Pennak, R. W., and Ward, J. V. (1985b). New cyclopoid copepods from interstitial habitats of a Colorado mountain stream. *Trans. Am. Microsc. Soc.* **104**, 216–222.

Pennak, R. W., and Ward, J. V. (1986). Interstitial faunal communities of the hyporheic and adjacent groundwater biotopes of a Colorado mountain stream. *Arch. Hydrobiol., Suppl.* **74**, 356–396.

Poff, N. L., and Ward, J. V. (1989). Implications of streamflow variability and predictability for lotic community structure: A regional analysis of streamflow patterns. *Can. J. Fish. Aquat. Sci.* **46**, 1805–1818.

Reid, J. W. (1994). *Acanthocyclops pennaki* n. sp. (Copepoda: Cyclopoida) from the hyporheic zone of the South Platte River, Colorado, U.S.A. *Trans. Am. Microsc. Soc.* (in press).

Ritter, D. F. (1987). Fluvial processes in the mountains and intermontane basins. *In* "Geomorphic Systems of North America" (W. L. Graf, ed.), pp. 220–228. Geol. Soc. Am., Boulder, CO.

Robson, S. G. (1987). Bedrock aquifers in the Denver Basin, Colorado—A quantitative water-resources appraisal. *Geol. Surv. Prof. Pap.* (*U.S.*) **1257**, 1–73.

Schminke, H. K., and Noodt, W. (1988). Groundwater Crustacea of the order Bathynellacea (Malacostraca) from North America. *J. Crustacean Biol.* **8**, 290–299.

Scott, G. R. (1963). Quaternary geology and geomorphic history of the Kassler Quadrangle, Colorado. *Geol. Surv. Prof. Pap. (U.S.)* **421-A**, 1–70.

Stanford, J. A., and Ward, J. V. (1988). The hyporheic habitat of river ecosystems. *Nature (London)* **335**, 64–66.

Strayer, D. (1988). Crustaceans and mites (Acari) from hyporheic and other underground waters in southeastern New York. *Stygologia* **4**, 192–207.

Terborgh, J. (1971). Distribution on environmental gradients: Theory and preliminary interpretation of distributional patterns in the avifauna of the Cordillera Vilcabamba, Peru. *Ecology* **52**, 27–40.

Tweto, O. (1980). Summary of Laramide orogeny in Colorado. *In* "Colorado Geology" (H. C. Kent and K. W. Porter, eds.), pp. 129–134. Rocky Mt. Assoc. Geol., Denver, CO.

Ward, J. V. (1977). First records of subterranean amphipods from Colorado with descriptions of three new species of *Stygobromus* (Crangonyctidae). *Trans. Am. Microsc. Soc.* **96**, 452–466.

Ward, J. V. (1986). Altitudinal zonation in a Rocky Mountain stream. *Arch. Hydrobiol., Suppl.* **74**, 133–199.

Ward, J. V., and Voelz, N. J. (1990). Gradient analysis of interstitial meiofauna along a longitudinal stream profile. *Stygologia* **5**, 93–99.

Ward, J. V., and Voelz, N. J. (1995). Interstitial fauna along an epigean-hypogean gradient in a Rocky Mountain river. *In* "Groundwater/Surface Water Ecotones" (J. Gibert *et al.*, eds.). Cambridge Univ. Press, Cambridge, UK, (in press).

Ward, J. V., Stanford, J. A., and Voelz, N. J. (1994). Spatial distribution patterns of Crustacea in the floodplain aquifer of an alluvial river. *Hydrobiologia* (in press).

Whittaker, R. H. (1975). "Communities and Ecosystems." Macmillan, New York.

Williams, D. D. (1989). Towards a biological and chemical definition of the hyporheic zone in two Canadian rivers. *Freshwater Biol.* **22**, 189–208.

16

Dynamics of Communities and Ecology of Karst Ecosystems: Example of Three Karsts in Eastern and Southern France

J. Gibert,* Ph. Vervier,† F. Malard,* R. Laurent,* and J.-L. Reygrobellet*

* *Université Lyon I*
U.R.A. CNRS 1451 "Ecologie des Eaux Douces et des Grands Fleuves"
Laboratoire d'Hydrobiologie et Ecologie Souterraines
43 Boulevard du 11 Novembre 1918
69622 Villeurbanne Cedex, France

† *Centre d'Ecologie des Ressources Renouvelables, CNRS*
29 Rue Jeanne Marvig
31055 Toulouse Cedex, France

I. INTRODUCTION

Earlier research works devoted to the study of the population dynamics of cave organisms (Ginet, 1960; Vandel, 1964; Mohr and Poulson, 1966) were indeed open to criticism in that they only considered some specific habitats of a karst aquifer. However, it was soon realized that there was a need to investigate the main physical parameters of the different karst biotopes that lead to the development of a particular community of organisms with adapted strategies [for a review see Rouch (1986)].

Because it clearly appeared that the interactions among the communities, the structural characteristics of the aquifer, and the groundwater flow pattern had to be perceived on the scale of the karst system, a holistic approach was developed in the seventies (Chapter 1). A new sense was given to the faunal results of cave explorations and spring studies because these sampling sites could be located within the framework of a groundwater ecosystem. It soon became evident that such a new approach offered some interesting perspectives, and numerous studies on the different karst areas in the northeast and southeast of France rapidly followed intensive surveys carried out by Rouch (1971, 1977, 1982; Rouch and Carlier, 1985) on the Baget system. Such an explosive increase in the number of studies and the variety of methodologies may at first be disturbing, because it has produced in a very short period of time a large body of ideas and concepts, which have simultaneously created a need for new data.

This chapter aims to show how different studies based on different methodological principles and sampling schemes have contributed to a better understanding of the interrelations existing between groundwater organisms and external factors (*i.e.*, the groundwater flow). First, we present the general situation and the hydrogeologic settings of the three karst areas under study: the Dorvan–Cleyzieu system in the Jura mountains, the Foussoubie system in the Ardèche plateau, and the Lez spring system in the stony hills of the Languedoc region. Second, using the results of these surveys and the findings of other authors, attention is paid to the main factors controlling the spatial and temporal distribution of organisms in the aquifer. Third, we emphasize new ideas and concepts raised by such studies and also discuss their usefulness for further progress in karstwater ecology.

II. GROUNDWATER FLOW MODELS

Ecological surveys carried out on karst systems must be developed within the framework of a groundwater flow conceptual model. Although the aim of the biological studies is not to validate such a model, they may in turn provide substantial information that could then be used to improve it. Indeed, the groundwater flow in karstified limestone is so complex that multidisciplinary conceptual models incorporating hydrogeologic, hydrochemical, and biological data are highly desirable prior to the development of simulation models.

Whatever the spatial scale considered in limestone aquifers, the heterogenity does not seem to exhibit any homogeneous properties and the fracture geometry cannot be accurately determined. Therefore, conceptual models have often used simplifying assumptions, but they have nevertheless greatly contributed to a better understanding of fundamental flow in fractured aquifers. Ecological spring studies carried out on the scale of an entire catchment area have so far relied on the flow model proposed by Mangin (1974, 1975, 1985). The hydrodynamic response of the karst system to rainfall events is analyzed and used to determine the structural characteristics of the reservoir (Chapter 2). The aquifer is then believed to be made up of a main drainage network, comprising conductive channels that isolate several physical subsystems connected to the drainage network (Fig. 1A).

On the site scale the double fissural porosity flow model developed by Drogue (1974, 1980) has been used to interpret the faunal data of an ongoing survey (Malard *et al.*, 1994a). The aquifer is idealized as a fracture continuum of high flow velocity, which separates and forms discontinuities between thinly fissured rock masses (Fig. 1B).

Emphasis is placed on the fact that whatever the spatial scale considered, the models developed have all concentrated on the interactions and relative flow rates of groundwater in the transmissive and capacitive zones of the aquifer: the drainage network and the subsystems on the catchment scale and the conductive fractures and the adjacent fissured matrix on the site scale. During low-water periods, the transmissive zones with a low pressure head drain much of the groundwater from the adjacent and capacitive areas. Then, during times of intensive groundwater recharge, the pressure head increases in the transmissive zones, in which groundwater flows rapidly and enters the adjacent capacitive parts of the aquifer.

It is also recognized that groundwater recharge occurs by means of a double infiltration mechanism: rapid infiltration through large subvertical openings and slow, double-phase (air and water) infiltration through the thin fissures of the unsaturated zone (Bakalowicz, 1981; Mangin and Bakalowicz, 1989). The epikarstic and infiltration zones may thus concentrate the downward groundwater flow at particular places, but they may also retain a significant amount of water.

III. THREE KARST SYSTEMS

Three karst aquifers with completely different relationships with surface water have been studied. The Dorvan–Cleyzieu system is an exsurgence karst mainly fed by diffuse infiltration. The Foussoubie system is a resurgence

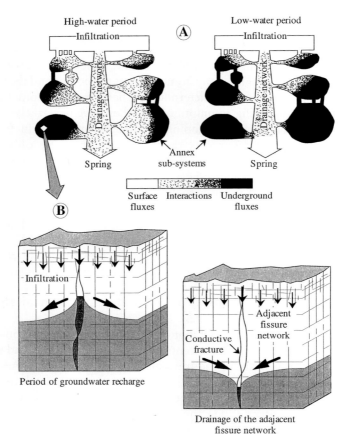

FIGURE 1   Conceptual groundwater flow models for the different ecological surveys carried out on the Dorvan–Cleyzieu and Foussoubie karst systems (A) and the Terrieu limestone site (B).

karst whose recharge is partlty due to a surface stream flowing through a sinkhole during rainy periods. The Terrieu site is recharged with sewage-polluted surface water during low- and high-water periods.

A. Study Areas

The Dorvan–Cleyzieu system is a 10-km² low-mountain karst whose average altitude and dislevelment are about 620 m and 126 m/km⁻¹, respectively. It is situated on the southwestern outer range of the Southern Jura mountains in middle Jurassic limestone (Bathonian and Bajocian) (Fig. 2A). Three different drainage levels can be distinguished. In the upper zone, an epikarstic aquifer is drained by a main subsurface channel, corresponding

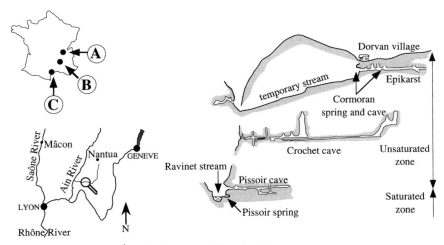

A - The Dorvan - Cleyzieu karst system

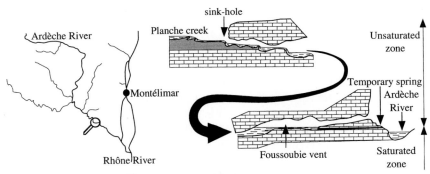

B - The Foussoubie karst system

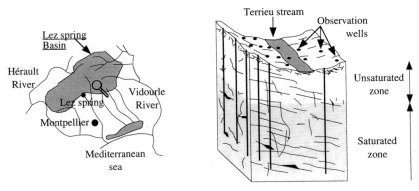

C - The Terrieu limestone site

FIGURE 2   Location and description of the three study karst aquifers.

to the stream and exsurgence of the Cormoran cave (steady flow and vadose regime; mean annual discharge, 4.2 l/sec^{-1}). In the intermediary and vertical transfer zone, a subterranean channel runs through the Crochet cave (steady flow and vadose regime). Finally, in the lower zone, a main channel emerges at the Pissoir spring and flows through the Albarine alluvial plain (semisaturated and saturated regime and temporary flow, mean annual discharge 76.5 l/sec^{-1}). The relationships between these different zones have been demonstrated by a multitracer experiment (Gibert *et al.*, 1982). The effective rainfall is about 64% (Gibert, 1986). The hydrologic cycle presents a high-water period during winter and a low-water period during which the main outlet of the massif, the Pissoir spring, is intermittent. Groundwater flow at this spring ranges from 2 to 2,000 l/sec^{-1} during autumn, winter, and spring, and it ceases during summer. Vegetation covers 80–90% of the ground surface (meadows and crops, 50%; forests, 40%; and moors and bushes, 10%). An impact of agriculture on the groundwater quality was detected only in the Cormoran stream, in which chloride and nitrate concentrations up to 8 and 14.7 mg/l^{-1}, respectively, were measured (Gibert, 1986; 1990). Because this contamination has remained slight and confined to a small area in the saturated zone, this karst system is still considered a pristine one.

The Foussoubie karst system is situated at the southeast boundary of the French Massif Central in a limestone plateau (Fig. 2B). This 400-km^2 plateau belongs to the lower Cretaceous formation, including Bedoulian limestone with an Urgonian facies made up of hard, compact, white zoogene limestone up to 300 m thick (Mazellier, 1971; Callot, 1978; Adam, 1979; Belleville, 1985; Balazuc, 1986). During the upper Miocene, the Ardeche River, which had cut through this plateau, creating a canyon, initiated the formation of karst systems, like that of the Foussoubie. This system lies in the upper part of the plateau to the east of the Ardèche River. A temporary stream (the Planche creek), draining a 14-km^2 catchment area dominated by clay, marl, and Cretaceous sandstone formations, forms the only point inlet to the karst through a sinkhole called "Goule of Foussoubie." The ensuing underground stream flows a linear distance of 5 km, with an elevation difference of 115 m, to two outlets: a cave, the Foussoubie vent that allows access to groundwater flows, and a temporary spring situated 10 m below the vent and 3 m above the mean level of the Ardèche River. During high-water periods, underground flow in the Foussoubie system is fed by the temporary stream (the Planche creek), by diffuse infiltrations of rainfall and also by karst storage (Vervier, 1990a). During these periods the Planche creek discharge ranges from 0.01 to 0.26 m^3/sec^{-1}, although it can reach 1 m^3/sec^{-1} during heavy rainfall events (Slama *et al.*, 1981). In this case, a tracer (fluorescein) was carried from the Planche creek to the outflow at the Foussoubie vent within 10 hr (Slama *et al.*, 1981). As the flow of the Planche creek is intermittent, increasing during winter and spring and ceasing during summer and autumn, it is clear that during low-water periods, water flowing from the outlet comes mainly from storage within the karst system.

The 200-km^2 Lez spring basin, which is located in the south of France, lies between the Hérault and the Vidourle Rivers (Fig. 2C). Much of the groundwater flowing through this hydrogeologic basin is drained toward the Lez spring, which is the main source of drinking water for the inhabitants of the town of Montpellier. Three research sites, from which many historical hydrodynamic, hydrochemical, and hydrothermic data are available and which constitute a dense network of observation wells, were selected in the downstream part of the basin. In this chapter we mainly use the results of an ongoing survey being carried out on the Terrieu site, whose groundwater is polluted with sewage effluent. This site is a 500-m^2 experimental area in which 21 uncased wells were installed in 1974 at regular 5-m intervals on each side of a sewage-polluted river. These wells are 60 m deep, and all reach the water table, which fluctuates between 15 and 40 m below the surface. The geology and hydrology of the site were described previously (Drogue, 1980). The rock intersected by the wells is solid Berriasian limestone whose dip is about 15–20° NW. The major fractures are oriented N 000–040 and N 100–120. Although a complex and dendritic system of interconnected joints and fissures was tapped by the different wells, active groundwater flow was recognized only in the western and more permeable part of the site, where karstification in depth has developed on bedding joints and on the main subvertical fractures. On the contrary, in the eastern and less permeable part of the site, the rock mass was intersected only by small aperture fissures, which do not seem to carry much water. The Terrieu River, which intersects the site, is dry for most of the year and it is flushed with storm water runoff during rainy periods. However, in 1991 and 1992, it received the secondary effluent overflow of a malfunctioning sewage treatment plant located 800 m upstream.

B. Sampling Methods

At the Dorvan–Cleyzieu system, five stations were set up to collect the fauna (Gibert, 1986). In the epikarst, inside the Cormoran cave, percolation and bottom and surface drift of the subhorizontal stream flow were filtered continuously through 100- and 150-μm mesh nets. Outside, 300-μm mesh nets were installed at the Cormoran and Pissoir springs to collect the aquatic fauna mainly during flood periods.

At the Foussoubie system, artificial substrata were used (Vervier, 1990b) to collect the groundwater fauna during the whole hydrologic cycle. They were composed of a 0.2-m-long and 0.1-m-wide PVC pipe containing 25 m of synthetic rope. They were placed on the bed sediments of the underground stream of the Foussoubie vent during one-month periods. During the floods, organisms were also collected by filtering the water of the Foussoubie spring through a 300-μm mesh net.

At the fractured limestone sites of the Lez spring basin, pumping in selected wells was conducted to collect the groundwater fauna. In small-diameter wells (<110 mm) a submersible pump could not be used because macroorganisms were considerably damaged as they passed through the turbines. On the contrary, we found that an air-lift pump was particularly suited to groundwater fauna sampling because it did not have any moving parts that could damage the organisms. At the Terrieu site, four wells were thus selected in order to sample both the conductive and the capacitive parts of the saturated zone. Well W8 was shown to intersect an enlarged bedding joint at a depth of 43 m, characterized by a high hydraulic change by "conductivity" of approximately 10^{-1} to 10^{-3} m/sec^{-1} (Drogue, 1985). On the contrary, wells W7, W10, and W16 were shown to intersect narrow and rather closed fissures at depths ranging from 25 to 45 m, characterized by low hydraulic conductivities of approximately 10^{-7} to 10^{-9} m/sec^{-1} (Drogue, 1985). Because we wanted to describe the assemblage characteristics of each well before and after the floods, the sampling surveys were conducted during low- and high-water periods. Because of the heterogeneous nature of fractured limestone, wells responded very differently to pumping, and the same amount of water could not be obtained from the different wells. In high-capacity wells W7, W10, and W16, pumping lasted 15 min and approximately 200 l of water was obtained and filtered through a 125-μm mesh net. In the highly transmissive well W8, pumping was carried out for 3 h at a constant discharge rate of about 0.55 l/sec^{-1}. Therefore, whatever the faunal groups, the numbers of organisms were related to a 200-l sample in high-capacity wells (W7, W10, and W16) and to a 6.000-l sample in well W8.

IV. DYNAMICS OF THE AQUATIC COMMUNITIES

Because the methodological principles and sampling techniques used during these three surveys were different, it has been possible to describe the community dynamics at different scales of space and time. At the Dorvan–Cleyzieu and Foussoubie karsts, the information provided by the drifting fauna at the outlets of the catchment was analyzed and used to define some of the main biological characteristics of the karst system. Such an approach is similar to that developed by Mangin (1983, 1986; see also Chapter 2) in hydrology and is called the black box approach. Finally, at the Terrieu site, we may consider that we are effectively trying to analyze some of the processes occurring within the black box, as we investigated only some smaller parts of the aquifer by means of deep observation wells.

A. Nonuniform Distribution Communities within Karst Systems

At the scale of habitat and within the same kind of habitat (*e.g.,* a drain), the distribution of fauna is not homogeneous. This could be observed at the Foussoubie vent, where artificial substrata were placed in the riffle and pool sections of the underground stream. Considering the "planktonic" organisms such as the cyclopoids, it clearly appeared that the low-water-velocity habitats were colonized preferentially, because a higher number of species and individuals were collected in the pools (Table I). These differences in community assemblages increased when different kinds of habitats were taken into account. For example, in the Terrieu site, relevant differences in the community structure of wells separated by only a few meters were observed (Fig. 3). Assemblages of the conductive fractures (well W8) were characterized by a high epigean component and the almost total absence of hypogean fauna, whereas samples collected from the slow-moving water parts of the saturated zone (wells W7, W10, and W16) contained both epigean and hypogean forms. For example, taking all the sampling surveys into account, *Tubifex tubifex* and *Macrocyclops albidus* represented 97 and 65%, respectively, of all oligochaetes and cyclopoids collected in well W8. In this highly transmissive well large swimming forms such as the Coleoptera *Haliplus lineatocollis, Agabus bipustulatus, Colymbetes fuscus,* and *Sticto-nectes sp.;* the Heteroptera *Notonecta;* and Corixidae larvae were also collected. On the contrary, assemblages from well W16 were dominated by small epigean or interstitial pollution-tolerant species (*Nais elinguis,* 28% and *Pristinella jenkinae,* 45%) and by stygophile taxa (*Trichodrilus sp.,* 7%, and *Paracyclops fimbriatus,* 95%).

The heterogeneous distribution of the fauna increases depending on the scale of the system. For the Dorvan–Cleyzieu karst it has been demonstrated that from the zone closest to the surface down to the saturated zone, communities changed in the numbers of both organisms and species. In the topmost layers of this massif, there are highly simplified communities (Table II). For example, in the percolation studies, only 13 aquatic species, including four hypogean crustaceans were collected. This low number of species can be linked to the fact that the sampled habitat is relatively small, because the catchment area is only 400 m^2 with a low discharge of 0.012 l/sec^{-1}. The discharge of the underground Cormoran stream, whose catchment area is about 0.14 km^2, is 350 times higher than that of the upper level percolations. In this stream 7 stygobite species were collected, and the total taxonomic richness more than tripled when compared with that of the percolations for the following groups: nematods, oligochaetes, gastropods, harpacticoids, ostracods, and isopods. When the epigean organisms collected at the outlets are taken into account, the total number of systematic units for the epikarstic aquifer is as high as 71. Such a difference in the number of species between the upper level of percolations and the underground stream demonstrates

TABLE 1 Spatial Distribution of Cyclopoids within the Foussoubie Stream

| | Species | | | | | | | | | | | | | | | | | |
| Sampling dates | Eucyclops serrulatus | | Paracyclops fimbriatus | | Megacyclops viridis | | Diacyclops bicuspidatus | | Acanthocyclops robustus | | Acanthocyclops venustus | | Diacyclops languidoides | | Tropocyclops prasinus | | Diacyclops languidus | |
	R	P	R	P	R	P	R	P	R	P	R	P	R	P	R	P	R	P
23/11/84	50	50	62	38	90	10	0	100							100	0		
27/12/84			0	100	0	100									0		0	
22/3/85	33	67	3	97	100	0							0	100				
26/4/85	0	100	42	58									0	100				
12/2/86			0	100	100	0	0	100					0	100				
12/2/86			100	0							100	0	0	100				
14/4/86	6	94	17	83					0	100	0	100	0	100			0	100
9/5/86	0	100	0	100	0	100					0	100						
9/5/86	50	50	0	100	0	100			0	100	0	100						

Note. R, riffle; P, pool.

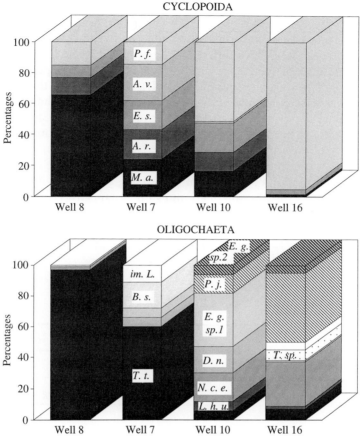

FIGURE 3 Percentages of different species of cyclopoids and oligochaetes collected in four sampled wells of the Terrieu site. Cyclopoids: P. f., *Paracyclops fimbriatus*; A. v., *Acanthocyclops venustus*; E. s., *Eucyclops serrulatus*; A. r., *Acanthocyclops robustus*; M. a., *Macrocyclops albidus*. Oligochaetes: E. g. sp. 2, *Enchytraeidae g. sp. 2*; P. j., *Pristinella jenkinae*; im. L., *immature Lumbriculidae*; T. sp., *Trichodrilus sp.*; B. s., *Bichaeta sanguinea*; E. g. sp. 1, *Enchytraeidae g. sp. 1*; D. n., *Dero nivea*; N. c. e., *Nais communis* and *Nais elinguis*; L. h. u., *Limnodrilus hoffmeisteri* and *Limnodrilus udekemianus*; T. t., *Tubifex tubifex*.

TABLE II Spatial Changes in the Number of Species Collected within the Dorvan–Cleyzieu Karst

Zone of the aquifer	Catchment (km²)	Number of species	Hypogean species
Epikarst			
Percolations	0.04	13	4
Cormoran stream	0.14	45	7
Cormoran spring	0.14	71	7
Saturated zone			
Pissoir spring	10	142	14

that the fauna is not uniformly distributed in the unsaturated zone and that percolation areas have different species assemblages. In the saturated zone the number of species is twice as high as that in the epikarstic zone, with a total of 14 stygobite species (Table III). Other biotopes that contained animals that we did not find in the upper layers of the karst have thus been flushed out. Due to its position and size, the saturated zone appears to be taxonomically richer and provides an overall view of the faunal diversity and density of the karst.

Finally, it clearly appears that, whatever the spatial scale considered in the different zones of a karst system (epikarstic, infiltration, and saturated zones), the spatial distribution of groundwater organisms is highly heterogeneous and may be related to the presence of diversified habitats (Barr, 1967; Barr and Kuehne, 1971; Delay, 1969; Lescher-Moutoué and Gourbault, 1970; Rouch, 1971; Rouch *et al.,* 1968; Bertrand, 1975; Henry, 1976;

TABLE III   Spatial Distribution of Hypogean Species within the Dorvan–Cleyzieu Karst

	Sampling sites		
	Cormoran		
Species	*Percolations*	*Spring*	*Pissoir spring*
Cyclopoida			
Speocyclops cf. *proserpinae* (Kieper, 1937)	×	×	×
Harpacticoida			
Ceuthonectes serbicus (Chappuis, 1924)	×	×	×
Elaphoidella cavatica (Chappuis, 1957)	×	×	×
Amphipoda			
Niphargus rhenorhodanensis (Shellenberg, 1937)	×	×	×
Isopoda			
Proasellus cavaticus (Leydig, 1871)		×	×
Hydrobida			
Mollessiera lineolata (Coutagne, 1881)		×	×
Hauffenia minuta globulina (Paladilhe, 1866)		×	×
Ostracoda			
Pseudocandona zschokkei (Wolf, 1919)			×
Pseudocandona herzogi (Klie, 1934)			×
Pseudocandona n. sp.			×
Cyclopoida			
Eucyclops graeteri (Chappuis, 1927)			×
Tubificida			
Haber turquini (Juget, 1979)			×
Hydrobida			
Bythiospeum diaphanum diaphanum (Michaud, 1831)			×
Bythiospeum diaphanum diaphanum bourguignati (Paladilhe, 1869).			×

Magniez, 1976; Culver, 1982; Howarth, 1983). Thus, we must consider a fluctuating mosaic of communities throughout very different biotopes whose interrelations are mainly dependent on the groundwater flow pattern. For example, organisms inhabiting the epikarstic zone may be flushed by percolation down to the saturated zone, but organisms of this latter zone may also reach some upper levels of the aquifer during periods of intense rising of the groundwater table.

B. Groundwater Flow: An Organizing Factor for Communities

A rapid change in water velocity at a particular location in response to rainfall input is probably one of the most striking characteristics of karst systems. These strong hydraulic dynamics earlier had been believed to be one of the main factors controlling changes in the community structure through time and space in caves (Hawes, 1939).

Concerning the karst as a whole, on the flood scale, the influence of the increase of the discharge can be noticed within the saturated zone even if the karst system is mainly fed by diffuse infiltration. At the outlet of the Dorvan–Cleyzieu system, the drift density of *Niphargus rhenorhodanensis* was studied during four floods (Fig. 4). The number of hypogean amphipods flushed out of the aquifer is higher at the beginning of the flood. Then fluctuations of the drift density during the flood change according to the type of flood. However, there are not always significant relationships between drift density or drift rate and discharge (VII flood, $r = 0.872$; VIII flood, $r = 0.716$; IX flood, $r = -0.213$; X flood, $r = 0.224$) (Gibert, 1986). Other studies have shown similar patterns (Rouch, 1979; Rouch and Carlier, 1985; Turquin 1984, 1986; Turquin and Barthélémy, 1985; Mathieu and Essafi, 1991).

When surveys are carried out inside the karst, they reveal that the groundwater flow pattern during a flood induces a spatial redistribution of organisms. Those inhabiting some transmissive zones of the aquifer are transported along with the current and may be carried out of the karst system or spread throughout some other adjacent areas of the saturated zone. The result of such a dispersion mechanism could be clearly observed at the Terrieu site, at which the floods induce a decrease in the number of epigean invertebrates in the conductive fractures and on increase in the adjacent and thinly fissured matrix (Fig. 5). In this case, the spatial distribution of organisms that could be observed a short time after the floods seems to be linked essentially to the reverse flow process occurring during the flood. At the Foussoubie karst, a similar observation was made because the stream in the Foussoubie vent was shown to contain a higher number of organisms and species after the floods (Fig.6) (Vervier and Gibert, 1991).

Although the groundwater flow is an important factor to consider in karst systems, it alone cannot explain the dynamics of aquatic communities.

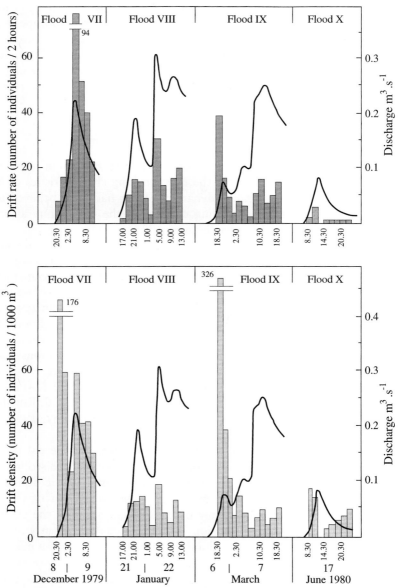

FIGURE 4 Evolution of *Niphargus rhenorhodanensis* rate and density during floods VII, VIII, IX, and X at the Pissoir spring.

FIGURE 5 Spatiotemporal distribution of cyclopoids and oligochaetes collected at the Terrieu site in 1991 and 1992. For abbreviations of species, see Fig. 3. *D. l., Diacyclops languidus; M. v., Megacyclops viridis.*

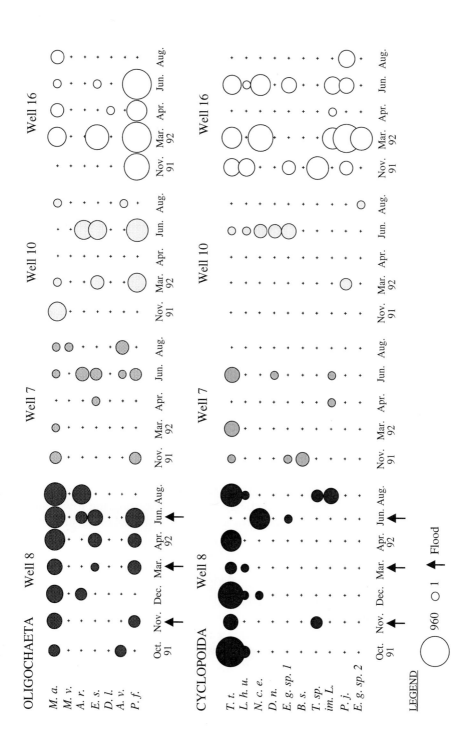

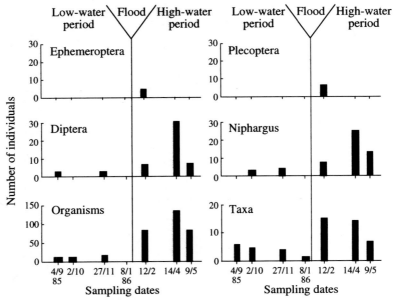

FIGURE 6   Number of organisms and species collected before and after a flood in the stream of the Foussoubie vent.

Indeed, we cannot consider the karst invertebrates only as inert particles that are simply displaced from one point to another, but we have to take into account the strategies developed by highly adapted organisms. As an example, Turquin (1981a,b, 1986) showed that the first flood events of the hydrologic cycle induced a high drift of juveniles of *Niphargus virei* outside the karst aquifer and thus a loss for the population. However, this author suggested that *N. virei* might effectively lay its eggs during the low-water periods and thus use the floods to allow an efficient dispersal of the juveniles throughout the aquifer.

C. Control of the Karst Invertebrate Assemblages

Because previous studies were often conducted on cave rivers and springs, which are usually located in the vicinity of large conductive channels, the organization of animal assemblages has sometimes been considered to be waterflow dependent. However, we know that karst habitats differ greatly in their hydraulic conductivity, creating a wide range of flow conditions, whatever the global hydrologic situation. It is probably one of the most important features to consider if we have to understand the heterogeneous distribution of fauna within the karst systems. This was demonstrated by Rouch *et al.* (1993) for the Baget karst system, in which the adjacent subsystems had faunal assemblages different from those of the main drain. Assemblages of the main channels were dominated by the harpacticoid *Parapseudo-leptomesochra subterranea,* whereas assemblages of the adjacent subsystems were dominated by the harpacticoids *Ceuthonectes gallicus* (Hillère sub-system) and *Nitocrella gracilis* (Peyrère sub-system). A similar scenario could be observed at the Terrieu site when a limited network of interconnected joints and subvertical fractures with different hydraulic conductivities was considered. Indeed, the most relevant result of this ongoing survey has been the differentiation of the conductive fractures from the adjacent small-sized fissure network on the basis of their assemblage characteristics. Such a difference between the faunal composition of fast-moving and that of slow-moving water parts of the saturated zone was also noticed in some other unpolluted limestone sites of the Lez spring basin, where as many as 34 hypogean species were collected from four wells, with estimated yields rang-ing from 2 to 150 m^3/hr^{-1} (F. Malard, unpublished data).

In Section II, we noticed that the interactions and relative flow rates of groundwater in the drainage network and its adjacent subsystems described by Mangin (1974, 1975) on the scale of a karst system show some similarities with those observed in the conductive fracture and in the adjacent fissured matrix described by Drogue (1974) on the scale of a fractured limestone site. Because biological similarities also appeared on these two different spatial scales, we may wonder if the organization of the fracture network of a karst aquifer is such that it allows the repetition of hydrodynamic, physicochemical, and biological features on different scales of space.

Because flow conditions may not fluctuate so much in some parts of the saturated zone, other factors have to be taken into consideration to explain the spatial and temporal distribution of groundwater organisms. In karst milieus, a few attempts have been made to correlate the faunal distribu-tion with changes in the quantity of organic matter. In the stream of the Foussoubie vent, four artificial substrata, immersed during the same period of time, did not contain the same quantity of organic matter (pieces of leaves and twigs and fine particulate organic matter). The richest one in organic matter also had the highest number of organisms and species (Table IV). Because this huge quantity of organic matter, compared with that found in

TABLE IV   Relation between the Quantity of Organic Matter and the Number of Organisms and Species Collected in the Foussoubie Stream

Habitats	Organic matter (g/substrata)	Number of organisms	Number of species
Riffle 1	0.8	334	31
Riffle 2	3.5	404	24
Pool 1	2.0	487	27
Pool 2	30.3	1469	43

the other substrata, was observed after a flood event, we may consider that the groundwater flow has led to some high organic matter concentrations at particular locations, toward which the animals migrate actively. This was also demonstrated at the karst/floodplain interface (Mathieu *et al.*, 1991).

In moderately populated and species-diversified habitats, interactions between organisms have also to be taken into consideration. Such interactions have been described extensively in the United States, especially by Culver (Culver and Ehlinger, 1980; Culver, 1982; Culver *et al.*, 1991; see also Chapter 10) from the results of cave studies in the southern Appalachians (United States) and laboratory experiments on different species of amphipods and isopods. The importance of competition (and amensalism) varies from one karst system to another. For example, competitive interactions dominated in the Thompson Cave and were undetectable in the Alpena Cave. In some situations both predation and trophic commensalism have been detected. Therefore trophic interactions appear to be much more complex than previously recognized. In the Mammouth Cave (Kentucky), the most studied "cave ecosystem" and probably the best understood in the world [see the review in Poulson (1992)], Poulson studied the species composition (fish predators, crayfish, isopods, amphipods, flatworms, etc.) of different habitats (seasonally wet, shallow streams, and medium to deep streams). He demonstrated that the differences in species composition were related to the way in which competition was modulated by food supply and differential abilities to escape seasonal drying and predation. Finally species interactions—not solely of one type—have major effects on microdistribution patterns.

Groundwater contamination may also lead to changes in community structure. This has been particularly well observed at the Terrieu site, at which the original assemblages, essentially made up of stygobites, are now dominated by epigean and ubiquitous species (*i.e.*, *Tubifex tubifex*, *Macrocyclops albidus*, *Attheyella crassa*, and *Paracyclops fimbriatus*). In this case, the spatial distribution of hypogean and epigean species was closely related to the infiltration and movement of sewage-polluted surface water toward and within the aquifer (Malard *et al.*, 1994b). It must be noted that, although the elimination of the contamination results in a decrease of polysaprobiont species such as *T. tubifex*, whose presence is organic matter dependent, ubiquitous organisms, once they have colonized the site, are probably able to accommodate new changes in environmental conditions. In this case, a time-limited source of pollution may induce long-term modifications of the community structure. A similar observation has been made by Rouch *et al.* (1993), who demonstrated that the community structure of the Peyrère subsystem of the Baget karst, which had been disturbed by a high-discharge and three-day-long pumping test, had not recovered to its original state one year after the pumping date.

Finally, it is clear that in many parts of a karst aquifer, the community organization is, to a certain extent, not so waterflow dependent. Therefore, a multifactor approach that considers the hierarchization and interactions of different constraining variables such as food availability, contaminant concentrations, and organism interactions, as well as the ability of groundwater organisms to migrate in order to meet their ecological tolerances and preferences, has to be developed.

V. ECOLOGICAL FUNCTIONING OF AQUATIC KARST SYSTEMS

Links between the surface environment and the karst aquifer can be quite varied from one karst system to another, leading to strong consequences for biological functioning. Interactions may be strongly influenced by the nature of the surface water/groundwater boundary. Thus, there is a considerable interest in determining the extent to which the boundary acts as a differentially permeable zone that facilitates some biological fluxes but impedes others (Gibert *et al.*, 1990; Vervier *et al.*, 1992). Moreover, because the karst is now viewed as a dynamic mosaic of invertebrate assemblages distributed throughout the different slow- and fast-moving water parts of the aquifer, biological interactions and exchanges that take place among these different elements are considered. Within the karst, interactions between elements confer a certain stability to the ecosystem on a given space and time scale.

A. Links between the Karst Aquifer and Surface Environment

From a hydrologic point of view, karst systems are very open, whereas the openness of these aquatic ecosystems in relation to the biological fluxes may be modulated. Based on the three study karsts, we can envisage the epigean invertebrate colonization from upstream and downstream.

Upstream, numerous individuals often penetrate the aquifer by means of large sinkholes. At the Foussoubie system, many epigean animals inhabiting the Planche stream are flushed out into the aquifer and are transported over a distance of 5 km, because they were collected at the spring. At the Baget system, Rouch (1979) studied the drift of epigean harpacticoids in a temporary stream that sinks at a point called the "Peyrère sinkhole" and demonstrated that 7 to 14 million individuals enter the aquifer each year (Rouch, 1979). Although we cannot say, in the case of streams flowing through sinkholes, that epigean organisms actively migrate into the underground environment, it is, as outlined by Rouch (1986), a typically anthropocentric point of view to consider that the epigean animals are passively displaced (passive drift) into the aquifer. At the Terrieu site, the infiltration of the effluent stream (average discharge, 5 l/sec^{-1}) occurs on a bed surface of approximately 4000 m^2. Therefore, the infiltration rates are relatively low, and we cannot consider that epigean organisms are passively displaced toward the groundwater zone. On the contrary, during low-water periods these organisms effectively migrated in the aquifer, especially when the stream started to dry up. Moreover, we demonstrated that these epigean organisms were spread throughout the whole adjacent fissure network during periods of groundwater recharge. Thus, the fact that surface organisms are able to establish some permanent populations in the aquifer mainly depends on their ecological tolerances and their ability to sustain the environmental conditions of the karst aquifer.

Downstream, it clearly appears that the epigean organisms actively migrate into the karst aquifer simply because it may offer some relatively new and free biotopes to colonize. In the Dorvan–Cleyzieu system, larvae of epigean insects, adult water beetles, and individuals of *Gammarus* were collected inside the karst in some places located 200 m upstream from the outlet. Becasue there is no sinkhole or losing stream at the karst surface, the presence of epigean organisms is due to an upstream migration process similar to that observed in surface streams (Kurek, 1967; Elliot, 1971; Hobbs and Butler, 1981) and that probably exists for all springs. However, in many cases, this migration from downstream is probably space limited.

Considering the three study karsts as examples, the biological functioning of a karst aquifer with regard to its relation with the surface environment can be analyzed (Fig. 7).

— The Dorvan–Cleyzieu case is probably the simplest one and the least frequent because no surface running water that could recharge the aquifer occurs at the land surface (Fig. 7A). The upstream and space-limited migration process at the spring is the only biological input to the system, which is thus distinctly autogenic (from a biological point of view).

— The Foussoubie karst aquifer is a semiopen biological system because the migration of epigean species occurs only during high-water periods (Fig. 7B). As demonstrated by Rouch (1986) for the Baget system, the underground community structure is disturbed during periods of groundwater recharge but the aquifer recovers its biological equilibrium during low-water periods.

— In an open biological system, epigean organisms may continuously enter the aquifer at different locations (Fig. 7C). For example, at the Terrieu site, epigean invertebrate migration occurs during low- and high-water periods all along the longitudinal profile of the stream. In this case, the invertebrate community in some parts of the aquifer may be dominated by epigean forms throughout the hydrologic cycle.

B. Connections and Exchanges within the Karst and Ecosystem Stability

Only a small percentage of the fractures forming the dendritic and three-dimensional network of a karst aquifer participates in the flow in the

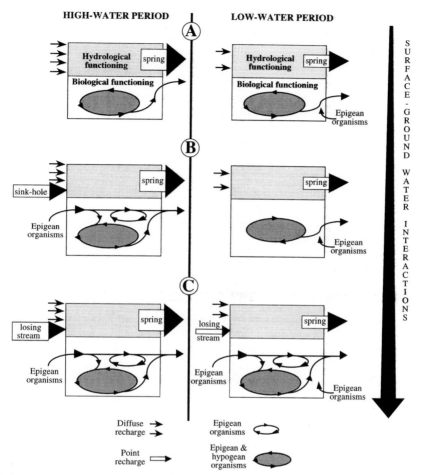

FIGURE 7 Epigean and hypogean invertebrate fluxes within the three study karst systems. (A) the Dorvan–Cleyzieu, (B) the Foussoubie, and (C) the Lez spring systems.

direction of the spring. This means that during a flood, active groundwater flow with high velocities probably takes place in a small part of the aquifer, whereas the remaining parts only undergo some rising of the water levels. The latter zones probably have a high "physical stability," comparable to that encoutered in porous aquifers, which made possible the development of permanent and diversified communities. This may also happen in huge and highly karstified openings because low flow does not imply small aperture but only that a particular area is barely connected to the conductive zones. For example, in a sinkhole connected to the Peyrère subsystem of the Baget karst, Rouch (1992) collected a new species of *Parastenocaris* that did not appear in any of the numerous filterings made at the main spring of the system. In the same ways, individuals of *Bathynella sp.* and *Microcharon sp.* were regularly collected from high-capacity wells located in the upstream part of the Lez spring catchment. These stygobites are known to be very sensitive to disturbances in alluvial systems (Dole-Olivier *et al.*, 1993).

As for the surface aquatic ecosystems, on a given space and time scale, the stability of a karst ecosystem is seen to be the result of an equilibrium between antagonist forces. Floods are often considered a destabilizing force that moves the system away from its equilibrium point. However, during periods of intensive groundwater recharge, food supply increases (*i.e.*, the entrance of organic matter and the drift of terrestrial animals), highly contaminated zones are diluted, and epigean communities that established themselves in the conductive channels (*i.e.*, the Terrieu site) are flushed out. During periods of low water, it is known that subsystems of a hydrogeologic catchment may be hydraulically disconnected from the drainage network (Rouch *et al.*, 1993; Karam, 1989). For example, in the well-studied Edwards Aquifer in Texas, which contains diverse communities of groundwater species (Holsinger and Longley, 1980), different areas are connected only when the piezometric levels are high (Longley, 1981, 1992). From a biological point of view, this means that during low-water periods the ecosystem is cut up into smaller parts, which may evolve in different ways. As an example, two species of blind catfishes, *Satan eurystomus* and *Trogloganis pattersoni*, live in the San Marcos part of the aquifer, whereas they are absent in the San Antonio pool (Longley, 1981). On the contrary, we may consider that groundwater recharge will allow the system to recover its biological continuum. At the Baget karst system, Rouch *et al.* (1993) pointed out that, if 10 to 20 floods occurred each year, during which a high drift of harpacticoids toward the oulets was observed, the community structure of the system would not be affected. Finally, he also noticed that perturbation of the community structure, caused by a high-discharge pumping test, tended to be resorbed with the succession of floods.

Another example also underlines the considerable resistance of the karst ecosystem. From the results of a 10-year survey carried out on a karst of the Jura (France), Turquin (1984, 1986) has demonstrated that the abundance and age distribution of a population of *N. virei* could undergo significant changes throughout the years. On the decade scale, the age distribution of the population was related to the rainfalls and hydrologic responses of the karst system. For example, during rainy decades, the reproduction activity was shown to be higher as an offset to the loss of individuals caused by intensive drift toward the outlet.

Finally, a karst ecosystem disposes of a variety of homeostatic mechanisms that, over a given period of time, attenuate the effects of floods as well as the effects of pollution. It appears likely that the karst ecosystem, as a whole, is more resistant and resilient to physical and chemical disturbances than previously recognized.

VI. CONCLUSION AND PERSPECTIVES

Several authors have reviewed some of the main concepts frequently used in groundwater ecology and demonstrated how critical examination of first ideas has greatly contributed to a better understanding of the functioning of the ecosystem (Chapter 1). Rouch (1986) emphasized all the advantages arising from the substitution of a single cave habitat approach for a multihabitat and karst system approach. Rouch and Danielopol (1987) have brought to light the active behavior of organisms in the colonization process of the underground environment. Gibert (1991, 1992) demonstrated how, with the change in basic conceptions and the growing appreciation of the scale and complexity of the underground environment, the ecological approach has become practical and not simply descriptive.

Although much research has been carried out these last 30 years and our knowledge of karst system properties has considerably increased, a global and conceptual model that could describe and explain the spatial and temporal distribution of karstwater organisms is still lacking. In milieus as complex as those of fractured and karstified aquifers which are known to be highly heterogenous, anisotropic, and sometimes discontinuous; in which the groundwater flow conditions at particular locations may change rapidly in response to rainfall events; and which contain both epigean and hypogean species as well as ubiquitous and highly adapted species, it is questionable that a simple model based on a continuous change of a few synthetic variables will effectively be able to render an account of the spatial and temporal dynamics of communities.

Because the theoretical aspect of karst water ecology is still very weak, future research, based on the multidisciplinary framework, on karst key units, such as the annex subsystems or capactive zones, must be encouraged. The fundamental properties of these zones are the life reservoir, buffer effect, probably zones in which self-purification is possible, and the starting point for restoration and restructuration after disturbance. Documentation from the standpoint of this zone structure and function is only beginning to be written. We need more synthetic biological models that will formalize knowledge and guide research directions through interactive biological and physical attributes.

ACKNOWLEDGMENTS

These case studies represent the collaborative efforts of many people, speleologists, and inhabitants of the different areas studied. Special thanks to D. Martin for her valuable assistance during the field and laboratory operations, to J. Mathieu and M. Lafont for identifying cyclopoids and oligochaetes, and to Glyn Thoiron for checking the English of a first draft of this manuscript. The bulk of this research was supported by the Interdisciplinary Research Programme on the Environment (PIR—Groundwater Ecosystems) of the URA-CNRS 1451 laboratory (Lyon I).

REFERENCES

Adam, C. (1979). Composition des eaux souterraines du département de l'Ardèche. Thèse, Univ. Lyon I.

Bakalowicz, M. (1981). Les eaux d'infiltration dans l'aquifère karstique. *Proc. Int. Congr. Speleol., 8th,* Bowling-Green, pp. 710–713.

Balazuc, J. (1986). "Spéléologie du département de l'Ardèche." Bouquinerie Ardéchoise, Aubenas.

Barr, J. C., Jr. (1967). Ecological studies in the mammoth cave system of Kentucky. I. The biota. *Int. J. Speleol.* 3, 147–204.

Barr, T. C., Jr., and Kuehne, R. A. (1971). Ecological studies in the Mammoth cave system of Kentucky. II. The ecosystem. *Ann. Spéléol.* 26, 48–96.

Belleville, L. (1985). Hydrogéologie karstique. Géométrie, fonctionnement et karstogenèse des systèmes karstiques des Gorges de l'Ardèche (Ardèche, Gard). Thèse, Univ. Grenoble.

Bertrand, J. Y. (1975). Etude d'un aquifère épikarstique des Corbières (Opoul, Pyrénées orientales). *Ann. Spéléol.* 30, 513–537.

Callot, Y. (1978). A propos des plateaux ardéchois: Karst, rapport fond-surface et évolution des paysages calcaires ou en roche perméable cohérente. Essai sur les paramètres influant dans la formation des paysages calcaires ou en rouch perméable. Thèse, Univ. Reims.

Culver, D. C. (1982). "Cave Life: Evolution and Ecology." Harvard Univ. Press, Cambridge, MA.

Culver, D. C., and Ehlinger, T. J. (1980). The effects of microhabitat size and competitor size on two cave isopods. *Brimleyana* 4, 103–114.

Culver, D. C., Fong, D. W., and Jernigan, R. W. (1991). Species interactions in cave stream communities: Experimental results and microdistribution effects. *Am Midl. Nat.* 126, 364–379.

Delay, B. (1969). Les circulations verticales d'eau et le peuplement de la zone de percolation temporaire. *Ann. Spéléol.* 23, 705–716.

Dole-Olivier, M. J., Creuze des Châtelliers, M., and Marmonier, P. (1993). Redundant gradients in subterranean landscape. Example of the stygofauna in the alluvial floodplain of the Rhône river (France). *Arch. Hydrobiol.* 127, 451–471.

Drogue, C. (1974). Structure de certains aquifères karstiques d'après les résultats de travaux de forages. *C. R. Acad. Sci.* 278, 2621–2624.

Drogue, C. (1980). Essai d'identification d'un type de structure de magasins carbonatés, fissurés. Application à l'interprétation de certains aspects du fonctionnement hydrogéologique. *Mém. Hors-Sér. Soc. Geol. Fr.* 11, 101–108.

Drogue, C. (1985). Geothermal gradients and groundwater circulations in fissured and karstic rocks: The role played by the structure of the permeable network. *J Geodyn.* 4, 219–231.

Elliot, J. M. (1971). Upstream movements of benthic invertebrates in a lake district stream. *J. Anim. Ecol.* 40, 235–252.

Gibert, J. (1986). Ecologie d'un système karstique jurassien. Hydrogéologie, dérive animale, transits de matrières, dynamique de la population de *Niphargus* (Crustacé Amphipode). *Mém. Biospéol.* 13, 1–379.

Gibert, J. (1990). Behavior of aquifers concerning contaminants: Differential permeability and importance of the different purification processes. *Water Sci. Technol.* 22(6), 101–108.

Gibert, J. (1991). Groundwater systems and their boundaries: Conceptual framework and prospects in groundwater ecology. *Verh. Int. Ver. Limnol.* 24, 1605–1608.

Gibert, J. (1992). Groundwater ecology in a perspective to environmental sustainability. *In* "Ground Water Ecology" (J. A. Stanford and J. J. Simons, eds.), pp. 3–13. Am. Water Resour. Assoc., Bethesda, MD.

Gibert, J., Guezo, B., Laurent, R., and Marchand, T. (1982). Expérience de traçage artificiel dans le Jura méridional. Mise en évidence de liaisons souterraines dans le massif de Dorvan (Torcieu, Ain, France). *Spelunca* 7, 19–26.

Gibert, J., Dole-Olivier, M.-J., Marmonier, P., and Vervier, P. (1990). Surface water-groundwater ecotones. *In* "The Ecology and Management of Aquatic-Terrestrial Ectones" (R. J. Naiman and H. Décamps, eds.), Man & Biosphere Ser., Vol. 4, pp. 199–255. Parthenon Publ., London.

Ginet, R. (1960). Ecologie, éthologie et biologie de *Niphargus* (Amphipodes Gammaridés hypogés). *Ann. Spéléol.* **15**, 1–254.

Hawes, R. S. (1939). The flood factor in the ecology of caves. *J. Anim. Ecol.* **8**, 1–5.

Henry, J. P. (1976). Recherches sur les Asellides hypogés de la lignée *cavaticus* (Crustacea, Isopoda, Asellota). Thése, Univ. Dijon.

Hobbs, H. H., and Butler, M. S. (1981). A sampler for simultaneously measuring drift and upstream movements of aquatic macroinvertebrates. *J. Crustacean Biol.* **1**, 63–69.

Holsinger, J. R., and Longley, G. (1980). "The Subterranean Amphipod Crustacean Fauna of an Artesian Well in Texas." Smithson. Inst. Press, Washington, DC.

Howarth, F. G. (1983). Ecology of cave arthropods. *Annu. Rev. Entomol.* **28**, 365–389.

Karam, Y. (1989). Essais de modélisation des écoulements dans un aquifère karstique—Exemple de la source du lez (Hérault-France). Thése, Univ. Montpellier.

Kurek, A. (1967). Uber die tagesperiodische Ausdrift von *niphargus aquilex schellenbergi* Karaman aus Quellen. *Z. Morphol. Oekol. Tiere* **58**, 247–262.

Lescher-Moutoué, F., and Gourbault, N. (1970). Etude écologique du peuplement des eaux souterraines de la zone de circulation permanente d'un massif karstique (recherches sur les eaux souterraines, 13). *Ann. Spéléol.* **25**, 765–848.

Longley, G. (1981). The Edwards aquifer: Earth's most diverse groundwater ecosystem? *Int. J. Speleol.* **11**, 123–128.

Longley, G. (1992). The subterranean aquatic ecosystem of the Balcones fault zone edwards aquifer in Texas—threats from overpumping. *In* "Ground Water Ecology," (J. A. Stanford and J. J. Simons, eds.) pp. 291–300. Am. Water Resour. Assoc., Bethesda, MD.

Magniez, G. (1976). Contribution à la connaissance de la biologie des Stenasellidae (Crustacea Isopoda Asellota des eaux souterraines). Thése, Univ. Dijon.

Malard, F., Reygrobellet, J. L., Gibert, J., Chapuis, R., Drogue, C., Winiarsky, T., and Bouvet, Y. (1994). Sensitivity of underground karst ecosystems to human perturbation-conceptual and methodological framework applied to the experimental site of Terrieu (Hérault-France). *Verh. Int. Ver. Limnol.* (in press).

Malard, F., Reygrobellet, J. L., Mathieu, J., and Lafont, M. (1994). The use of invertebrate communities to describe groundwater flow and contaminant transport in a fractured rock aquifer. *Arch. Hydrobiol.* **131**, 93–110.

Mangin, A. (1974). Contribution à l'étude hydrodynamique des aquifères karstiques. *Ann. Spéléol.* **29**, 283–332; 495–601.

Mangin, A. (1975). Contribution à l'étude hydrodynamique des aquifères karstiques. (troisième partie) *Ann. Spéléol.* **30**, 21–124.

Mangin, A. (1983). L'approche systémique du karst, conséquences conceptuelles et méthodologiques. *Karst-Larra, Reun. Monogr. Karst-Larra, 1982*, pp. 141–157.

Mangin, A. (1985). Progrès récents dans l'étude hydrogéologique des karsts. *Stygologia* **1**, 239–257.

Mangin, A. (1986). Réflexion sur l'approche et la modèlisation des aquifères karstiques. *Jorn. Karst Euskadi, Donostia-San Sebastian, Spain* **2**, 11–31.

Mangin, A., and Bakalowicz, M. (1989). Orientation de la recherche scientifique sur le milieu karstique. Influences et aspects perceptibles en matière de protection. *Spelunca* **35**, 71–79.

Mathieu, J., and Essafi, K. (1991). Changes in abundance of interstitial populations of *Niphargua* (stygobiont amphipod) at the karst/floodplain interface. *C.R. Acad. Sci.* **312**, 489–494.

Mathieu, J., Essafi, K., and Doledec, S. (1991). Dynamics of particulate organic matter in bed sediments of two karst streams. *Arch. Hydrobiol.* **122**, 199–211.

Mazellier, R. (1971). Contribution à l'étude géologique et hydrogéologique des terrains crétacés du Bas-Vivarais. Thèse, Univ. Montpellier.

Mohr, C. E., and Poulson, T. L. (1966). "The Life of the Cave." McGraw-Hill, New York.

Poulson, T. L. (1992). The Mammoth Cave ecosystem. *In* "The National History of Biospeology" (A. I. Camacho, ed.), Monogr. 7, pp. 568–611. Nac. Mus. Cienc. Nat., C.S.I.C., Madrid.

Rouch, R. (1971). Recherches sur les eaux souterraines. XIV. Peuplement par les Harpacticides d'un drain situé dans la zone de circulation permanente. *Ann. Spéléol.* **26**, 107–133.

Rouch, R. (1977). Le système karstique du Baget. VI. La communauté des Harpacticides. Signification des échantillons récoltés lors des crues au niveau de deux exutoires du système. *Ann. Limnol.* **13**, 227–249.

Rouch, R. (1979). Le système karstique du baget. VIII. La communauté des Harpacticides. Les apports au sein du système par dérive catastrophique. *Ann. Limnol.* **15**, 243–274.

Rouch, R. (1982). Le système karstique du Baget. XIII. Comparaison de la dérive des Harpacticides à l'entrée et à la sortie de l'aquifère. *Ann. Limnol.* **18**, 133–150.

Rouch, R. (1986). Sur l'écologie des eaux souterraines dans le karst. *Stygologia* **2**, 352–398.

Rouch, R. (1992). *Parastenocaris mangini* N. sp., nouvel Harpacticoïde (copépodes) stygobie des Pyrénées. *Crustaceana* **63**, 306–312.

Rouch, R., and Carlier, A. (1985). Le système karstique du Baget. XIV. La communauté des Harpacticides. Evolution et comparaison des structures de peuplement épigé à l'entrée et à la sortie de l'aquifére. *Stygologia* **1**, 71–92.

Rouch, R., and Danielopol, D. (1987). L'origine de la faune aquatique souterraine, entre le paradigme du refuge et la modèle de la colonisation active. *Stygologia* **3**, 345–372.

Rouch, R., Juberthie-Jupeau, L., and Juberthie, C. (1968).Essai d'étude du peuplement de la zone noyée d'un karst. *Ann. Spéléol.* **23**, 717–733.

Rouch, R., Pitzalis, A., and Descouens, A. (1993). Effets d'un pompage à gros débit sur le peuplement des Crustacés d'un aquifère karstique. *Ann. Limnol.* **29**, 15–29.

Slama, P., Leroux, P., Chedhomme, J., and Cheilletz, E. (1981). Etat actuel des recherches dans la goule de Foussoubie et dans diverses autres cavités au confluent de l'Ardéche et de l'Ibie. Vallon-Pont d'Arc. Ardèche. *Spelunca* **2**, 28–31.

Turquin, M. J. (1981a). The tactics of dispersal of two species of *Niphargus* (Perennial, troglobitic Amphipoda). *Proc. Int. Congr. Spéléol., 8th,* Bowling-Green, pp. 353–354.

Turquin, M. J. (1981b). Profil démorgraphique et environnemnt chex une population de *Niphargus virei* (Amphipode troglobie). *Bull. Soc. Zool. Fr.* **106**, 457–465.

Turquin, M. J. (1984). Age et croissance de *Niphargus virei* (Amphipode Perennant) dans le système karstique de Drom: Méthodes d'estimation. *Mém. Biospéol.* **11**, 37–49.

Turquin, M. J. (1986). Mortalité et stock de l'Amphipode *Niphargus virei* dans le système karstique de Drom. *Bull. Soc. Linn. Lyon* **55**, 293–304.

Turquin, M. J., and Barthélémy, D. (1985). The dynamics of a population of the troglobitic Amphipod *Niphargia virei* Chevreux. *Stygologia* **1**, 109–117.

Vandel, A. (1964). "Biospéléologie: La biologie des animaux cavernicoles." Gauthier-Villars, Paris.

Vervier, P. (1990a). Hydrochemical characterization of the water dynamics of a karstic system. *J. Hydrol.* **121**, 103–117.

Vervier, P. (1990b). A study of aquatic community dynamics in a karstic system by the use of artificial substrates. *Arch. Hydrobiol.* **119**, 15–33.

Vervier, P., and Gibert, J. (1991). Dynamics of surface water/groundwater ecotones in a karstic aquifer. *Freshwater Biol.* **26**, 241–250.

Vervier, P., Gibert, J., Marmonier, P., and Dole-Olivier, M. J. (1992). A perspective on the permeability of the surface freshwater-groundwater ecotone. *J. North Am. Benthol. Soc.* **11**, 93–102.

17

Organ Cave Karst Basin

D. C. Culver,* W. K. Jones,† D. W. Fong,*
and T. C. Kane‡

*Department of Biology, American University
Washington, D.C. 20016

†Karst Waters Institute
P.O. Box 490
Charleston, West Virginia 25414

‡Department of Biological Sciences
University of Cincinnati
Cincinnati, Ohio 45221

I. INTRODUCTION

Rouch and his coworkers have pioneered the concept of the karst basin as the unit of ecological analysis (Rouch, 1977; Mangin, 1985). The utility of this approach has been amply demonstrated by Rouch's extensive series of papers on the Baget Basin (e.g., Rouch, 1982), and by Gibert's work on the Dorvan Massif (Gibert, 1986). Ford and Williams (1989), arguing from a hydrogeologic perspective, have put forward a similar idea of looking at the karst basin as a hydrologic gray box.

In this chapter we take a preliminary look at a North American karst basin—the Organ Cave basin—from this perspective. The Organ Cave basin differs strikingly from the better studied Baget and Dorvan basins (Rouch, 1982; Gibert, 1986), primarily due to the presence of over 65 km of cave passage in the 8.1-km² basin. Over 60 km of passage is in Organ Cave, the

6th longest cave in the United States and the 14th longest cave in the world (Courbon *et al.,* 1989). It is indicative of both the complexity of the cave and the absence of an ecosystem perspective for karst that, although the cave has been known since at least the 18th century and actively explored and mapped since the late 1940s, the sole resurgence to the system was discovered only in 1973 (W. K. Jones, 1988). The presence of extensive cave stream passage has also led biologists studying the fauna of the Organ Cave karst drainage basin to focus on population and community ecology, rather than the ecosystem.

In this chapter, we take a preliminary look at the Organ Cave karst drainage basin from an ecosystem perspective. We begin with a consideration of the regional, environmental, geologic, and hydrologic setting of the karst area, move to a detailed description of the hydrogeology of the basin, describe the fauna of the cave and resurgence, and finish with a description of a plausible scenario for cave colonization.

II. THE REGIONAL SETTING

Organ Cave is situated in an extensive karst plateau drained by the Greenbrier River in southeastern West Virginia (Fig. 1). The area is in the eastern part of the Appalachian Plateau geomorphic province. A mature fluvial–doline karst has developed on lower Carboniferous (Mississippian) limestones, which range in thickness from 350 m in southern West Virginia to 100 m in central West Virginia. The karst is characterized by numerous dolines, blind valleys, the absence of surface drainage, and well-defined subsurface drainage basins (W. K. Jones, 1973). The karst plateau is sur-

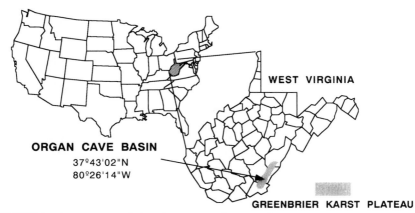

ORGAN CAVE BASIN
37º43'02"N
80º26'14"W

WEST VIRGINIA

GREENBRIER KARST PLATEAU

FIGURE 1   Locator map of Greenbrier Karst Plateau and Organ Cave Basin. Latitude and longitude refer to the entrance to Organ Cave.

rounded by clastic ridges and hills that channel aggressive water onto the carbonate outcrop area at discrete recharge points. Discharge from the karst plateau is generally concentrated at base-level springs along the Greenbrier River or its major tributaries. The springs respond within hours to storm events, and the discharge characteristics of the springs are typical for conduit flow regimes.

This karst region of about 500 km^2 ranges from Sinks Grove in the south to Droop Mountain in the north and has over 1200 known caves (William Balfour, personal communication). A total of 8 of these caves are longer than 20 km (Courbon *et al.*, 1989). The caves of this region have seen extensive exploration and mapping activity since the 1940s.

The area has a humid–temperate climate with a mean temperature of 11°C, a mean annual rainfall of approximately 100 cm, and an average snowfall of approximately 60 cm. Although precipitation is more or less evenly distributed throughout the year, there are relatively dry falls. Stream levels in and out of the cave are highest in the spring due to snow melt, ground thawing, and spring precipitation. Jones (1973) calculated the total annual evapotranspiration by the Thornwaite method at 66 cm, with evapotranspiration exceeding precipitation from May through September. High infiltration rates through the karst land surface and the greatly reduced evaporation rate of channeled subsurface flow could cause a lower than predicted evapotranspiration rate and a higher than predicted runoff rate.

The Organ Cave drainage basin is part of the Organ Cave Plateau (Fig. 2). The surface of the plateau is generally 670 m above sea level with a few residual hills at 700 m and sinkholes at 630 m. It is bounded on the north and west by the Greenbrier River, on the south by Second Creek, and on the east by Carpenter Creek and White Rock Mountain. The base level at the Organ Cave resurgence on Second Creek is 540 m. White Rock Mountain to the east rises to 950 m in elevation. The Organ Cave drainage basin is one of seven karst basins on the plateau. The mature karst developed on the plateau has obliterated all traces of the original surface stream channels. The drainage divides were largely determined by tracer tests and the position of surveyed cave passage (Jones, 1988). There is no evidence that the present catchment area has changed very much from the time just prior to the onset of karstification.

III. HYDROGEOLOGY OF ORGAN CAVE BASIN

The karst basin is dominated by Organ Cave, over 60 km in length (Fig. 3). The cave has 10 known entrances and has been the subject of intensive exploration and mapping since the late 1940s, with a detailed map on the scale of 1:1200 (Stevens, 1988). Most entrances are at the downstream ends of short blind valleys at the contact of the limestone and the underlying

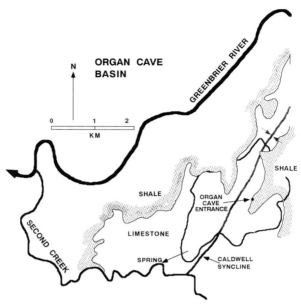

FIGURE 2 Map of Organ Cave Plateau, showing the hydrologic setting. The basin is outlined by the solid line of medium thickness. The basin is confined on the east and west by the underlying Maccrady Shale as the Caldwell syncline exposes the basal units of the Greenbrier limestone in the center of the basin. Several of the entrances are located at or near the shale–limestone contact.

Maccrady Shale, and they contain wet weather streams. There are no permanent surface streams in the basin.

Passage development has a complex pattern (Deike, 1988a), but largely follows the axis of the southwest plunging Caldwell Syncline (Fig. 4). Most passages are long shallow stream channels, with 10×10 m cross-sections not uncommon. Stream channels are typically littered with breakdown and gravel fill. The most common sedimentary deposits are silty and sandy gravels between 1 and 10 cm in diameter (Deike, 1988a). Although the limestone is over 250 m thick, most cave development is within the lower 30 m of the Greenbrier limestones and the top 5 m of the underlying Maccrady shales. The sole resurgence of the cave is a spring on Second Creek.

The synclinal axis forms a structural trough with the master drain (Organ/Hedricks streams), flowing southwest down the plunge until base level is reached about 1 km from the resurgence. From the last sump the water rises about 150 m through the stratigraphic section and may follow a phreatic conduit that loops as much as 40 m below base level before rising at the resurgence on Second Creek (Deike, 1988b). Joints appear to control the development of most of the cave. Both limbs of the syncline have locally steep monoclinal folds and drag folds with reverse faults parallel to the

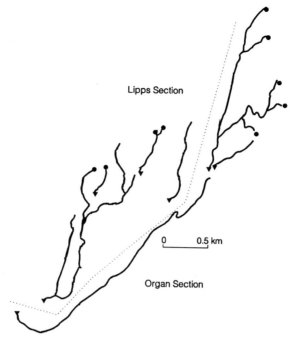

FIGURE 3 Planar map of major stream passages in Organ Cave, adapted from Jones (1988). Circles are entrances and triangles are sumps. The dotted line indicates the drainage divide in the cave. Note that most passages are not shown (Stevens, 1988).

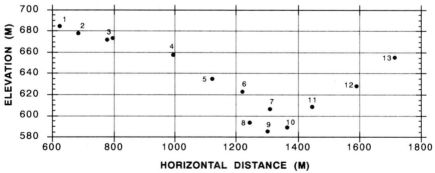

FIGURE 4 Organ Cave passages displayed in an east–west cross-section of the cave through the Organ entrance. Numbers refer to the following sites in the cave (see Fig. 3 for planar map): 1, Near Deems entrance; 2, Lipps entrance; 3, Lipps Maze; 4, Skid Row (near Jones Canyon); 5, Treasure Room; 6, Fun Room Passage; 7, North Upper Level Organ (NULO); 8, Nine Foot Bat Passage; 9, Hedricks Stream; 10, Organ Stream; 11, Discovery Passage; 12, Rotunda Room; 13, Organ entrance. Hedricks Stream (No. 9) underlies the North Upper Level Organ (No. 7) stream in the trough of the Caldwell syncline. Adapted from Jones (1988).

strike. Connections between stratigraphic levels within the cave are generally at these deformed zones. An east–west cross section (Fig. 4) through the center of the caves shows the positioning of passages on the limbs and trough of the syncline.

The passages in Organ Cave are primarily developed in two horizons of the limestone. Most passages are developed at the base of the limestone, and streams have eroded up to 5 m into the underlying shale at a few locations. The shale beds are relatively incompetent and include varied calcareous rocks and breccia. Passage development in these beds probably began with phreatic flow in cherty biosparite 3 to 7 m above the shale contact (Deike, 1988b). Higher level passages, some with perched streams showing phreatic features, are developed in micrite and sparite about 25 m stratigraphically above the shale. Thin clastic and dolomitic beds, as well as some chert horizons, cause the perching of the north upper level Organ (NULO) stream above the main Organ/Hedricks stream for as much as 1.5 km (Fig. 3).

Passages of the northern part of the cave form a complex dendritic pattern with three subsurface streams draining the upper cave (Fig. 3). The eastern part of the cave is drained by the Organ and Hedricks streams. The central section is drained through Jones Canyon, which is a tributary to the Organ/Hedricks stream in the southern part of the cave system. The western part of the cave drains through the lower Lipps Canyon. The two main drains of the southern part of the cave merge before they reach the resurgence on Second Creek. The stream passage gradients are much stronger in the upstream reaches of the cave and become gentler in the downstream direction (Fig. 5).

Horton (1945) and Strahler (1964) have suggested that the arrangement of surface stream tributaries is highly ordered. In the Strahler ordering

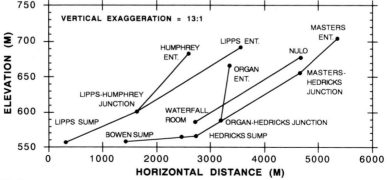

FIGURE 5   Stream passage gradients in Organ Cave. For planar location of passages, refer to Fig. 3. Note that stream passage gradients are gentler downstream.

system, all unbranched tributaries are of first order and a junction of two streams of order u becomes a stream segment of order $u + 1$. Horton and Strahler suggest that

where N_u is the number of segment of order u, k is the order of the trunk stream, and

$$R_u = N_u/N_{u+1}.$$

The branching ratio, R, is typically between 3 and 5 for surface streams (Strahler, 1964). White (1988), citing Baker's (1973) work on several New York caves, indicated that the branching ratio is lower in caves, between 2.4 and 3.3. However, our analysis of the stream order in Organ Cave (and the hydrologically connected Foxhole Cave) indicates a ratio typical of surface basins ($R = 4.2$; Fig. 6). The lower ratios in caves found by Baker indicate a relative scarcity of first-order streams, and the analysis here of the more extensive Organ Cave streams suggest that his results may be an artifact of the inaccessibility of parts of the subterranean drainage of the New York caves he studied.

A short-term temporal picture of both the cave stream and the spring habitat is available for two streams near the Organ entrance and the resurgence spring. Based on temperature recorded at 15-min intervals for 49 days during the summer of 1993, both cave streams were more variable than the spring. In spite of a diurnal temperature cycle at the spring, its coefficient of variation was smaller (Table I).

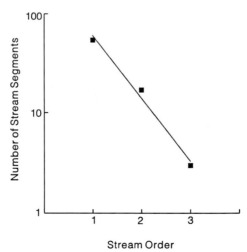

FIGURE 6 Relationship between stream order and frequency of stream segments for Organ Cave basin caves. The resurgence of Organ Cave is a fourth-order stream, but no accessible passages in the cave are fourth-order streams.

TABLE I   Variability of Temperature (°C) in Two Cave Streams in
Organ Cave and the Resurgence of Organ Cave

Location	Mean	Maximum	Minimum	Coefficient of variation
1812 stream	11.5	11.7	11.2	0.027
Organ Stream	8.9	9.0	8.7	0.011
Resurgence	8.4	9.0	8.2	0.008

Note. Organ Stream is fed largely by vertically percolating water and the 1812
stream is fed by direct input. Data are based on 4601 measurements taken at
15-min intervals during the late summer of 1992 (Julian Day 219 to 267).

Relative water levels in the two cave streams responded differently to
a storm event (Fig. 7). In one cave stream with an open input from the
surface, water rose quickly and fell quickly following a rain. In the other
cave stream, with diffuse inputs from the surface, the water level rose more
slowly and fell very slowly. The spring showed a bimodal response, which
lagged the storm response in the two cave streams by several hours. Thus, the

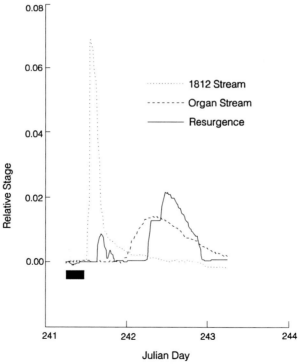

FIGURE 7   Response of two streams in Organ Cave and Organ Cave resurgence to a storm
event of approximately 1 cm. For each site, the water level (a correlate of discharge) is scaled
so that the total area under the curve is identical (1). Thus, only information about shape of
the response is available. The solid bar indicates the duration of the rain.

spring is less responsive to fluctuations caused by storms, and the response represents an averaging of the subterranean tributaries.

The discharge from the resurgence is difficult to measure directly because the spring is at base level and is back-flooded by Second Creek. A comparison of the water budget and drainage area with those of nearby springs suggests about 380 mm of runoff from the Organ Cave basin, for a mean discharge of 200 l/sec. The minimum discharge is about 9 l/sec, and the maximum is estimated to be greater than 3000 l/sec. Extreme flood discharges for the Organ and Hedricks streams were estimated using the threshold of motion criteria (Graf, 1971) to calculate the flow required to move bedload cobbles through various stream segments in the cave. The measured discharge of the Organ/Hedricks stream was 28 l/sec, but the calculated flood discharge was 7900 l/sec with a velocity of 4.7 m/sec (Jones, 1988).

A quantitative tracer test was conducted in November 1990. A total of 0.23 kg of the fluorescent dye sodium fluorescein (CI generic name, Acid Yellow 73) was added to the Organ/Hedricks stream and 0.038 kg of Rhodamine WT (CI generic name, Acid Red 388) was added to the Lower Lipps stream near the Humphrey entrance (see Fig. 3). The travel time from the Organ Hedricks stream was 30 hr to the first appearance of the dye and 42 hr to the maximum dye concentration. The travel time from the Humphreys stream was 66 hr to the first appearance of the dye and 82 hr to the maximum dye concentration. The discharge was calculated at 125 l/sec by integrating the area under the time–concentration recovery curve (Fig. 8) and dividing the quantity of dye injected by the result of the integration.

IV. THE FAUNA

A. The Subterranean Fauna

The aquatic fauna of Organ Cave is listed in Table II. Although the macroscopic fauna is well known, the microscopic fauna, especially micro-crustaceans, has been little studied. In the single study of this fauna, Starr [summarized in Holsinger *et al.* (1976)] found two *Bryocamptus* species in stream gravels. The macroscopic aquatic cave fauna is typical of that found in caves throughout the Greenbrier River valley and is dominated by amphipods and isopods. Five obligate subterranean species (*i.e.,* troglobites) are present (one snail, one flatworm, two amphipods, and one isopod), none endemic to Organ Cave (Holsinger, 1988). The morphologically distinct cave form of *Gammarus minus* is also present in Organ Cave.

There are several distinct habitats and communities within the cave. Although a few small surface streams (*e.g.,* the stream at the Organ entrance) enter the cave directly, many of the small first-order streams in the cave

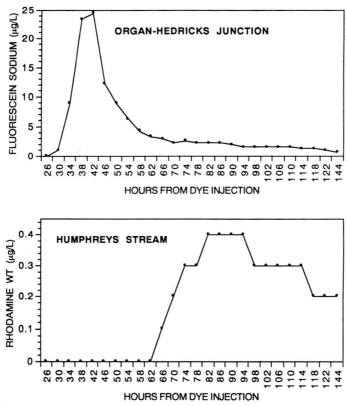

FIGURE 8 Time–concentration recovery curves for sodium fluorescein (upper) injected at Organ–Hedricks Junction and Rhodamine WT (lower) injected in Humphreys Stream. See Fig. 3 for locations. The combined discharge from these two subdrainages was 125 l/sec.

have their origin in vertically dripping and seeping water from the epikarstic zone. The drip pools and seeps form a distinct habitat in the cave (Fig. 9). The other major aquatic cave habitat is the extensive dendritic stream system in the cave (Fig. 10). The small permanent first-order streams are often little more than trickles of water with base flow rates of approximately 0.02 l/sec. The largest streams in the cave are third-order streams with base flow rates of between 15 and 30 l/sec.

Many species in the amphipod genus *Stygobromus* are rarely found in streams, but are found in seeps and drip pools, although their existence in any particular set of seeps or pools is often quite ephemeral. The primary habitat of these species is almost certainly the subcutaneous zone, *i.e.,* epikarst (Williams, 1983; Mangin, 1985). As is true for most North American karst areas, we have not sampled the epikarstic zone directly. Although none of these obligately epikarstic species occurs in the Organ basin, *Stygobromus*

TABLE II Aquatic Species Known from Organ Cave

Phylum Plathelminthes
  Order Tricladida
    *Macrocotyla hoffmasteri**
Phylum Mollusca
  Order Mesogastropoda
    *Fontigens tartarea**
    Physa sp.
Phylum Arthropoda: Subphylum crustacea
  Order Harpacticoida
    Bryocamptus nr. morrisoni
    Bryocamptus nivalis
  Order Amphipoda
    Crangonyx gracilis
    Gammarus minus
    *Stygobromus emarginatus**
    *Stygobromus spinatus**
  Order Isopoda
    *Caecidotea holsingeri**
  Order Decapoda
    Cambarus bartonii
Phylum arthropoda: Subphylum Uniramia
  Order Plecoptera
    Taeniopteryx
Phylum Chordata
  Order Urodela
    Eurycea bislineata
    Eurycea lucifuga
    Gyrinophilus porhyriticus

Note. Cave-limited species (i.e., troglobites) are indicated by an asterisk.

spinatus and to a certain extent the isopod *Caecidotea holsingeri* are common in drip pools and seeps.

The most diverse fauna occurs in first-order and small second-order streams, which are the primary habitat for *Macrocotyla hoffmasteri, Fontigens tartarea, Stygobromus emarginatus,* with *S. spinatus, G. minus, Caecidotea holsingeri,* larvae of *Eurycea lucifuga* and *E. bislineata,* and *Crangonyx gracilis* being also common. Except for *C. gracilis,* which is known only from stream pools in one first-order stream in the cave, and *Eurycea* larvae, which sporadically occur in stream pools, all of the other species are primarily found in riffles (shallows) of stony bottomed streams. Within these riffles, amphipods and isopods are quite abundant, as indicated by the data in Table III. In optimum riffle habitats, densities of *C. holsingeri* reach nearly 50/m^2, those of *G. minus* reach over 100/m^2, and those of *Stygobromus* reach over 30/m^2.

FIGURE 9   Drip pool habitat near Jones Canyon (see Fig. 4).

The larger third-order streams in the cave have the lowest diversity of troglobites. Of the troglobites, only *C. holsingeri* is regularly reported from these streams. The cave form of *G. minus* predominates, reaching densities of between 5 and 25/m² in the Organ main stream (R. Jones, 1990). Both *G. minus* and *C. holsingeri* are found primarily in riffles rather than pools. *Cambarus bartonii* and larvae of *Gyrinophilus porphyriticus* are sometimes found in pools, but neither is common.

FIGURE 10   Organ main stream, a typical third-order stream passage.

The overall population size in Organ Cave is likely to exceed 10^5 for *G. minus* and 10^4 for *C. holsingeri, S. emarginatus,* and *S. spinatus*. Although large populations are common in optimum habitats, the optimum habitat itself is highly heterogeneous in its occurrence. Pools rather than riffles take up the majority of space in such streams (up to 90%), and for reasons not entirely clear (see below) not all riffles have high densities of animals. No quantitative data are available for *F. tartarea* or *M. hoffmasteri,* but at least

TABLE III   Density/m^2 for Amphipod and Isopod Species in Riffles in Five
Streams in Organ Cave

Stream	Square meters sampled	Width	Species	Density
Siveley No. 2	2.88	0.33	*Gammarus minus*	8.3
			Stygobromus emarginatus	11.5
			Caecidotea holsingeri	0.4
Siveley No. 3	2.03	0.53	*Gammarus minus*	117.2
			Stygobromus emarginatus	1.0
			Caecidotea holsingeri	1.5
Deems	1.12	0.20	*Stygobromus emarginatus*	30.4
			Caecidotea holsingeri	48.2
Jones Canyon	0.96	0.20	*Stygobromus spinatus*	6.2
			Caecidotea holsingeri	13.5
Lipps Main	1.72	0.22	*Stygobromus emarginatus*	5.2
			Caecidotea holsingeri	44.2

Note. All streams are first-order streams, except Lipps Main, which is a second-order
stream. Twelve riffles or portions of riffles were sampled in each stream. Width is the mean
riffle width in meters.

F. tartarea shows a pattern similar to that of amphipods and isopods, locally
common but highly sporadic.

Based on the presence in or absence from 23 stream segments in Organ
Cave, Fong and Culver (1994) analyzed the habitat niche of the five amphipod and isopod species in terms of stream order. The pie diagrams depicted
in Fig. 11 indicate the niches of the five amphipod and isopod species, which
emphasize the points made above. *C. gracilis, S. emarginatus,* and *S. spinatus*
are primarily in tiny first- and second-order streams; *G. minus* is primarily
but not exclusively in larger streams; and *C. holsingeri* is ubiquitous.

The microgeographic pattern of occurrence of the five species is shown
in Fig. 12. The most striking is that, whereas *G. minus* occurs in all but 1
of the 12 streams located in the eastern subbasin of the cave system, it is
absent from all of the 11 streams located in the western (Lipps) subbasin.
Collecting trips to the far downstream sections of the Lipps subbasin confirm
the absence of *G. minus* from the subbasin. The reason for the absence of
this species from the Lipps subbasin apparently is a failure of upstream
invasion by *G. minus,* due to historical reasons, competition from other
species, or both. The other striking pattern is the restriction of the troglophile
C. gracilis to one small first-order stream in the Lipps subbasin. This restriction is apparently the result of a failure of downstream invasion, due to
historical reasons, competition from other species, or both. Based on our
current knowledge of the biology of *G. minus* and *C. gracilis,* it is unlikely
that the reason for their absence from parts of the cave system is the lack
of a suitable habitat.

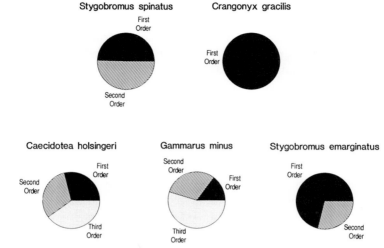

FIGURE 11 Habitat niches based on stream order for the five amphipod and isopod species found in Organ Cave. Each of the three stream orders was weighted equally in determining proportions of the pie diagrams.

In general, patterns of distribution also reflect the direction of invasion of the cave. That is, *S. emarginatus* and *S. spinatus,* which are mostly found in headwater streams, probably invaded the cave from other subsurface habitats in the epikarstic zone. This is not to imply that these species do not have true cave populations, but rather that their ancestors lived in epikarstic rather than surface habitats (springs and spring runs). Although *C. holsingeri* is ubiquitous, its occurrence in drips and seeps suggests that

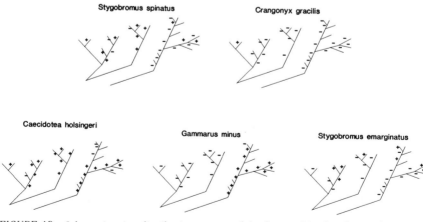

FIGURE 12 Schematic microdistribution pattern of the five amphipod and isopod species in 22 stream segments in Organ Cave. The orientation of the schematic diagram is the same as that of the line map in Fig. 3. See Fong and Culver (1994) for details.

it originally invaded caves from the epikarstic zone as well. The association of *G. minus* with larger streams, and its occurrence in springs (see below) suggests that it invaded upstream rather than downstream. The distribution of *C. gracilis* and its direction of invasion are somewhat enigmatic. Its limitation to one headwater stream largely formed by vertically moving water would suggest an epikarstic origin. On the other hand, the species is widespread in cold-water streams in the eastern United States and is found in the resurgence of Organ Cave as well. The cave population, although not morphologically distinct, may be genetically distinct and may be the result of a rare upstream colonization. Downstream colonization is also possible. The species occurs near an entrance to the cave, and it may be a remnant of a population that previously occurred in a former surface stream now captured by the cave.

The distribution of the number of amphipod and isopod species in different cave streams is summarized in Fig. 13. The mean numbers of species found in first-, second-, and third-order streams are very similar, ranging from 2.0 in third-order streams to 2.18 in first-order streams. What is different is the variability in species richness (numbers). The standard deviation for third-order streams is 0.0 ($n = 3$), for second-order streams is 0.33 ($n = 9$), and for first-order streams is 1.08 ($n = 11$). Based on 500 resamples (bootstraps) of standard deviation, the variability among the three stream orders significantly differs. The decreasing variability of species number with increasing stream order is in part the result of the averaging effect of stream order; *e.g.,* each second-order stream must be formed by at least two first-order streams. But the great variability of first-order streams is a

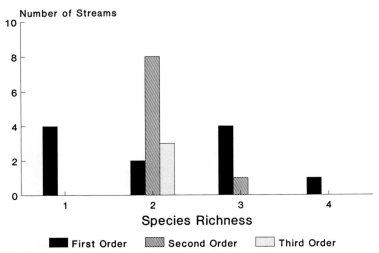

FIGURE 13   Frequency distribution of number of amphipod and isopod species found in Organ Cave stream segments of different order.

reflection of the spatial variability of the epikarstic zone. This echoes Rouch's (1988) finding of great heterogeneity in harpacticoid copepod distribution and abundance in the underflow of small streams in karst regions in the Pyrenees.

B. The Resurgence Fauna

Although Organ Cave offers extensive opportunities for sampling and study of the subterranean fauna, the resurgence of the Organ Cave basin has proved difficult to study and monitor because it is at the base level of Second Creek and subject to back-flooding. Under low-flow conditions the water rises from several openings in the bottom of the spring pool, which is surrounded by a dilapidated containment wall (Fig. 14). The spring pool has a mud bottom, with extensive clumps of a leafy liverwort (probably *Chiloscyphus polyanthus*). Most of the species listed in Table IV have been found in clumps of the leafy liverwort.

Except for the two species of amphipods, *G. minus* and *C. gracilis*, the stone fly *Taeniopteryx,* and the snail *Physa,* there is no overlap of species composition with the cave fauna. The stone fly and snail are rare in the cave. *C. gracilis* is rare in both the cave and the spring (see above). The dominant species in both the cave and the resurgence is *G. minus.* The abundance of *G. minus* in the resurgence exceeds 10^4. However, cave and resurgence populations are highly distinct, both genetically and morphologically (see below). The major predator of *G. minus* is the sculpin *Cottus bairdii* (Man, 1991), which, due to size-selective predation, reduces the average size of individuals in the population. The planarian *Phagocata* is also a predator of *G. minus.*

Similarly, the fauna of the resurgence is distinct from that of Second Creek, although there is some faunal overlap due to back-flooding and physical proximity. *G. minus, C. gracilis,* and *C. scrupulosus* have never been found in Second Creek. The plecopteran genus *Taeniopteryx* is more abundant in the spring than in the creek. Other common spring species, especially *Physa, Brachycentrus, Baetis,* and *Dixa,* are common in both habitats. The distinctness of spring and creek fauna is probably the result of the great physicochemical constancy of the spring, lower food levels in the spring, and greater predation in the creek. For example, *G. minus* is probably absent from the stream due to its susceptibility to elevated temperatures (Glazier *et al.,* 1992).

V. COLONIZATION OF THE CAVE BY *GAMMARUS MINUS*

G. minus is the dominant animal of both the cave and the resurgence, but there are profound morphological differences between the cave and the

FIGURE 14 The resurgence to Organ Cave. Water rises in a spring pool immediately in front of the person in the photograph.

TABLE IV   Aquatic Species Known from the
Resurgence of Organ Cave

Phylum Plathelminthes
  Order Tricladida
    *Phagocata**
Phylum Mollusca
  Class Gastropoda
    Physa
    Lymnaea
    Hydrobiidae
Phylum Arthropoda: Subphylum Crustacea
  Order Amphipoda
    *Crangonyx gracilis**
    *Gammarus minus**
  Order Isopoda
    *Caecidotea scrupulosus**
Phylum Arthropoda: Subphylum Uniramia
  Order Coleoptera
    Haliplus
  Order Diptera
    Dixa
    Simulium
  Order Ephemeroptera
    Baetis
    Heptagenia
    Isonychia
  Order Plecoptera
    Acroneuria
    Leuctra
    *Taeniopteryx**
  Order Trichoptera
    Brachycentrus
    Lepidostoma
    Molanna
Phylum Chordata
  Class Osteichthyes
    Cottus bairdi
    Lepomis

Note. Species whose primary habitat is the spring
(rather than Organ Cave or Second Creek) are indicated by an asterisk.

spring populations, which is typical of the populations of large caves in this area. At maturity (approximately one year) individuals from caves are larger, have relatively longer antennae, and have smaller eyes. The difference in ommatidia number is particularly striking. At maturity, individuals from Organ Spring had on average 16.2 ($\pm$1.0) ommatidia, whereas individuals from the cave had an average of 5.8 ($\pm$0.5) ommatidia (Culver, 1987).

In an extensive study of quantitative genetics of this species Fong (1989) found that, based on full-sib analysis of 60-day-old individuals, nearly all eye and antennal traits showed significant heritability (Table V). Further evidence of the genetic differences are the morphological differences between cave and spring individuals raised in the same environment (Table V). The only anomaly in Table V is the smaller head length (and overall body size) of 60-day-old cave individuals relative to spring individuals.

Kane and coworkers (Kane and Culver, 1991; Kane *et al.*, 1992) analyzed the resurgence population and populations from six streams in the cave for electrophoretic variation at 18 presumptive gene loci for 13 enzyme systems. As is clearly indicated in Fig. 15, the resurgence population is distinct from the cave population. This is not a distance effect. The cave population sampled nearest to the resurgence is 3 km away, and the cave population sampled farthest from the resurgence is 5.5 km away. Based on Wright's F_{ST}, a measure of population differentiation (Wright, 1978), with the assumption of the selective neutrality of electrophoretic variation, the average number of migrants per generation ($N_m = [1 - F_{ST}]/4F_{ST}$) is 0.85, indicating considerable differentiation ($F = 0.228$). If only cave populations are considered lower differentiation and higher migration are obtained ($F = 0.032$, $N_m = 7.56$).

Nei's unbiased standard genetic distance (Nei, 1987), with the assumptions of selective neutrality and the equality of mutation rates in different populations, increases linearly with time and thus can be used to measure

TABLE V   Mean and Standard Error of Size and Heritabilities for a Series of Morphological Characters from 60-Day-Old *G. minus* from the Organ Cave Resurgence and from the Organ Main Stream in the Cave, Reared under Identical Conditions in Darkness in the Laboratory

Character	Cave		Spring	
	Size	Heritability	Size	Heritability
Head length	4.46 ± 0.05	1.34 ± 0.26	4.96 ± 0.03	0.85 ± 0.12
Ommatidia No.	3.22 ± 0.11	0.79 ± 0.27	19.67 ± 0.21	0.46 ± 0.11
Eye area	4.32 ± 0.12	0.79 ± 0.27	16.61 ± 0.23	0.11 ± 0.10
Antenna 1				
Peduncle length	6.18 ± 0.11	0.57 ± 0.26	5.06 ± 0.04	0.50 ± 0.11
Flagellar length	15.35 ± 0.35	0.91 ± 0.27	13.05 ± 0.11	0.50 ± 0.11
Flagellar segment No.	12.87 ± 0.23	0.31 ± 0.26	10.34 ± 0.08	0.37 ± 0.11
Antenna 2				
Peduncle length	5.39 ± 0.11	1.18 ± 0.26	4.22 ± 0.04	0.63 ± 0.12
Flagellar length	4.59 ± 0.11	0.92 ± 0.27	3.54 ± 0.04	0.68 ± 0.12
Flagellar segment No.	5.15 ± 0.11	0.86 ± 0.27	4.12 ± 0.02	0.38 ± 0.11

Note. Lengths are in millimeters × 10 and areas in square millimeters × 100. Data are from Fong (1989).

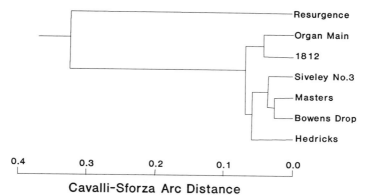

Cavalli-Sforza Arc Distance

FIGURE 15   UPGMA dendrogram of Organ Cave populations of *Gammarus minus*, based on electromorphs from 18 loci and 13 enzyme systems. See Kane and Culver (1991) for details.

the time since invasion of the cave (Kane and Culver, 1991). For the standard mutation rate of 10^{-7}, Nei's genetic distance, D, can be scaled in years:

$$t = 5 \times 10^6 \, D.$$

Although there is considerable controversy concerning the validity of such "molecular clock" equations (Gillespie, 1986), we believe that, at the very least, it has considerable heuristic value. The time since isolation, based on the six samples from Organ Cave, is $6.7 \pm 0.1 \times 10^5$ years. Although it is even more difficult to estimate the age of the basin itself, Ford and Williams (1989), on the basis of the U/Th ratios of speleothems, suggest an age of 2.5×10^6 years for the Friar's Hole karst system immediately to the north of the study area. Thus, the invasion of *G. minus* is likely to have occurred after karstification.

    G. minus probably invaded the Organ Cave system during the interglacial periods of the Pleistocene. Note that this region was not glaciated. In the laboratory at least, *G. minus* is very sensitive to elevated temperatures, and a temperature increase is a possible cause of the migration. Ecological and genetic data indicate that effective population sizes are large and have not gone through a bottleneck. This suggests an active migration (Rouch and Danielopol, 1987) into caves rather than the stranding of a few individuals. We have no direct information on the early stages of migration, but the lack of genetic continuity and the possibility of a physical separation (due to a water-filled phreatic loop) suggest early isolation of cave populations from the spring population. Until more is known about the nature of the subterranean habitat immediately upstream of the resurgence, it is difficult to speculate further.

VI. FINAL PERSPECTIVES

    From a hydrogeologic point of view, the Organ Cave basin is relatively well understood. There is of course much more information that could be obtained, and long-term monitoring of the resurgence is an especially critical gap in our knowledge. Nonetheless, the outlines are clear. It is a system dominated by conduit flow, with two major subsurface streams finally converging at the resurgence. Relative to the Dorvan and Baget ecosystems, conduit flow in Organ Cave is at least more accessible and probably more important. The real gap is in our knowledge of the ecosystem dynamics. The microcrustacean fauna, which Rouch and others have used with such great success to unlock the secrets of ecosystem function, is a mystery. Is it less diverse, perhaps the result of a more diverse macrocrustacean fauna or perhaps because of greater conduit flow? Or is it simply unknown? These problems await groundwater ecologists.

REFERENCES

Baker, V. R. (1973). Geomorphology and hydrology of karst drainage basins and cave channel networks in east-central New York. *Water Resour. Res.* **9**, 695–706.

Courbon, P., Chabert, C., Bosted, P., and Lindsley, K. (1989). "Atlas of the Great Caves of the World." Cave Books, St. Louis, MO.

Culver, D. C. (1987). Eye morphometrics of cave and spring populations of *Gammarus minus* (Amphipoda: Gammaridae). *J. Crustacean Biol.* **7**, 137–147.

Deike, G. H. (1988a). Geology. *In* "Caves of the Organ Cave Plateau, Greenbrier County, West Virginia" (P. J. Stevens, ed.), Bull. No. 9, pp. 13–39. West Virginia Speleol. Surv., Barrackville.

Deike, G. H. (1988b). Geological factors in the hydrology of the Organ Cave Plateau, West Virginia, USA. *Proc. Int. Assoc. Hydrol. Congr., 21st*, pp. 394–399.

Fong, D. W. (1989). Morphological evolution of the amphipod *Gammarus minus* in caves: Quantitative genetic analysis. *Am. Midl. Nat.* **121**, 361–378.

Fong, D. W., and Culver, D. C. (1994). An analysis of the distribution of the crustacean fauna within a subterranean drainage basin. *Hydrobiologia* (in press).

Ford, D. C., and William, P. W. (1989). "Karst Geomorphology and Hydrology." Unwin Hyman, London.

Gibert, J. (1986). Ecologie d'un système karstique jurassien. Hydrogéologie, dérive animale, transits de matières, dynamique de la population de *Niphargus* (Crustacé Amphipode). *Mém. Biospéol.* **13**, 1–376.

Gillespie, J. H. (1986). Natural selection and the molecular clock. *Mol. Biol. Evol.* **3**, 138–155.

Glazier, D. S., Horne, M. T., and Lehman, M. E. (1992). Abundance, body composition and reproductive output of *Gammarus minus* in ten cold springs differing in pH and ionic content. *Freshwater Biol.* **28**, 149–163.

Graf, W. H. (1971). "Hydraulics of Sediment Transport." McGraw-Hill, New York.

Holsinger, J. R. (1988). Biology. *In* "Caves of the Organ Cave Plateau, Greenbrier County, West Virginia" (P. J. Stevens, ed.), Bull. No. 9, pp. 49–50. West Virginia Speleol. Surv., Barrackville.

Holsinger, J. R., Baroody, R. A., and Culver, D. C. (1976). "The Invertebrate Cave Fauna of West Virginia," Bull. No. 7. West Virginia Speleol. Surv., Barrackville.

Horton, R. E. (1945). Erosional development of streams and their drainage basins: Hydrophysical approach to quantitative morphology. *Geol. Soc. Am. Bull.* **56**, 275–370.

Jones, R. (1990). Evolution of cave and surface populations of the Amphipod *Gammarus minus*. Ph.D. Dissertation. Northwestern University, Evanston, IL.

Jones, W. K. (1973). Hydrology of limestone Karst in Greenbrier County, West Virginia. *W. Va., Geol. Econ. Surv., Bull.* **36**. Morgantown.

Jones, W. K. (1988). Hydrology. *In* "Caves of the Organ Cave Plateau, Greenbrier County, West Virginia" (P. J. Stevens, ed.), Bull. No. 9, pp. 40–48. West Virginia Speleol. Surv., Barrackville.

Kane, T. C., and Culver, D. C. (1991). The evolution of troglobites—*Gammarus minus* (Amphipoda: Gammaridae) as a case study. *Mém. Biospéol.* **18**, 3–14.

Kane, T. C., Culver, D. C., and Jones, R. (1992). Genetic structure of morphologically differentiated populations of the amphipod *Gammarus minus. Evolution (Lawrence, Kans.)* **46**, 272–278.

Man, Z. (1991). Life History Variation Among Spring Dwelling Populations of *Gammarus minus* Say. Thesis, American University, Washington, DC.

Mangin, A. (1985). Progrès récents dans l'étude hydrogéologique des karsts. *Stygologia* **3**, 239–257.

Nei, M. (1987). "Molecular Evolutionary Genetics." Columbia Univ. Press, New York.

Rouch, R. (1977). Considérations sur l'écosystème karstique. *C. R. Hebd. Seances Acad. Sci., Ser. D* **284**, 1101–1103.

Rouch, R. (1982). Le système karstique du Baget. XIII. Comparaison de la dérive des Harpacticides a l'entrée et à la sortie de l'aquifère. *Ann. Limnol.* **18**(2), 133–150.

Rouch, R. (1988). Sur la répartation spatiale des Crustacés dans le sous-écoulement d'un ruisseau des Pyrénées. *Ann. Limnol.* **24**, 213–234.

Rouch, R., and Danielopol, D. (1987). L'origine de la faune aquatique souterraine, entre le paradigme du refuge et le modèle de la colonisation active. *Stygologia* **3**(4), 345–372.

Stevens, P. J., ed. (1988). "Caves of the Organ Cave Plateau, Greenbrier County, West Virginia," Bull. No. 9. West Virginia Speleol. Surv., Barrackville.

Strahler, A. N. (1964). Quantitative geomorphology of drainage basins and channel networks. *In* "Handbook of Applied Hydrology" (V. T. Chow, ed.) Ch. 4, pp. 39–76. McGraw-Hill, New York.

White, W. B. (1988). "Geomorphology and Hydrology of Karst Terrains." Oxford Univ. Press, New York.

Williams, P. W. (1983). The role of the subcutaneous zone in karst hydrology. *J. Hydrol.* **61**, 45–67.

Wright, S. (1978). "Evolution and Genetics of Populations," Vol. 4. Univ. of Chicago Press, Chicago.

SECTION *FOUR*

*HUMAN INTERFERENCE
AND MANAGEMENT
IMPLICATIONS*

18

Groundwater
Contamination and Its
Impact on Groundwater
Animals and Ecosystems

J. Notenboom,* S. Plénet,† and M.-J. Turquin†

Laboratory of Ecotoxicology
National Institute of Public Health and Environmental
Protection
P.O. Box 1
3720 BA Bilthoven, The Netherlands

† *Université Lyon I*
U.R.A. CNRS 1451, "Ecologie des Eaux Douces et des Grands
Fleuves"
Laboratoire d'Hydrobiologie et Ecologie Souterraines
43 Boulevard du 11 Novembre 1918
69622 Villeurbanne Cedex, France

I. Introduction
II. Groundwater Contamination
III. Ecological Impact of Groundwater Contamination
    A. Effects on Organisms
    B. Effects on Ecosystem Structure and Function
    C. Ecological Adaptations and Community Structure of
        Groundwater Invertebrates
IV. Stygobites as Biomonitors
V. Groundwater Management and Ecology
VI. Conclusion
    References

I. INTRODUCTION

Groundwater contamination is a socioeconomic problem that receives considerable attention in modern industrialized societies (Khondaker *et al.,*

1990). The disciplines involved in the study of contamination are, in particular, hydrology, hydrogeology, hydrochemistry, and environmental engineering, which focus mainly on the fate and transport of pollutants and the remediation of polluted groundwaters. Most of the regulatory standards for contaminants in groundwater are derived from mammalian toxicity studies and aim to safeguard groundwater as a drinking water resource (Piver, 1993).

The importance of groundwater as part of the biosphere is now widely recognized (Ghiorse and Wilson, 1988; Stanford and Simons, 1992), but many groundwaters now contain toxic wastes (Page, 1981; Piver, 1993). Greater attention has been given to the adverse effects of contaminants on the ecological properties of groundwater systems. Results stimulated discussions of the need to develop a rational basis for optimizing groundwater management, including protection of its ecological integrity (Nachtnebel and Kovar, 1991; Stanford and Ward, 1992).

In this chapter we review studies of pollution ecology and ecotoxicology related to groundwater ecosystems and organisms, especially invertebrates. We emphasize ecological responses to contamination in relation to the unique adaptations of groundwater organisms and the peculiarities of their communities. However, studies are limited, and information in this chapter should be considered a very first attempt to summarize responses of groundwater ecosystems to contamination. The focus is mainly on shallow groundwater zones (*e.g.*, alluvial aquifers), because pollutants enter these zones before reaching deeper layers. Moreover, invertebrates in shallow systems have been studied in greater detail and seem to be more abundant and diverse in comparison with very deep (phreatic) groundwaters.

Assessment of the ecological impact of contamination should always be preceded by assessment of the physical, chemical, and biological processes controlling the fate and behavior of pollutants in the groundwater pathways (Suter, 1993). Such exposure assessments allow estimation of biologically available concentrations of polluting chemicals. An exhaustive discussion on this subject falls outside the scope of this chapter. Our objective is to summarize what is known about the ecological consequences of groundwater pollution at various levels of biological organization and how this knowledge can be used to advance groundwater management.

II. GROUNDWATER CONTAMINATION

The special nature of groundwater pollution in terms of occurrence and duration and problems related to identification, measurement, and prediction (models), have provoked considerable growth in the field of groundwater quality research [Fried, 1976; Edworthy, 1987; Galceran *et al.*, 1990; see a review in Mayer *et al.* (1993)]. Extensive literature on the environmental chemistry of groundwater and hydrologically related soils and surface waters exists (Freeze and Cherry, 1979; Edworthy, 1987; Page, 1987; Mayer *et al.*, 1993; Allen *et al.*, 1993). In general, groundwater contamination originates from soils or surface waters that have been polluted by human activities. Pollutants from some point sources (*e.g.*, disposal sites) often enter the subsurface environment directly, whereas those from diffuse sources (*e.g.*, atmospheric deposition and agriculture) pass through soil or sediment layers before reaching saturated zones (Table I).

Vertical transport of solutes and particulates from soils into the groundwater is principally driven by percolating precipitation. These flows pass through unsaturated (vadose) layers that are multiphasic and contain immiscible fluids. Water-saturated single-phase systems exist only when direct hydrologic connection routes exist between surface water and groundwater (*e.g.*, through alluvial sediments). In both situations, dispersal of solutes and particulates is essentially determined by interactions between system heterogeneity and hydrodynamics (Mayer *et al.*, 1993). Filtration capacity is determined by the structure of the interstitial matrix and the size of the voids (Gibert, 1990). Along flow paths, the fate of pollutants (*i.e.*, filtration, retention, or alteration) is further determined by physical, chemical, and biological processes (Table II), depending on the characteristics of the system and the type of pollutants. In general, sorption and degradation processes

TABLE I Groundwater Pollutants and Some Sources

Category	Source type	Sources
Heavy metals	Diffuse	Fertilizers, manure, atmospheric deposition, sewage sludge, pesticides containing metals, and line sources (e.g., motorways, railways, and sewerage systems)
	Point	Industrial sites, urban areas, landfills, mining disposal sites, waste and sludge disposal sites, and hazardous waste sites
Pesticides	Diffuse	Agriculture and atmospheric deposition
Natural organics and xenobiotics	Diffuse	Atmospheric deposition
	Point	Industrial sites, urban areas, landfills, mining disposal sites, waste and sludge disposal sites, hazardous waste sites, leaking storage tanks, and line sources
Acid and aluminum	Diffuse	Atmospheric deposition
Fertilizers (nitrogen and phosphate)	Diffuse	Agriculture and wastewater infiltration
Sludge and manure	Diffuse	Agriculture
	Point	Wastewater infiltration

TABLE II Chemical, Physical, and Biological Processes that Influence the Mobility, Dilution, Chemical Speciation, and Degradation of Pollutants in Subsurface Environments

Category	Process
Physical process	Solubility, dilution, sorption and desorption, and evaporation
Chemical process	Oxidation and reduction, hydrolysis, complexation, ion exchange, and precipitation
Biological process	Decomposition and accumulation

are more prominent in superficial soil and sediment layers than in deeper subsurface compartments, which usually contain less organic matter. In porous systems the "filter" phenomenon is generally considerable. Locally, however, because of the presence of preferential flow paths, there may be a strong reduction in this phenomenon.

Fissured and karst systems are considered to be very vulnerable to pollution (Muller, 1981). However, vulnerability varies with the structure of the karst (Mangin and Bakalowicz, 1989). In karst the dispersivity of pollutants can be very high, and "filter" phenomena can be deficient or even totally absent. Moreover, adsorption and sites for biological and chemical degradation are reduced in karst, giving these systems a limited self-purification potential (Gibert, 1990; Turquin, 1989). Another peculiarity of karst is the accumulation of sediment and polluted materials in lateral compartments (Maire, 1979; see also Chapter 5).

Physical, chemical, and biological properties regulating the retention of pollutants in soils and sediments (Table III) are dynamic in space and time and can be affected by human activities. Stigliani (1988) showed that small changes in these regulating or capacity-controlling properties, for example, pH, can lead to a large change in storage capacity of soils and sediments (Ter Meulen *et al.*, 1993).

Once pollutants have reached aquifers, sorption and degradation processes are generally low and residence times, long. This makes groundwater environments vulnerable and difficult to rehabilitate (Travis and Doty, 1990). A key parameter in assessing the ecological risks associated with contaminants is bioavailability, which is determined by the characteristics of the compound and the physical and chemical properties of the environment, as well as the biology of the organism(s) in question. Bioavailibility is commonly simplified by the assumption that organisms are exposed only to the chemicals dissolved in pore water (Dickson *et al.*, 1987; Van Gestel and Ma, 1988). So, in multiphase systems, the partition of the pollutant between aqueous and solid phases has an impact on the bioavailable fraction. Physical and chemical knowledge of this partition is a basis for the estimation of bioavailable concentrations of pollutants.

TABLE III   Linkage between Human Activities and Changes in Capacity-Controlling-Properties (CCPs) [from Stigliani (1993)]

CCP	Human activities affecting CCPs
Cation exchange capacity (CEC)	Affected indirectly by activities that affect pH, organic matter content, and salinity
pH	Atmospheric acid deposition, agricultural practices, climate change altering the nitrogen cycle, and seasonal patterns of precipitation and evapotransportation
Redox potential (Eh)	Drainage of wetlands or flooding of dry lands either as deliberate land-use policy or as a consequence of climate change
Organic matter content (OM)	Changes in vegetation, agricultural practices, climate change can increase or decrease OM
Salinity	Irrigation with salty groundwaters, salt water intrusion of groundwaters due to sea level rise or overexploitation
Microbial activity	Affected indirectly by activities that affect redox potential and pH and directly by synergetic effects of multiple toxins and by climate change

Groundwater contaminants can be grouped into three categories: (1) inorganic and organic toxic compounds (heavy metals, pesticides, and other xenobiotic organics); (2) nitrogen and phosphorous compounds, nutrients for primary producers, and other microorganisms, and (3) organic matter from sewage sludge, manure, and other sources that may serve as a carbon and energy source for heterotrophic organisms. Table IV classifies

TABLE IV   Groundwater Contaminants and Potential Adverse Effects on Groundwater Microorganisms and Invertebrates

	Ecological effect	
Pollutant	Microorganisms	Fauna
Heavy metals	Inhibition activity	Bioaccumulation and adverse effects
Pesticides	Inhibition activity, biodegradation, and adaptation	Bioaccumulation, biotransformation, and adverse effects
Natural organics	Stimulation activity	Increased density and distress by anoxia
Xenobiotics	Inhibition activity, biodegradation, and adaptation	Adverse effects
Acid (aluminum)	Inhibition activity	Adverse effects
Fertilizers	Stimulation activity (when carbon is not limiting)	

groundwater contaminants with respect to the potential effects on microorganisms and invertebrates.

III. ECOLOGICAL IMPACT OF GROUNDWATER CONTAMINATION

Groundwater ecotoxicology examines the ways toxic contaminants affect the health, reproduction, and survival of organisms, or inhibit microbial processes, in groundwater ecosystems, in relation to abiotic conditions and the adaptation of the organisms for life in natural groundwaters. The study of the impact of nontoxic chemical, physical, or biological agents on organisms, populations, and communities is covered by pollution ecology. The organism is the primary focus in ecotoxicology (Suter, 1993), because toxic compounds affect organisms before manifesting themselves at population and community levels.

A. Effects on Organisms

1. Toxicant Effects

Acute toxicity experiments with groundwater animals from natural populations have been performed in only a few studies. Toxicological end points are lethality or immobility, which are not always easy to discriminate. Sublethal end points in chronic toxicity studies have not yet been used in groundwater ecotoxicology, mainly because of the difficulties in culturing animals in an experimental design. A summary of published LC_{50} or E_iC_{50} values for groundwater invertebrates is given in Table V. One of the first groundwater ecotoxicological experiments was performed by Mathews *et al.* (1977) on the blind cave crayfish *Orconectes australis australis,* collected from Merrybranch Cave (Tennessee). They exposed acclimated (three days at sublethal concentrations of $0.21-1.50$ mg/l^{-1} total residual chlorine) and unacclimated crayfish to cave water chlorinated with sodium hypochlorous solutions over 24 hr. A clear dose–response relationship was documented, with the unacclimated crayfish a little more sensitive ($p < 0.10$) than the acclimated ones. Bosnak and Morgan (1981a) performed 96-hr acute toxicity experiments with cadmium and copper on the stygobite isopods *Caecidotea stygia* and *C. bicrenata* and with hexavalent chromium only on *C. stygia*. Again test animals came from Merrybranch Cave. For cadmium they found *C. stygia* to be four times more sensitive than *C. bicrenata*.

The influence of environmental factors (pH and oxygen) on the sensitivity of groundwater invertebrates to toxicants has been the subject of a few studies. Meinel and Krause (1988) and Meinel *et al.* (1989) studied the acute toxicity of cadmium and zinc at different pH conditions (5, 6, 7, and 8) on the stygobites *Trichodrilus tenuis* (Oligochaeta), *Proasellus cavaticus*

TABLE V   Published Data on Acute Toxicity of Groundwater Invertebrates

Species	LC_{50} (96 hr) (mg/l)	Test conditions	References
Crustacea			
Amphipoda			
Niphargus aquilex	Zinc: 180	pH 7; $T = 12°C$; $H = 103$ mg $CaCO_3$/l	Meinel and Krause (1988)
	Cadmium: 4.5		Meinel *et al.* (1989)
Isopoda			
Proasellus cavaticus	Zinc: 127	pH 7; $T = 12°C$; $H = 103$ mg	Meinel and Krause (1988)
	Cadmium: 0.5		Meinel *et al.* (1989)
Caecidotea bicrenata	Zinc: 20	pH 7; $T = 13°C$; $H = 220$ mg	Bosnak and Morgan (1981b)
	Cadmium: 2.2		
	TRC: 0.11		
	Cadmium: 1.2	pH 7.2; $H = 83$ mg	Bosnak and Morgan (1981a)
	Copper: 2.2	pH 6.7; $H = 82$ mg	
Caecidotea stygia	Cadmium: 0.29	pH 7.5; $H = 70$ mg	Bosnak and Morgan (1981a)
	Chromium(VI): 2.4	pH 7.2; $H = 86$ mg	
	Copper: 2.3	pH 7.0; $H = 70$ mg	
Decapoda			
Orconectes a. australis	TRC: 2.7	pH 8.9; $T = 10.5°C$; $H = 242$ mg	
Copepoda			
Parastenocaris germanica	Zinc: 1.7		
		pH 6.8; $T = 10.5°C$	Notenboom *et al.* (1992)
	Cadmium: 2.2		
	PCP: 0.036		
	3,4-DCP: 4.6		
	Aldicarb: 2.9	pH 7; $T = 13°C$	Notenboom and Boessenkool (1992)
	Thiram: 0.003		
Annelida			
Oligochaeta			
Trichodrilus tenuis	Zinc: 8.25	pH 7; $T = 12°C$; $H = 103$ mg	Meinel and Krause (1988)
	Cadmium: 1.05		Meinel *et al.* (1989)

Note. TRC, total residual chlorine; PCP, pentachlorophenol; 3,4-DCP, 3,4-dichlorophenol.

(Isopoda), and *Niphargus aquilex* (Amphipoda). The pH dependency of zinc and cadmium toxicity, often reported in aquatic ecotoxicology (Gerhardt, 1993), was also found in these experiments as a general trend. The pH-dependent toxicity is also influenced by biological and chemical processes, including metal complexation and biologically mediated sorption and desorption. Notenboom *et al.* (1992) used the stygobite *Parastenocaris germanica* (Copepoda) in acute toxicity experiments with cadmium, zinc, pentachlorophenol, and 3,4-dichlorophenol under normoxic (10 mg/l^{-1} O_2) and hypoxic (0.1 mg/l^{-1} O_2) conditions. Differences in the response of the animals to these chemicals under both conditions were almost insignificant; only the log-logistic dose–response curve of pentachlorophenol was slightly, but significantly, steeper under hypoxic than under normoxic conditions. The difference in response was explained by assuming a lower metabolic rate under hypoxic conditions (low concentrations of pentachlorophenol uncouple the oxidative phosphorylation, causing ATP deficiency, but in *P.*

germanica ATP occurs less rapidly at hypoxic than at normoxic conditions). Hence, on the basis of these studies, ambient oxygen concentration seems to have negligible influence, at least within the 0.1–10 mg/l^{-1} O$_2$ range, on the sensitivity to toxicants of *P. germanica* and probably also other similarly adapted stygobite species.

2. Relative Sensitivity of Groundwater Invertebrates Compared with Epigean Species

Bosnak and Morgan (1981b) compared the acute toxicities of cadmium, zinc, and total residual chlorine for the hypogean *C. bicrenata* and for the epigean *Lirceus alabamae*, both asellid isopods. For cadmium and zinc the epigean isopod was, respectively, 14 and 2 times more sensitive than the stygobite. For total residual chlorine, no significant differences in sensitivity were found.

Variations in the percentage mortality of two amphipod crustaceans, when exposed to zinc and copper, also revealed that the epigean amphipod *Gammarus fossarum* was more sensitive than the hypogean amphipod *Niphargus rhenorhodanensis* (Plénet, 1993). In contrast, Barr (1976) mentioned that a hypogean isopod, *C. stygia,* was more sensitive to nickel, cadmium, and chromium than the epigean species *Lirceus fontinalis*.

M. J. Turquin and C. Boucheseiche (unpublished data) examined sublethal toxic effects on amphipod embryos in an attempt to develop a less time-consuming and more precise bioassay than that of considering the whole life cycle. The work was done in winter, when the range of temperatures was low; *G. fossarum,* easier to collect than the associate stygobite amphipods, was used. Thirty rivers were divided into five groups according to pollution load. The production of eggs *in situ* in the five groups of rivers differed significantly (Fig. 1) These data show that populations may over-

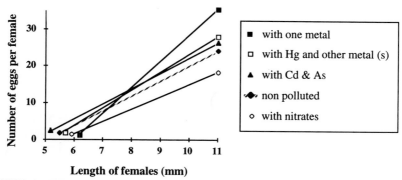

FIGURE 1 Regression lines of number of eggs plotted against females length. *Gammarus fossarum* originating from five groups of rivers with different pollution loads (M. J. Turquin and C. Boucheseiche, unpublished data).

come unsuitable environments: the genetic strains of the species should be taken into account to explain the success of reproduction, whatever the water quality. It remains to be seen if the underlying stygobite amphipods will reproduce as well.

3. Bioaccumulation of Heavy Metals

Bioaccumulation is the net accumulation of a chemical by an organism as a result of uptake from all routes of exposure. Biomagnification is the tendency of some pollutants, in particularly persistent ones, to accumulate within food webs. Exact knowledge of bioaccumulation processes is needed in order to perform valid ecological risk assessments (Timmermans *et al.*, 1989).

Information on the bioaccumulation of pollutants by groundwater animals is very scarce. The few studies performed concern only heavy metals. Dickson *et al.* (1979) collected stygobitic (*O. a. australis*) and stygophilic (*Cambarus tenebrosus*) crayfish from the same Merrybranch Cave mentioned above and determined tissue concentrations. For almost all heavy metals except zinc, tissue concentrations were significantly higher in *O. a. australis* than in *C. tenebrosus* (Fig. 2). Both nonessential metals (Cd and Pb) accumulated because they are immobilized by protein binding (metallothioneins) rather than excreted. The distribution of the metals among tissues

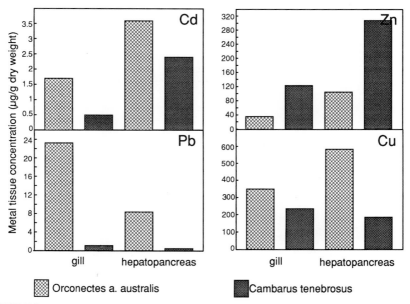

FIGURE 2   Accumulation of cadmium and lead (nonessential) and zinc and copper (essential) in the gill and hepatopancreas of *Orconectes a. australis* (troglobite) and *Cambarus tenebrosus* (troglophile), both from Merrybranch cave (Virginia), [after Dickson *et al.* (1979)].

of the two crayfish species differed and could result from temporal differences (*e.g.*, moulting cycle), specific physiological differences, or differences in behavior and biology (*e.g.*, food source).

Dickson *et al.* (1979) suggested that the higher tissue metal concentrations in *O. a. australis* might be explained by its greater longevity. Increased longevity, lower metabolic rates, and low reproductive output are well-known phenomena that have been recorded in many different stygobite animals. But the consequences of these adaptations for the bioaccumulation patterns of heavy metals are not well understood. Three amphipod species (*G. fossarum*, *N. rhenorhodanensis*, and *Niphargopsis casparyi*) from the Rhône and Ain Rivers in France demonstrated various accumulation patterns for two essential metals, zinc and copper (Plénet, 1994) (Fig. 3). At all sites mean metal concentrations in surface and interstitial sediments were approximately the same (zinc, 97.2 and 99.3 $\mu g/g^{-1}$ dry weight, respectively; copper: 37 and 37.4 $\mu g/g^{-1}$ dry weight, respectively; concentrations of both metals in the water were always <50 $\mu g/l^{-1}$). The epigean species had lower zinc and copper concentrations than the hypogean ones, a finding that corroborates the results of Dickson *et al.* (1979). In laboratory experiments (12-day exposure), the epigean species showed a net accumulation of zinc, which increased with increasing ambient concentration. The hypo-

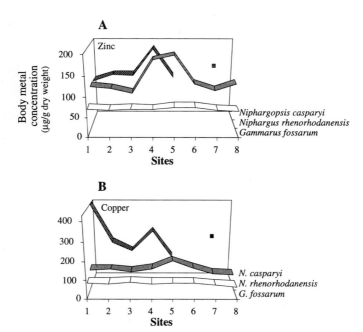

FIGURE 3   Profiles of mean zinc (A) and copper (B) concentrations in epigean and hypogean amphipods in the superficial and the interstitial zones of the Rhône (sites 1–6) and the Ain (sites 7 and 8) Rivers during five sampling periods [after Plénet (1994)].

gean *N. rhenorhodanensis* can apparently regulate its total body zinc and copper concentration (Plénet, 1993).

B. Effects on Ecosystem Structure and Function

1. Pollution with Organic Matter

Organic contamination interferes with the oligotrophic and hypoxic nature of groundwater habitats, which are generally carbon limited. Under reductive conditions, however, nitrogen may serve as an alternative electron acceptor for biochemical processes (Korom, 1992). The appearance of oxygen depletion is a result of the interaction between the oxygen diffusion rate (Enckell, 1968) and assimilatory oxygen consumption by microbial decomposition of organic matter. The oxygen diffusion rate depends on many characteristics of the system, but, in general, the rate is lower in porous systems than in fissured or karst systems. Assimilatory oxygen consumption depends on the quantity of organic matter and its biological degradation potential (*i.e.,* some organic matter may have a low biological oxygen demand). Changes in capacity-controlling properties, in particular redox conditions, are often associated with organic pollution and eutrophication (Danielopol, 1983). Hence, groundwaters polluted by organic matter are difficult, if not impossible, to restore.

Most studies of organic pollution in groundwater deal with microbes (degradation of pollutants) and the distribution of microbes associated with contaminant plumes from point sources (Armstrong *et al.,* 1991). Organic contamination can eliminate metazoans from groundwater because dissolved oxygen is often reduced to zero by oxidation and biodegradation of organic matter. Such complete anoxic conditions are often more easily reached in porous systems of low oxygen diffusion capability. When dissolved oxygen is available at only very low levels many stygobites can probably survive, but as some field observations show (see below), densities remain small (Chapter 8).

Infiltration of sewage sludge is a potential source of organic material. Bacteria and macroinvertebrates were surveyed using an existing array of wells in an alluvial aquifer below a sewage irrigation field (Sinton, 1984). Three crustacean species (amphipods and isopods) composed about 98% of fauna abundance and had the highest biomass. No significant correlations to chemical or microbial patterns were evident. In general, crustacean abundances and biomass were highest in more polluted wells, except where oxygen stagnation occurred (Fig. 4). Aquifer macroinvertebrates were theoretically responsible for assimilating approximately 20% of organic matter in the sewage reaching the aquifer; this estimate was based on extrapolation of assimilation data for the epigean amphipod *Hyalella azteca.*

In the Johannes flood plain of the Fulda River (Germany), the impact of organic pollution on a phreatic system was studied (Husmann, 1975;

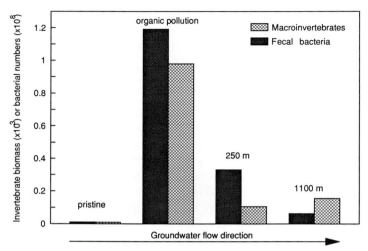

FIGURE 4 Effect of sewage sludge infiltration into a phreatic alluvial aquifer on biomass (mg/d.wt) of dominant (98%) macroinvertebrates and numbers of fecal bacteria ($\times 10^5$) per cubic meter of well water. Upstream of, below, and at distances downstream of a disposal area [after Sinton (1984)].

Marxsen, 1981a,b; DWVK, 1988). Organically enriched groundwater (very low in oxygen) flows into the plain from a lateral valley. In general, metazoan abundance was positively correlated with oxygen concentrations and elevated levels of organic substances (as COD). Where oxygen concentration was high and no organic pollution was present, lower fauna abundances were found. (Fig. 5). The term COD*O_2 (min) explained 49% of total

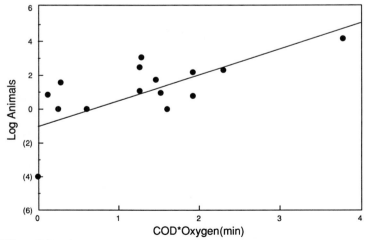

FIGURE 5 Relationship among total number of animals, organic matter (COD), and minimum oxygen in the period before sampling in shallow boreholes of the Johannes flood plain (Fulda, Germany) [after DWVK (1988)].

variance in metazoan abundance. Of course, no direct causal relationship existed between COD and metazoan abundance. The quantity and quality of organic matter determine microbial activity and biomass. Microbial biomass forms the food base for the metazoans. The highest microbial biomass was found at the border of the organically contaminated plume. In the center of the plume, bacterial numbers were very low, similar to numbers in the unpolluted part of the flood plain. Deleterious oxygen conditions and the occasional occurrence of toxic levels of NH_4^+ and H_2S occurred in the center of the plume. Moreover, Husmann (1975) postulated that a decrease of fauna-activity could lead to the clogging of pore spaces, subsequently altering the infiltration process.

Ward *et al.* (1992) studied the interstitial macroinvertebrate communities of a sewage-polluted river site in comparison with upstream sites at which pollution was not present. They distinguished three different habitat types: (1) "surface gravel" (top 10 cm of surficial bed sediments); (2) "hyporheic" habitats (30 cm below the stream bed); and (3) "phreatic" habitats (30 cm below the water table and 10–20 m from water's edge). Total fauna-densities were similar at reference and treatment sites in phreatic habitats, similar or higher at treatment sites in hyporheic habitats, and consistently higher at treatment sites in the surface gravels (benthic) (Fig. 6). Various density responses were shown by individual taxa, including reduced abundance (*e.g.*, harpacticoid copepods), little or no change in abundance (*e.g.*, cyclopoid copepods), and increased abundance (*e.g.*, nematods and oligochaetes). Some groups, such as ostracods, were resistant to low dissolved

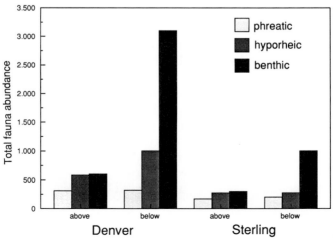

FIGURE 6   Total fauna abundance in phreatic, hyporheic, and benthic habitats of the South Platte River at sites upstream (above) and downstream (below) from the sewage effluents from Denver and Sterling. Abundances indicate mean numbers of animals per 5-l sample (phreatic and hyporheic) or per liter of gravel (benthic) [slightly modified after Ward *et al.* (1992)].

oxygen concentrations associated with the pollution, which other researchers have also oberved (Creuzé des Châtelliers *et al.*, 1992; Schmidt *et al.*, 1991).

Organic pollution of karst waters and its impact on macroinvertebrate communities have been the subject of various studies. Increases in densities of cave macroinvertebrates in response to sewage contamination were observed by Holsinger (1966). He compared the aquatic macroinvertebrate communities of two caves, of which one was polluted by septic tank leachates. Both caves contained the same isopod and planarian species, but they were considerably more abundant in the polluted cave. This was thought to be caused by invertebrate feeding on an abundant microbial flora or direct feeding on the sewage-associated particulate organic matter.

Culver *et al.* (1992) recorded that, after a severe organic contamination, the original and diverse aquatic fauna, consisting of stygobite isopods and amphipods, with occasionally a few *Gyrinophilus porphyriticus* salamanders, had completely disappeared. Moreover, after the event an increasing number of surface organisms were observed, especially chironomid larvae and tubificid worms. Very similarly, a previously rich cave fauna (*Niphargus* sp., *Synurella* sp., and *Proteus anguinus*) was replaced by epigean detrivorous species, such as *Tubifex tubifex* and *Ancylus fluviatilis*, after acute organic pollution (Sket, 1973).

Environmental conditions in inland marine caves on Bermuda support unique faunas that are extraordinarily vulnerable to even very slight organic pollution. Under pristine circumstances very low oxygen conditions commonly occur in these caves, owing to a highly stratified water column and the long residence time of cave waters. A slight increase of organic matter in these systems can rapidly produce anoxia and loss of the highly adapted fauna (Iliffe *et al.*, 1984). These caves often contain an endemic fauna (Iliffe *et al.*, 1983).

Organic pollution can change groundwater food webs. Hypogean populations can increase as discussed above, but organic pollution also creates more favorable conditions for colonization by detritivorous epigean organisms (Malard *et al.*, 1994). Organic pollution may favor the appearance of detrivorous stream invertebrates (*e.g.*, Oligochaeta, Chironomidae, and epigean Gammaridae) at the portal of cave systems, and stygobite species may be reduced or pushed farther into the deeper parts of the karst (Sket, 1979; Dumnicka and Wojtan, 1990). Apparently, under these conditions the unique adaptation of many stygobites to oligotrophic conditions is overridden by the pollution, and they are replaced by epigean species.

2. Toxic Chemicals

Studies on the impact of metal pollution on interstitial invertebrate communities on the Rhône River (Plénet, 1993) supported the findings of many investigators. Macroinvertebrate communities were selectively influenced, quantitatively and qualitatively, by the concentrations of zinc and copper in water and sediments (Fig. 7). For example, interstitial station 5,

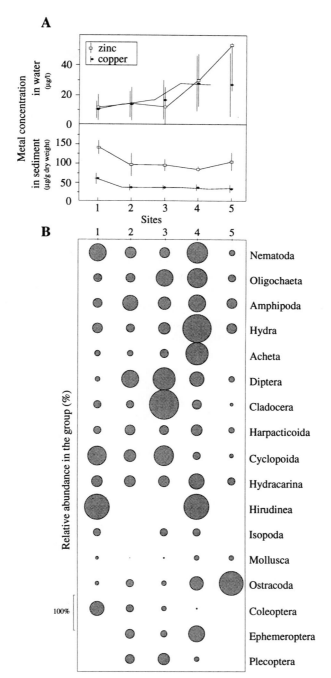

FIGURE 7   (A) Mean zinc and copper concentrations in water and in sediments (with standard deviation) in the upper Rhône river (sites 1–5). (B) Longitudinal distribution of different groups of invertebrates collected in the benthos and in the interstitial zone [modified from Plénet (1993)].

characterized by high metal levels in water, displayed a low total abundance and a decrease in species diversity compared with those of the other stations. Some groups were more resistant to pollution than others. Ostracods were the most abundant group at polluted station 5 (37.5%); 98.3% were hypogean species such as *Pseudocandona zschokkei* and *Cryptocandona kieferi*. However, it is likely that distribution or abundance was determined by numerous chemical or physical factors (*e.g.*, substrate type and stability, water velocity, available food resources, and predation), as well as metal contamination.

If we compare the amphipod fauna of pumping wells at two catchment areas upstream and downstream of Lyon, supplied mainly by the Rhône river, the number of amphipod species is greater in groundwater (7) than in the river (1). But the Rhône river, which provides the groundwater for the towns south of Lyon, is deeply polluted downstream of the city; the biological result is to be seen with the vanishing of the amphipods between two stations only 17 km apart (Table VI). Schmidt *et al.* (1991) also discussed the low taxonomic richness (9 species) of the interstitial fauna of the Grand Gravier sector, which is highly impacted by industrial pollution, compared with other segments situated 20 km upstream of Lyon (Miribel area; 19 species) and 170 km downstream of Lyon (Donzère segment; 17 species). Moreover, where the interstitial fauna is still present, its composition has changed: insect larvae and Amphipoda, *e.g.*, *Gammarus* sp., are no longer present, but microcrustaceans (Cladocera and Ostracoda) remain abundant and diversified.

Karstic areas provide opportunities for ecologists to trace changes in groundwater over very long distances *in situ*. In the French Jura region, for instance, half the water is more than seven years old at the outflow springs (Blavoux, 1981). The vadose zone is accessible by means of caving and contains a legacy of water resource (mis-)management. The terrestrial fauna

TABLE VI   The Occurrence of Amphipods in an Unpolluted and in a Polluted Sector in the Interstitial Zone of the Rhône River [Data from Schmidt *et al.* (1991) and M. J. Turquin, unpublished data]

Species	Crépieux sector (upstream)	Grand Gravier sector (downstream)
*Niphargus kochianus**		O
*Niphargus pachypus**	O	
*Niphargus renei**	O	
*Niphargus gallicus**	O	
*Niphargopsis casparyi**	O	
Salentinella sp.*	O	
Gammarus pulex	O	O

Note. Asterisks indicate stygobite species.

(troglobites), which provides food for aquatic biota in the limestone aquifer (Turquin, 1980), reflects changes in the percolation input over 20 years of increasing pollution by wastes, fertilizers, and toxic chemicals (Fig. 8). Different populations of troglobite species were, directly or indirectly, exposed to the pollution, according to the zone of the cave. The true stygobites were also exposed to predation by an epigean staphylinid beetle, *Deleaster dichrous,* that had spread through the infiltration pathways. Eutrophication of the reservoir has also followed increasing exposure to pollutants. Celi (1991) related copper bioaccumulation in troglobites in the caves of the Venetian Alps to munition dumps from World War I, which continue to leach metals through fissures in the limestone. As a consequence "certain springs seem to be small mines" for copper ore.

C. Ecological Adaptations and Community Structure of Groundwater Invertebrates

1. Sensitivity of Groundwater Invertebrate Populations

In general, the true groundwater environment is characterized by the absence of primary production, relatively stable physical and chemical condi-

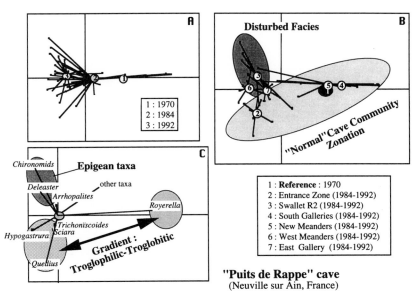

FIGURE 8 Impact of progressive pollution of water (1970–1992) on the unsaturated zone of the karstic system of Puits de Rappe (Ain, French Jura). Ordination diagrams corresponding to (A) evolution of the stations with time suggesting that the cave community is unstable; (B) major contributing taxa of the cave community; and (C) abnormal biotopes along the percolation routes of polluted water, characterized by new epigean taxa [after Turquin and Crague (1994)].

tions, detrivorous food chains, and oligotrophic and hypoxic (or anoxic) conditions. Remarkably, independent evolutionary lineages of organisms that have conquered this environment have developed very similar morphological, physiological, and life-history adaptations, indicating the presence of similar selection forces (Chapter 1). Are stygobite populations by the nature of their adaptations more or less sensitive to toxicants? This question can hardly be answered owing to the absence of relevant studies on the toxicology, physiology, and population dynamics of these organisms. Necessary information on the relationships between life history and physiology on the one hand and among the bioaccumulation pattern, the sensitivity of lethal and sublethal end points, and population dynamics after long-term exposure on the other is not available for groundwater organisms. Such information is available for only a very few aquatic and terrestrial species, and then it is mostly incomplete. The only data for groundwater organisms come from acute studies with mortality or immobility as toxic end points. These studies indicate that stygobite animals are not more sensitive than epigean relatives. On the contrary, it seems that hypogean species are somewhat less sensitive in acute toxicity tests. This is demonstrated in Fig. 9 for cadmium and zinc (see also Table V).

The relative sensitivity of a species to a toxicant depends also on the type of toxicological end point concerned. For example, a greater LC_{50} value for one species will not necessary mean that a particular EC_{50} value (*e.g.*,

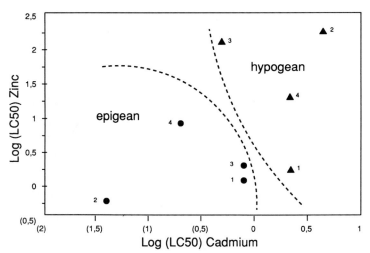

FIGURE 9   Comparison of acute toxicity data (LC_{50}) for zinc and cadmium between epigean and hypogean crustaceans. Epigean species: 1, *Nitocra spinipes; 2, Gammarus pulex; 3, Allorchestes compresa; 4, Lirceus alabamae.* Hypogean species: 1, *Parastenocaris germanica; 2, Niphargus aquilex; 3, Proasellus cavaticus; 4, Caecidotea bicrenata.* Data are from the AQUIRE database (1993), Bosnak and Morgan (1981b), Meinel and Krause (1988), Meinel *et al.* (1989), and Notenboom *et al.* (1992).

for growth or reproduction) is also greater when compared with that of species. Similarly, differences are seen when individual toxicological parameters are compared with population parameters. Van Straalen *et al.* (1989) observed a greater effect of cadmium on the population growth of *Platynothrus peltifer* compared with *Orchesella cincta* (both soil microarthropods), even though the LC_{50} value for *O. cincta* was lower. In an ecotoxicological study of two *Brachionus* species (Rotifera), Halbach (1984) found the population parameters more sensitive than the individual parameters. In contrast, Meyer *et al.* (1987) studied *Daphnia pulex* (Cladocera) exposed to cadmium and copper and asserted that results were similar using either population or individual measures. Crommentuijn *et al.* (1993) concluded that general rules cannot be given and that the effect of toxic substances on population parameters is species and substance specific. To accurately explain population responses to pollutants, it is necessary to determine how individuals in different life stages respond in addition to long-term trends in abundance. In other words, extrapolating results of LC_{50} bioassays to predict population dynamics is not recommended because stygobites typically have long life cycles characterized by low reproductive rates, and they often exist in isolated populations with probably little gene flow among them (Chapter 1). Therefore, pollutants may have greater effects at the population level than those inferred from ecotoxicological studies of individuals.

Clearly additional research on the effects of long-term exposure to pollutants on stygobite populations is needed before generalizations can be made. But, *a priori*, it appears that stygobites will have high ecotoxicological risk profiles, as do some soil arthropods with similar life characteristics and that have been studied in more detail (Posthuma and Van Straalen, 1993).

2. Susceptibility of Groundwater Communities

Do communities of groundwater organisms share ecological characteristics (*i.e.*, with respect to adaptive traits for life in groundwater, food web interactions, and responses to food and energy resources)? And, if so, what does this mean for the susceptibility of groundwater communities when they experience pollution-induced stress? Overall, groundwater ecosystems are food limited, except in some very shallow situations and karst sinks. Hence, densities of the fauna are low. The current literature suggests that food webs are simple with few trophic levels and low connectivity. Patchy distribution of food resources, especially in porous systems, is another characteristic feature of groundwater ecosystems. Moreover, high levels of endemicity are characteristic, especially in karst, and many species represent very old phylogenetic lineages. Indeed, a few taxonomic orders are almost entirely restricted to groundwater (Botosaneanu, 1986). Although the number and densities of species are generally low, the biodiversity of groundwater biotopes has a special nature.

Hence, possibilities for recolonization and recovery after some pollution appear to be limited, because of the low reproductive output of many groundwater invertebrates and the fact that the groundwater environment is often heterogenous with patchy distributed resources. Due to the special nature of groundwater biodiversity, the probability of complete species extinction seems high. However, cave systems can be resilient, if the pollution comes from a point source, because repopulation of the network can occur from refugia situated laterally from the principal drainage axis (Turquin, 1981). Another consideration is that the functional redundancy of groundwater ecosystems (*i.e.*, most groundwater organisms are generalists or omnivores) greatly enhances the effects of pollutants. The low functional and structural diversity of many groundwater communities probably makes them relatively unstable when they experience pollution-induced stress (Elton, 1927). Moreover, Kooijman (1985) suggested that food-regulated systems are more sensitive than predator-regulated systems.

IV. STYGOBITES AS BIOMONITORS

Biomonitoring is a complex concept that always seems to include discussions about cost effectiveness, interpretation of results, and its value in comparison with conventional pollution monitoring techniques (Hellawell, 1989). Biomonitoring mostly involves repeated analysis of specific bioindicator organisms over time. A bioindicator is generally considered an organism that, because of its presence and its appearance, gives information about the biophysical conditions of the environment in which it lives.

Mainly with respect to potability of groundwaters, Husmann (1971) advocated the notion of stygobites as indicators of a negligible influence of surface water. The presence of stygophiles may point to connection with surface waters but may not necessarily suggest an unacceptable situation. More intense infiltration of meso- or polysaprobic surface water may be indicated by the presence of abundant stygoxenes, a condition that should be considered undesirable for drinking water. Husmann (1971) clearly stated that the presence of these different ecological groups in drinking water distribution networks gives only a very rough indication of its potability. A more refined assessment is possible when the ecology of the indicator species is known, along with information on bacteria and protozoa. For example, Danielopol (1981) used the composition and phenology of the ostracod fauna in shallow wells to differentiate epigean and hypogean influences.

Groundwater organisms have been used as describers (Bournaud and Amoros, 1984) for hydrologic connectivity and stability of aquifers. The origin of groundwater and its dynamics in alluvial plains were described by the presence of *Salentinella sp.* (Amphipoda) or by *P. zchokkei* (Ostracoda)

(Marmonier, 1988; Creuzé des Châtelliers and Marmonier, 1990; Dole-Olivier and Marmonier, 1992). Marmonier and Dole (1986) suggested that the spatial distribution of amphipod crustaceans in the hyporheic zone of the river was related to the groundwater circulation patterns and the intensity of the water exchanges between the river and the interstitial zone. A good describer of the intensity of exchange between a river and an adjacent aquifer appeared to be the oligochaete worm *Phallodrilus sp.* (Lafont and Durbec, 1990). The occurrence of animals, such as *Troglochaetus beranecki* (Archannelida), *Microcharon reginae* (Isopoda), and *Pseudocandona triquetra* (Ostracoda), appeared to be restricted to those parts of the alluvial aquifer characterized by physical stability. Mostly these conditions are encountered in the relatively deeper phreatic zones, in which the very "fragile" organisms are able to survive (Dole and Coineau, 1987).

Groundwater organisms are used more and more as indicators of biophysical conditions in aquifers. Danielopol (1989) showed that, in an abandoned channel of the Danube, the distribution of two ostracod crustaceans was related to the oxygen levels and the percentage of fine sediment. For example, *C. kieferi* appeared to be indicative of zones with low oxygen content and little fine sediment. Some harpacticoid crustaceans proved to be good indicators too. For example, *Parapseudoleptomesochra subterranea* was a descriptor for the well-oxygenated zones in the sediment of a small Pyrenean river (Rouch, 1988). Organisms were used as indicators of available biological fractions of pollutants in hyporheic and phreatic environments (Plénet, 1993). In the hyporheic environment some stygobites may reflect saprobic conditions (Mestrov and Latinger-Penko, 1977/1978, 1981; Rejic, 1973) (oligosaprobic, *Hauffenia subpiscinalis*; oligosaprobic-β mesosaprobic, *Niphargus puteanus spoekeri*, and *Asellus aquaticus cavernicolus*.

Barthélémy (1984) selected a series of polluted and nonpolluted karst springs in a 100-km^2 area of French Jura and sampled organisms four times a year with a drift net. Spatial and temporal comparisons of abiotic factors clearly demonstrated the incidence of organic pollution in half of the springs. A sophisticated analysis (Carrel *et al.*, 1986) failed to show a relation between taxa and the water quality of the springs. The number and types of drifting organisms may be too few to allow differentiation of pollution effects, or the differences between polluted and unpolluted karst aquifers were not great enough to affect the fauna. In spite of these seemingly negative results, seven crustaceans were indicators of water quality along the altitudinal gradient (Fig. 10).

V. GROUNDWATER MANAGEMENT AND ECOLOGY

Aquifers form the most extended array of freshwater ecosystems on our planet. It is remarkable that considerations about the intrinsic ecological

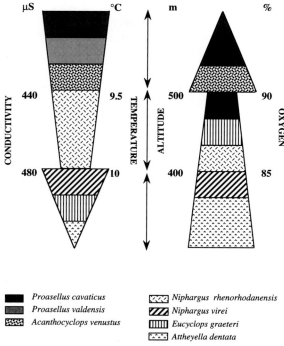

μS   °C   m   %

CONDUCTIVITY

440   9.5   500   90

TEMPERATURE

ALTITUDE

OXYGEN

480   10   400   85

■ *Proasellus cavaticus*   *Niphargus rhenorhodanensis*
Proasellus valdensis   *Niphargus virei*
Acanthocyclops venustus   *Eucyclops graeteri*
                                  Attheyella dentata

FIGURE 10   Environmental requirements of sensitive stygobites in kartsic aquifers in the French Jura. Only 7 species out of 19 stygobites and 14 stygophiles are discriminated by a statistical multivariate analysis [after Barthélémy (1984)]. The crustacea community is not affected by NH_4, PO_4, NO_3, or Cl (no data here). The geographical situation of the stations is related to the specific physicochemical conditions of each station. For example, *Proasellus cavaticus* is present in caves for which high altitude is characterized by cooler, less mineralized, and more oxygenated water.

properties of these systems play only a tiny role—if at all—in the development of water resource management. Most groundwater management strategies are based on simple calculations and the monitoring of quantity. Water quality, as reflected by ecological conditions, is rarely included in management agendas, although aspects of microbial ecology clearly influence decisions with respect to groundwaters used for drinking water. A more synthetic approach to the sustainability of groundwater resources should be based on the conservation of ecological functions so that pollution can be detected early and be reversed before the self-cleansing or degradation capacity of natural food webs is compromised. Moreover, groundwaters have an intrinsic value as refugia for rare species and unique biodiversity patterns (Nachtnebel and Kovar, 1991; Stanford and Simons, 1992).

Unfortunately, management instruments used to deal with the protection of the intrinsic ecological values of groundwater are barely developed. The assessment of ecological risks associated with various human activities should be the foundation of management approaches (Suter, 1993). Currently, the most prominent studies are related to riverine aquifers and environmental problems caused by surface water pollution. These studies focus mainly on changes in hydrologic exchanges between the river and its aquifer, on the impact of organic pollution, and to a limited extent on pollution by toxic chemicals (*e.g.,* heavy metals). Studies devoted to the ecological implications of groundwater pollution are very scarce, even though aquifers of alluvial and coastal plains are strongly influenced by changes in soil characteristics and by patterns associated with land-use practices worldwide. Standards for chemical substances in groundwater must be based on ecotoxicological information (Notenboom and Van Gestel, 1992). For example, the acute toxicity studies of selected pesticides for the stygobite *Parastenocaris germanica* showed that current drinking water standards for some pesticides in groundwater will probably not sufficiently protect the ecological properties of groundwater (J. Notenboom and J.-J. Boessenkool, unpublished data).

A research agenda aimed at the improvement of the ecological basis of groundwater management should include the study of effects of pollutants on biodegradation processes; effects on the viability of populations of groundwater invertebrates after chronic exposure to low environmental concentrations of toxicants; the ecological recovery of groundwater ecosystems after remediation; the impact of land-use practices on structural and functional aspects of shallow and karstic groundwater ecosystems; and the impact of hydrologic interventions (*i.e.,* groundwater mining) on the functioning of groundwater ecosystems.

VI. CONCLUSION

Information on the impact of pollutants on invertebrates and their communities in groundwater is scarce. Most information is related to riverine aquifers and their interaction with rivers. Fewer studies exist for karst and phreatic groundwater. Most existing field observations concern the incidence of organic pollution, especially in karst systems. Field studies on hyporheic communities in reference sites and sites exposed to heavy metals provide some indication of toxic effects, but these studies may be biased by other, nontoxic effects. The responses of groundwater communities to the different types of pollution are complex, and response models are needed.

It is fairly well known that organic pollution interferes strongly with the mostly oligotrophic and hypoxic nature of groundwater systems. An organic subsidy may stimulate a population increase, because food is often a limiting factor. However, organic pollution can also allow colonization of the subterranean realm by epigean animals, thereby pushing stygobites into the oligotrophic refugia. Extreme organic pollution can produce complete oxygen depletion and cause the disappearance of all metazoans.

Toxic chemicals are much more problematic. The few available groundwater ecotoxicological studies mainly address the sensitivity of a few stygobites and stygophile species in short-term laboratory experiments. Unfortunately, assessment of the ecological risks associated with the chemical pollution of groundwater is limited by the paucity of acute toxicity data. Information on the response of populations to chronic exposure would be even more relevant. However, ecotoxicological studies of stygobite growth, reproduction, and population dynamics are very difficult because stygobite phenologies are complex and poorly known and stygobites are very difficult to culture. Owing to long life histories, low reproductive rates, and low gene flow among populations, we expect that stygobites will be very sensitive to all classes of pollution, even though the few species studied in LD_{50} bioassays were rather insensitive. Almost nothing is known of the impact of toxicants on *in situ* community structure and food web dynamics.

Groundwater invertebrates appear to be good biomonitors, at least where pollution is severe. As more ecological information on the stygobite distribution, abundance, life histories, food web dynamics, and ecotoxicology becomes available, a general framework for biomonitoring and risk assessment may emerge. Comparisons with benthic studies may be very useful, especially with respect to coupling hydrologic, hydrochemical, and ecological effects models. However, biophysical models describing the fate and behavior of contaminants in groundwater ecosystems require a much more synthetic understanding of basic groundwater ecology, as summarized in other chapters in this volume.

That groundwater resources are being pervasively polluted without consideration of the resiliency of the self-cleansing capacity associated with groundwater food webs, seems in conflict with the strategic importance of aquifers in all countries of the world. Clearly, greater investment in groundwater ecology with respect to the influences of pollutants is warranted.

ACKNOWLEDGMENTS

The Institute of Public Health and Environmental Protection is thanked for the opportunity given to J. Notenboom to continue his research in groundwater ecology and ecotoxicology. Support was provided in part by the Interdisciplinary Research Program on the Environment (PIR—Groundwater Ecosystems) within the URA–CNRS 1451 laboratory (Lyon I). Special thanks to R. Laurent and D. Martin for their valuable assistance during the field and laboratory operations. We thank Professors J. Gibert and J. Stanford for comments on and improvements of the manuscript.

REFERENCES

Allen, H. E., Perdue, E. M., and Brown, D. S. (1993). "Metals in Groundwater." Lewis Publishers, Boca Raton, LA.

Armstrong, A. Q., Hodson, R. E., Hwang, H.-M., and Lewis, D. L. (1991). Environmental factors affecting toluene degradation in ground water at a hazardous waste site. *Environ. Toxicol. Chem.* **10**, 147–158.

Barr, T. C. (1976). "Ecological Effects of Water Pollutants in Mammoth Cave", Final Tech. Rep. Contract N° CX 500050204. U.S. National Park Service.

Barthélémy, D. (1984). Impact des pollutions sur la faune stygobie karstique: Approche typologique sur seize émergences des départements de l'Ain et du Jura. Thesis, Univ. Lyon 1.

Blavoux, M. B. (1981). L'eau dans le karst: Les différentes composantes de l'écoulement et leur participation au débit à l'exutoire en crue et en étiage. *Actes Colloq. Nat. Prot. Eaux Souterraines, 1st,* Besançon 1980, pp. 59–81.

Bosnak, A. D., and Morgan, E. L. (1981a). Comparison of acute toxicity for Cd, Cr, and Cu between two distinct populations of aquatic hypogean isopods (Caecidotea sp.). *Proc. Int. Congr. Speleol., 8th,* Bowling Green, Vol. 1, pp. 72–74.

Bosnak, A. D., and Morgan, E. L. (1981b). Acute toxicity of cadmium, zinc and total residual chlorine to epigean and hypogean isopods (Asellidae). *NSS Bull.* **43**, 12–18.

Botosaneanu, L., ed. (1986). "Stygofauna mundi." E. J. Brill/Dr. W. Backhuys, Leiden.

Bournaud, M., and Amoros, C. (1984). Des indicateurs biologiques aux descripteurs de fonctionnement: Quelques exemples dans un système fluvial. *Bull. Soc. Ecol.* **15**(1), 57–66.

Carrel, G., Barthélémy, D., Auda, Y., and Chessel, D. (1986). Approche graphique de l'analyse en composantes principales normée: Utilisation en hydrobiologie. *Acta Oecol. [Ser.]: Oecol. Gen.* **7**(2), 189–203.

Celi, M. (1991). Biospeleologia. *Barbastrijo* **3**, 1–18.

Creuzé des Châtelliers, M., and Marmonier, P. (1990). Macrodistribution of Ostracoda and Cladocera in a by-passed channel: Exchange between superficial and interstitial layers. *Stygologia* **5**(1), 17–24.

Creuzé des Châtelliers, M., Marmonier, P., Dole-Olivier, M. J., and Castella, E. (1992). Structure of interstitial assemblages in a regulated channel of the Rhine River (France). *Regulated Rivers* **7**, 23–30.

Crommentuijn, T., Brils, J., and Van Straalen, N. M. (1993). Influence of cadmium on life-history characteristics of Folsomia candida (Willem) in an artificial soil substrate. *Ecotoxicol. Environ. Saf.* **26**, 216–227.

Culver, D. C., Jones, W. K., and Holsinger, J. R. (1992). Biological and hydrological investigation of the Cedars, Lee County, Virginia, an ecologically significant and threatened karts area. *In* "Ground Water Ecology" (J. A. Stanford and J. J. Simons, eds.), pp. 281–290. Am. Water Resour. Assoc., Bethesda, MD.

Danielopol, D. L. (1981). Distribution of ostracods in the groundwater of the North Western coast of Euboa (Greece). *Int. J. Speleol.* **11**, 91–103.

Danielopol, D. L. (1983). Der Einfluss organischer Verschmutzung auf das Grundwasser-Ökosystem der Donau im Raum Wien und Niederöstereich. "Gewässerökologie," Beitr. 5. Budesminist. Gesundh. Umweltschutz, Wien, Austria.

Danielopol, D. L. (1989). Groundwater fauna associated with riverine aquifer. *J. North Am. Benthol. Soc.* **8**(1), 18–35.

Dickson, G. W., Briese, L. A., and Giesy, J. P., Jr. (1979). Tissue metal concentrations in two crayfish species cohabiting a Tennessee cave stream. *Oecologia* **44**, 8–12.

Dickson, K. L., Maki, A. W., and Brungs, W. A. (1987). "Fate and Effects of Sediment-Bound Chemicals in Aquatic Systems." Pergamon, New York.

Dole, M.-J., and Coineau, N. (1987). L'isopode *Microcharon* (Crustacea, Isopoda) abondant dans les eaux interstitielles de l'est lyonnais, *M. reginea n. sp.,* écologie et biogéographie. *Stygologia* **3**(3), 200–217.

Dole-Olivier, M.-J., and Marmonier, P. (1992). Effects of spates on the vertical distribution of the interstitial community. *Hydrobiologia* **230**, 49–61.

Dumnicka, E., and Wojtan, K. (1990). Differences between cave water ecological systems in the Krakow-Czestochowa upland. *Stygologia* **5**, 241–247.

DWVK (1988). "Bedeutung biologischer Vorgänge für die Beschaffenheit des Grundwassers. Deutscher Verb. für Wasserwirtschaft u. Kulturbau e. V. (DVWK)." Verlag Paul Parey, Hamburg and Berlin.

Edworthy, K. J. (1987). Groundwater contamination—A review of current knowledge and research activity. *Stygologia* 3/4, 279–295.

Elton, C. (1927). "Animal Ecology." Sidgwick & Jackson, London.

Enckell, P. H. (1968). Oxygen availability and microdistribution of interstitial mesofauna in Swedish fresh-water sandy beaches. *Oikos* 19, 271–291.

Freeze, A. R., and Cherry, J. A. (1979). "Groundwater." Prentice-Hall, Englewood Cliffs, NJ.

Fried, J. (1976). "Ground Water Pollution." Elsevier, Amsterdam.

Galceran, M. T., Rubio, R., Rauret, G., and Alonso, L. (1990). Assessment of groundwater contamination subsequent to an environmental release. *Waste Manage.* 10, 261–268.

Gerhardt, A. (1993). Review of impact of heavy metals on stream invertebrates with special emphasis on acid conditions. *Water, Air, Soil Pollut.* 66(3/4), 289–314.

Ghiorse, W. C., and Wilson, J. T. (1988). Microbial ecology of the terrestrial subsurface. *Adv. Appl. Microbiol.* 33, 107–172.

Gibert, J. (1990). Behavior of aquifers concerning contaminants: Differential permeability and importance of the different purification processes. *Water, Sci. Technol.* 22(6), 101–108.

Halbach, U. (1984). Population dynamics of rotifers and its consequences for ecotoxicology. *Hydrobiologia* 109, 79–96.

Hellawell, J. M. (1989). "Biological Indicators of Freshwater Pollution and Environmental Management" Pollut. Monit. Ser. Elsevier, London.

Holsinger, J. R. (1966). A preliminary study on the effects of organic pollution of Banners Corner Cave, Virginia. *Int. J. Speleol.* 2, 75–89.

Husmann, S. (1971). Die gegenseitige Ergänzung theoretischer und angewandter Grundwasser-limnologie; mit Ergenissen aus Wasserwerken Wiesbadens. *In* "Die Sicherstellung der Trinkwasserversorgung Wiesbadens." Stadtwerke Wiesbaden A G, Wiesbaden.

Husmann, S. (1974/1975). Versuche zur Erfassung der vertikalen Verteilung von Organismen und chemischen Substanzen im Grundwasser von Talauen und Terrassen. Methoden und erste Befunde. *Int. J. Speleol.* 6, 271–302.

Husmann, S. (1975). Die Schotterufer des Niederrheins bei Krefeld: Abwasserkranke Biotope mit gestörter Uferfiltration. *Wasser Abwasser* 57/58, 7–26.

Iliffe, T. M., Hart, C. W., Jr., and Manning, R. B. (1983). Biography and the caves of Bermuda. *Nature (London)* 308, 141–142.

Iliffe, T. M., Jickells, T. D., and Brewer, M. S. (1984). Organic pollution of an inland marine cave from Bermuda. *Mar. Environ. Res.* 12, 173–189.

Khondaker, A. N., Al-Layla, R. I., and Husain, T. (1990). Groundwater contamination studies. The state of the art. *CRC Crit. Rev. Environ. Control* 20(4), 231–256.

Kooijman, S. A. L. M. (1985). Toxicity at the population level. *In* "Multispecies Toxicity Testing" (J. Cairns, Jr., ed.), pp. 143–164. Pergamon, New York.

Korom, S. F. (1992). Natural denitrification in the saturated zone: A review. *Water Resour. Res.* 28(6), 1657–1668.

Lafont, M., and Durbec, A. (1990). Essai de description biologique des interactions entre eau de surface et eau souterraine: Vulnérabilité d'un aquifere à la pollution d'un fleuve. *Ann. Limnol.* 26(2/3), 119–129.

Maire, R. (1979). Comportement du karst vis à vis des substances polluantes. *Ann. Soc. Géol. Belg.* 102, 101–108.

Malard, F., Reygrobellet, J.-L., Mathieu, J., and Lafont, M. (1994). The use of invertebrate communities to describe groundwater flow and contaminant transport in a fractured rock aquifer. *Arch. Hydrobiol.* 131, 93–110.

Mangin, A., and Bakalowicz, M. (1989). Orientation de la recherche scientifique sur le milieu karstique. Influences et aspects perceptibles en matière de protection. *Spelunca* 35, 71–79.

Marmonier, P. (1988). Biocénoses interstitielles et circulation des eaux dans le sous-écoulement d'un chenal aménagé du Haut-Rhône français. Thèse, Univ. Lyon.

Marmonier, P., and Dole, M.-J. (1986). Les Amphipodes des sédiment d'un bras court-circuité du Rhône. Logique de répartition et réaction aux crues. *Rev. Fr. Sci. Eau* 5, 461–486.

Marxsen, J. (1981a). Bacterial biomass and bacterial uptake of glucose in polluted and unpolluted groundwater of sandy and gravelly deposits. *Verh. Int. Ver. Limnol.* 21, 1371–1375.

Marxsen, J. (1981b). Fluoreszensmikroscopische Untersuchungen der Bakterienflora und Bestimmung ihrer heterotrophen Aktivität in organisch belastem und unbelastem Grundwasser sandig-kiesiger Ablagerungen. *Int. J. Speleol.* 11, 173–201.

Mathews, R. C., Bosnak, A. D., Tennant, D. S., and Morgan, E. L. (1977). Mortality curves of blind cave crayfish (*Orconectes australis australis*) exposed to chlorinated stream water. *Hydrobiologia* 53(2), 107–111.

Mayer, A. S., Rabideau, A. J., Mitchell, R. J., Imhoff, P. T., Lowry, M. I., and Miller, C. T. (1993). Groundwater quality. *Water Environ. Res.* 65(4), 486–534.

Meinel, W., and Krause, R. (1988). Zur Korrelation zwischen Zink und verschiedenen pH-Werten in ihrer toxischen Wirkung auf einige Grundwasser-Organismen. *Z. Angew. Zool.* 75, 159–182.

Meinel, W., Krause, R., and Musko, J. (1989). Experimente zur pH-Wert-Abhängigen Toxizität von Kadmium bei einigen Grundwasserorganismen. *Z. Angew. Zool.* 76, 101–125.

Mestrov, M., and Latinger-Penko, R. (1977/1978). Ecological investigations of the influence of a polluted river on surrounding interstitial underground waters. *Int. J. Speleol.* 9, 331–355.

Mestrov, M., and Latinger-Penko, R. (1981). Investigation of the mutual influence between a polluted river and its hyporheic. *Int. J. Speleol.* 11, 159–171.

Meyer, J. S., Ingersoll, C. G., and McDonald, L. L. (1987). Sensitivity analysis of population-growth rates estimated from Cladoceran chronic toxicity tests. *Environ. Toxicol. Chem.* 6, 115–126.

Muller, I. (1981). Quelques aspects de la pollution bactériologique et chimique des sources karstiques du Jura Neuchatelois. *Actes Colloq. Nat. Prot. Eaux Souterraines, 1st,* Besançon 1980, pp. 263–273.

Nachtnebel, H. P., and Kovar, K. (1991). Hydrological basis of ecologically sound management of soil and groundwater. *IAHS Publ.* 202.

Notenboom, J., and Boessenkool, J.-J. (1992). Acute toxicity testing with the groundwater copepod *Parastenocaris germanica* (Crustacea). *In* "Ground Water Ecology" (J. A. Stanford and J. J. Simons, eds.), pp. 301–309. Am. Water Resour. Assoc., Bethesda, MD.

Notenboom, J., and Van Gestel, K. (1992). Assessment of toxicological effects of pesticides on groundwater organisms. *In* "Ground Water Ecology" (J. A. Stanford and J. J. Simons, eds.), pp. 311–317. Am. Water Resour. Assoc., Bethesda, MD.

Notenboom, J., Cruys, K., Hoekstra, J., and Van Beelen, P. (1992). Effect of ambient oxygen concentration upon the acute toxicity of chlorophenols and heavy metals to the groundwater copepod *Parastenocaris germanica* (Crustacea). *Ecotoxicol. Environ. Saf.* 24, 131–143.

Page, W. G. (1981). Comparison of groundwater and surface water for patterns and levels of contamination by toxic substances. *Environ. Sci. Technol.* 15(12), 1475–1481.

Page, W. G. (1987). "Planning for Groundwater Protection." Academic Press, Orlando, FL.

Piver, W. T. (1993). Contamination and restoration of groundwater aquifers. *Environ. Health Perspect.* 100, 237–247.

Plénet, S. (1993). Sensibilité et rôle des invertébrés vis a vis d'un stress métallique à l'interface eau superficielle/eau souterraine. Thèse, Univ. Lyon-1.

Plénet, S. (1994). Freshwater amphipods as biomonitors of trace-metal pollution in surface and interstitial aquatic systems. *Freshwater Biol.* (in press).

Posthuma, L., and Van Straalen, N. M. (1993). Heavy-metal adaptation in terrestrial invertebrates: A review of occurrence, genetics, physiology and ecological consequences. *Comp. Biochem. Physiol. C* 106C(1), 11–38.

Rejic, M. (1973). Biological pollution indicators in underground waters. *Biol. Vestn.* 21(1), 11–15.

Rouch, R. (1988). Sur la répartition spatiale des Crustacés dans le sous-écoulement d'un ruisseau des Pyrénées. *Ann. Limnol.* **24**(3), 213–234.

Schmidt, M., Marmonier, P., Plénet, S., Creuzé des Châtelliers, M., and Gibert, J. (1991). Bank filtration and interstitial communities. Example of the Rhône River in a polluted sector (downstream of Lyon, Grand Gravier, France). *Hydrogéologie* **3**, 217–223.

Sinton, L. W. (1984). The macroinvertebrates in a sewage-polluted aquifer. *Hydrobiologia* **119**, 161–169.

Sket, B. (1973). Gegenseitige beeinflussung der wasserpollution und das höhlenmilieus. *Proc. Int. Congr. Speleol., 6th*, Vol. 5, pp. 253–262.

Sket, B. (1979). The cave fauna in the triangle Cerknica- Postojna-Planina (Slovenia, Yugoslavia); its conservational importance. *Varstvo Narave (Nature Conservation)* **12**, 45–59.

Stanford, J. A., and Simons, J. J., eds. (1992). "Ground Water Ecology." Am. Water Resour. Assoc., Bethesda, MD.

Stanford, J. A., and Ward, J. V. (1992). Emergent properties of ground water ecology: Conference conclusions and recommendations for research and management. *In* "Ground Water Ecology" (J. A. Stanford and J. J. Simons, eds.), pp. 409–415. Am. Water Resour. Assoc., Bethesda, MD.

Stigliani, W. M. (1988). Changes in valued "capacities" of soils and sediments as indicators of non-linear and time delayed environmental effects. *Environ. Monit. Assess.* **10**, 245–307.

Stigliani, W. M. (1993). Overview of the chemical time bomb problem in Europe. *In* "Delayed Effects of Chemicals in Soils and Sediments," Foundation for Ecodevelopment "Stichting Mondial Alternatif," Hoofddorp, The Netherlands.

Suter, G. W. (1993). "Ecological Risk Assessment." Lewis Publishers, Boca Raton, FL.

Ter Meulen, G. R. B., Stigliani, W. M., Salomons, W., Bridges, E. M., and Imeson, A. C. (1993). Chemical time bombs. *In* "Delayed Effects of Chemicals in Soils and Sediments" Foundation for Ecodevelopment "Stichting Mondial Alternatif," Hoofddorp, The Netherlands.

Timmermans, K. R., van Hattum, B., Kraak, M. H. S., and Davids, C. (1989). Trace metals in a littoral foodweb: Concentrations in organisms, sediment and water. *Sci. Total Environ.* **87/88**, 477–494.

Travis, C. C., and Doty, C. B. (1990). Can contaminated aquifers at Superfund sites be remediated? *Environ. Sci. Technol.* **24**, 1467–1468.

Turquin, M. J. (1980). La biocénose terrestre cavernicole: Apport énergétique pour la communauté aquatique souterraine. *Mém. Biospéol.* **8**, 9–16.

Turquin, M. J. (1981). La pollution des eaux souterraines: Incidence sur les biocénoses souterraines. *Actes Colloq. Nat. Prot. Eaux Souterraines, 1st*, Besançon 1980, pp. 341–347.

Turquin, M. J. (1989). Faut-il protéger la faune souterraine? *Spelunca* **35**, 91–93.

Turquin, M. J., and Crague, G. (1994). Impact de l'anthropisation sur la biologie du système karstique de Rappe de la zone non saturée à l'émergence (Neuville-sur-Ain, France). *Bull. Soc. Linn. Lyon* **63**, 3.

Van Gestel, C. A. M., and Ma, W.-C. (1988). Toxicity and bioaccumulation of chlorophenols in earthworms, in relation to bioavailability in soil. *Ecotoxicol. Environ. Saf.* **15**, 289–297.

Van Straalen, N. M., Schobben, J. H. M., and De Goede, R. G. M. (1989). Population consequences of cadmium toxicity in soil microarthropods. *Ecotoxicol. Environ. Saf.* **17**, 190–204.

Ward, J. V., Voelz, N. J., and Marmonier, P. (1992). Groundwater faunas at riverine sites receiving treated sewage effluent. *In* "Ground Water Ecology" (J. A. Stanford and J. J. Simons, eds.), pp. 351–365. Am. Water Resour. Assoc., Bethesda, MD.

19

Groundwater Operations and Management

J. Margat

BRGM
Avenue de Concyr
BP 6009
45060 Orléans Cedex 2, France

I. INTRODUCTION

Because groundwater is "invisible" and difficult to study, it was a source of mythology and conjecture long before it was a subject of serious study. Even as late as the seventeenth century, accounts of the hydraulic cycle turning in the "wrong" direction were advanced by the great minds of the time to account for "visible" springs. Miners were perhaps the first group to study and manipulate groundwater. Their need to drain mines in order to reach mineral resources not only confronted them with the reality of the widespread occurrence and abundance of groundwater, but also stimulated their thinking on ways to use groundwater resources. Thus, the concept of groundwater evolved from that of an obstacle to be overcome to a valuable resource with multiple uses. One of the earliest attempts at using groundwater was probably the effort made, in response to desert encroachment toward the end of the neolithic era, to recover receding springs. This led to the invention of irrigation and the extensive proliferation of drainage galleries in the arid and semiarid regions of the ancient world, such as the "kanats" of Iran.

Methods of locating and pumping groundwater have become increasingly varied and sophisticated with increased technology. New methods make it possible to prospect increased depths and provide greater yields in many cases. But in spite of this, freshwater is a finite resource, and the constant demand for more and more freshwater should act as a warning of the growing need to wisely evaluate and manage our groundwater resources.

The view of groundwater as a natural resource that must be managed and used wisely is a recent one. This view grew from the adverse effects of intensive water pumping operations, including depletion of artesian wells; drawdowns, resulting in lower production; and deterioration in groundwater quality. This view became linked with the idea of reasonable limits for groundwater use. However, confusion and conflicts over what constitutes reasonable use and what is "overexploitation" are exacerbated by the lack of clearly defined criteria for making judgments.

The concept of groundwater resource management evolved from two viewpoints. One viewpoint is the focus on the physical parameters of aquifers (hydrogeologic and hydrodynamic), often with an emphasis on the mean flow of each water-bearing stratum as the basis for estimating the volume of renewable resources. A second viewpoint seeks to extend the criteria for evaluating exploitability to external conditions related to the possible interaction between groundwater and surface water. This direction developed from a better understanding of the continuum between groundwater and surface water, as well as the indivisibility of water resources.

The wise management of water resources needs to be approached from both viewpoints, *i.e.,* from the impact of groundwater manipulation on the natural environment (including groundwater, surface water, and riparian ecosystems) and from the volume of water resources available for sustainable use. Groundwater operations vary considerably in different locations depending on what emphasis is placed on the above viewpoints, as well as the importance of groundwater to an area's water supply, either for drinking water or for other purposes.

The need to manage groundwater resources as an integral part of overall available water resources and in recognition of their place in the equilibrium of the natural environment is better understood today by scientific experts and water resource managers. However, the concepts of enlightened management are still far from being fully grasped by those directly or indirectly involved in the day-to-day management operations of groundwater and who, by their actions, can modify the groundwater regime or groundwater quality.

A strong public education program is needed to emphasize the need for a better understanding of the nature of groundwater resources and of their complexity and diversity and to emphasize how this understanding must form the basis for operating conditions and constraints. This is the only way to positively influence for the long term the attitude of the different actors involved.

II. IMPORTANCE OF GROUNDWATER IN WATER MANAGEMENT

Groundwater supplies a significant proportion of total water for the three major uses: urban and rural drinking water, industrial, and agricultural. The dependence on groundwater varies with geographic location. Whereas the urban and rural community sectors constitute the major operators in the developed nations of Europe (including the former USSR), the industrial sector is the number one operator in some industrialized countries like the former East Germany, South Korea, Japan, the Netherlands, Norway, and the former USSR (at least until 1980). The industrial sector also ranks number two in some others: Belgium, France, the United Kingdom, Czechoslovakia, and Yugoslavia. The pattern is somewhat reversed in those developed nations in which large-scale, irrigated agriculture is practiced: some Mediterranean countries (Spain and Greece), Australia, and parts of the United States. In these countries the agricultural sector constitutes the major operator. This is also true for most developing and some developed nations outside the humid tropical regions. Available figures are quite revealing in this regard: Greece (85%), the United States (62%), South Africa (84%), Tunisia (85%), India (96%), China (90%), Saudi Arabia (80%), Argentina (79%), Mexico (64%), etc.

As indicated above, groundwater has different degrees of importance depending on available sources of water. In most developed nations and in many developing ones, groundwater represents the major source of drinking water. In some cases it is virtually the only available source: Austria and Denmark (97%), Italy (more than 90%), Hungary (78%), Germany, Belgium, Switzerland, Russia, and Yugoslavia (between 70 and 75%), and France and the Netherlands (more than 60%). For the whole of the European community the average is about 70%. In many industrialized countries, groundwater accounts for a significant source of water available to industries not located close to watercourses: Denmark (65%), Japan (41%), France (40%), Germany (26%), the United States (18%), and the former USSR (15%).

Groundwater also supplies significant water for agricultural use. Quite naturally this is the case for countries in arid regions: Saudi Arabia (86%),

Libya (100%), and Argentina (70%). Less arid countries also use significant groundwater resources: India (40%), the United States and Mexico (33%), and Pakistan and some Mediterranean countries like Spain, France, Greece, and Italy (25% or more). Finally, groundwater is also an important source for meeting the water needs of extensive animal husbandry, especially in the arid and semiarid regions.

A. Characteristics of Aquifer Systems

Groundwater refers first and foremost to the physical unit of the aquifer system. The proper evaluation of resources should therefore begin with an analysis of aquifer structure and functioning (dynamic, hydrodynamic, and even hydrobiological). Considerations pertaining to water use (social, economic, and public health) as well as the need to protect the natural environment (groundwater ecosystems and related ecosystems) should be considered. The following features of aquifer systems emphasize this need and help define the objectives and modalities of groundwater resource management.

1. Aquifer Size and Volume

Aquifers are depicted in cross-sectional illustrations as relatively thick compared with their width. In reality, aquifers are more like pancakes, with relatively little thickness for the area they cover. Variations in size are quite impressive and may range from a few square kilometers of area to millions of square kilometers. Thickness is measured in tens of meters, sometimes in hundreds of meters, and only very rarely reaches beyond the 1000-m mark.

2. Nonhierarchical Structure

Compared with hydrographic networks, which are organized in tree structures, aquifer heterogeneity does not usually allow for such structuring. Karst terranes and more transmissive layers may be the exception to this.

3. Inseparability of Aquifer Conservation–Storage and Flow

An aquifer is both a reservoir and a conductor. Groundwater conservation–storage volume may vary significantly on the basis of aquifer size, thickness, and porosity. It may be in millions, billions, or thousands of billions of cubic meters. The total aquifer discharge is usually equivalent to the flow received and depends on climatic and surface conditions. This discharge also varies significantly (between one millimeter and one meter per year per surface area).

4. A Wide Variety of Renewal Rates and Times

Renewal rates and times measure the relationship between mean storage and flow in aquifers. Because conservation–storage and flow are independent of each other (they are not determined by the same conditions), renewal rates vary greatly: conservation–storage may be lower or higher than the mean annual flow, which is usually several times higher than flow (expressed in tens, hundreds, or thousands).

5. Flow Regimes

Two flow regimes are clearly distinguishable: first, the unconfined groundwaters (aquifers) with varying water and conservation–storage levels, usually topped by a water-free zone and fed mainly by seepage from the topsoil, and second, the confined groundwaters, similar to pressure conduits and which are fed and discharged solely through interaction with other unconfined or confined groundwaters.

6. Influence Propagation Field

The hydrodynamic inertia of this field varies with the nature of the groundwater, but is usually higher for confined groundwaters. The effects of all localized actions accumulate over time and affect all sections of this field.

7. Different Types of Groundwater/Surface Water Interface

The interface is more pronounced for unconfined groundwater. Operations affecting groundwater quantity or quality may affect rivers and riparian zones in different ways. Reciprocally, groundwater quality and the integrity of the groundwater system may be adversely affected by recharge from contaminated surface water.

8. Variations in the Sensitivity of the Regime to Pollution and Alterations in the Surface Soil State

Unconfined groundwaters are more vulnerable to changes in land uses and pollution, whether concentrated or diffuse (point source or nonpoint source).

B. Implications for Groundwater Management Based on Physical Conditions

There are far fewer constraints in the choice of the location of groundwater operations than in that of surface water operations. The location of wells is limited only by the need to find water of adequate quality and quantity in the transmissive (and productive) layers, especially the fractures of discontinuous aquifers. However, wells may be delocalized by constructing a network of drains, especially if the groundwater is shallow. Groundwater operations do not normally require the construction of regulatory devices (underground dams, gates, etc.), although they may prove necessary and useful in some situations.

Groundwater management may be combined to varying degrees with aquifer flow and reserve management. The degree of combination may range from the long-term withdrawal of flow alone to the short-term withdrawal of reserves alone.

Groundwater operations may enjoy varying degrees of freedom of action depending on requirements, such as maintaining a surface water discharge rate or regime or even maintaining levels in natural ponds for specific protected uses. The conservation of groundwater resources, from both a quantity and a quality standpoint, may impose strict requirements on water and land uses and planning.

Groundwater management and watershed management of a basin overlying an aquifer can be performed independently of each other in the case of deep, confined aquifers. However, management of unconfined, more shallow aquifers should always be an integral part of local watershed planning and land uses.

The diversity of situations discussed above reflects the diversity of hydrologic conditions and necessitates entirely different approaches to groundwater management. However, these approaches may be grouped under a few headings, as in Table I.

III. RENEWABLE AND NONRENEWABLE GROUNDWATER RESOURCES

For the sake of simplicity, let us begin by distinguishing between renewable groundwater resources, with the possibility of withdrawing aquifer flow or part of its natural yield without disrupting mean, long-term, multiannual equilibrium, and nonrenewable groundwater resources, with the possibility of emptying aquifer reserves (usually measured in maximum exploitable unit volumes, which can be expressed as the mean annual flow for a given length of time). Such an operation necessarily disrupts equilibrium and is therefore time limited. "Possibility" is used here to stress the point that estimates should consider exploitability criteria that may vary with the operator's approach and objectives, as well as with other conditions, such as the proportion of conservation yield reserved for receiving watercourses. Estimates are not simply derived from gross flow and reserves.

Groundwater resources are more complex than many other fixed resources, such as mineral deposits. Regardless of the volume and renewable potential of aquifer reserves, they are never fixed in the manner of mineral deposits, but are part of aquifer dynamics. This is the case even when the aquifer is a large reservoir containing essentially fossil waters. Aquifer flow and reserves are interdependent sets of resources and should not be managed separately.

Withdrawal of a significant part of the mean total flow of a aquifer requires major drawdowns to rechannel internal flow to waterworks. Such measures are always to the detriment of natural discharge channels and constitute permanent cuts on reserves during the initial phase of a search for equilibrium. Every intensive operation, even if in equilibrium, is always at the expense of the groundwater reserve levels. The deliberate withdrawal of a significant portion of aquifer reserve, on the other hand, leads inevitably to the gradual depletion of the natural discharge of the aquifer and the reduction of the mean natural flow to total water output. This is why renewable and nonrenewable water resources do not in any way constitute two independent sets of resources to be managed separately. They simply offer two choices: either we match maximum withdrawal with mean aquifer flow for long-term sustainable yield or we withdraw in excess of the mean natural flow and progressively exhaust reserves.

In practice, however, an aquifer may be considered to yield only renewable resources if the difference between reserves and the mean annual flow is within the two-figure range. It would be possible to offer nonrenewable resources if the difference between reserves and the mean annual flow is in the three- or four-figure range. In other words, the conventional distinction between aquifers offering *a priori* renewable or nonrenewable resources is equal to a renewal rate in the range of 0.01 or to a renewal time of about one century.

A. Using Aquifer Reserves

Because aquifer reserves are important to the above conditions, their level should be considered in evaluating the extent of groundwater resources, although in a different manner for each case. In the first situation of a renewable resource, reserves allow a certain amount of freedom in the choice of a withdrawal regime. In contrast to variations in the natural flow regime, variations in aquifer reserve levels may be artificially increased. This is often the case with unconfined groundwater resources, especially if withdrawals have to be adjusted to consider fluctuations in demand, *e.g.,* for irrigation water. In the second situation of a nonrenewable resource, the major part of the production needed to meet temporary increases in demand is obtained by withdrawing aquifer reserves.

Based on the desired use of aquifer reserves, three strategies for resource management can be identified. (1) Maintain the short-term dynamic equilibrium of the aquifer system layer (managing flow alone). Because of the rapid side effects of such operations, this objective is best suited to the pumping of low-capacity, unconfined groundwater aquifers. This is especially true of operations in alluvial aquifers that readily generate "new" waters and make it possible to restrict operations to drawing on the surplus over normal yield without disrupting equilibrium (operating in overequilibrium). (2) Maintain long- or medium-term equilibrium (relying on the multiannual regulatory potentials of aquifer reserves in flow management). This objective is often

TABLE I Types of Groundwater and Management Conditions

Types of groundwater	Connection with surface water	Sensitivity	Appropriate operating strategy (cf. Infra 2)	Limiting factors
Alluvial groundwaters (essentially unconfined) of limited size and reserve	Closely interfaced; groundwater/river exchanges possible in both directions	Sensitive to river works and soil use; vulnerable to pollution in surface soil and neighboring watercourses	(a) Short-term equilibrium; overequilibrium possible because of river recharge; flow management	External constraints; bank clogging; maintaining surface compensation flow
Low regulatory potential Plain and plateau subsoil waters and karst networks; huge reserve (usually >10 × annual flow)	Generally one way; river drainage (except in the case of possible river losses in plateau karsts); varying densities, depending on hydrographic structure	Varying yields; sensitive to drought; sensitive to land use; vulnerable to pollution, especially nonpoint	(b) Pluriannual equilibrium; flow management; regulatory storage management	External constraints; maintaining spring and watercourse compensation flows
High regulatory potential; high hydrodynamic inertia				
Interconnected subsoil and semiconfined waters (multilayer systems); huge partly regulatory	Varying degrees of connection with high unconfined waters	Sensitive to soil use (high unconfined waters); varying degrees of sensitivity to pollution;	(b) Final equilibrium after a possible phase of prolonged depletion of reserve; flow and reserve management	External constraints (high unconfined waters); maintaining spring and watercourse

reserve (>100 to 1000× annual flow); varying inertia	risk of interexchange between groundwaters of different quality as a result of drilling	evolving toward flow management alone	compensation flows; internal constraints (semiconfined groundwaters); maximum acceptable drawdown
Unconfined groundwaters with huge reserve (>100× annual flow) in arid regions; high inertia	Insignificant (no permanent hydrographic network)		
	Low sensitivity to climatic changes; sensitive to pollution, especially nonpoint	(c) Overall disequilibrium; longer lasting reserve after stabilization of withdrawals reserve management	Internal constraints; maximum acceptable drawdown; risk of attracting brackish waters
Deep unconfined groundwaters; low primary reserve; huge "secondary" reserve constituted by associated aquitards or by contiguous unconfined groundwater aquifers (>1000 and >10,000× annual flow); nonregulatory reserve; low hydrodynamic inertia	Independent (except in the case of artesian springs)		
	Not sensitive to climatic conditions; not sensitive to surface pollution; risk of interexchange between groundwaters of different quality as a result of drilling	(c) Overall disequilibrium (final equilibrium in some sectors); longer lasting reserve after stabilization or through the modulation of withdrawals; reserve management and, sometimes, flow management in particular areas (near the open limits)	Internal constraints; maximum acceptable drawdown; risk of attracting brackish waters

sought after an initial phase of prolonged imbalance due to increasing withdrawals. This situation is the case in most extended unconfined and semiconfined groundwater or multilayer systems. (3) Accept long-term disequilibrium even after draw-off has been stabilized. This involves the management of conservation–storage alone and is feasible only with deep unconfined groundwaters as well as unconfined groundwater with huge conservation–storage in arid regions.

It should be noted that the search for equilibrium in strategies 1 and 2 nearly always implies a reduction in the natural aquifer flow, and this flow is normally equal to the volume of draw-off. Targeted dynamic equilibrium cannot be fixed on the basis of overall aquifer yield, because this would mean the complete depletion of all flow sources, but rather on the basis of an acceptable reduction in this yield coupled with restrictions as may be imposed by the targeted level of conservation of the resource (conservation yield).

B. Is Overexploitation a Case of Disruption of Equilibrium or One of Excessive Pumping?

From a strictly semantic point of view, overexploitation denotes pumping in excess of a maximum established criterion. Pumping would be considered excessive if it causes negative, long-term effects. Whether or not an operation is considered overexploitation depends solely on the criterion chosen for such a judgment. This criterion is usually chosen to reflect resource operation objectives for a particular aquifer. Some of the objectives that may contribute to determining criteria for such judgments include: preserving existing favorable operating conditions, increasing the operation as part of a development project in which water supply is important, and reducing, to the barest minimum possible, negative impacts of operations on surface waters or land uses.

The criterion may be purely physical and quantitative, to maintain hydrodynamic equilibrium and guarantee recharge of resources; qualitative, to meet the need to guard against deterioration in water quality resulting from operations; economic, to maintain profitability, especially in comparison with other sources of water supply, or, more broadly speaking, to reduce direct costs (borne by the operator) and indirect costs (borne by others) that together may exceed overall benefits to all parties involved (however, indirect costs may not easily translate into monetary terms, making them difficult to compare with other costs); social, to reduce problems resulting from conflicting standards of operators of different social standing or damages caused to third-party users of surface water and groundwater; or environmental, to prevent damage to the natural environment, particularly to fragile aquatic ecosystems.

For the water resource manager, overexploitation is essentially a function of physical and hydraulic criteria, especially disruption of equilibrium (demand in excess of supply). For the operator or the analyst, socioeconomic criteria alone count, and reference is constantly made to undesirable or unacceptable operation consequences (water scarcity).

Maintaining favorable operating conditions may be more or less tantamount to maintaining production in a state of dynamic equilibrium, but not necessarily maintaining a feasible maximum operating state without disrupting equilibrium. The need to strike a balance between intensive operations and dynamic aquifer equilibrium by adjusting draw-off to the mean flow yield or the famous "safe yield" arises only in those situations in which operation objectives are both to maximize production and to have it run for as long as possible. Only then would overexploitation refer to a case of inequality in which withdrawals would be greater than the mean flow yield. In other words, overexploitation would correspond to an operation index of greater than 100% and to an operation not only in disequilibrium but also in excess of flow.

Making *a priori* judgments or predictions of overexploitation requires that the analyst has access to accurate data on the mean flow yield and total withdrawal, which are necessary for useful comparisons. Given the required precision of data, this condition is difficult to meet for either the mean flow or total withdrawal. Water budgets are often based on estimates of data and are thus only approximate measures of disequilibrium. In practice, overexploitation diagnoses are made *a posteriori* on the basis of observed symptoms of prolonged imbalance—continued drawdown, possible effects on boundary flows, and water quality.

This is not a refusal to admit that the problem of overexploitation is real or a refusal to take public outcry seriously. Rather, it is saying that when judgments and the generalizations they give rise to are based on the disruption of dynamic aquifer equilibrium alone, they are of limited use. At best, they lead to an overly simplistic account of the phenomenon by comparing operations based on safe yields as determined by the safety objectives of the operator. At worst, they lead to unfounded overexploitation diagnoses and, in practice, to a Malthusian policy of "underexploitation."

C. Equilibrium and Disequilibrium in Groundwater Operations

An operation in disequilibrium, as characterized by drawdowns and a reduction in the conservation–storage level, is not necessarily one in which withdrawal is in excess of the aquifer natural yield. Instead, disequilibrium may result from a mere increase in withdrawal, whether or not it is higher than the natural flow. It may also result because the effects of rising or stabilized withdrawal on boundary flows (possible factors for a return to equilibrium) have not been completely attained. In large aquifers with high hydrodynamic inertia, intensive withdrawals, even when stable, may require considerable time to return to equilibrium. In addition, operation effects add to the natural dynamics of an aquifer, and any "water balance" judgments must consider a sufficiently long period of time. In the case of unconfined aquifers, this may be several years. Proving that disequilibrium results from groundwater operations requires the analyst to work on a longer time scale than that necessary to determine the aquifer natural mean flow. Such disequilibrium is not simply a result of the inequality

$$P > Q_a \tag{1}$$

but rather the inequality

$$P > Q_d + Q_b, \tag{2}$$

where P is the sum of mean flow withdrawals, Q_a is the mean flow yield, Q_d is the induced reduction in the mean discharge flow (spontaneous aquifer yield), and Q_b is the induced increase in the mean recharge flow. Both cases imply a reduction in aquifer storage.

Balancing groundwater withdrawals is feasible only in the long term. Short- or medium-term disequilibrium does not always result from operations in excess of the mean yield. If we reduce the diagnosis of overexploitation to the single criterion of disequilibrium, then, in the above-identified strategies, strategies 1 and 2 would give inaccurate diagnoses and strategy 3 would not fit the criterion.

Thus, the concepts of "excessive withdrawals" and "unbalanced withdrawals" must be kept separate. An operation in disequilibrium is not necessarily one that is excessive and vice versa. If overexploitation is defined as operations that are both excessive and in disequilibrium, this would overrestrict its meaning. On the other hand, generalizing the definition to include all cases of operation in disequilibrium would be the other extreme (unless we gave the term a new meaning without the negative connotations of withdrawing aquifer reserves). But, as explained above, this is but one aspect of the role assigned to reserves in aquifer operations. Therefore, there is no way of overcoming the semantic dilemma posed by the use of these terms (Table II). Is overexploitation a case of $C + D$, $B + D$, or simply D?

From the viewpoint of groundwater resource management seeking to strike a fair balance between operator and economic interests and the need to conserve surface and groundwater resources, overexploitation would designate a situation in which withdrawal causes serious negative consequences for users or for the environment, *i.e.*, excessive under user and environmental criteria $B + D$.

TABLE II   The Semantic Dilemma

Operations	Not excessive	Excessive
In equilibrium	A	B
In disequilibrium	C	D

IV. GROUNDWATER MANAGEMENT

A. Socioeconomic Considerations

Aside from the physical conditions and constraints outlined above, a number of socioeconomic considerations must also be balanced in groundwater management. If groundwater resources are substantial and accessible, there are always influences (financial, political, and legal) that are felt in groundwater operations. These operations include private uses (farming and drinking water), industrial uses (manufacturing and underground injection), and public uses (drinking water and power utilities). There are many interested parties that can exert a variety of influences on groundwater.

• Land users, in addition to their political influence on groundwater use, cause the more direct effects of effluent discharges, agricultural practices, dredging, mining, etc.
• Projects related to watercourses that are connected to groundwater, including construction of dams, flood control structures, embankments, etc.

All groundwater is, in a sense, the common property of several different parties. The different individual operations of these parties, as well as any other actions capable of having importance for groundwater quality and quantity, often have different objectives. Although their dependence on the same groundwater provides a reason for them to work together, these parties are usually not aware of their participation in the *de facto* management of this common groundwater (*i.e.,* coproprietorship neither found in law nor governed by any form of mutual agreement). When the various uses of groundwater cause conflicts for other water and land uses, groundwater operators do not show a collective sense of responsibility, nor a willingness to accept it. They do not see their use in terms of overall adverse impacts to the aquifer.

Opposition arises between the indivisibility of the aquifer and the multiple parties with different and sometimes conflicting interests, and this leads to the need for groundwater management. Management of groundwater by an ad hoc authority invested with decision-making powers can be achieved only by influencing the actions of that authority with different forms of pressure and inducements. In contrast to the management of dam waters or mineral deposits by the operators and project managers with the required technical expertise, groundwater management is necessarily indirect in nature. This suggests that groundwater management is often constrained by the need from within to create conditions for the peaceful coexistence of groundwater users, those involved in withdrawals and in wastewater discharge, and the need from without to reconcile the rights and interests of other users of surface waters and various land users for whom groundwater operations may have negative consequences.

Finally, groundwater management cannot be conceived of separately from management of surface waters. It cannot be said too strongly that surface waters and groundwater *together* form the indivisible water resource system, as illustrated by the hydrologic cycle. Because of the interdependence of surface water and groundwater, operations on any part of the system have consequences for the other parts. It should be remembered that both groundwater and surface water are used as a source of drinking water and as a receptacle for waste disposal. The uses of groundwater, which influence management objectives, depend largely on the degree of choice offered operators in terms of their location and targeted water quantity and quality. These choices are usually made following microeconomic criteria defined by the different agents and sometimes management and/or environmental macroeconomic criteria.

There is, therefore, an interplay among conditions related to abundance, accessibility, and quality of groundwater and surface water and the variety of demands for water and activities that are capable of affecting water in various ways. This creates a great variety of situations that require management.

B. *De Facto* Management Limitations

The need for management barely arises as long as groundwater operations do not involve more than the extensive withdrawals of water, the resultant side effects cause minimum conflicts among operators, and the groundwater regime and quality are not seen to be adversely affected. This *de facto* management goes hand in hand with the lack of awareness that groundwater is the collective property of all involved parties. *De facto* management can continue only as long as conflicts between interested parties, including operators, land users, and those in favor of maintaining the surface water and groundwater quality and regime, are not a problem.

The need for deliberate and concerted management of groundwater resources arises when conflicts exist or are looming ahead; the extent of the physical aquifer system is known and recognized as a separate management unit; or groundwater operators come to realize that their fate is tied to activities in a watercourse because the groundwater directly interacts with the surface water.

C. Management Objectives

Groundwater management objectives vary and reflect varying degrees of compatibility. It is the task of the manager to define and prioritize these objectives.

1. Prevention and Arbitration in Conflict Situations

The first step in groundwater management is to create conditions conducive to the peaceful coexistence of the different groundwater users. When the direct and indirect demands for groundwater increase and diversify, conflicts will likely increase, and the manager can arbitrate only with reference to clearly defined priorities that also reflect socioeconomic objectives. The same can be said for the use of groundwater that impacts surface water and vice versa.

Generally speaking, a primary management objective should be to prohibit activities as necessary to prevent damage to the environment, including groundwater, surface water, and riparian zones.

2. Conservation of Productivity and Resources

Conservation of productivity relates to accessibility and production conditions at a particular stage of operation. This can be accomplished by prohibiting increases in the production level, which may entail allowing established users to continue operations and refusing withdrawals by new users.

Resource conservation, in terms of quality and quantity, is an objective with a less pronounced Malthusian dimension. Quantity-wise, it means preventing overexploitation in the sense of long-term unbalanced operations. This includes any form of excessive operations, including those that transform surface water and land conditions. Quality-wise, resource conservation means protecting groundwater from all forms of pollution or the side effects of intensive operations, such as saltwater intrusion or intrusion of polluted waters. In summary, conservation means recognizing the status of groundwater as a common heritage and the need to protect it for sustainable yields and long-term uses.

3. Resource Allocation

The selective allocation of groundwater resources to high-priority sectors established on the basis of general public interests (*e.g.*, public drinking water supply) is a rather interventionist objective. It may necessitate a system of zoning similar to a resource distribution chart and, if need be, establishing water quantity zones, as is the case with hydromineral basins. It may also reserve part of the resources as well as protect their quality for future needs, including the needs of receiving springs or rivers, or even their protection as aquatic ecosystems (ecological objective). In the case of aquifer conservation—storage management (deep confined groundwater), there is the need to arbitrate between immediate/short-term needs and long-term future needs.

4. Development

Where resources are thought to be underutilized, intensifying operations to meet the needs of certain sectors, especially as an alternative to less economically viable sources, may constitute a macroeconomic objective, *i.e.*, not neglecting any factor of economic development. Allocation would then be based on the argument of the most advantageous use, given the specificity of the particular groundwater in comparison with other available sources of water supply and the existing economic conditions.

Naturally, conservation-oriented management is not compatible with management that seeks to stimulate a development project necessitating major transformations and redistribution of available resources. Such compatibility problems only underscore the critical need for management to identify priority objectives on which to base decisions.

D. The Future of Groundwater Management

Groundwater management is here to stay. Future progress seems to lie in its complete integration into the overall management of other water resources and into urban and country planning.

1. A Shift from Potential Management to Real Management

Groundwater management has barely passed the stage of policy statements and guidelines, rather than actual implementation with enforcement capability. It can be said that those who have taken up the issue of groundwater management have until now been content with suggesting what these guidelines should be and who should be responsible for implementing them. That, however, is still not groundwater management, it is potential management.

In the future it is hoped that groundwater management will be more concerned with real management and that it will involve all the affected and concerned parties. This will require a higher level of awareness on the part of private individuals as well as other users of groundwater. It also will require clearly defined general objectives and priorities and the institution of appropriate management' authorities.

2. A More Integrated Management Style

It will no longer be sufficient to define management objectives to satisfy only the operators and direct users of groundwater. Consideration must also be given to broader objectives that recognize groundwater as a resource that serves many diverse functions, including an ecological function. In some cases, this may require a complete reordering of priorities.

• Maximizing intensive groundwater withdrawal instead of drawing from receiving watercourses, especially where the surface waters are of poor quality.

• Producing water of high quality by imposing strict observances on land users.

• Maintaining, on a sustainable basis, the surface conservation yield and the quality of surface waters interacting with groundwater. This may mean including groundwaters that form the upstream part of a watercourse in the conservation yield and defining water quality standards.

• Allowing only those land uses in selected areas that have acceptable effects on groundwater regime and quality.

• Dewatering particular zones of subsoil for other uses (mining, quarrying, underground construction projects, etc.), necessitating the transformation of groundwater distribution or the groundwater regime.

These priorities can be achieved only through proper consultation with the involved parties and a thorough explanation of the need for any proposed actions. In some cases, government intervention or arbitration may be required.

3. New Challenges

Future groundwater management is likely to evolve toward more total aquifer management. This management will not only oversee pumping of groundwater, but also be involved in conservation–storage and aquifer recharge, including limitations on activities in recharge areas that could contaminate critical sources of groundwater. Future activities may include the following.

• Extension of operating strategies, as described in strategy 2, and reliance on the regulatory potentials of some reservoirs to make up for the deficit in surface seepage during drought periods (substituting resources and compensation flow). This may be achieved through the use of springs with different drawdown levels or through the construction of underground dams. The regulatory functions of the enormous potential of aquifers, not only as an alternative but also as a complement to surface retention, will have to be relied on more often in the effort to manage the irregularity of surface water flow. This will be especially true in those countries in which the available surface storage is in constant and rapid decline due to the accumulation of silt.

• Increased use of artificial recharge within the framework of the strategy outlined in the preceding paragraph.

• Use of the self-purification potentials of some aquifer systems in the management of water quality, especially in wastewater reuse.

• Reservation of some areas exclusively for the production of high-quality water. This implies limiting land uses, such as agriculture, and will have to be developed as part of an integrated policy on agricultural, urban, and rural development. In the case of the European community, such a measure could help address the problems posed by the "land freeze" used to regulate agricultural production. Put differently, such a policy would mean extending the sanitary zone of drinking water wells and seeking to create what G. De Marsily (1991a,b) so aptly describes as "hydrogeologic natural reserves." This does not mean, however, sacrificing other groundwaters, but rather viewing them within the framework of negotiable water quality objectives.

• A nonrenewable resource management style, as described in strategy 3, may become necessary for countries in arid regions in which renewable water resources are becoming increasingly scarce. Long-term objectives should reinvest profits accruing from existing water operations for future development in water management.

ACKNOWLEDGMENT

The author greatly appreciates the time that J. Simons (Environmental Protection Agency, Washington, D.C.) took to improve and edit the manuscript.

REFERENCES

Bodelle, J., and Margat, J. (1980). "L'eau souterraine en France," Les objectifs scientifiques de demain. Masson, Paris.

Collective book (1983). "Groundwater in Water Resources Planning." UNESCO, IAH, IAHS, Natl. Comm. Fed. Rep., Germany.

Collective book (1990). "L'eau souterraine, un patrimoine à gérer en commun," Vol. 1, Doc. du BRGM, N°. 195. Colloque Paris-Orléans.

Collective book (1991a). "Pollution des nappes d'eau souterraine en France," Rapp. N°. 29. Académie des Sciences, Paris.

Collective book (1991b). Les eaux souterraines et la gestion des eaux. *Soc. Hydrot. Fr., C. R. Journ. Hydraul., 21st,* Sophia-Antipolis.

Collective book (1991c). "La gestion des eaux souterraines: Préparation d'un séminaire Ministériel." Inst. Eur. de l'Eau, Como, Paris.

Collective book (1991d). "Sustainable Use of Groundwater: Problems and Threats in the European Communities." RIVM, RIZA / Pays-Bas.

De Marsily, G. (1991a). La gestion des eaux souterraines. *Soc. Hydrot. Fr., C. R. Journ. Hydraul., 21st,* Sophia-Antipolis, Question III.

De Marsily, G. (1991b). Création de "parcs naturels hydrogéologiques." Minist. Environ., REED/SRETIE.

Margat, J. (1973). L'utilisation des eaux souterraines. *In* "Encyclopédie de la Pléiade," Vol. II, pp. 1057–1092. Gallimard, Paris.

Margat, J., and Saad, K. F. (1983). Concepts for the utilization of non-renewable groundwater resources in regional development. *Nat. Resour. Forum* 7(4), 377–383.

Margat, J. (1985). Groundwater conservation and protection in developed countries. *Mém. Congr. Hydrogeol. Serv. Man, 18th,* Part 1, pp. 270–301.

Margat, J. (1987). "La ressource en eau souterraine revisitée. De sa définition à son évaluation et à sa gestion," Doc. BRGM 87 SGN 524 EAU.

Margat, J. (1989). Défense et illustration des eaux souterraines en Europe. *Hydrogéologie* 2, 75–91.

Margat, J. (1990). Les gisements d'eau souterraine. *Rech. Spéc. Eau* 221, 590–596.

Margat, J. (1991a). Les eaux souterraines dans le monde. Similitudes et différences. *Soc. Hydrot. Fr., C. R. Journées Hydraul. 21st,* Sophia-Antipolis, Conf. N°. 1.

Margat, J. (1991b). La surexploitation des aquifères. *Proc. Congr. Int. Assoc. Int. Hydrogéol., 22nd,* Puerto de La Cruz, Canarias, Espagne, pp. 1–12.

Ollagnon, H. (1979). Propositions pour une gestion patrimoniale des eaux souterraines: L'expérience de la nappe d'Alsace. *Bull. Interminist. Rationalis. Choix Budgét.* 36.

20

Ecological Basis for Management of Groundwater in the United States: Statutes, Regulations, and a Strategic Plan[1]

C. A. Job and J. J. Simons

U.S. Environmental Protection Agency
Groundwater Protection Division
Mail Code 4602
401 M Street, S.W.
Washington, D.C. 20460

[1] The views expressed in this chapter are solely those of the authors and may or may not be endorsed by the U.S. Environmental Protection Agency.

I. INTRODUCTION

Groundwater quality management requires balancing human use of the resource with protection of the natural self-purifiation processes of groundwater ecosystems. Reliance on groundwater as a drinking water source is growing; in the United States it has risen from 50% in 1984 to 53% in 1986. Moreover, 40% of the average annual stream flow in the United States is groundwater discharge (U.S. Geological Survey, 1986). In humid areas of the United States, 90% or more of stream flow may be groundwater discharge. With increasingly stringent statutes and requirements and the escalating costs of monitoring groundwater quality, managers are searching for indicators and metrics that will enable them to monitor changes in the quality of the groundwater environment in a more comprehensive and economical way. Biota, chemical constituents, and physical characteristics of the relatively stable groundwater environment likely can be effective variables for monitoring groundwater quality and influences on surface water (Intergovernmental Task Force on Monitoring Water Quality, 1992).

In this paper we provide an overview of U.S. statutes designed to protect and enhance groundwater resources. We also discuss monitoring needs in the context of ecological indicators of groundwater quality. Finally, we present a strategic plan for integrating principals of groundwater ecology into a management framework. Our discussion is primarily based on experience in the United States. However, owing to global pressures and increasing reliance on groundwater (Chapter 19), we believe the rationale and recommendations for dealing with these pressures also have worldwide application.

II. OVERVIEW

A. Scientific Basis and Direction

A fundamental aspect of groundwater management is the recognition of flow pathways, rates of flow, and the differential retention of materials contained in the water. These are key characteristics of the porous media of aquifers. Certain pollutants, such as pesticides, exhibit mobility and persistence in groundwater. For example, five pesticides, Atrazine, Simazine, Cyanazine, Alachlor, and Metolachlor, commonly found in groundwaters are also associated with serious toxicological effects (40 CFR 152 Subparts I and J, proposed regulation, U.S. Environmental Protection Agency, Washington, DC, 1993). Furthermore, certain petroleum-based products, such as organic solvents, called dense nonaqueous phase liquids (DNAPLs), are gravity driven and not flow driven. They may move in directions opposite to groundwater flow and may even migrate vertically (e.g., through fractures in bedrock; Huling and Weaver, 1991a,b). Once pesticides, DNAPLs, and other hazardous chemicals reach the saturated zone (e.g., due to accidents or poor disposal practices), they may resist washout and breakdown and can cause significant, long-term contamination in groundwater that may be impossible or very expensive to clean up (Environmental Protection Agency, 1993a). Indeed, in the United States fewer than 200 of more than 2000 priority hazardous ("Superfund") sites have been remediated and declared safe, in spite of over $30 billion spent for this purpose in the last two decades. Groundwater contamination is a major problem at most of these sites.

Groundwater environmental research has to a great extent focused on determination of safe sources of drinking water and, if the sources were contaminated, how they could be made safe to drink. This predominant emphasis on human health and the possible ingestion of harmful chemicals has, until recently, diverted attention from the larger issues of how the subsurface environment functions and how the quality of that environment affects other important resources. Groundwater too often is viewed as a physically isolated resource, even though hydrologists and ecologists have long recognized the importance of groundwater and surface water interactions and the vital role of groundwater in the hydrologic cycle.

Humans have long known that bacteria can exist deep below the surface of the Earth (Chapter 7), and many animals have been found in cave streams (Chapter 1). Over the last 20 years, researchers in the United States and Europe have identified numerous macroscopic invertebrates in shallow groundwater, particularly in the hyporheic zone (the interstitial volume penetrated by surface water) of streams, (Chapters 12–15). Physical, chemical, and biological processes in the hyporheic zone are now thought to maintain or, in some cases, improve the quality of stream water (Duff and Triska, 1990; Gibert *et al.*, 1990). Because groundwater and its biochemical constituents typically move slowly and because microbial metabolism is usually low, groundwater quality remains relatively constant over the long term. Therefore, the presence or absence of different assemblages of organisms may indicate groundwater quality. That is, these organisms could serve as "indicators" of the quality of this resource in specific aquifers.

Additionally, many river ecologists now think of rivers as including both the surface water and the zones under and along rivers in which surface water and groundwater exchange occurs repeatedly (Amoros *et al.*, 1987;

Ward, 1989). This concept should be of great interest to managers because land-based disposal methods (e.g., settling ponds), which are commonly used in the United States and elsewhere, may often pollute the zone of groundwater that is in constant flux with the river. This is analogous to directly polluting the river itself. Management of river corridors requires an updated perspective that includes recognition of hyporheic zones. In many cases setback zones to protect rivers may not be adequate, and some activities that currently are permitted in protected areas may need to be reexamined.

B. Statutory and Regulatory Directives in the United States

The U.S. Environmental Protection Agency's role in water resource protection was created by the U.S. Congress. In 1972 the U.S. Congress enacted the Federal Water Pollution Control Act (33 U.S.C 466 *et seq.*), which was later renamed the Clean Water Act as amended by the Water Quality Act of 1987 (Public Law 100-4). The Clean Water Act is the major U.S. law relating to protection of water resources. Through the years the court's interpretation of this law has consistently upheld water quality measures required by the Act. The Act's stated objective is "to restore and maintain the physical, chemical, and biological integrity of the Nation's waters," but the focus is on surface water. It is clear from the legislative history of the Clean Water Act that Congress never intended the Act to set standards for groundwater. The added complexity of many different state groundwater laws discouraged this. For example, states make distinctions among various types of groundwater based on characteristics such as flow, and in some cases, for legal purposes, groundwater is considered surface water. In some states in which groundwater is important in maintaining the quality and quantity of the stream base flow, groundwaters are considered surface waters. However, the treatment of groundwater as it relates to surface water in state water law overall is not consistent and does not recognize the important interstate nature of the groundwater resource, which will be addressed below (Goldfarb, 1988).

Under the Clean Water Act, Section 304, the Environmental Protection Agency was given authority and direction to produce water quality criteria "(A) for the kind and extent of all identifiable effects on health and welfare including, but not limited to, plankton, fish, shellfish, wildlife, plant life, shorelines, beaches, esthetics, and recreation which may be expected from the presence of pollutants in any body of water, *including groundwater;* (B) for the concentrations and dispersal of pollutants, or their byproducts, through biological, physical, and chemical processes; and (C) for the effects of pollutants on biological community diversity, productivity, and stability, including information on the factors affecting the rates of eutrophication and rates of organic and inorganic sedimentation for varying types of receiving waters" (emphasis added). Although groundwater has been addressed in changes to some regulations, further attention is needed to fully capture the current knowledge of groundwater and surface water interactions.

Although it appears that the current statutory and regulatory framework addresses ecological effects, many aspects of these ecological relationships in fact are not considered. The primary deficiency is that groundwater and surface water are treated as distinct resources with no physical or biological connections. This is particularly noticeable in the Clean Water Act, which defines "waters of the United States" as "navigable waters." Although this definition has as its basis the "commerce clause" of the Constitution, Section 8, relating to Federal authority over interstate commerce, current understanding in the field of hydrology indicates that groundwater is responsible for a significant portion of stream flow, and, during drought conditions in some locations, groundwater may actually maintain the entire flow (Environmental Protection Agency, 1993b).

For example, during the Midwestern drought of 1988, a significant portion of the flow of the Mississippi River and its tributaries was maintained by groundwater discharge. This discharge allowed interstate commerce (barge and recreational traffic) on these streams past the time when it otherwise would have been possible if maintained only by surface flow and runoff. Thus, groundwater and the condition and extent of the saturated zone have a significant effect on the navigable waters of the United States, as well as the sustainability of intermittent and ephemeral streams, associated wetlands, and lakes. This groundwater discharge also affected the quality and ecology of these surface waters. Yet, this relationship is largely unrecognized in the laws and regulations directly affecting these resources, and the legal concept of navigable waters, including groundwater connected to surface water, is not clearly resolved. Furthermore, in many cases, actions taken to mitigate nonpoint sources of pollution through best management practices (BMPs) may only delay surface water impacts and intensify local groundwater quality conditions. This is because most BMPs are meant to reduce surface water pollution by increasing the residence time of water on the land and thus increasing infiltration and possible groundwater contamination.

The current direction in protecting ecological systems associated with groundwater and groundwater interaction with surface water is in the Environmental Protection Agency's policy, "Protecting the Nation's Groundwater: EPA's Strategy for the 1990's" (Environmental Protection Agency, 1991). This strategy indicates that both public and private drinking water supplies should be protected to prevent adverse health risks for present and future generations, and groundwater closely hydrologically connected to surface waters should be protected so as not to interfere with the attainment of surface water quality standards, which is necessary to protect the integrity of associated ecosystems. The Strategy recognizes the states as having the primary role in groundwater protection.

A major step in implementing the Strategy was the issuance of the "Final Comprehensive State Groundwater Protection Programs Guidance"

(Environmental Protection Agency, 1992). Under Section 106 of the Clean Water Act, the Environmental Protection Agency provides funds to states for water quality programs, including comprehensive groundwater protection programs. The guidance for these state programs emphasizes an integrated approach, including groundwater remedial programs. Based on the Guidance, state programs address six strategic elements: (1) establishing a groundwater protection goal; (2) establishing priorities based on resource characterization (including groundwater's relationship to surface water quality), sources of contamination, and program needs; (3) defining roles, responsibilities, resources, and cross-program coordinating mechanisms; (4) implementing efforts necessary to accomplish the state's goals; (5) coordinating information collection and management, including monitoring; and (6) improving public education and participation. Although these state programs can incorporate groundwater and surface water information and their ecological relationships, the states need research and technical assistance in identifying the significance of these relationships for their water management programs.

We conclude that the role of groundwater in the hydrological cycle, including groundwater/surface water interactions (e.g., in hyporheic zones and wetlands), is not effectively addressed in principal U.S. water resource protection laws. It remains to be seen to what extent these deficiencies will be addressed in reauthorization of the Clean Water Act in 1994. Recognition of groundwater and surface water interaction in the Clean Water Act would augment the ability of public water systems to be in compliance with the maximum contaminant levels at the point of drinking water use established under the Safe Drinking Water Act. Several other laws affecting water resource protection will be discussed below.

III. CURRENT REGULATORY FRAMEWORK

A. Clean Water Act

1. Overview

Even though the Clean Water Act does not mandate groundwater quality control nor incorporate groundwater and surface water interaction in its regulatory program, the Act does recognize the significance of the groundwater resource. Section 102 of the Clean Water Act states that the Environmental Protection Agency will work with other federal, state, and interstate agencies and municipalities and industries to "prepare or develop comprehensive programs for preventing, reducing, or eliminating the pollution of the navigable waters and groundwaters and improving the sanitary condition of surface and undergroundwaters. In the development of such comprehensive programs due regard shall be given to the improvements which are necessary to conserve such waters for the protection and propagation of fish and aquatic life and wildlife, recreational purposes, and the withdrawal of such waters for public water supply, agricultural, industrial, and other purposes." This language points the way for water quality management that recognizes groundwater and surface water relationships, including uses that are affected by their related biotic communities, which respond to as well as maintain water quality.

2. Water Quality Control and National Pollution Discharge Elimination System

The focus of the Clean Water Act has been primarily to address surface water quality. Section 302 of the Act provides for a National Pollutant Discharge Elimination System (NPDES), which identifies the public water supply, agricultural and industrial uses, the propagation of fish and wildlife, and swimming as the major activities affected by water quality. The basic thrust of the limited case law is that groundwater contamination and surface water pollution from diffuse (nonpoint) groundwater sources are not covered by the Clean Water Act (Smith, 1991). Including the quality of groundwater discharged to streams will ultimately affect the management of both point and nonpoint sources of pollution to surface waters. Although benthic organisms are used as comprehensive indicators of surface water quality (Resh and Jackson, 1992; Rosenberg and Resh, 1993), the equivalent is not true of groundwater biota. Riverine invertebrates have been found in alluvial aquifers over 2 km from a stream channel in Montana. (Stanford and Ward, 1988). Other investigations indicate that this likely is a widespread phenomenon in river systems in North America and elsewhere (Stanford and Ward, 1993; see also Chapter 14). These organisms have the potential to be used as indicators to delineate the extent of groundwater/surface water interaction. They also have the potential to be used as predictors of the future base flow quality of the stream. Bioassays of various classes of pollutants using groundwater organisms also could help establish more effective evaluations of groundwater quality.

The Clean Water Act also requires states to identify those waters for which technology-based effluent limitations are not sufficiently stringent to attain specific water quality standards. The technology-based limits are mandated under Sections 301 and 307 of the Act. The Environmental Protection Agency promulgates industry-specific effluent guidelines. The states must each rank their water-quality-limited waters and set total maximum daily loads (TMDL) of pollutants that will meet the standard. Finally, the TMDLs are to be used to compute load allocations. Best-management practices tailored for a stream or watershed, to be effective, should account for the extent of overland and groundwater flow to streams. If groundwater is not integrated in this process, circumstances in which nonpoint source loads to streams, in the longer term, may increase because groundwater was contaminated as a result of using best-management practices may exist.

3. Nonpoint Source Pollution Control Program

The Environmental Protection Agency's nonpoint source program, under Section 319 of the Clean Water Act, is the primary vehicle by which the Agency provides technical and financial assistance to states to implement programs to control nonpoint source pollution. This program addresses both groundwater and surface water, with about 25–30% of grant funds being used to address groundwater problems each year.

In 1990 Congress enacted the "Coastal Zone Act Reauthoriazation Amendments of 1990," which requires states with approved coastal zone management programs to develop coastal nonpoint pollution control programs, including enforceable policies and mechanisms, to protect coastal waters from nonpoint source pollution. Congress directed the Agency to publish guidance specifying management measures for sources of nonpoint pollution in coastal waters. Although the focus of this legislation was on protecting coastal surface waters, the Agency recognized that hydrological connections between surface water and groundwater necessitated the inclusion of measures that would protect groundwater as well. Thus, the guidance contains a variety of measures that protect groundwater, including, for example, appropriate management of animal wastes; the proper application of fertilizers, pesticides, and irrigation water to farmland; and the control of septic tanks.

Nonetheless, the state of knowledge concerning the potential effects of some management practices on groundwater quality is not well developed at this time. Potential influences of surface water regulations (e.g., allowing specified pollutant loads) on groundwater organisms are particularly problematic. Further research on the relationship of surface water and groundwater quality and on the relative impact of management practices on each of these media is needed. Otherwise, the efforts to protect one medium may have unintended adverse effects on the other. Further, contamination of surface water may occur when polluted groundwater provides significant stream flow, in which case the intended regulations for surface water fail. Research could help identify which hydrogeologic environments are in need of protection from various chemicals and wastes, thereby guiding implementation of management practices to fit specific local conditions (Weitman, 1994).

4. Storm Water Controls

Storm water management typically promotes retention of water in ponds and reservoirs that allow infiltration of water through the soil to groundwater to reduce impacts to surface water quality. Storm water controls (Clean Water Act, Section 402 p), although recognizing possible groundwater impacts in regulation implementation guidance, have a problem similar to nonpoint source control when carried out in that more information may be needed on the trade-offs between groundwater and surface water pollution prior to the implementation of the controls. Therefore, these practices need the same type of research as nonpoint source controls.

5. Wetlands Protection

Under Section 404 of the Clean Water Act, wetland areas cannot be dredged and filled without the necessary permits. The implementation of this law principally revolves around the definition and delineation of wetland areas, which may include permanently or seasonally inundated areas. The Section 404 regulations define wetlands based on three factors: vegetation, soils, and hydrology. Most wetland areas result from groundwater discharge (upwelling) during a significant portion of the year, although wetlands may occur in zones of either groundwater discharge at an up-gradient point or groundwater recharge at a down-gradient point (Winter, 1989). Comprehensive protection of wetlands should include groundwater food webs because they can influence the quality of the water within the wetlands over time (Gibert *et al.*, 1990).

6. Other Clean Water Act Relationships

Under Section 304, the Environmental Protection Agency has developed bioassays intended to show the effect of pollutants on biological communities and serve as a more holistic indicator of surface water quality. Such bioassays could be applied to groundwater organisms in areas of significant groundwater discharge, affecting surface water quality.

Under Section 303, the states are required to set water quality standards that protect the public health or welfare, protect the environmental integrity, and serve the purposes of the Act for all waters in the state. These standards employ water quality criteria developed by the Environmental Protection Agency for the states and designate uses for each water body.

B. Safe Drinking Water Act

The Safe Drinking Water Act as amended in 1986 (Public law 99-339, June 1986) sets maximum contaminant levels for 83 chemicals in drinking water. These apply to water at the potable outlet (tap), which originates from either groundwater or surface water. The Act also requires States to establish wellhead protection programs for public water supply systems relying on groundwater. The Act also establishes a regulatory program for controlling injection of wastes underground. Healthy groundwater ecosystems should be sources of safe drinking water, but solid evidence is needed that demonstrates that protection of these ecosystems helps promote safe sources of drinking water.

The Wellhead Protection Program, established under Section 1428, provides guidance to states to establish programs to control contaminant sources, including microbial contamination, around public water supply wells used for drinking water. It is possible that organisms in groundwater could serve as valuable indicators of water quality before the water was allowed to enter the public water supply system.

Under Part C of the Safe Drinking Water Act, the Underground Injection Control Program assures that underground injection of fluids, including various types of wastewater, will not endanger drinking water sources and will be done in a manner that is protective of the public health and the environment. Underground injection activities include, among others, deep injection of hazardous and nonhazardous waste into isolated geologic environments; injection of solutions for subsurface mining; injection of fluids associated with oil and gas recovery; disposal of oil and gas production wastes (mostly brine); and shallow discharge of wastewater through drains, pipes, or septic systems. Groundwater organisms could be used to monitor the water quality of aquifers affected by these shallow discharges, especially when wastes may contain pathogenic organisms. The Sole Source Aquifer Program, under Section 1427, protects critical aquifers that are used as a sole or principal source of drinking water by a community. Any project receiving federal financial assistance and potentially affecting a sole-source aquifer must ensure that the source of drinking water will not be contaminated. If the project is not compatible with protection of the aquifer, it must be modified or it is not allowed. Again, groundwater organisms could be valuable tools to assist with this determination and to serve as indicators of the water quality of these aquifers. Groundwater organisms could perhaps be effectively used to indicate a safe drinking water source as well as monitor conformance to pollution standards after the project is developed.

Two regulations address treatment requirements for minimizing pathogen concentrations in drinking water. The draft Groundwater Disinfection Regulation would establish vulnerability criteria for drinking water wells with regard to sources of viruses and other pathogens. Source water not meeting the vulnerability criteria must be disinfected. Suggested criteria are set-back distances, hydrological factors, groundwater time of travel, and virus time of travel. The Surface Water Treatment Rule defines groundwater that contains pathogenic protozoa as surface water. These waters may have to be filtered as well as disinfected. The existence of surface water organisms in water pumped from the well is used as an indication of the quality of the water and whether treatment of the water is needed to make the well a safe source of drinking water.

C. Federal Insecticide, Fungicide, and Rodenticide Act

This law as amended in 1988 requires manufacturers of pesticides to perform a variety of tests to demonstrate effects of pesticides on the physical, chemical, and biological environment and to report the results to the Environmental Protection Agency. Although physical and chemical effects, along with impacts on vegetation and laboratory test animals, are routinely reported, effects on subsurface organisms are not recognized nor researched. For persistent pesticides, the impacts on these organisms may be critical. Because these chemicals have been used extensively, groundwater organisms in some areas may have been eradicated, not permitting an accounting of the natural subsurface biota. Under State Management Plans for use of pesticides in different hydrogeological settings, the role of organisms is not considered.

D. Other Laws Affecting Groundwater

Several other laws affect groundwater quality and ecology. The Resource Conservation and Recovery Act (RCRA) established programs to control and monitor for chemical releases from hazardous and municipal solid waste treatment, storage, and disposal sites, including underground storage tanks. When chemical and waste sites have contaminated groundwater—thereby affecting the local subsurface ecology—monitoring is done to determine when corrective action for active RCRA-regulated sites or remediation under the Comprehensive Environmental Response, Compensation, and Liability Act (i.e., the Superfund cleanup program) has achieved its goals. Various laws and regulations related to radioactive waste disposal also provide for monitoring near waste sites. All of these programs could benefit from research on the use of organisms and other groundwater parameters as indicators of water and ecosystem quality. Bioremediation efforts would also be enhanced through this proposed work.

IV. MONITORING

A. Federal, State, and Local Government

Humans' increasing use of chemicals combined with increasing use of groundwater suggests that additional monitoring of groundwater quality is necessary to adequately manage and provide safe drinking water. Most increases in the use of groundwater for drinking water occur near the fringe of urban or developing areas, the same areas in which chemical use and disposal are also increasing. Additionally, as populations increase, more groundwater is needed to support greater agricultural production. Groundwater resources are finite, and, clearly, accurate knowledge of trends in the quality and quantity of groundwater is of strategic importance to countries worldwide. Traditional methods of monitoring groundwater quality not only are expensive, but also have significant limitations. For example, values often change in the time between field sampling and laboratory analysis, and analytical methods may be insufficient to allow detection of trace amounts of pollutants that are harmful at very low concentrations. In the United States federal, state, and local governments spent an estimated $71 million in 1990 on ambient water quality monitoring (principally for surface water;

Intergovernmental Task Force on Monitoring Water Quality, 1992). Under the Clean Water Act, the purpose of monitoring the quality of surface water is to ensure that in-stream or lake water quality is within standards that are based on designated uses for the respective waters. Monitoring the quality of drinking water is typically done after water treatment and must indicate that the water is below "maximum contaminant levels." These are levels judged to be safe for human consumption under the Safe Drinking Water Act or by state statutes in the rare cases in which more stringent local standards must be met.

Typically, federal maximum contaminant levels are used to establish the concentrations to which contaminant levels must be lowered at sites of groundwater remediation to ensure that water is safe for human consumption. Maximum contaminant levels are based on human risk factors. However, the effects of these or higher levels on groundwater organisms are not known. Groundwater organisms may be more or less sensitive to contaminants than humans. From analysis of abandoned hazardous waste sites (Superfund sites) in the United States, it is known that hypogean food webs, over a long period of time, can break down organic chemicals. The relationship of contaminant levels to the distribution and abundance of groundwater biota must be defined in the context of the extent to which these organisms can function to break down pollutants and maintain groundwater quality.

B. Groundwater Organisms and Water Quality

1. Indicators of Water Quality

Ecological relationships involving groundwater biota and their physical/chemical environment should be used to monitor groundwater quality in a more synthetic manner than is currently encouraged by management. Homeostasis mechanisms in groundwater ecosystems are highly evolved, owing to the relative biophysical constancy of the environment (Chapter 1). Therefore, the flux and fate of pollutants in groundwater may provide very direct metrics in terms of adverse impacts on affected flora and fauna. The subsurface environment may have a limited ability to moderate stress mediated by groundwater pollution. In general it is thought that because the groundwater environment is a stable system, the organisms are not highly adaptable to change. On the other hand, the hyporheic zone of streams is subject to frequent changes, and biotic assembledges are highly dynamic, which suggests ecological resiliency and resistance (Chapter 1). A better understanding of the biophysical interactions that control the flux and fate of contaminants in groundwater is needed so that managers can help design effective groundwater protection strategies. However, some biota are ubiquitous in groundwater and should serve as fine biomonitors, once a thorough understanding of their life histories and resilience to pollutants is achieved.

2. Use across Federal and State Protection Programs

Because groundwater is addressed by nine major U.S. Environmental Protection Agency programs as well as those of other federal agencies, and similar state programs, groundwater organisms and chemical/physical parameters could be very useful monitoring tools in integrating and coordinating these many programs. The distribution and abundance of biota may provide a very practical and integral way to evaluate groundwater quality and, where hydrologically connected, surface water quality. Such use of groundwater organisms, similar to testing for standard chemical constituents in water, may provide relatively quick assessments of future water quality for public water supplies, for which testing for each suspected contaminant is extraordinarily costly. Changes in the distribution and population dynamics of organisms in groundwater may also suggest the degree of protection needed.

V. POLLUTION PREVENTION AS A MANAGEMENT FRAMEWORK

A. Legal Hierarchy

The hierarchy of the Pollution Prevention Act in the United States provides a useful approach for handling chemicals and wastes (Pollution Prevention Act, 1990). Simply stated, this hierarchy attempts to

1. reduce or eliminate pollution,
2. recycle residuals,
3. stimulate proper treatment, and
4. mediate safe disposal.

The goal is to reduce the transfer of pollution to future generations and preserve the quality of all waters for future use. However, this hierarchy could be used to emphasize the importance of groundwater ecology in the management of groundwater resources. The first two steps, to reduce or eliminate pollution and to recycle residuals, mandate that engineering and construction activities focus on minimizing the production of wastes. The third and fourth steps directly concern subsurface systems. For example, proper treatment currently includes injection of significant volumes of wastes of various kinds deep into the Earth, below any useful aquifers. Although subsurface disposal units are selected so no migration of waste occurs, no monitoring of these deep zones is required. Without focus on pollution prevention, residuals management may inadvertently allow contaminants to infiltrate into groundwater. Research is needed to determine the extent that biophysical processes may detoxify these contaminants in different types of aquifers and in groundwater/surface water interaction zones. Where groundwater moves most slowly, organisms may be able to degrade some chemicals but not others, and the risk of massive change in or destruction of hypogean ecological systems is great. In situations in which groundwater moves rapidly (e.g., porous alluvial or well-developed karstic aquifers), determination of the extent of the aquifer and, hence, the management area becomes problematic and often negates protection efforts. The distribution and abundance of biotic assemblages and various metrics associated with biophysical interactions (e.g., rates of microbial respiration, denitrification, and nitrification) can be used to define aquifer boundaries and sensitive interaction zones with adjacent systems (ecotones), as well as provide protocols for sound risk assessments of potential pollutants. But, as noted by Notenboom *et al.* (Chapter 18), process–response information on various waste disposal activities and chemicals, and their amounts, is badly needed. This approach ultimately may lead to a hierarchy of protection and management based on predictions of ecosystem responses to stress induced by groundwater pollution. In the interim, the development of criteria based on the toxicology of groundwater biota would be very helpful.

B. Relation to Current Laws and Regulation

The Water Quality Standards Program under the CWA, Section 303, currently uses the response of organisms to stress to set water quality criteria standards for surface waters. These standards are the basis for writing permits that establish the legally allowable pollutant concentrations in discharges. Because a significant volume of stream flow can come from groundwater discharge, pollutant concentrations in groundwater are important, not only to maintain the quality of the groundwater and associated groundwater organisms, but also to maintain the quality of the water in the stream. The precedent and framework that already exists in the Clean Water Act could be expanded to groundwater biota, if a better understanding of toxicology and the biomonitoring of hypogean biota existed. Minor modifications to regulations would be required to recognize groundwater and surface water interaction and the linkages to protecting groundwater for drinking water and for ecological benefits. However, obstacles remain with respect to the regional significance of groundwater discharge to surface water, because in the United States groundwater has historically been regulated by individual states, not the federal government.

VI. THE ENVIRONMENTAL PROTECTION AGENCY'S PROGRAM FOR PROTECTION AND ENHANCEMENT OF GROUNDWATER ECOSYSTEMS

A. The Strategic Plan

1. Overview

The Environmental Protection Agency is responding to the need to address the physical, chemical, and biological relationships that affect the quality and ecology of groundwater and the associated effects of groundwater and surface water interaction in its programs by developing a strategic plan for groundwater ecology. This activity began as a new initiative in the Agency's Groundwater Protection Division in 1990. Prior to that time, the primary interest was related to the enhancement of contaminant-degrading microbial populations in remediation efforts. Other aspects of groundwater ecology were largely unknown or unrecognized. Since that time, the Environmental Protection Agency has become actively involved in a variety of groundwater issues, including cosponsoring a biannual international conference on groundwater ecology and developing a strategic plan to incorporate groundwater ecology issues into relevant Environmental Protection Agency programs and other federal and state programs. The primary management tool needed for implementing this plan is the development of a methods manual to delineate areas of groundwater and surface water interaction. This manual is in progress.

2. Goal

The primary goal is to provide management protocols for the protection of surface water and groundwater interaction zones, because attainment of surface water quality standards in many instances is contingent on high-quality groundwater. As noted above, current statutes do not adequately recognize this important relationship. Hence, the plan focuses on near-surface groundwater ecosystems and their influences on terrestrial ecosystems (surface waters and wetlands). Environmental Protection Agency programs that can be responsive to this goal include the Watershed Program and the Wetlands Program. The plan is intended to promote integration of these programs and others in order to protect the integrity of groundwater and associated ecosystems.

3. Objectives

1. Provide technical assistance to water resource managers, especially at the state and local level, so that they may delineate and prioritize areas of surface and groundwater interactions that need protection.

2. Continue to work to incorporate or amend the use of existing statutory–regulatory authorities, policies, and guidance to better protect groundwater and related ecosystems.

3. Cooperate and coordinate with other public, private, and international organizations involved in groundwater ecology.

4. Increase the understanding of groundwater ecology.

4. Implementation

The plan identifies specific and practical recommendations to fulfill the objectives of the plan. Each objective has several recommendations, phased in over a three-year period, along with a budget to achieve the recommendations. For example, under the objective "to increase understanding of groundwater ecology," specific research priorities are listed.

B. Research Priorities

1. Physical and hydrological processes relevant to groundwater ecosystem structure and function. Topics include
   — hydrological connectivity between groundwater and surface water;
   — role of physical and chemical variables in determining the distribution and abundance of groundwater biota;
   — geochemical linkages among groundwater, surface water, and terrestrial ecosystems; and
   — ecosystem responses to environmentally rare and/or catastrophic events.
2. Biochemical and microbial processes relevant to groundwater ecosystem structure and function. Topics include
   — identification of dominant biogeochemical processes and pathways;
   — elucidation of energy transfer and nutrient cycling; and
   — understanding of microbial degradation of contaminants in the natural system.
3. Ecological processes relevant to groundwater ecosystem structure and function. Topics include
   — identification and classification of groundwater organisms;
   — study of population, community, and ecosystem dynamics and interspecific interactions;
   — elucidation of food web relationships;
   — identification of adaptive traits of groundwater organisms;
   — study of temporal and spatial patterns;
   — quantification of biodiversity and other descriptive indices; and
   — understanding the ecological relationships among groundwater and related ecosystems.
4. Anthropogenic influences on the structure and function of groundwater ecosystems. Topics include
   — fate and transport of contaminants in groundwater ecosystems;
   — effects of induced global climate change on groundwater and surface water systems;
   — ecological toxicity and risk assessment methods for groundwater ecosystems;
   — effects of stream management practices on groundwater ecosystems; and
   — development of a monitoring strategy for groundwater.

VII. CONCLUSION

Advances in groundwater ecology have unraveled important physical, chemical, and biological relationships in the subsurface aquatic realm. The research infers a basic message for management: Protect groundwater quality by protecting ecosystem functions. It appears that groundwater biota effectively detoxify at least some contaminating stressors, thereby maintaining and improving groundwaters as potable resources as well as reducing impacts on associated ecosystems, such as rivers, wetlands, and estuaries. This is a strong argument for implementing principles of groundwater ecology into water resource management activities. However, a better understanding of groundwater toxicology and the efficacy of using groundwater biota as biomonitors is needed. Biological, chemical, and physical indicators can be more effectively used to quantify and prioritize protection strategies and management areas (e.g., strategic groundwater supplies for potable and agricultural use, and zones of high bioproductivity or refugia for endangered species). In many situations, the current regulatory framework in the United States could address the conservation and protection of groundwater ecological processes, especially with respect to influences on surface water quality. However, such a practice currently is not routine. The strategic plan of the U.S. Environmental Protection Agency for protection and enhancement of groundwater ecosystems provides a basis for more effectively applying regulatory processes to groundwater pollution problems. A significantly expanded ecological research effort, as prioritized in this plan, will be necessary as an information base to resolve the plethora of management considerations that are problematic in the United States and elsewhere.

REFERENCES

Amoros, C., Roux, A. L., Reygrobellet, J.-L., Bravard, J. P., and Pautou, G. (1987). A method for applied ecological studies of fluvial hydrosystems. *Regulated Rivers* 1, 17–36.

Duff, J. H., and Triska, F. J. (1990). Denitrification in sediment from the hyporheic zone adjacent to a small forested stream. *Can. J. Fish. Aquat. Sci.* 47, 1140–1147.

Environmental Protection Agency (1991). "Protecting the Nation's Groundwater: EPA's Strategy for the 1990's," Rep. No. 21Z-1020. EPA, Washington, DC.

Environmental Protection Agency (1992). "Final Comprehensive State Groundwater Protection Program Guidance," Rep. No. 100-R-32-001. EPA, Washington, DC.

Environmental Protection Agency (1993a) "Evaluation of the Likelihood of Dense Nonaqueous Phase Liquids Presence at National Priority List Sites, National Results," Rep. No. 540-R-93-073. EPA, Washington, DC.

Environmental Protection Agency (1993b). "National Assessment of Groundwater Discharging to Surface Water" (unpublished report). EPA, Groundwater Protect. Div., Washington, DC.

Gibert, J., Dole-Olivier, M.-J., Marmonier, P., and Vervier P. (1990). Surface water-groundwater ecotones. *In* "The Ecology and Management of Aquatic-Terrestial Ecotones" (R. J. Naiman and H. Décamps, eds.), Man & Biosphere Ser., Vol. 4, pp. 199–225. Parthenon Publ., London.

Goldfarb, W. (1988). "Water Law." Lewis Publishers, Chelsea, MI.

Huling, S. G., and Weaver, J. W. (1991a). "Dense Nonaqueous Phase Liquids," Rep. No. 540/4-91-002, pp. 25–45. U.S. Environ. Prot. Agency, Washington, DC.

Huling, S. G., and Weaver, J. W. (1991b). "Dense Nonaqueous Phase Liquids—A Workshop Summary," Rep. No. EPA/600/-92/030. U.S. Environ. Prot. Agency, Washington, DC.

Intergovernmental Task Force on Monitoring Water Quality, Interagency Advisory Committee on Water Data, and Water Information Coordination Program (1992). "Ambient Water-Quality Monitoring in the United States: First Year Review, Evaluation, and Recommendations." Am. Water Resour. Assoc., Bethesda, MD.

Pollution Prevention Act (1990). 42 U.S. Code 13101, Sec. 6602 B.

Resh, V. H., and Jackson, J. K. (1992). Rapid assessment approaches to biomonitoring using benthic macroinvertebrates. *In* "Freshwater Biomonitoring and Benthic Macroinvertebrates" (D. M. Rosenberg and V. H. Resh, eds.), pp. 192–229. Chapman & Hall, New York.

Rosenberg, D. M., and Resh, V. H., eds. (1993). "Freshwater Biomonitoring and Benthic Macroinvertebrates." Chapman & Hall, New York.

Smith, K. (1991). Memo from Environmental Protection Agency Assistant Regional Counsel, Region IV to Anne Heard, November 1, 1991, Atlanta, GA (unpublished).

Stanford, J. A., and Ward, J. V. (1988). The hyporheic habitat of river ecosystems. *Nature (London)* **335**, 64–66.

Stanford, J. A., and Ward, J. V. (1993). An ecosystem perspective of alluvial rivers: Connectivity and the hyporheic corridor. *J. North Am. Benthol. Soc.* **12**(1), 48–60.

U.S. Geological Survey (1986). "National Water Summary, Groundwater Quality—Overview. *Geol. Surv. Water-Supply Pap. (U.S.)* **2325**, 3.

Ward, J. V. (1989). The four-dimensional nature of lotic ecosystems. *J. North Am. Benthol. Soc.* **8**, 2–8.

Weitman, D. (1994). Memo to Chuck Job, February 18, 1994. Environmental Protection Agency, Washington, DC.

Winter, T. (1989). "Hydrologic Function of Wetlands." U.S. Geological Survey Yearbook, Reston, VA.

EPILOGUE

Conclusions and Perspective

Jack A. Stanford* and Janine Gibert†

*Flathead Lake Biological Station
The University of Montana
Polson, Montana 59860

†Université Claude Bernard-Lyon 1
U.A./C.N.R.S. 1451, Ecologie des Eaux Douces et des Grands Fleuves
Laboratoire d'Hydrobiologie et Ecologie Souterraines
43 Boulevard du 11 Novembre 1918
69622 Villeurbanne Cedex, France

The chapters in this book provide a foundation for a new direction in groundwater science. Although we recognize the importance of the traditional emphasis on quantification of water volume and movement within aquifers of the lithosphere, the works presented herein greatly expand groundwater science by demonstrating biological and physical linkages between epigean and hypogean components of the Earth's aquatic realm. Moreover, the dynamic nature of groundwater ecosystems in time and space and in terms of biogeochemistry was articulated throughout. All of the fundamental ecological principals relating processes and responses in surface waters apply to groundwaters and must be included in an ecosystem context in future studies. The unifying conclusion of this volume is that groundwater aquifers, whether alluvial, fissured, or karst, are spatially and temporally dynamic in structure and function and that epigean/hypogean boundaries (ecotones) are sites of complex processes that produce rapid transformations of matter and energy that are vital to the self-renewal capacity (resilience) of the biosphere.

However, several important and less generalized conclusions also emerged from the contributions in this volume. We believe these will emerge as seminal concepts as the science of groundwater ecology matures.

First, the biotic attributes of groundwaters are arrayed within a continuum of geohydrological units. For example, a karstic system may drain a mountain region, and groundwaters may feed directly into a river via large springs. Or water from a river may penetrate the alluvium at the upstream end of a floodplain, flow through unconfined aquifers within the floodplain, and upwell back into the channel via a network of spring brooks at the downstream end of the floodplain. In either case surface water becomes groundwater, followed by reentry into surface systems. The number and diversity of geohydrological units may be better viewed as an interconnected mosaic (underground landscape or stygoscape) of different patches. This is readily apparent in large, wet caves that can be entered so that the different units may be viewed directly. The stygoscape is less apparent but nonetheless present in any aquifer system. Regardless, in terms of groundwater ecology, the primary physical features are the dynamics of water volume and residence time within units.

Second, as water and materials flux from one unit to the next, significant biogeochemical transformations usually occur. Types and concentrations of solutes change in relation to flow volume and rate as the geomorphology of the media changes within the aquifers. Likewise, the composition of the media itself changes in response to the attributes (e.g., temperature, solute load, and recharge rate) of flow. Biotic species composition and abundance may change significantly in response to these dynamic geohydrological gradients or patches within the aquifer system. Hence, physical and biological heterogeneity (e.g., in flow, temperature, redox, ion concentration, biodiversity, and bioproduction), not uniformity, fundamentally characterizes groundwater ecosystems. Moreover, biological and physical gradients may be very steep (i.e., conditions change very quickly) at boundaries or ecotones between the different geohydraulic units.

Third, virtually all groundwaters have some sort of biotrophic (food) web composed of microbes (bacteria and protozoa) and metazoans, except in situations in which oxygen is insufficient to support any organisms other than chemotrophic bacteria. As in surface waters, the food web is based on the energy transformations (bioproduction) through a microbial loop involving the use of dissolved or particulate organic matter as the primary energy source. Because photosynthesis cannot occur in groundwaters, the supply of organic matter is critical. Flux of dissolved organic matter or entrainment of particulate organic matter from geohydrologic units upstream (e.g., soils and surface waters) influence groundwater food webs. Generally, groundwaters are much less productive than surface waters because organic matter to drive microbial production is in short supply. Nonetheless, groundwaters can support speciose food webs, including in some cases even vertebrates, that play a vital role in the transformations of solute concentrations as waters flux through the ground between successive surface environments.

Fourth, flux of water through groundwater systems and associated biogeochemical transformations add complexity to our view of aquatic ecosystems in general. For example, we now examine river ecosystems in four dimensions: upstream–downstream (longitudinal), channel–riparian (lateral), channel–groundwater (vertical), and temporal (time). The penetration of river water into floodplain gravels defines the groundwater–surface water ecotone (sometimes referred to as the hyporheic zone). The ground/surface water ecotone in alluvial ecosystems may be expected to contain a speciose community of organisms having both ground and surface water affinities that greatly influence local and regional biodiversity patterns. Water and materials flux may change in time and direction as the piezometric head on the aquifer changes in association with spates and droughts. Chemical transformations within the ecotone and the dynamics of materials flux via the water vector may actually control bioproduction within the river channel and the aquifer as well as the distribution and abundance of riparian vegetation. We cannot observe corollaries for karst and fissured systems because the extent of exchange zones between surface and groundwater is limited upstream to sink streams and downstream to springs and different kinds of outlets. However, it is much easier for karst to define the system as a unit of drainage, as a whole, which contributes to the structuration and the persistence of surface ecosystems.

Finally, even though we recognize heterogeneity as a fundamental feature of aquifers, most groundwater systems are naturally more benign and biophysically constant (ecologically stable) in comparison with surface water environments. Groundwater organisms and their environment are rather nonresilient in that they have evolved in conditions that in many cases have weakly changed for long periods of geologic time. Therefore, groundwater organisms are not likely to be resistant to environmental change. Yet we are well aware of massive groundwater pollution worldwide and many examples of threats to native groundwater biota exist. Indeed, it may be that many groundwater systems have been destroyed or severely damaged by abstraction and pollution, either for irrigation and potable use or for disposal of wastes, before their natural integrity was recognized. Moreover, where groundwaters are connected components of river, lake, and marine ecosystems vital ecosystem linkages may have been disconnected by pervasive and mismanaged human activities.

Future research must be responsive to these general conclusions. Future ecosystem-level approaches will have to involve interdisciplinary teams of geohydrologists, geochemists, microbial ecologists, systematists, zoologists, and limnologists.

An interdisciplinary approach will bring more methodology to bear on groundwater problems. For example, limnologists working on groundwaters find it difficult, if not impossible, to accurately estimate population size (number per unit area or volume) of fauna, owing to uncertainties about the aquifer volume sampled by pumping monitoring wells. Innovative interaction with geohydrologists likely will provide new approaches to this fundamental problem. New drilling methods, including use of hand-installed piezometers to collect organisms in addition to analysis of hydraulic gradients, and other innovative devices that are responsive to the needs of limnological inquiry, as well as geohydrology, will allow the knowledge of groundwater ecology to expand rapidly. Also, more realistic models of the physical aspects of groundwater transport, from solute breakout curves to full-blown mass balance simulations, by geohydrologists and geochemists (physical scientists) will help microbial ecologists and limnologists (biological scientists) to more accurately identify ecological connectivity between water and materials flux, transformations of organic matter, and the distribution and abundance of groundwater biota. Grids of monitoring wells and other sampling sites outfitted with electronic sensors and data loggers that can accumulate and download massive quantities of monitoring data to detect times series changes that occur either naturally or to detect movement of pollutants or experimental tracers will be needed to provide more complete empirical input to these models. Eventually, more synthetic models, formalizing knowledge and guiding the direction of research for the many interactive biological and physical attributes, should contribute in a major way toward the goal of a predictive understanding of groundwater ecology at the ecosystem level of organization.

Particular attention should be given to the ecology of the microbial loop, because it is rather paradoxical that the food webs in many of the groundwater systems described to date could be supported entirely on bacterial production derived from limited sources of organic carbon. Moreover, microbes mediate ammonia, nitrate, and other solute transformations that apparently determine the concentrations of bioavailable solutes; concentration of solutes clearly will influence autotrophy in surface waters receiving groundwater discharge, especially at the groundwater–surface water interface. The microbial loop is also a critical component of bioremediation of contaminated groundwater supplies.

Inhabitation of groundwaters by metazoans may have occurred as early as the Triassic time, and therefore much about biogeography as related to plate tectonics is inherent in these biota. Much more syntheses of these potential relations are attempted. Isotopic aging of waters in association with biota may offer great insights. Groundwater biota also offer interesting and potentially ideal models for evolutionary studies, particularly regressive and convergent evolution and the role of metapopulations. Different relevant works have been provided in this field and for a long time. In terms of the usefulness of groundwater organisms in biomonitoring, which is a major theme proposed for future regulation and management of groundwaters, concern for how well lab experiments relate with field conditions in the assessment of contamination exists. We observe variability in the environmental tolerances of groundwater organisms, which is likely explained by physiological adaptations that are poorly understood. Refinement of experimental approaches and utilization of designs that allow manipulations in a field setting will greatly augment basic ecological studies.

Finally, researchers need to clearly define the geohydrologic unit(s) and scales in which their work is being conducted. For example, the ground–surface water ecotone (such as the hyporheic zone of rivers or the karst spring) is often very large and difficult to define. Understanding how processes, such as flux of materials, manifest themselves at different scales of organization (i.e., microhabitat to entire aquifer) is essential to properly framing ecological inquiry in groundwaters. Perhaps greater emphasis on conceptualization is needed because so much new information about groundwater ecology is rapidly forthcoming.

We suspect that major new discoveries of groundwater communities and interactions involving connections between hypogean and epigean biotopes will eventuate over the next few years, especially if research on groundwater ecology can proliferate in the manner inferred by the work summarized in this book and other recent colloquia (e.g., the First International Symposium on Groundwater Ecology in the United States and the International Symposium on Surface and Groundwater Ecotones in France). We close with a reiteration of an introductory comment. Groundwater ecology is a young, synthetic science that merits much greater research investment. The return on investment portends to be very great, owing to the strategic importance of groundwater resources in all countries of the world.

ACKNOWLEDGMENTS

This paper is based in part on text contained in a recent paper [Stanford, J. A., and Ward, J. V. (1992). Emergent properties of groundwater ecology: Conference conclusions and recommendations for research and management. *In* "Proceedings of the First International Conference on Groundwater Ecology" (J. A. Stanford and J. J. Simons, eds.), pp. 409–415. American Water Resources Association, Bethesda, MD] and is used here by permission of the American Water Resources Association. We thank J. V. Ward for his perspective in preparing these concluding comments.

Index